T0260489

ON THE LIFE OF GALILEO

FRONTISPIECE. Ottavio Leoni (1578–1630). Portrait of Galileo; engraving. Inscription: "Galileus Galileus Florentinus | Superiorum licentia 1624. | Eques Octauius Leonus Romanus pictor fecit." // "Galileo Galilei, Florentine | With the superiors' permission, 1624 | Made by Cavalier Ottavio Leoni, painter of Rome." Rome: Istituto Nazionale per la Grafica, Gabinetto Stampe (by kind permission). See *OGA* 1, D9, pp. 41–43; see also ibid., D9a-c, D45-D46a and D48, pp. 44–45, 112–114 and 116, respectively.

ON
THE
LIFE
OF
GALILEO

VIVIANI'S *HISTORICAL ACCOUNT*
& OTHER EARLY BIOGRAPHIES

EDITED, TRANSLATED, AND ANNOTATED BY
STEFANO GATTEI

PRINCETON UNIVERSITY PRESS
PRINCETON AND OXFORD

Published by Princeton University Press
41 William Street, Princeton, New Jersey 08540
6 Oxford Street, Woodstock, Oxfordshire OX20 1TR

press.princeton.edu

Library of Congress Control Number: 2019939551
ISBN 978-0-691-17489-1

British Library Cataloging-in-Publication Data is available

Editorial: Ingrid Gnerlich, Jessica Yao, and Arthur Werneck
Production Editorial: Karen Carter
Jacket and Text Design: Chris Ferrante
Jacket art: Ottavio Leoni, engraved portrait of Galileo (1624).
    Rome: Istituto Nazionale per la Grafica, Gabinetto Stampe
Production: Erin Suydam
Publicity: Katie Lewis and Matthew Taylor
Copyeditor: Barbara Liguori

This book has been composed in Arno Pro

Printed on acid-free paper. ∞

Printed in the United States of America

10 9 8 7 6 5 4 3 2 1

"[...] da iungere dextram,
da, genitor, teque amplexu ne subtrahe nostro."
Sic memorans largo fletu simul ora rigabat.
Ter conatus ibi collo dare bracchia circum,
ter frustra comprensa manus effugit imago,
par levibus ventis volucrique simillima somno.

"Let me take your hand, father,
let me take it: do not hold back from my embrace."
As he recalled this, waves of tears washed over his cheeks.
Three times he tried to reach arms around his father's neck,
three times the image eluded his hands, stretched out in vain—
like the puff of a breeze, like a dream that flits away.

VIRGIL, *Aeneis*, VI, 697–702

CONTENTS

SHAPING THE MYTH

WRITING THE LIFE OF GALILEO IN THE SEVENTEENTH CENTURY

I saw two-horned Venus
navigating suavely in the clear sky.
I saw valleys and mountains on the Moon,
and three-bodied Saturn–
I, Galileo, first among humans.
I saw four stars wander around Jupiter,
and the Milky Way break up
into infinite legions of new worlds.
I saw—not believed—foreboding spots
taint the face of the Sun.
I built this telescope myself,
a man of learning, but with shrewd hands:
I polished its lenses, I aimed it at the Sky
the way you would aim a bombard.
I was the one who broke Heaven open
before the Sun burned my eyes.

    Before the Sun burned my eyes
    I had to yield and say
    I did not see what I saw.
    He who chained me to the Earth
    did not unleash earthquakes or thunderbolts:
    he had a soft, flat voice–
    he had a face like everyman.
    The vulture that gnaws at me every evening
    has the face of everyman.

PRIMO LEVI, *Sidereus Nuncius* (April 11, 1984)[1]

---

1 "Ho visto Venere bicorne / Navigare soave nel sereno. / Ho visto valli e monti sulla Luna / E Saturno trigemino / Io Galileo, primo fra gli umani; / Quattro stelle aggirarsi intorno a Giove, / E la Via Lattea scindersi / In legioni infinite di mondi nuovi. / Ho visto, non creduto, macchie presaghe / Inquinare la faccia del Sole. / Quest'occhiale l'ho costruito io, / Uomo dotto ma di mani sagaci: / Io ne ho polito i vetri,

With these lines, the Italian chemist and writer Primo Levi (1919–1987) beautifully and effectively presented Galileo's scientific breakthrough and the complex interplay of science, religion, and power that affected the second half of his life. The turning point was March 13, 1610, when the first few hundred copies of the *Sidereal Messenger*, the terse record of his telescopic discoveries, were published in Venice. News spread all over Europe: Galileo's observations offered decisive evidence against the established Ptolemaic system. The "great, and very wonderful sights" (*magna, longeque admirabilia spectacula*) announced on the title page of the book—and the important steps taken shortly thereafter—would prove crucial for the eventual success of the heliocentric theory, but the path was much longer and much more difficult than Galileo would have ever anticipated.[2]

Some four centuries have passed, and still the Galileo story has not lost its fascination. The literature on Galileo's life and works is huge and ever growing and can possibly be compared with only that on Charles Darwin or Albert Einstein. Independent of the relevance of these giants' contributions to science, the proliferation of all sorts of works on them, scholarly and popular, is the outcome of their becoming more than men of science—in fact, of their becoming myths, either in their own day or shortly after their demise. The Galileo myth, in particular, is multifarious and has varied through the ages, adapting itself to the varying historical and cultural transformations, as is also testified by the different styles, functions, and messages of his portrait.[3]

Galileo's pupils and followers played a major role in this transformation. Chief among them was Vincenzo Viviani, who appeared to be, after the death of Benedetto Castelli, Bonaventura Cavalieri, and Evangelista Torricelli (all within five years of Galileo's demise), Galileo's leading follower, and himself a scientist of great repute.[4]

---

io l'ho puntato al Cielo / Come si punterebbe una bombarda. / Io sono stato che ho sfondato il Cielo / Prima che il Sole mi bruciasse gli occhi. / Prima che il Sole mi bruciasse gli occhi / Ho dovuto piegarmi a dire / Che non vedevo quello che vedevo. / Colui che m'ha avvinto alla terra / Non scatenava terremoti né folgori, / Era di voce dimessa e piana, / Aveva la faccia di ognuno. / L'avvoltoio che mi rode ogni sera / Ha la faccia di ognuno": in Primo Levi, *Ad ora incerta*, Milan: Garzanti, 1984, p. 73 (originally published in *La stampa*, 13 April 1984, p. 3); English translation based on that of Jonathan Galassi, in *The Complete Works of Primo Levi*, edited by Ann Goldstein, Introduction by Toni Morrison, New York-London: Liverlight Publishing Corporation, 2015, vol. 3, pp. 1970–1971. The title of the collection in which the poem was first included—*Ad ora incerta*, "at an uncertain hour"—is from Samuel Taylor Coleridge, "The Rime of the Ancyent Marinere," in *Lyrical Ballads, with a Few Other Poems*, London: John & Arthur Arch, 1798 (Part VII, p. 48, line 13; the whole stanza reads: "Since then, at an uncertain hour, / That agony returns: / And till my ghastly tale is told, / This heart within me burns"). Levi was not the first to compare Galileo to Prometheus: see the anonymous poem, written shortly after the trial of 1633, in Vaccalluzzo, *Galileo Galilei nella poesia del suo secolo* (1910), pp. 102–103, especially lines 21–30.

2 Galileo did not provide decisive evidence in favor of the Copernican hypothesis in the *Sidereal Messenger*, nor did he offer any conclusive arguments in its support later on, in the *Dialogue*. What Galileo thought was a piece of crucial evidence in favor of the earth's movement—the argument of the tides, in the Fourth Day of the *Dialogue*, was wrong, and seems to have convinced very few readers. In fact, Galileo succeeded in showing only that the Aristotelian explanation was no longer acceptable. Galileo's clear advocacy of Copernicanism notwithstanding, after the suspension of *De revolutionibus* in 1616 and Galileo's trial of 1633, the Tychonic system remained a viable hypothesis until the triumph of Newtonianism.

3 See *OGA*, vol. 1, and Tognoni, *Galileo: il mito tra Otto e Novecento* (2014), and *I volti di Galileo* (2018).

4 As Leibniz would later refer to him, Viviani was "the single person in which the spirit of Galileo still lives" ("[. . .] in quo uno adhuc vivit et spirat Galilaeus"): Gottfried W. Leibniz to Francesco Bianchini, 18 March 1690, in *Sämtliche Schriften und Briefe*, Series III, vol. 4, no. 244, p. 482.

The *Racconto istorico* (*Historical Account*), first drafted in 1654 but published only in 1717, was included in all subsequent editions of Galileo's collected works (1718, 1744, 1808–1811, 1842–1856), as well as in other major collections, and remained the main source for later biographers for nearly two centuries, until the publication of the *Edizione Nazionale* (*National Edition*) of Galileo's works and correspondence by Antonio Favaro and his collaborators (1890–1909), which is still the standard reference edition for Galileo scholars around the world.

Today, three centuries after its first publication, Viviani's account of Galileo's life and achievements—its limitations, errors, and constraints notwithstanding—is still an important and widely read document. And although we now know much more about Galileo, and have a much more comprehensive picture of the historical, scientific, religious, political, and intellectual milieu in which he lived, we cannot do without the information it provides and ignore the role it played in all subsequent accounts of the events. However, even though the *Racconto istorico* overshadowed all other period biographical sketches (including a few additional texts by Viviani himself), the shaping of the Galileo myth started much earlier, while he was still alive, and was spread throughout Europe by a number of other early biographical accounts.

Together the *Racconto istorico* and the other biographical notices not only contribute to forming the public image of Galileo across the centuries, in Italy and abroad, but also provide historians with important elements for understanding the wide-ranging repercussions of what has since become known as the "Galileo *affaire*"—the sequence of events that began with the publication of the *Sidereal Messenger* and culminated with the condemnation of Galileo by the Roman Inquisition. Each author's peculiar approach and tone in coping with the 1633 events—either by ignoring the very existence of the trial (or the *Dialogue* itself), or by diminishing its significance and impact, or else by dealing with its consequences in ways we may now find objectionable—needs to be contextualized within the complex intertwining of historical and religious elements of the day.

This volume presents fourteen accounts of Galileo's life and work written and published between 1633 and 1702. Four were written by Viviani, starting with the 1654 manuscript version of his *Racconto istorico* and culminating in his *Grati animi monumenta* (*Testimony of a Grateful Soul*), a published version of the inscriptions Viviani had painted on scrolls decorating the façade of his palace in Florence, in 1693. Some were written by people equally close to Galileo—his son, Vincenzo; the priest Niccolò Gherardini, who first met Galileo because of the trial and followed him back to Florence; and the artist Joachim von Sandrart, who recalled meeting Galileo at Palazzo Medici during the trial, many years after the fact. Others are products of the political and cultural life of the Roman court—Leo Allatius's hastily edited *Apes Vrbanæ* (1633) and Gian Vittorio Rossi's more inflammatory *Pinacotheca imaginvm illvstrivm* (1643), the source of information about Galileo's allegedly illegitimate birth. Still others are biographies written in the context of the Europe-wide passion to write the lives of famous men. Together they form an important corpus of source material that is essential for understanding what people thought they knew about Galileo and how their perceptions began to shape early mythologies about his life, his work, and his trial and condemnation. The appendix at the end of this volume discusses Galileo's and Campanella's relationship with Pope Urban VIII and contextualizes the *Adulatio perniciosa*, a poem celebrating Galileo's astronomy the pope wrote when still a cardinal, in 1620.

These texts document the different—at times repetitive, at time contrasting—ways early modern authors dealt with the problem of writing the life of Galileo in the aftermath of the trial. Authors of these biographical accounts were Galileo's pupils, friends, and critics, most of whom were not at all acquainted with Galileo's scientific activity and works, and not even with the actual proceedings of the trial. Their understanding of the events was based on their attitude toward one of the most important institution of the day, the Roman Catholic Church. More than anybody else, Viviani and the Medici realized the urgent need to capture the living memory of Galileo before his image was seized by others, who tried to appropriate him or wished him forgotten. Viviani felt the need to come to a compromise with the Church, struggling to transform the sentence, consign the clash to the past, and move forward, in the hope of having Galileo's reputation rehabilitated and the ban on his works lifted. Others were caught in between and opted for passing the trial over in silence, out of careless negligence or careful prudence, hoping to save appearances and please everybody (or those in power, at least). Still others openly sided with authority, burying Galileo with superficial, diminishing, and at times sarcastic language, resting assured that a miserable page of history had been turned once and for all. Their sneaky accounts turned into epitaphs that, however filled with scattered words of appreciation, would serve as warning for future similar attempts—or so they thought. Whatever their approach to what is undoubtedly one of the central episodes in the history of science, the early generations of scholars could not distance themselves from the events, the individuals, or the institutions involved; authors knew what was at stake and attempted at reconstructions that provided the meaning they intended to offer. Shaping the public image of Galileo was the collective work of a community of scholars who cared for the continuation of Galilean science or for the vindication of the Church's authority in the Counter-Reformation and its aftermath. For centuries to come, through all the transformations of his image, Galileo would remain a symbol for his independence—one that would resist any attempts at taming or framing it.

Vincenzo Viviani (1654) and Niccolò Gherardini (1654)

"About two years ago," wrote Viviani to Élie Diodati in 1656, "the Most Serene Prince Leopoldo was informed that a publisher in Bologna had undertook the task of reprinting in one volume all of Galileo's works (except those that in the present season bring with them a sort of *noli me tangere*). And Viviani decided to have an account of the life of such a great author added to the volume [. . .]".[5] In 1654, twelve years after Galileo's death (1642), Prince Leopoldo de' Medici—son of Grand Duke

---

5 Viviani to Élie Diodati, 23 February 1656, in *ODG Carteggio*, vol. 2, no. 679, p. 302 ("Sono presso a due anni che il Ser.ᵐᵒ Principe Leopoldo hebbe notizia che un libraro a Bologna aveva intrapreso a ristampar in un corpo tutte l'opere del Sig.ʳ Galileo (eccettuate però quelle che in questo clima portano seco quasi il *noli me tangere*), onde s'invogliò di farvi aggiugnere appresso un disteso della vita di così grande autore [. . .]").

6 We know little about Gherardini's life: see Salvini, "Elogio del Can. Niccolò Gherardini" (1768).

Cosimo II, and founder, together with his brother, Grand Duke Ferdinando II, of the Accademia del Cimento—asked Vincenzo Viviani (1622–1703), Galileo's last and most devoted pupil, and Niccolò Gherardini (1607–1678),[6] a close friend of Galileo's since 1633, to write biographical sketches that may serve Carlo Rinaldini (1615–1698), a "very learned man"[7] and cofounder of the Accademia del Cimento, to write a proper biography of Galileo. This was meant to be included in the first edition of Galileo's collected works, published in Bologna by Carlo Manolessi, in 1655–1656. Things, however, did not turn out as planned: Galileo's patrons and followers were not happy with the overall quality of Manolessi's edition,[8] and Rinaldini's biography never appeared in print. If it was ever written, it did not survive; Viviani's and Gherardini's outlines, by contrast, did. We owe the preservation of Gherardini's account to Viviani, who not only annotated it, and treasured it along with many other letters and documents, but kept revising and tinkering with his own biographical account for several years, contemplating a new and much better edition of Galileo's works. At the same time Viviani, a very hesitant man—subject also to the constraints imposed by the Church—and a perfectionist, never entirely satisfied with what he had written or the material he had gathered never managed to bring his grand project to completion.[9]

A member of a noble Florentine family, Vincenzo Viviani de' Franchi, from Colle Val d'Elsa, was born in Florence on April 5, 1622.[10] A child prodigy, he studied humanities with the Jesuits, in Florence, and logic with the Franciscan friar Sebastiano da Pietrasanta. His early education in geometry was under the Scolopian Father Clemente Settimi, a follower of Galileo's, who taught mathematics at the Pious Schools. Settimi was a student of Famiano Michelini's, a court mathematician and teacher to Grand Duke Ferdinando II's younger brothers. In 1638, after Viviani began to show great potential in his studies, Michelini mentioned Settimi's promising student to the Grand Duke, who insisted on meeting with Viviani in Livorno, where the Medici court was temporarily staying. When they met, Viviani offered an excellent demonstration of his mathematical talent, and the Grand Duke was so impressed that he offered Viviani a salary to help him pursue his studies, also giving him the opportunity to meet Galileo, who was living under house arrest at Arcetri.

Galileo was 74 years old, and blind. He needed an amanuensis but most of all a capable and intelligent assistant who could help him with his works and provide the critical feedback he required to develop his ideas; he needed someone who could write the letters he would dictate, illustrate his works with diagrams, and make

---

7 Viviani to Élie Diodati, 23 February 1656, in *ODG Carteggio*, vol. 2, no. 679, p. 302: "soggetto litteratissimo."

8 See, for example, Viviani to Diodati, 23 February 1656, in *ODG Carteggio*, vol. 2, no. 679, p. 305; and Viviani to Cosimo Galilei, 21 March 1656, ibid., no. 691, p. 321. The whole story is documented in *ODG Carteggio*, vol. 2, pp. 301–308.

9 In fact, it took centuries to realize Viviani's idea: the *National Edition* of Galileo's works and related documents (1890–1909), under the supervision of Antonio Favaro, may well be seen as the fulfillment of Viviani's ambitious project. However, Favaro's edition could hardly have been published had it not been for Viviani's tireless work and collecting activity.

10 For more comprehensive accounts of Viviani's life, see Favaro, "Amici e corrispondenti di Galileo Galilei: XXIX. Vincenzio Viviani" (1912–1913); Bonelli, "L'ultimo discepolo: Vincenzo Viviani" (1972); and Boschiero, "Post-Galilean Thought and Experiment in Seventeenth-Century Italy" (2005).

observations and experiments, if necessary. Viviani seemed to fit the role perfectly: not only did Galileo find the assistant he needed, but Viviani himself had the opportunity to improve his mathematical skills in the company of Galileo. The *Two New Sciences* had just been published in Leiden, and Viviani immediately took interest in Galileo's attempt to demonstrate the acceleration of falling bodies, discussing the topic with Galileo at length. Their collaboration eventually resulted in the scholium Viviani added to the Third Day of the *Two New Sciences*.

In 1641, Evangelista Torricelli joined Viviani and Galileo in Arcetri, and when the latter died, in 1642, Grand Duke Ferdinando II appointed Torricelli as Galileo's successor to the position of Court Mathematician. Torricelli, however, died in October 1647, and two other close pupils of Galileo's, Bonaventura Cavalieri and Vincenzo Renieri, passed away shortly thereafter (in November 1647); and Vincenzo Galilei, Galileo's only son, died two years later. Another pupil and collaborator of Galileo's, Benedetto Castelli, had died in 1643. This made Viviani, age 27, the only surviving pupil of Galileo's, and possibly the only natural philosopher in a position to carry on Galileo's work. Thus began the second part of Viviani's life, during which he not only pursued his own mathematical interests but took on greater responsibilities at the Medici court and dedicated himself to preserving and promoting Galileo's thought and works.

Viviani gradually rose to be appointed Court Mathematician, taught mathematics at the Accademia del Disegno (Academy of Drawing, founded by Cosimo I de' Medici in 1563, under the influence of Giorgio Vasari), and tutored the pages of the Medici family. His duties at the Medici court varied according to the desires of the Grand Duke, but he devoted most of his time to fulfilling his obligation as engineer, which required substantial travel (often on horseback) and led to frequent periods of illness. His repute led to various appointments to foreign scientific academies,[11] and several European princes offered him positions, which he always declined. He remained in Florence, at the Medici court, until he died on September 22, 1703, having published a few mathematical works, and leaving more than one hundred volumes of unpublished manuscripts.

Foremost among Viviani's many duties for the Medici court was the responsibility to be Galileo's successor. He took it upon himself to reorder Galileo's papers, and throughout his life he continued to translate and restore Galileo's unpublished works and gather his correspondence by sending out requests to Galileo's contacts in Italy and abroad. He planned to publish his own collection of Galileo's works, together with a more comprehensive biography than the report for Prince Leopoldo he had prepared in 1654, under the title *Racconto istorico*. The project, however, was never accomplished, as Viviani never felt satisfied with the material he gathered and was disappointed with the Catholic Church's refusal to grant permission to reprint the *Dialogue* and the other Copernican works, or to erect a proper funeral monument in a prominent position in the Santa Croce Basilica, in Florence. This personal failure notwithstanding, his huge efforts would prove crucial in the establishment of the Galileo collection at the National Library of Florence, and, as mentioned, in the

11 Among them, the Royal Society of London (1696) and the Académie Royale des Sciences of Paris (1699).

publication of the *National Edition* of Galileo's works, eventually edited by Antonio Favaro in 1890–1909.

Viviani's original plan for a new edition of Galileo's works, in both Latin and the vernacular, is still preserved in the collection of his papers at the National Library of Florence. Viviani's comprehensive biography of Galileo (of which the *Racconto istorico* was somewhat of a preliminary draft) was meant to introduce the set. Viviani described the plan in the following terms:

Order to be followed in the book with the Life of Galileo, etc.
Title page, in the Dutch manner[12]—*Life of Galileo*
Portrait of the King of France.
Engraved title page, with ornaments referring to the King of France, with the full portrait of the King, in his armor; in the white background, at the center: *Life of Galileo Galilei, Florentine Nobleman, Lyncean Academician, Chief Philosopher and Mathematician of the Most Serene Grand Dukes of Tuscany, Cosimo II and Ferdinando II, now reigning—Written by Vincenzo Viviani, Member of the Accademia della Crusca, Mathematician of the said Most Serene Highness, dedicated to the Catholic Majesty of the King of France and Navarre.*
Copper portrait of Galileo, aged approximately 68 years, on a large folio, within an oval or a square, with ornaments of various devices, and Latin mottoes referring to his major inventions and discoveries, and the coat of arms of the Galilei family standing below it.—Around the portrait: *Galilaeus Aet. suae 68.* Below the portrait, within a wreath, a couplet.
Poem explaining the devices.
Latin or Tuscan poems in praise of Galileo, especially that by the Pope.[13]
Dedicatory letter to the King.
Preface to the readers.
Preface to Galileo's life, in three books: 1) from the birth to the invention of the telescope; 2) from the telescope to Galileo's death; 3) on his health conditions, his illnesses, his habits, his sayings and what he enjoyed, his opinions, etc., and untold or unwritten judgments. About his friends and pupils. Testimonies of very eminent people.—Letters to Galileo from important people.
Pictures of his published works.[14]

---

12 It is not entirely clear what Viviani had in mind. Most likely, as he says in the next few lines, he intended the volume to have a title page incorporating both an engraving and letterpress: a full-page engraving, with reserved white space in the middle where the printed title wording would go; or else, he planned that the entire title-page would be engraved, with the words of the long title engraved on the plate. Engraved title pages were a Dutch speciality, and although other countries had them, too, there were more specialist draughtsmen and engravers who made them in Holland.

13 See the appendix.

14 "Ordine da tenersi nel Libro della vita &c. | Frontespizio all'Olandese.—*Vita di Galileo.* | Ritratto del Re di Francia | Frontespizio in rame con ornam.*to* alludente al Rè di Francia, col ritratto di questo da capo, e con l'Arme da piedi & nel bianco di mezzo sia scritto: *Vita di Gal.º Gal.ⁱ Nobil Fior. Accad.ᶜᵒ Linceo, Filosofo, e Matematico Primario dei Ser.ᵐⁱ G. D.ᶜʰⁱ di Toscana Cosimo I[I] e Ferdinando II regnante.—Descritta da Vincenzio Viviani Accad. della Crusca, Mat.ᶜᵒ della med.ᵐᵃ Ser.ᵐᵃ Altezza, alla M.ᵗà Crist.ᵃ del Re di F.ᶜⁱᵃ e di Nau.ʳᵃ* | Ritratto in rame del Galileo in età d'anni 68 inc.ᵃ in foglio grande in ovale, o in quadro, con ornamenti di uarie imprese, con motti latini alludenti alle sue principali invenzioni, e scoprim.ᵗⁱ et in piedi per di sotto con l'arme della famiglia del Galilei.—Attorno al ritratto sia scritto:—*Galileus Æt. suæ 68.* | In

Possibly, the "Preface to the readers" is the text now included in the appendix to the *Racconto istorico*. And a portion of what was likely meant to be the "Preface to Galileo's Life" may still be read in Viviani's papers at the National Library of Florence:

> Ever since its origin—ever since its childhood, so to say—the city of Florence has always been privileged, by Nature and by God, to have men eminent to the utmost in every art, occupation, science, or discipline. By the universal consensus of all nations, they deserved the title of Heroes, or Divine. And, leaving aside the names of so many leaders who showed the highest valor in war, etc., suffice it to mention here some of those who achieved the highest level in the nobler arts and letters, and perhaps even exceeded the most renown heroes of antiquity: Dante Alighieri, nobleman and poet; Petrarch; Marsilio Ficino, nobleman, Platonic philosopher; Amerigo Vespucci, nobleman; Filippo Brunelleschi, sculptor and architect; Donatello, sculptor; Giotto, through whom painting was revived; Michelangelo Buonarroti, nobleman, unparalleled painter, sculptor, and architect—Michael, less man than divine angel. But God's gracious Providence wished to preserve the noble privilege so abundantly enjoyed by my dear homeland in the past centuries, and granted it to restore the loss it endured after the death of such a great citizen, sung by the noble Ferrara swan [i.e., Ludovico Ariosto]: "Michael, less man than divine angel."[15]

As already mentioned, Viviani never managed to realize his grand project of a complete new edition of Galileo's works, prefaced by a comprehensive biography. He was still working on it in 1668, as is testified by Lorenzo Magalotti.[16] At the time

---

fondo in cestelletta sia un distico | Poesia esplicante imprese | Poesie latine, e Toscane in lode del Galileo e particol.^te quella del Papa | Lettera dedicatoria al Re | Proemio ai Lettori | Proemio alla Vita diuisa in tre libri: Il p.^mo dalla nascita fino all'inuenzione dell'occhiale.—Il 2.do dall'occhiale fino alla morte.—Il 3.º delle abitudini del suo corpo, delle malattie, de costumi, de detti, de i diletti, e di dogmi &c. opinioni non dicchiarate ne scritte; degli Amici e Scolari. Clarissimorum Virorum Testimonia.—Lettere di personaggi al Gal.^o | Illustrazioni dell'Opere stampate": *Gal.* 97, f. 82r. A detailed plan and list of Galileo's works to be included in the volumes is presented on ff. 75r-81r.

15 "Che la Città di Firenze è stata fin dalla sua origine e per così dire dalla sua infanzia, privilegiata sempre dalla Natura e da Dio d'Uomini in ogni arte, in ogni esercizio, in qualunque scienza, e professione eminenti in supremo grado di eccellenza, e tali che per concorso universale delle Nazioni ancora che straniere si sono meritati il titolo di Eroi e del Diuino. E tralasciando di far menzione di tanti e tanti gran Capitani di preclaro ualore nell'armi &c. basterà qui far registro di alcuni di quelli che nell'arti più nobili e nelle lettere sono peruenuti al maggior colmo, e traboccato forse sopra i più rinomati eroi dell'antichità: Dante Alighieri, Nobile, Poeta. Petrarca. Marsilio Ficino, Nobile, filosofo Platonico. & Amerigo Vespucci, nobile. Filippo di Ser Brunellesco Scultore, e Architetto. & Donatello scultore; un Giotto, *per quem pictura reuixit*. Un Michelagnolo Buonarroti, Nobile, Pittore, Scultore, Architetto senza pari. *Michel più che mortal, Angel diuino*. Ma che la benefica prouidenza dello sommo Dio uolendo conservare alla mia cara patria così nobile priuilegio da essa goduto così ampiamente ne secoli andati, si compiacque di ristorare la perdita che fatta aueua per la morte di cosi gran Cittadino, dal nobile cigno di Ferrara cantato: *Michel più che mortal Angel diuino*": *Gal.* 11, f. 132r-v. The sources of the quotations are the first line of Angelo Poliziano's epitaph for Giotto, in Vasari, "Vita di Giotto" (1568), p. 233: "Ille ego sum, per quem pictura extincta revixit"; and Ariosto, *Orlando fvrioso* (1536), XXXIII.12.

16 See Magalotti to Viviani, 20 May 1668, in Magalotti, *Delle lettere familiari* (1769), vol. 1, p. 19.

of Viviani's death, in 1703, both his and Gherardini's accounts were still unpublished; they appeared posthumously, in 1717 and 1780, respectively.

Its overall accuracy and comprehensiveness notwithstanding, Viviani's *Racconto istorico* is responsible for a number of historiographical clichés that proved to be very influential in later historiography. For example, the reader's attention is repeatedly called to the empirical and practical aspects of Galileo's work, whom Viviani presents as the first true empiricist. It was Galileo who first discovered the isochronism of pendulum oscillations after observing a lamp swinging in the Cathedral (Duomo) of Pisa, Viviani reports, and it was he who showed Aristotle's theory of free fall wrong by having objects of different weights fall from the Leaning Tower of Pisa (one of the most famous anecdotes in the whole history of science). Both these anecdotes were disputed by modern historians, as were other undocumented elements offered by Viviani, such as the ways in which Galileo was introduced to the study of mathematics; the reasons behind his decision to give up his appointment at the University of Pisa, in the early 1590s; his lectures on Euclid's geometry at the University of Padua, allegedly crowded by students; and the presence of Gustav II Adolf, King of Sweden, among Galileo's students in Padua.[17] Most important, Galileo's birthdate is incorrectly indicated as February 19, 1564 (instead of February 15, as correctly reported in the first printed version of the *Racconto istorico*, edited by Salvino Salvini), which was, in fact, the day on which Galileo was baptized. But it was close enough to the date on which Michelangelo Buonarroti died in Rome (February 18, 1564) that the temptation to have the two events very close to each other, and therefore to establish a sort of continuity between them, was too strong to resist. And indeed, years later, in the inscriptions on the façade of his palace in Florence, Viviani would indicate February 18 as his master's birthdate, also having the very hour of Michelangelo's death and Galileo's birth almost coincide—further and undisputable evidence of God's design.[18]

---

17 Many of these controversial events are repeated in Gherardini's account of Galileo's life, which was prepared at the same time as Viviani's and presents a similar structure but was written independently from it (as can be seen from Viviani's critical annotations). Like Viviani, for example, Gherardini reports the events that eventually led to Galileo's taking up an appointment at the University of Padua, but explicitly mentions the name Giovanni de' Medici as the high-ranking personality in the court with whom Galileo clashed. Gherardini also reports that Gustav II Adolf was one of Galileo's pupils in Padua, thereby suggesting that the information came to both Viviani and Gherardini from Galileo himself. Gherardini does not report Galileo's alleged experiments on free fall from the top of the Leaning Tower, nor his observation of a pendulum in the Duomo of Pisa. On the inaccuracies of the *Racconto istorico* see, for example, Favaro, "Vincenzo Viviani e la sua *Vita di Galileo*" (1902–1903), "Sulla veridicità del *Racconto istorico della Vita di Galileo* dettato da Vincenzio Viviani" (1915), and "Di alcune inesattezze nel *Racconto Istorico della Vita di Galileo* dettato da Vincenzio Viviani" (1917); and Wohlwill, "Über einen Grundfehler aller neueren Galilei-Biographien" (1903), "Galilei-Studien" (1905–1907), and *Galilei und sein Kampf für die copernicanische Lehre* (1909–1926), passim. See also Cooper, *Aristotle, Galileo, and the Tower of Pisa* (1935), and Segre, "Galileo, Viviani and the Tower of Pisa" (1989).

18 Just like for other anecdotes in the *Racconto istorico* (the oscillating lamp in the Duomo, the experiments from the top of the Leaning Tower, or King Gustav's attending Galileo's lectures in Pisa), the story about Galileo's birthdate coinciding with that of Michelangelo's death might have originated from Galileo himself: Viviani possibly got it from Galileo's own recollections, and 'worked' on it trying his best to have the two events perfectly coincide.

However, such inaccuracies or—at least in the case of Viviani's playing with Galileo's birthdate—(slight) alterations of facts, major or minor as they might seem nowadays, are to be considered within the wider historical and intellectual context in which Viviani was working. The writing of history and biography, in Viviani's day, was very different from our present ideas about it. Biographies, and histories in general, were literary, possibly more than scholarly, works: they were supposed to offer moral lessons and provide readers with ethical models, and as such abided by a preestablished pattern. Moreover, following in the tradition of his day, Viviani addressed a very specific audience, largely made up of readers with little interest in science. Most important, he turned Galileo from a sentenced heretic into a pious man, obedient and willing to repent as soon as he was shown his errors. In so doing, Viviani was hoping to come to terms with the Roman Catholic Church, convince the Inquisition and the pope to lift the ban on Galileo's works, have them translated and reprinted, and eventually be granted the permission to move Galileo's remains to a more prominent place in the Santa Croce Basilica.

The *Racconto istorico* is an eloquent account of Galileo's life that presents the various stages of his career: from Galileo's youth and early studies in Florence and Pisa, his appointments at the universities of Padua and Pisa, his telescopic discoveries and return to the Tuscan court, his controversies, the trial of 1633, and his later life and work; it ends with some general remarks about Galileo's personal traits. Following Galileo's own example, Viviani's choice of the vernacular indicates that the volumes it was meant to preface were intended to reach a much broader audience than that of professional scholars, whose lingua franca was Latin. Also, whereas Latin scientific treatises usually paid little to no attention to expository style, clarity was essential to reaching the general educated public, who used to meet at the various literary academies, such as the Florentine Academy, the Accademia della Crusca and, later, the Accademia dell'Arcadia. Most likely, as Michael Segre has convincingly argued, Viviani was writing with such an educated audience in mind.[19] And although there is no evidence that the *Racconto istorico* was ever circulated, or that Viviani ever lectured on Galileo at these meetings, Viviani's account was eventually published in something like a proceedings volume of the Florentine Academy—a collection of the lives and works of the academy's consuls.[20]

19 See Segre, "Viviani's Life of Galileo" (1989), pp. 214–217.
20 See Salvini, *Fasti consolari dell'Accademia Fiorentina* (1717), pp. 397–431. Galileo was elected consul of the Florentine Academy, although he was never active; he resigned in 1622. Torricelli's *Lezioni accademiche*, posthumously published in 1715, were originally delivered at the Accademia della Crusca between July 1642 and September 1643. In 1642, Cavalieri advised him about how to deliver his lectures: "I understand they expect physical rather than mathematical things, and perhaps rightly so, for I would liken the former to the bran, whereas the latter are the best flour, the true food and nourishment of the mind. It is nonetheless advisable to meet their inclinations, and indeed everybody's inclinations, as nobody holds mathematics in high esteem, unless they can see some material applications of it" ("Sento che vogliono cose piuttosto fisiche che mattematiche, e forsi con ragione, poiché quelle assomiglierei io piuttosto alla crusca, e queste al fior di farina, vero cibo e nutrimento dell'intelletto. Nondimeno conviene accomodarsi al loro genio, anzi al genio universale, che non istima punto le mattematiche se non ne vede qualche applicazione alla

Viviani had to meet further requirements, too, and of a rather different kind. As mentioned, at the time he was writing the *Racconto istorico*, biographies were meant to offer exemplary models. As such, they were a specific literary genre still very much influenced by patterns established in antiquity. The classic example is Giorgio Vasari's *The Lives of the Most Excellent Painters, Sculptors, and Architects*, first published in 1550 and reprinted, in a revised and enlarged edition, in 1568.[21] There are a number of striking similarities between the overall structure of Vasari's biographies of artists and Viviani's *Racconto istorico*. Artists, for instance, are often presented as child prodigies by Vasari, and the expressions Viviani uses to describe Galileo's precocious talent very much resemble Vasari's words about young Giotto or Michelangelo:

Vasari	*Viviani*
[...] Still a child, [Giotto] showed an extraordinary vivacity, and mental alertness in all his actions.[22]	In the early years of his childhood, Galileo began to display the vivacity of his mind.[24]
The old Lorenzo de' Medici, the Magnificent, who welcomed Michelangelo, still a young boy, in his gardens [...], saw samples of his work: blossoms as they were, they hinted at the fruits that would later bountifully sprout from the vivacity and greatness of his mind.[23]	

Another recurrent theme is that of the young prodigy triumphing over obstacles put in the way of his chosen profession, often by those nearest to him:

---

materia [...]": Bonaventura Cavalieri to Evangelista Torricelli, 14 July 1642, in *ODG Carteggio*, vol. 1, no. 19*, p. 18). Just like the *Lezioni accademiche*, the *Racconto istorico* was written to meet the expectations of a specific audience—that of academies. Indeed, in the *Racconto istorico* Viviani repeatedly calls attention to the empirical aspects of Galileo's work, to Galileo's craftsmanship, to his ability to put mathematics to practice and solve specific problems.

21 Interestingly, Carlo Manolessi, the editor of the first edition of Galileo's collected works, for which Galileo's biography was originally intended, published an edition of Vasari's *Lives* in 1647, in three volumes.

22 Vasari, "Vita di Giotto" (1568), p. 119: "[...] mostrando in tutti gli atti, ancora fanciulleschi, vna viuacità, & prontezza d'ingegno straordinario"; a similar expression may be found already in the 1550 edition, p. 139.

23 Vasari, "Vita di Michelagnolo Buonarruoti" (1568), p. 787: "Il Magnifico Lorenzo vecchio de' Medici, che riceueua nel suo giardino [...] Michelagnolo fanciullo, hauendo ueduti certi saggi di lui, che accennauano in que primi fiori, i frutti che poi largamentc sono usciti della uiuacità, e grandezza del suo ingegno"

24 *Racconto istorico*, pp. 3–4.

Vasari	*Viviani*
[...] Being poor, and with little income, [Michelangelo's father] set his children, and particularly Michelangelo, who had already grown up, to work in the wool and silk trade. Michelangelo was put under the guidance of Francesco da Urbino, a grammar teacher, and since his inclination led him to amuse himself with drawing, he secretly devoted to drawing as much time as possible. As a consequence, his father and elder relatives scolded him, and at times beat him. Possibly, they believed that devoting time to that art, which they ignored, was a low-level activity, something unworthy of their ancient family.[25]	[...] his father decided to send him to the University of Pisa despite the financial burden. He did hope, however, that one day Galileo would relieve him of the burden by becoming a physician, which Vincenzo wanted him to become, since medicine was the most suitable and fastest way to ease the situation.[26] As Ricci continued his lessons Galileo's father realized that his son was increasingly ignoring medicine as he was growing increasingly fond of geometry. Thus fearing, as time went by, that Galileo would abandon the former, from which he could have benefited more, from the point of view of utility and opportunity, given the tight financial circumstances, Vincenzo repeatedly scolded Galileo (pretending not to know the reason [for his lack of interest in medicine]), but always in vain, for the more Galileo grew fond of mathematics, the more he turned away from medicine.[27]

Further evidence of Galileo's heroic struggle with obstacles that appeared to be preventing him from following his calling is, for instance, Viviani's account that Galileo acquired his first mathematical notions by hiding himself outside the room in which Ricci was tutoring the pages of the Grand Duke.[28] Similar patterns of Renaissance biography also appear in Gherardini's account:

25 Vasari, "Vita di Michelagnolo Buonarruoti" (1568), pp. 716–717: "[...] & essendo male agiato, & con poche entrate, ando accomodando all'arte della Lana, & Seta i figlioli, & Michelagnolo, che era gia cresciuto; fu posto con maestro Francesco da Vrbino alla scuola di gramatica; & perche l'ingegno suo lo tiraua al dilettarsi del disegno, tutto il tempo, che poteua mettere di nascoso lo consumaua nel disegnare, essendo percio, & dal padre, & da suoi maggiori gridato, & tal uolta battuto, stimando forse che lo attendere a quella virtu non conosciuta da loro, fossi cosa bassa, & non degna della antica casa loro."

26 *Racconto istorico*, pp. 5–6.

27 *Racconto istorico*, pp. 9–10.

28 *Racconto istorico*, pp. 10–11.

Finding himself in some hardship, given the very limited amount of goods fortune provided him with, he decided to have his elder son, Galileo, work at the manufacturing of wool. Once he realized that Galileo was exceptionally bright and willing to study, he graciously allowed him to study grammar. Galileo went to study with a rather mediocre teacher who gave lectures at his own house, in via de' Bardi. Here, after a very brief time, once he had overcome the usual youthful disgust for such boring work, he learned the basic rules of the Latin language; then, by reading a few books, which he did most greedily, he learned the most beautiful secrets of that language, to the unspeakable astonishment of his teacher. Bewildered, he told Galileo's father he was no longer able to teach anything to the young boy. After this, Vincenzo decided to enroll Galileo in Pisa, so that he might take up the study of medicine. He hoped that would enable Galileo to earn money much more quickly and in larger amounts, and so help him out of his difficulties.[29]

[Galileo] went to the room where Master Ricci lectured on mathematics. He had no right to speak and was even less sure that he had a right to attend those lectures, as they were meant to be for the pages only, or for those who were in service at the court; so he remained in the room, in a place from which he could hardly hear what was being taught.[30]

As these few (out of many more) examples show, there are close parallels between Vasari's lives of artists, and Viviani's and Gherardini's biographies of Galileo. These works belonged to the same genre, were addressed to very similar audiences, and abode by equally stringent requirements, whose roots were in classical antiquity and were reintroduced during the Renaissance. In their groundbreaking 1934 study of the cross-cultural links between the legends and myths about artists, Ernst Kris and Otto Kurz argued that from the time artists made their appearance in historical records, certain stereotyped anecdotes and legends have been linked with their works and personality. Their best example of a biographical formula is the tale of Giotto, the son of a poor peasant farmer, who drew superb pictures of the animals on a rock while tending his father's flock of sheep. By a happy coincidence, Cimabue happened to be passing, immediately detected the boy's extraordinary talent, and took him to Florence, where he rapidly blossomed into the great genius of Italian Renaissance art we know.[31] According to the authors, Giotto-type legends show a

---

29 Gherardini, pp. 140–141.
30 Gherardini, pp. 142–143.
31 See "Vita di Giotto" (1568), p. 119. In fact, Vasari's biography of Giotto opens and closes with one such anecdote. Toward the end of it, Vasari reports an equally fictitious and telling story: "They say that when Giotto was only a boy with Cimabue, he once painted a fly on the nose of a face that Cimabue had drawn; and the fly was so natural that when the master returned to continue his work, he tried more than once to drive it away with his hand, thinking it was real, before he realized his mistake" ("Dicesi, che stando Giotto, ancor giouinetto con Cimabue, dipinse una uolta in sul naso d'una figura, che esso Cimabue hauea

striking resemblance to sagas and myths, especially to those that have come to be known as "myths of the birth of the hero" in psychoanalytic literature.[32]

In the numerous accounts of the lives of artists encompassed by Kris and Kurz, we encounter themes that typically recur with little or no variation: the belief that an artist possesses a more profound knowledge of nature than the layman;[33] the role played by chance that enables the youth to choose his career and thence to rise in social standing;[34] and interest in everything reported about the artist's childhood and youth. Not only are childhood events believed to have a decisive impact on the future development of man, but the earliest available details are taken to be sorts of premonitory signs or evidence of the early completion of uniqueness. Anecdotes, in particular, do not relate to one specific artist but are to be found in the lives of many typical artists. Most important, the question whether such anecdotes are true in any particular case is irrelevant: "The only significant factor is that an anecdote recurs, that it is recounted so frequently as to warrant that it represents a typical image of the artist."[35] It is a fixed and recurring biographical theme, a rhetorical model to which Viviani and Gherardini were bound, too.[36] In particular, Viviani and Gherardini were writing in the tradition renewed and exemplified by Vasari in the sixteenth century, as further testified by their calling attention to Galileo's equally profound knowledge of art and science.[37] Certainly, both Viviani and Gherardini embellished facts, but their stories remained very close to the truth, and it would be an error to dismiss them altogether, as Emil Wohlwill and other historians suggested doing, not only for the truth their accounts actually contain but for a proper understanding of the genre they belong to, and of the birth of the Galileo myth.

After the sentence, Galileo was afraid that his name might have been "deleted from the book of living men," and his works might have gone forgotten.[38] As mentioned previously, Viviani took on the responsibility of having Galileo's works collected and published. Soon after Galileo's death, in 1642, he began to ask friends and correspondents of Galileo's for letters, and began a lifelong search for documents and other

---

fatta, una mosca tanto naturale, che tornando il maestro per seguitare il lauoro si rimise piu d'una uolta a cacciarla con mano, pensando che fusse uera, prima, che s'accorgesse dell'errore": ibid., p. 133).

32 See Kris and Kurz, *Die Legende vom Künstler* (1934/1979), especially pp. 13–38.

33 See, for example, *Racconto istorico*, pp. 6–7: "Galileo, whom nature had elected to unveil to the world part of those secrets that had been buried in the dense darkness of human minds for so many centuries".

34 See, for example, Gherardini, pp. 142–143: "I do not know how, but Galileo befriended him. He told me it was by chance: Galileo went to talk with him a few times, and always found him teaching and explaining Euclid."

35 Kris and Kurz, *Die Legende vom Künstler* (1934/1979), p. 11.

36 See Segre, "Viviani's Life of Galileo" (1989), and *In the Wake of Galileo* (1991), pp. 107–126.

37 From the literary point of view, scientists (natural philosophers) and artists were considered very close to one another: with different tools, both were trying to offer representations of nature that were intended to be as close as possible to their subject. Viviani repeatedly calls attention to Galileo's literary and artistic tastes and skill, which competed with those of celebrated writers and artists. See, for example, *Racconto istorico*, pp. 6–7: Galileo's "bent for drawing was so natural and innate, and he acquired such an exquisite taste as time went by, that his judgment on paintings and drawings was preferred to those of eminent masters by the masters themselves." And Gherardini, pp. 150–151: "Galileo was an expert in all sciences and arts, just like those who taught or wrote about them. He thoroughly enjoyed music, painting, and poetry".

38 Maria Celeste Galilei to Galileo, 15 October 1633, in *OG* 15, no. 2747*, p. 302: "depennato *de libro viventium*."

material. Throughout his life, preserving the memory of his master's life, discoveries, and achievement became Viviani's major concern and focus. Most important, he had to deal with the enormous pressure of censorship: after 1633, the publication or republication of any work by Galileo was banned in all countries with an Inquisition, and all Viviani did to come to terms with the Church hierarchies must be considered in light of his effort to have the ban on Galileo's works lifted.[39] In his eyes, this was the greater good against which everything else might be conceded.[40]

The erection of a public funeral monument in Santa Croce was an integral part of this project, for a proper monument would bear the double meaning of recalling the importance of Galileo's works, highlighting their superiority with respect to traditional scientific and philosophical ideas, and of claiming the liberty to build on them, extending them to other contexts. The image of Galileo heretic, as it sprang from the 1633 trial, was a clear obstacle to this—hence the key strategic, as well as psychological and political, relevance of a proper funeral monument for Galileo in Italy's most celebrated pantheon: the Santa Croce Basilica, in Florence.

For Viviani, however, redeeming his master's reputation did not mean only defending Galileo from his detractors; it meant, first and foremost, claiming his true and profound *pietas*. In the *Racconto istorico*, he claimed that Galileo did not present the arguments for and against the Ptolemaic and the Copernican systems impartially but, rather, "showed himself closer to the Copernican hypothesis, which the Holy Church had damned as contradicting the Divine Scriptures."[41] As a consequence, Viviani continues, the Church rightfully called attention to his mistake and asked him to recant his statements. In so doing, however, the pope and the Inquisitors did not treat him harshly. As soon as Galileo arrived in Rome,

> by the greatest mercy of that Tribunal and of the Sovereign Pope Urban VIII (who otherwise knew he was most meritorious in the republic of letters) [he] was confined in the very beautiful palace in Trinità dei Monti, the residence of the Tuscan ambassador. And having been shown his mistake, Galileo promptly retracted his opinion, as a true Catholic. As a punishment, however, his *Dialogue* was banned, and after five months (during which the plague infected the city of Florence) he was dismissed from Rome, and by an act of generous compassion

39 As Fulgenzio Micanzio reported to Galileo, Inquisitor Clemente da Iseo told him in no unclear terms that the prohibition to reprint did not concern only the *Dialogue*, or other works on the Copernican hypothesis, but it was a "general prohibition, concerning all of Galileo's published works and those he might want to publish in the future": Micanzio to Galileo, 10 February 1635, in *OG* 16, no. 3075, p. 209 ("divieto generale *de editis omnibus et edendis*"). In another letter, Micanzio also indicated that the prohibition held in Italy as well as in other countries: see Micanzio to Galileo, 17 March 1635, ibid., no. 3095, p. 237. See also Galileo to Nicola-Claude Fabri de Peiresc, 21 February 1635, ibid., no. 3082, p. 216; and Micanzio to Galileo, 10 March 1635, ibid., no. 3088, p. 230, and 24 March 1635, ibid., no. 3098, p. 239.

40 Viviani not only tried his best to prevent distorted characterizations of Galileo (such as, for example, Rossi's claim that he was born out of wedlock) from spreading but was also committed to preventing the circulation of depictions of Galileo as being uncompromising, thereby reinforcing the Church's decision to prohibit his works. Viviani also went so far as to suppress any correspondence between Galileo and Paolo Sarpi, the Servite friar and polymath who never refrained from displaying his frankly polemical and highly critical views of the Catholic Church and its Scholastic tradition. As the *Racconto istorico* makes very clear, Viviani carefully presented Galileo as a pious scholar that immediately and gratefully acknowledged his errors and willingly atoned for them.

41 *Racconto istorico*, pp. 34–35.

was sent, for his house arrest, to the residence of Archbishop Monsignor Piccolomini, his dearest and most esteemed friend in Siena. Galileo enjoyed such tranquility and satisfaction in their conversations that he resumed his studies and found and proved most of his mechanical propositions on the resistance of solids and other conjectures. After approximately five months, the plague having completely come to an end in his homeland, at the beginning of December 1633 His Holiness turned the confinement to the archbishop's house into the freedom of the countryside, which Galileo greatly enjoyed. Thus, he went back to his villa in Arcetri, in which he had spent much of his time, as it had the benefit of good air and was conveniently located close to Florence. Friends and members of his family often visited him there, and their visits were always of great comfort and solace to him.[42]

Furthermore, having received news that some of his works and manuscripts, which he had written while still wrongly "convinced of the truth of Aristarchus's and Copernicus's opinion—an opinion that he had abandoned shortly before, as a Catholic"—were being circulated, translated, and published abroad, "Galileo was greatly mortified," as he "did not wish to prove ungrateful for the beneficial gift infinite Providence kindly decided to give him by releasing him from such a terrible mistake."[43] And eventually, at the time of his death, "with philosophical and Christian perseverance he returned his soul to his Creator."[44]

Scholars tend to read Viviani with some benevolence, arguing that he "adopted this position for purely instrumental reasons."[45] In their eyes, Viviani—in agreement with the Grand Duke Ferdinand II and his brother Cardinal Leopoldo—attempted a conciliatory approach that, while giving in on the crucial issue of the status of the new science, aimed at neutralizing Galileo's alleged heresy, thereby removing what was regarded as the major obstacle to the circulation and understanding of Galileo's thought and works. Indeed, as Giovanni Targioni Tozzetti reported,

Viviani, as a disciple of Galileo's, was also being monitored, and was forced to live with great caution [...]. He was obliged [...] to work discreetly, and, as was reasonable, he was afraid of being searched; accordingly, he kept all of Galileo's works and those of his pupils and correspondents, in a grain hole in the ground.[46]

42 *Racconto istorico*, pp. 34–35. In Viviani's account, house arrest for life becomes a sort of pleasure retreat in the countryside, where Galileo could enjoy well-deserved rest in the company of friends.

43 *Racconto istorico*, pp. 36–37.

44 *Racconto istorico*, pp. 46–47. Incidentally, Galileo's death came after "he was struck by a long-lasting fever" ("sopragiunto da lentissima febbre": ibid., pp. 46–47); similarly, Gherardini reports: "after a few days of a persistent fever" ("dopo alcuni giorni di lenta febbre": Gherardini, pp. 160–161). Once again, it is a loan from Vasari, who says "Michelangelo became ill with a persistent fever" ("ammalatosi Michelagnolo di una lente febbre": Vasari, "Vita di Michelagnolo" (1568), p. 774).

45 Galluzzi, "I Sepolcri di Galileo" (1993/1998), p. 421.

46 Targioni Tozzetti, *Notizie degli aggrandimenti delle scienze fisiche accaduti in Toscana* (1780), vol. 1, pp. 123–124: "Esso Viviani, come Discepolo del Galileo, era anche osservato, ed era costretto a vivere con gran cautela [...]. Si trovava [...] costretto il Viviani a lavorare alla sordina, e sul ragionevole timore di qualche perquisizione, teneva nascosti entro a una Buca da Grano di Casa sua tutti gli scritti del Galileo, e dei suoi Discepoli, e Corrispondenti."

Viviani was torn between his resolution to rehabilitate Galileo's reputation, reprint his works, and recover his legacy, and his having to deal with the Church's relentless determination not to give in on the terms of the 1633 sentence and its consequences. He spent his whole life keeping a low profile, in the hope of coming to terms with the Church hierarchy. As Cardinal Michelangelo Ricci (a pupil of Benedetto Castelli's and a supporter of Galileo's) advised Viviani about the *Racconto istorico*,

> I believe it would be cautious for you to spend just a few words on the accident suffered by Galileo as a result of his *Dialogue on the System [of the World]*, and go into more details about the other astronomical and geometric subjects, as well as the many beautiful experiments and sensible deliberations he made, many of which are still unpublished.[47]

In the *Racconto istorico*, it is not the Inquisitors who sentenced Galileo; it is the Divine Providence that granted him the opportunity to realize his errors and repent—something that Galileo, a true and faithful Catholic, was very happy to do, and thankfully so. That is why, in Viviani's account, Galileo's revolution gets fragmented into more or less innocuous individual elements: the invention of the telescope, the astronomical discoveries and their consequences mingle with the observation of isochronism of the pendulum (which is later applied to clocks, as described in Viviani's 1659 letter to Prince Leopoldo), the youthful invention of a balance to solve Archimedes's problem, the compass, the magnet, and then mechanics, thermometers, microscopes, the long-awaited solution to the problem of longitude. Viviani's portrait of Galileo is that of a genius who was able to foster our understanding of the world and advance our ability to change it to our advantage, not that of a radical revolutionary who discarded a well-established worldview. In sharp contrast with the image of Galileo as a martyr to *libertas philosophandi*, which spread throughout Europe, and particularly in France, Viviani's image of Galileo as presented in the *Racconto istorico* is that of the promoter of a new philosophy, which, however, Viviani did his best to present in a nondistressing way. However new Galileo's science was, its inventor had always deferred to religious authorities and was fully aware of the limitations and temporary nature of every human endeavor. Accordingly, Viviani promoted the publication of the first collection of Galileo's works so as to make them once again available to scholars and laymen alike. Particularly important was the inclusion of the *Dialogue*, which would be read in a new and reassuring light, not as the authoritative statement of the truth of the Copernican hypothesis, that is, but as an important, although human and thereby intrinsically fallible attempt at a novel understanding of cosmology. In this way, Viviani hoped to present a new image of Galileo as a Christian hero of science: restored to faith by the acknowledgment

---

47 Michelangelo Ricci to Viviani, 13 March 1668, in Favaro, "Vincenzo Viviani e la sua *Vita di Galileo*" (1902–1903), p. 693: "Stimo prudente il consiglio di V. S. in passarsela con poche parole quando tratta il particolare dell'accidente dal Galileo patito per i Dialoghi del Sistema e diffondersi tanto più nell'altre materie astronomiche, geometriche ed in tante bell'esperienze e giudiziose considerazioni ch'egli fece, delle quali molte non saranno mai state pubblicate con le stampe."

of his mistakes, purged by a sincere act of contrition, and intimately and fervently obedient to the Church.

As years went by, however, Viviani realized that his efforts were unlikely to succeed. In September 1674—some twenty years after Manolessi's edition of Galileo's works—Viviani seemed to be giving up his original plan to move Galileo's body to a more prominent place in the Santa Croce Basilica and conspired with Gabriele Pierozzi, a friar of the basilica, to decorate the humble burial place by adding a celebratory inscription (the same printed in the Manolessi edition, next to Pope Urban VIII's poem in praise of Galileo, the *Adulatio perniciosa*),[48] as well as a bust of Galileo drawn from a sculpture (now lost) by the famous artist and architect Giovanni Caccini (1556–1613). On December 7, 1689, age 67, Viviani drew up his will, with which he bequeathed his heirs the money and goods he thought would be required to erect a proper funeral monument for Galileo, facing Michelangelo's, adorned with an inscription celebrating his discoveries and achievements. As to his own, body, Viviani added, he wished it to be buried

> in the [...] church of Santa Croce, below the above-mentioned statue and monument to the Great Galileo, or else to the side or below his bones, whenever they are moved there. In the meanwhile, until the above-mentioned plan is accomplished, he [the testator] wants and orders that his body temporarily be laid next to that of Galileo [...].[49]

Viviani never lost hope, and kept polishing the inscription to be carved in the final monument. He grew increasingly aware, though, that his original plan would not be fulfilled in the few years he was likely to live.[50] If no public monument was to be erected, he eventually resolved to create a private one—one which, however, could be enjoyed, and benefited from, publicly.

Having bought a residence in what is now via S. Antonino, in proximity to the church of Santa Maria Novella (a short walking distance from what is now Florence's Central Railway Station), he entrusted his friend and well-known architect Giovan Battista Nelli to restore the palace and turn its façade into a monument to Galileo's life and achievements. Nelli designed the façade so as to present, at its very center, below a short inscription, a bust of Galileo (by eminent artist Giovan Battista Foggini, 1652–1725); next to it, two bas-reliefs celebrating Galileo telescopic discoveries and his achievements in physics and mechanics. What was meant to impress and arouse the attention of passers-by, however, were two enormous scagliola scrolls (hence the name by which the palace is referred to: "Palazzo dei Cartelloni," namely, "Palace of the Large Scrolls") on each side of the façade.

48 The inscription (first published in *Opere 1655–1656*, Vol. I, p. 22) was reproduced by Lorenzo Crasso, at the end of his biographical account: see Crasso, pp. 196–197.

49 Viviani, *Testamento* (1735), p. 4: "eleggendo la Sepoltura per il suo proprio Cadavere nella detta Chiesa di S. Croce sotto alla detta statua, e memoria del predetto Gran Galileo, ed accanto, o sotto alle di lui ossa, quando saranno ivi trasportate; ed intanto, che non sarà adempito il suddetto suo concetto, vuole, ed ordina, che il suo Cadavere si ponga in deposito, vicino a quello del medesimo Sig. Galileo [...]."

50 See the appendix to Viviani 1702, pp. 274–275.

In the two long inscriptions celebrating Galileo's achievements and discoveries in astronomy and natural philosophy—prudently written in Latin, and not easy to read from the street level—no reference can be found to the trial of 1633, no mention to Galileo's abjuration, or to the prohibition to publish new works or reprint old ones. The Copernican theory is mentioned only in passing, disguised as "Philolaus's system," and the reader's (or the wayfarer's) attention is repeatedly called to Galileo's *pietas*:

> [Galileo was] so inflamed by the desire to contemplate the truth that he went far beyond the accepted views of both ancient and more recent philosophers, and, disregarding weaker opinions of human minds, relying only on the help of geometry (which he called the truth's guide to heaven), he was the first to teach others a more certain way to truth, and happily pursued it, accompanied throughout his arduous way by piety; so that his discoveries about the ocean tides and Philolaus's system (which he demonstrates especially in the *Letter* to Christina of Lorraine, and which he thought out only to engage his mind), he gladly sacrificed to religion.[51]

Viviani presents himself as Galileo's loyal disciple, in both scientific and religious matters: he, too,

> cared to choose the way of truth and pursued it to the utmost of his ability, following the golden commandments of civil, moral, and Christian wisdom, and the example of Galileo's life. He was never oblivious to God's will. He brought some truths to light out of the infinite secrets, from the immense treasuries of geometry, and discerned that by means of them men draw closer to God Himself.[52]

And, finally, Galileo returned his soul to the Creator in peace, accompanied by his closest relatives, his pupils, "the parish priest, and two others of extraordinary doctrine and piety, long before selected by Galileo so that they could purify his soul."[53] Just like the *Racconto istorico*, the inscriptions had a moral aim, too: they were written to preserve Galileo's memory, "for the education of Christian philosophers."[54]

The final section of the second scroll of inscriptions is addressed to Florence, *cara Deo prae aliis urbibus* ("dearer to God than other cities"). Divine Providence was kind enough to present the city with two of the greatest geniuses in history, Michelangelo and Galileo. Fully aware of its privileged fate, Florence must express its gratitude in a proper way. It is in Viviani's 1691–1692 drafts for the inscriptions on the scrolls that we first find the explicit statement of the fateful coincidence of the dates of Michelangelo's death and Galileo's birth. Giving up the parallel between Galileo and Amerigo Vespucci drawn in the *Racconto istorico*, Viviani brings his appeal to

51 Viviani 1702, pp. 258–259.
52 Viviani 1702, pp. 264–265.
53 Viviani 1702, pp. 270–271.
54 Viviani 1702, pp. 270–271.

Florence to a climax, declaring the divinely designed continuity between Michelangelo and Galileo almost seamless.

The incredible timing proved irresistible to Viviani and to the many who repeated his claim after him (from Nelli to Kant, who added Newton to the metempsychosis of genius): the disciples' veneration of their great master, the princes' desire to publicly trumpet their generous support of the arts and the sciences, and the need for Florence to call the world's attention to the continuity of the city's intellectual leadership, under the Medicis's patronage, of the grand cultural tradition from the Renaissance to present times—all summed up in the perfect handover between Michelangelo and Galileo and hence the necessity of a proper funeral monument in the Santa Croce Basilica for Galileo, facing Michelangelo's.

Viviani never saw his much-desired wish come true. He died in 1703 and was laid to rest beside his master, in the vestry of the Chapel of the Novitiate. His lifelong efforts were not in vain, though, as they triggered a mechanism that led eventually to the erection of the longed-for funeral monument for Galileo. The first step was the publication of Galileo's *Dialogue* (1710), followed by the publication of Torricelli's *Lezioni Accademiche* (1715), and the Florentine reprint of Gassendi's *Opera Omnia* (1727); in between the latter two was the publication of the second edition of Galileo's collected works (1718), which still lacked the *Dialogue*. After the death of Cosimo III, in 1721, things started to change, however slowly. With the support of a group of eminent intellectuals, his successor, Grand Duke Gian Gastone, undertook the task of limiting the power and influence of the Church and restoring to the state its full rights and independence. In 1737, Galileo's body was exhumed and moved to the base of the new funeral monument, which was well under way and was eventually completed on June 6.[55]

GIROLAMO GHILINI AND PAUL FREHER

The terse inscriptions on the façade of his palace in Florence bring Viviani's lifelong effort to recover Galileo's legacy to a close. The *Racconto istorico* is the central piece of Viviani's strategy and—together with its complements, the *Letter to Prince Leopoldo de' Medici on the application of pendulum to clocks* (1659) and the account of Galileo's last works (1674), partially based on Vincenzo Galilei's notes—undoubtedly represents the most important biographical document about Galileo, which was to inform all subsequent biographical accounts for centuries.

However, the *Racconto istorico* is not the first biography of Galileo. Indeed, the first two sketches of his life, very different in tone and publication history, were both completed in the midst of the 1633 trial. Their relevance lies not so much in their content, in the information about Galileo they provide (for, from this point

---

55 During the transportation of Galileo's remains from its original burial to the left nave of the Santa Croce Basilica, three fingers of Galileo's right hand, one vertebra, and one tooth were removed from his body. As years and centuries went by, they became true "relics," and testify to the metamorphosis of Galileo from myth, to hero, to martyr. The story is worth telling, but it goes beyond the limits of the present study. For a reconstruction, see Galluzzi, "I Sepolcri di Galileo" (1993), and Gattei, "From Banned Mortal Remains to the Worshipped Relics of a Martyr of Science" (2016).

of view, they add nothing to Viviani's or Gherardini's, and were rightly disregarded by successive scholars), as in the way they dealt with the problem "of writing about a man whose list of scientific accomplishments was as impressive as the Catholic Church's harsh response to the religious implications of his astronomy."[56] Just like Viviani, who spent his entire adult life keeping a low profile in the effort to come to terms with the Church's hierarchy, Galileo's earliest biographers shared with Viviani the same intellectual, political, and religious milieu and could not distance themselves from the events or from the individuals and institutions which set them in motion. As Paula Findlen has observed, "With far less evidence at hand than any modern historian but with far more tacit knowledge of what was at stake, Galileo's early modern biographers attempted to reconstruct his life to make it meaningful for their own times."[57]

The first account of Galileo's life ever to be published is Girolamo Ghilini's (1589–1668). Born in Monza, he first studied humanities, rhetoric and philosophy in the Brera Jesuit College, in Milan, and jurisprudence in Parma; he did not complete his studies, though, owing to illness and the sudden death of his father, and returned to Alessandria, where his family lived. On Christmas 1630 he was ordained a priest, after a profound crisis he underwent after the death of his wife and one of his brothers, whose lives were claimed during the outbreak of bubonic plague that ravaged northern and central Italy in 1629–1630 (the Great Plague of Milan, the backdrop for several chapters of Alessandro Manzoni's *The Betrothed*). While in Milan, Ghilini resumed the studies he had had to abandon years earlier in Parma, and earned a degree in theology and canon law. He devoted the rest of his life to literary studies.

Ghilini's biographical sketch of Galileo was included in the *Teatro d'hvomini illustri*, first published in Milan, with no date, and later reprinted in Venice, in two volumes, in 1647. The Milan edition was almost certainly printed in late 1632 (or early 1633), as the entry includes reference to the *Dialogue* (published in February 1632), and the volume is dedicated to Pope Urban VIII. As an ordained priest, Ghilini would not have dedicated a book including an entry on Galileo to the pope who sentenced Galileo as vehemently suspected of heresy. Hence, the book must have been published between February 1632 (most likely, later that year) and June 1633 (when Galileo was sentenced by the Inquisition). The Venice 1647 reprint presents only minor typographical differences but no significant additions; in particular, there is no reference to the trial, the publication of Galileo's *Two New Sciences* (1638), nor Galileo's death (1642).

Other than being the first-ever printed biographical sketch of Galileo, Ghilini's entry is not particularly significant. He offers a fairly complete list of Galileo's works up to the *Dialogue* and refers to "other works," not yet published, "on motion and the resistance of bodies to being broken" (that is, *Two New Sciences*)—which is further evidence of the fact that the entry was written long before 1647.

Ghilini's entry on Galileo in the *Teatro d'hvomini illustri* is explicitly referred to by Paul Freher in his own entry on Galileo, in the second volume of his *Theatrvm virorum eruditione clarorum* (1688). However different in content and scope, Ghilini's and

56 Findlen, "Rethinking 1633" (2012), pp. 205–206.
57 Findlen, "Rethinking 1633" (2012), p. 206.

Freher's works share the same aim: both the *Teatro* and the *Theatrvm* were source-books containing short biographies of the most eminent personalities. More than half a century separates Ghilini's original entry and Freher's,[58] and yet no attempt was made at improving the text, or at complementing it with more recent information, possibly drawn from other sources. Perhaps Freher was still working on the second volume of the *Theatrvm* at the time of his death, in 1682.

LEO ALLATIUS

Leo Allatius's (ca. 1586–1669) biobibliographical entry on Galileo in the *Apes Vrbanae* (1633) was likely written approximately at the same time as Ghilini's. The book was published not long after Ghilini's, though, and the short interval had remarkable effects on its content, especially concerning the entry on Galileo. Indeed, the trial of 1633 could not but change the situation completely, forcing Allatius to substantially revise the text he had written before the trial and transform the entry on Galileo from a standard entry about a very eminent personality into quite an ambiguous one, with a rather understated tone. The contrast between the totally inadequate entry on Galileo—which dryly reports some of his achievements and lists some of his works up to *The Assayer*, published ten years before Allatius's book was compiled—and Galileo's eminence as one of the foremost scientists of his time—the Chief Mathematician and Philosopher of the Grand Duke of Tuscany, and a personal friend of Pope Urban VIII's, Galileo was one of the foremost examples of the glory of the Barberini pontificate and patronage—can hardly go unnoticed.

Born on the Greek island of Chios, Allatius was a graduate of the Pontifical Greek College of Saint Athanasius, in Rome, where he spent his career as professor of Greek and devoted his life to the study of classics and theology. He was made *Scriptor* in the Vatican Library in 1616, and found a patron in Cardinal Alessandro Ludovisi, who was crowned Pope Gregory XV in 1620. Two years later, when Maximilian of Bavaria presented the Palatinate library to the pope in return for subsidies, Allatius supervised its transport across the Alps to Rome, where it was incorporated into the Vatican Library on June 24, 1623. Pope Gregory had promised Maximilian a canonry, but he died shortly after Allatius's return to Rome. Allatius soon fell victim to slander and backbiting from members of the Curia and became the librarian of Bishop Lelio Biscia (1575–1638). Ten years later, to regain the favor of Pope Urban VIII, Allatius decided to compile a biobibliographic catalogue—a sort of *Who's Who* of Barberini Rome, assessing the second Roman Renaissance and the first decade of the Barberinis' enlightened patronage—that listed all works (either published, forthcoming, or in preparation) by authors who visited Rome between 1630 and 1632.[59] Formally dedicated to the pope's nephew, Cardinal Antonio Barberini (papal legate in Avignon and protector of the Greek College of Rome), the *Apes Vrbanae* presents,

58 In fact, if we consider the Venice 1647 edition, which was most likely the source of Freher's *Theatrvm*, just four decades.

59 As such, the *Apes Vrbanae* is a sourcebook that belongs to the literary genre of the *elogium*, like Lorenzo Crasso's and Isaac Bullart's.

in fact, a second dedication to Urban VIII, followed by Allatius's own translation into Greek of one of the pope's Latin poems.[60] The title itself celebrates the pope and the new Rome under Urban VIII's reign, during which a great number of scholars, artists, scientists, and men of letters (the "bees" in Allatius's book title) enjoyed the sweetness of the nourishing honey of the Barberini patronage, in a city that prided itself on being the capital not only of Christianity but also of the Republic of Letters.[61]

The original manuscript of the *Apes Vrbanae* is still preserved at the Vatican Library (*Vat. lat.* 7075) and provides important information about the publication history of the book.[62] As the manuscript clearly shows, the book was written in two main stages: the first included a series of 330 entries, organized in alphabetical order (by a person's first name); later, these were complemented by additional information and new entries, for a total of 456 (including one on Allatius himself). The latter phase seems to have been mainly driven by Allatius's wish to pay homage to his patron, Lelio Biscia, who was made a cardinal by Urban VIII in 1626. The supplementary material was likely added to build or strengthen Allatius's network of alliances. Whereas, for instance, the entry on Galileo—who did not belong to Allatius's network of friends and patrons—presents complementary information added only in the margins, the ones on the literary and art scholar Ferdinando Carli (pp. 92–96) and the Jesuit astronomer Christoph Scheiner (pp. 68–71) are two extensive additions. Most interestingly, the *Vat. lat.* 7075 shows that problems arose when the book was already in press: a certain number of pages (two entire quires, D and F, each consisting of 16 pages, and 6 other pages from quires H and I, for a total of 38 pages) were substantially revised, so that the corresponding entries—some of which refer to key figures in Allatius's Roman network—are the most extensively reworked texts of the entire collection. Moreover, these revisions were likely made by people other than Allatius himself, as if the published version of the book was, in fact, more the work of a small group of editors rather than of a single author. And if the whole book was subjected to a very scrupulous reading by censors, the detailed revisions of quires D, F, H, and I all received independent imprimaturs, each personally signed by Niccolò Riccardi, Master of the Sacred Palace (also known as Father Monster).

The first draft of Galileo's entry (*Vat. lat.* 7075, f. 68v) mentions only three of his works: the *Sidereal Messenger* (1610), *On Sunspots* (1613), and *The Assayer* (1623), demonstrating Allatius's superficial knowledge of Galileo's production. Not only does he omit the *Compass* (1606), the *Difesa* (1607), *On Floating Bodies* (1612), and most important, the *Dialogue* (1632), but he seems not to be aware of any works Galileo was writing at the time. The list is followed by two quotes, from Paolo Gualdo's *Vita Ioannis Vincentii Pinelli* (1607), and Giovanni Battista Lauri's *Theatri Romani Orchestra*

---

60 See Allatius, *Apes Vrbanae* (1633), pp. [3]–[6] (official dedication to Antonio Barberini), 14–15 (unofficial dedication to Urban VIII, with praises of his poetical talents, described as *divina*), and 15–19 (poem). The poem is from the 1631 Jesuit edition of the pope's *Poemata*, pp. 230–234 (*editio maior*): "Ad Franciscvm Caroli fratris filivm."

61 *Apes Vrbanae*, namely, "Urban bees" but also "the bees of the city (*urbs*, in Latin) of Rome" (which was regarded as the city par excellence), the bees being the symbol of the Barberini family on its crest. His learned and flattering work notwithstanding, it would take Allatius six more years to enter the close circle of Cardinal Francesco Barberini, in the capacity of his librarian.

62 On this whole issue, see Cerbu and Lerner, "La disgrâce de Galilée dans les *Apes Urbanae*" (2000), on which I rely for my reconstruction. See also Lerner, "Premessa" (1998).

(1625). Despite that both quotes express admiration and appreciation of Galileo, Allatius must have felt that Gualdo's and Lauri's words could not suffice to pay a proper tribute to the Grand Duke's Chief Mathematician and Philosopher. Indeed, Allatius openly compares them to gooses' honks and compares them—after a personal appreciation for Galileo's high consideration of Plato—with the cry of an eagle, using a few carefully selected lines from the *Adulatio perniciosa*, a poem none other than Pope Urban VIII (then still a cardinal) wrote in praise of Galileo, in 1620. No bee could hope for a more prestigious seal to his own entries in the *Apes Vrbanae*.[63]

Later on, as in the case of other entries, Allatius obtained more information about Galileo, possibly from Francesco Stelluti (a member of the Academy of the Lynxes and a close friend of Galileo's) or Giovanni Ciampoli (himself a Lyncean academician, a former pupil, and a good friend of Galileo's, as well as chamberlain to Pope Urban VIII). The list of Galileo's works is supplemented by the *Compass*, the *Difesa*, *On Floating Bodies* (erroneously dated to 1623), and the *Dialogue*; and mention is made of Galileo's work in progress: on mechanics, on the new science of local motions (both natural and violent), and a rejoinder to Orazio Grassi's *Ratio pondervm libræ et simbellæ* (1626). Allatius also adds a personal remark, saying that he read a *Discourse on the Motion of the Earth*, in Italian. This may refer to Galileo's *Letter to Christina* (1615), which had been widely circulated; the *Discorso del flusso e reflusso del mare* (written for Cardinal Orsini, on January 8, 1616);[64] or else Galileo's rejoinder to Ingoli's *De situ et quiete Terrae contra Copernici systema disputatio* (1616), written in 1624.[65] There follow, before Gualdo's and Lauri's quotes, a few compliments to Galileo.[66]

Allatius sent this second, enlarged draft of the entry on Galileo to the press; it is printed in quire H, pp. 118–120. In the midst of publication, however, what was going to be an uncontroversial text, meant to please everyone, turned into a highly controversial biography. In July 1632, while the book was being printed, the Inquisition decided to prevent the circulation of Galileo's *Dialogue*, and appointed a special commission to examine the text.[67] In October, Galileo was summoned to Rome,[68] where he arrived on February 13, 1633, the very same day on which Allatius obtained the imprimatur

---

63 *Vat. lat.* 7075, f. 68v (in Allatius's handwriting).

64 In *OG* 5, pp. 376–395.

65 In *OG* 5, pp. 403–412; Galileo's rejoinder to Ingoli (in the form of a letter) remained unpublished at the time of his death: see *OG* 6, pp. 509–561. After the suspension ("until corrected," *donec corrigatur*: see *OG* 19, p. 323) of Copernicus's work, in 1616, Francesco Ingoli (1578–1649) was appointed to officially revise *De revolutionibus*; he submitted his report on April 2, 1618 (*De emendatione sex librorum Nicolai Copernici De revolutionibus*: in *Barb. lat.* 3151, ff. 58r–60v; English translation in Gingerich, "The Censorship of Copernicus's *De Revolutionibus*" (1981), pp. 51–56). Ingoli's *Disputatio* caused Johannes Kepler to write a rejoinder: *Responsio ad Ingoli disputationem de systemate* (1618, in *KGW* 20.1, pp. 168–180), to which Ingoli replied in the *Replicationes De situ et motu Terrae contra Copernicum ad Ioannis Kepleri Caesarei mathemathici Impugnationes contra disputationem de eadem re ad D. Galilaeum de Galilaeis Gymnasii Pisani mathematicum celeberrimum scriptam* (in Bucciantini, *Contro Galileo* (1995), pp. 207–209). Later, Ingoli drew from the latter text a report on Kepler for the *Index* of prohibited books, which led to Kepler's works' being banned.

66 *Vat. lat.* 7075, f. 68v, in the margin (in Allatius's handwriting).

67 See, for example, Filippo Magalotti to Mario Guiducci, 7 August 1633, in *OG* 14, no. 2285, pp. 368–369; and Francesco Niccolini to Andrea Cioli, 15 August 1632, ibid., no. 2287, p. 372, and 5 September 1632, ibid., no. 2298, pp. 383–385.

68 See *OG* 19, pp. 331–332, as well as Francesco Niccolini to Andrea Cioli, 11 September 1632, in *OG* 14, no. 2302, pp. 388–389, and Francesco Barberini to Giorgio Bolognetti, 25 September 1632, ibid., no. 2311*, pp. 397–398.

for the *Apes Vrbanae*. A few months later—on June 22, 1633—Galileo was found ve-hemently suspected of heresy and forced to abjure; the *Dialogue* was banned.[69]

At the very last minute, with the book already in print, Allatius had to make drastic changes, which affected only the last two pages of the entry; the originally printed pages, and the handwritten revisions penned on them, are preserved in *Vat. lat. 7075*, ff. 69r–v. The changes consisted of various deletions and a short addition. On p. 119, the reference to the *Dialogue*, without any mention of its censure, became totally inappropriate and was crossed out; and so was Allatius's allusion to reading a man-uscript on the same subject, namely, Galileo's *Discourse on the Motion of the Earth*.[70] Likewise, on p. 120, after the quote from Giovanni Battista Lauri, the last part of the entry—including Allatius's personal appreciation for Galileo's high consideration of Plato, and the lines from Urban VIII's *Adulatio perniciosa*—were crossed out. Had they remained, Allatius's efforts to gain the pope's favor would have been almost certainly frustrated. In their stead, Allatius mentioned Giulio Cesare Lagalla's (1571–1624) *De phœnomenis in orbe Lvnæ novi telescopii vsv nvnc itervm svscitatis Physica disputatio* (1612), an early critical reaction to Galileo's *Sidereal Messenger*.[71] This is the version of the entry that eventually came to light in 1633.

The radical changes in the entry on Galileo were not the only changes made in the book, though. Indeed, not only did Allatius delete any references to Galileo's works in support of the Copernican opinion (as already mentioned, Galileo's last recorded work is *The Assayer*, published in 1623, and dedicated to Pope Urban VIII) but took the opportunity to insert scathing and trenchant critiques of his Copernican works in the entry on Christoph Scheiner, one of Galileo's archenemies.

In summarizing Scheiner's achievements, not only does Allatius take sides with Scheiner about the controversial claim of priority over the discovery of sunspots but repeatedly and emphatically calls the reader's attention to Galileo's mistakes, also charging him with plagiarism. Most notably, it is while presenting Scheiner's interpretation of sunspots that we find the only explicit mention of the *Dialogue*. And whereas in the *Dialogue* Galileo tried to demonstrate the earth's motion from certain appearances connected with the movement of sunspots, in a forthcoming work

> Scheiner will show that from the course of that sunspot we cannot derive any conclusions Galileo wants to draw. But Galileo learned the true movement of sunspots from the *Rosa Vrsina*, and shrewdly disguised it, and likewise imposed it on the reader, and did violence to the heavens, the Sun, the *Rosa Vrsina* and its author. This work will be a sort of prelude, which will soon be followed by his *Preliminary Discourse for the Stability of the Earth against the Same Author of Dialogues*, in which will be summarized Galileo's logical errors, physical errors, mathematical errors, ethical errors, theological, and scriptural errors. Finally, from all these,

---

69 See *OG* 19, pp. 402–406 and 406–407.

70 *Vat. lat.* 7075, f. 69r.

71 *Vat. lat.* 7075, f. 69v. A former pupil of Lagalla's in Rome, Allatius held him in high esteem: "Giulio Cesare Lagalla, a man with a very sharp mind, glory of the honored Aristotelian philosophy" ("Iulius Cæsar Lagalla acerrimi ingenij vir, & Peripatethicæ Philosophiæ gloria celebris": *Apes Vrbanae* (1633), p. 96); see also Allatius's own entry, ibid., p. 177. Allatius edited Lagalla's *De cœlo animato dispvtatio* (Heidelberg, 1622) and would publish his biography in 1644.

once the mask is dropped, it will be clear that his doctrine is shown false because of his [that is, Galileo's] ignorance.[72]

Allatius's praise of Scheiner's achievements—the entry is nearly twice as long as Galileo's, and, just like Galileo's, was extensively reworked under the close scrutiny of censorship during the months of the trial—seems to be, in fact, just an excuse for a direct and open attack on Galileo and the *Dialogue*. By pure coincidence and ironic history, Galileo was being tried by the Inquisition in the very same weeks in which Allatius was submitting his work to censorship. The very same authorities that were keen to prevent the *Dialogue* from being read were eager to grant the circulation of the *Apes Vrbanae*—formally dedicated to Cardinal Antonio Barberini but openly addressed to Urban VIII—and the protection of its author. The publication history of the *Apes Vrbanae*, as documented and reconstructed by Cerbu and Lerner from the material in the *Vat. lat.* 7075, shows "by what complex steps this result has been obtained, and especially by what mechanism the work of censorship (and probably also self-censorship) resulted in transforming the initial praise of an eminent man, exalted by Pope Urban VIII himself, into a judgment which, without being patently negative, brings Galileo into disrepute."[73]

GIAN VITTORIO ROSSI

The first biography of Galileo published after his death, by Gian Vittorio Rossi (Ianus Nicius Erythraeus, 1577–1647), turned out to be a very controversial one. Born in Rome to a well-to-do family, Rossi was educated by Jesuits at the Roman College and graduated in law at La Sapienza, in Rome, when he was only 19 years old. His short biographical entry on Galileo was published in the first volume of Rossi's principal work, the *Pinacotheca imaginvm illvstrivm* (1643, reprinted in 1645 and supplemented by two additional volumes in 1645–1648). The *Pinacotheca* was meant to be a collection of 161 short biographies, or portraits (*imagines*), of eminent people who had lived in Rome and were part of the city's political and religious circles, and intellectual networks. Despite its purporting to belong to the literary genre of the *elogium*—just like Allatius's work (and, later on, Crasso's and Bullart's; see below)—the *Pinacotheca* did not, in fact, belong to encomiastic literature. It was not even a somewhat balanced collection of exemplary lives based on historical

---

72 Allatius, pp. 108–109. Shortly before this passage, with reference to Johann Georg Locher's (a pupil of Scheiner's in Ingolstadt) *Disqvisitiones mathematicæ, de controversiis et novitatibvs astronomicis*, Scheiner is credited for ruling out "the Copernican motion of the earth straight away, against all fallacious arguments of the renovators [*Nouatorum Paralogismos*]": ibid., pp. 104–105.

73 Cerbu and Lerner, "La disgrâce de Galilée dans les *Apes Urbanae*" (2000), p. 608: "Mais les documents publiés et analysés ici montrent sur le vif par quelles étapes complexes ce résultat a été obtenu, et surtout par quel mécanisme le travail de la censure (et sans doute aussi de l'autocensure) a abouti à transformer l'éloge initial d'un homme illustre porté aux nues par le Pape Urbain VIII lui-même en un jugement qui, sans être ouvertement négatif, fait tomber sur celui qui en est l'objet la disgrâce venue d'en haut." Another very interesting entry in the *Apes Vrbanae* is that on Tommaso Campanella, which underwent a process of revision similar to that on Galileo, as the pope's displeasure with the former grew more or less simultaneously with his rage at the latter (in the appendix, I will argue that, indeed, the two reactions were not independent from each other). See Lerner, "Le panégyrique différé" (2001).

researches, one that would provide moral and ethical models to follow, or negative examples to avoid. Unlike Allatius, Rossi was not trying to offer a learned account of eminent personalities, with bibliographic references. Certainly, the *Pinacotheca* includes standard, exemplary biographies, but its originality lies in the fact, so to speak, that the *imagines* present themselves as objects scattered in a *Wunderkammer* rather than portraits tidily hung on the walls of an art gallery. The entries describe eccentric personalities from the recent past, whose memories were still vivid in the minds of the readers. They are presented in the manner of caricatures, following the classical model of Theophrastus's *Characters*, rather than that of Tacitus's *Histories*, or *Annals*. These two features certainly contributed to the immediate and vast success enjoyed by the *Pinacotheca*, which may be seen as a companion to Rossi's *Evdemia* (1637 and 1645), a satirical work in which characters and events are disguised under fictitious names and clever allegories and which paints a faithful—though at times possibly a bit too unconventional—picture of Rome at the time of the Barberini.[74]

All this sets the background against which we have to read and understand Rossi's biographical sketch of Galileo. As Rossi himself declared in 1645, both the *Pinacotheca* and the *Evdemia* were written "as a joke" (*per ludum jocumque*)[75] and are imbued with gossip and iconoclastic humor. The *Pinacotheca* is a gallery of *imagines* of eminent personalities, portrayed in an unconventional way and, at times, with a witty malice. The *Evdemia* is, in its own way, a gallery as well, but of ill-famed personalities, who parade deprived of any trait of nobility or virtue and are almost invariably presented mercilessly, in their utter sordidness. Rossi's caustic criticism of the corruption of the political and intellectual establishment of Urban VIII's Rome was a scholar's disdainful reaction to the moral collapse of the expectations and promises of renovation of the pope-poet.[76]

---

74 See Gerboni, *Un umanista nel Seicento* (1899), especially pp. 75–86; Croce, *Nuovi saggi sulla letteratura italiana del Seicento* (1931), pp. 124–133; Giachino, "Cicero libertinus" (2002); and Herklotz, *Apes Urbanae* (2017), pp. 159–197. In 1603, Rossi became a member of the newly founded Accademia degli Umoristi (Academy of the Humorists), with the nickname "Aridus." The Accademia was a learned society of intellectuals, mainly noblemen, who significantly influenced the cultural life of seventeenth-century Rome. It began as a place for writers and intellectuals to celebrate burlesque and mock-heroic poetry but soon attracted some of the most prominent figures and patrons of the arts in Rome. For a full biography of Rossi, see Johann C. Fischer, "Vita Ioannis Victorii Roscii vvlgo Iani Nicii Erythraei," in Rossi, *Epistolae ad Tyrrhenvm et ad diuersos* (1749), pp. I-CLVIII (with supplements, pp. CLIX-CLXXXIIII).

75 Rossi, *Epistolæ ad Tyrrhenvm* (1645), p. 52: "But I would not want to be known and recognized by you only for the *Evdemia*, the *Pinacotheca*, and other similar trifles. Indeed, I wrote them as a joke, and the more easily I was prepared to belittle myself with this playful joke, since I realized that playful and light works are highly regarded and respected. They call everybody's attention and find someone willing to buy them; important and serious works, by contrast, are almost invariably turned down, as they are regarded as painful, boring, hideous, and impossible to sell" ("Sed nolim me tibi, de Eudemia, Pinacotheca, aliisque ejusmodi nugis, penitus esse notum atque perspectum. Etenim hæc per ludum jocumque composui; atque eo facilius ad hanc animi hilaritatem ludumque descendi, quod viderem, ludicras res ac leves in honore & gratia esse, omnium ad se voluntates allicere, emtorem invenire; graves vero & serias, tanquam tristes, molestas, odiosas, atque invendibiles merces, fere ab omnibus rejici"). Tyrrhenus was Rossi's nickname for his patron, Cardinal Fabio Chigi, who would later become Pope Alexander VII. Despite being officially dedicated to Maximilan Henry of Bavaria, the Archbishop Elector of Cologne, by printer Cornelius von Egmondt, the *Pinacotheca* was in fact dedicated to Chigi, who was then papal nuncio in Cologne: see Rossi's dedicatory letter in *Pinacotheca* (1633), ff. 6r–7r.

76 This explains the choice of publisher and place of publication: officially printed in Cologne by Cornelius von Egmondt, the *Pinacotheca* was, in fact, printed in Amsterdam by Joan Blaeu. The stratagem allowed authors working in Catholic countries to profit from the more liberal Dutch censorship, which would grant an imprimatur that would have been much more difficult to obtain in the Rome of Pope Urban VIII.

We may understand, then, why Rossi's portrait of Galileo is much less detailed and less carefully worded than Allatius's. Rossi was no librarian, keen to provide exhaustive and precise information about authors and works, both published and in progress. Nor was his aim to document the intellectual support of Urban VIII's pontificate. Accuracy was not the most important feature of Rossi's style—to the point that he often did not care to distinguish historical truths from widespread rumors or deliberate falsifications. For instance, the case of Galileo's illegitimate birth is claimed by Rossi at the very beginning of his biographical sketch (and later reported by other cut-and-paste biographers).[77]

However, whereas Rossi's diminishing of Galileo might be read as due to the overall caricatural tone of the *Pinacotheca*, Rossi's taking the side of the Church over the hot issue of the trial was undoubtedly serious:

> And while he was observing such enormous objects, very far removed from our eyes, if only he would have taken into account the weakness of these very eyes, which often get fooled by objects close to them and very clear! Certainly he would not have declared to see what he never actually saw, nor would he have attempted, by publishing books, to overthrow and shake things that rest upon the testimony of the Holy Scriptures, the concord of the Holy Fathers, and the truth of the Catholic faith. Furthermore, summoned to Rome by the Inquisitors, he would not have been forced to recant, and publicly state that the heavens move and the earth is at rest, against what he had taught before a wide audience.[78]

---

77 See Rossi, pp. 114–115; and Viviani 1702, pp. 268–269. In the *Evdemia*, Rossi poked fun at Campanella (who had supported Galileo in print, when Cardinal Bellarmino had admonished him, in 1616: see the appendix) and Galileo: "There were two men, the most learned of all: one of them, named Crepitaculus [i.e., "rattle": *campanella* means "little bell"], was a philosopher of the Convent and Order of the Dominicans. They said amazing things about him, almost supernatural. Certainly, he had completed his course of studies and stood out more as a teacher than as a student, and when still a child, almost a young man, he had learned everything, and memorized the doctrines of Aristotle and all philosophers. Furthermore, they said he had long and vigorously battled the waves of a bitter fortune and all sorts of hardships. Finally, pitiful and prostrated, thanks to the support of a powerful king [i.e., the pope], he had landed in a harbor, so to speak, and led a safe and peaceful life, devoting himself to literary studies [possibly, the *Commentaria* on Urban VIII's Latin poems: see the appendix]. The other man [i.e., Galileo] was said to be especially skilful in mathematics, even to an extraordinary degree, and from this ability he had earned patronage and friendships. He has left for posterity the memory of his name, for, with the aid of an eyepiece he invented, which is usually referred to as optical tube, things that are almost infinitely remote from us, and even distant from our understanding, become close to our eyes and look much bigger" ("aderant viri duo, omnium doctissimi, alter, ejusdem quo philosophus ille instituti atque ordinis, Crepitaculo nomine; de quo mira quædam narrabantur, ac pene extra naturæ ordinem; nimirum eum, cursum suum transcurrisse, atque adeo magistrum antea, quam discipulum, extitisse: nam etiam tum adolescentem, ac pene puerum, omnia habuisse percepta, ac memoria comprehensa, quae Aristotelis omniumque Philosophorum placitis continentur. addebant eundem, adversis fortunæ fluctibus diu multumque jactatum, & ad varios ærumnarum scopulos, miserandum in modum, afflictum, demum Regis potentissimi patrocinio, tanquam portu exceptum, tutam ac quietam in litterarum studiis vitam exigere. alter vero, qui aderat, mathematicis disciplinis ad miraculum excultus esse dicebatur, unde etiam sibi opes & amicitias peperisset; verum ex oculari, quod tubum opticum vocitant, à se invento, quo res, infinito propemodum intervallo à nobis disjunctæ, atque adeo ab aspectus judicio remotæ, oculis subjiciuntur vel aliquanto majores repræsentantur, nominis sui memoriam posteris reliquisset": Rossi, *Evdemiæ libri decem* (1645), Book X, p. 239.

78 Rossi, pp. 114–115.

The *Dialogue* is mentioned, too, but only to call attention to the "abominable opinion on the revolution of the earth around the heavens that are at rest."[79] Few words, tersely describing Galileo's last years in Arcetri, lead to the "epitaph" that eventually seals the entry, which sounds very much like an admonishment to any future similar attempts:

> He had been sentenced by the Inquisitors not to cross the boundaries of his villa as a punishment for demanding to be wiser than Wisdom itself, according to which the world was set up from the beginning.[80]

Vittorio Siri

Francesco Siri (1608–1685) studied at the Benedictine convent of San Giovanni Evangelista, in Parma, where he pronounced his vows on December 25, 1625, choosing the name Vittorino, although he later always used Vittorio.[81] At first, he specialized in geometry, and taught mathematics in Venice. There he befriended the French ambassador and took a liking to political matters. In 1640, under the pseudonym Capitano Latino Verità Monferrino, he published a book on the occupation of Casale Monferrato (*Il politico soldato Monferrino*), in which he defended the French position. This earned him the patronage of Cardinal Richelieu, who granted Siri access to his own archives. Based on what he found in the archives, Siri set up to publish *Del Mercvrio Ouero Historia De' correnti tempi*, a monumental work in 17 books, which was published in Venice between 1644 and 1682.[82] The key feature of the work is Siri's weaving the historical narrative with letters, private reports, and secret documents he managed to get from ambassadors, papal nuncios, and ministers from various European courts. Readers felt as if they could actually take a behind-the-scenes look at historical events, and this ensured *Del Mercvrio*'s great success.[83]

The biographical sketch of Galileo is appended at the end of Book III, followed by that of the Italian artist Guido Reni. Siri offers no special reason for their inclusion;

---

79 Rossi, pp. 116–117.

80 Rossi, pp. 116–117.

81 See Affò, *Memorie degli scrittori e letterati parmigiani*, vol. 5 (1797), no. CCXLVII, pp. 205–236.

82 In the preface to vol. 2, Siri defends himself against the allegations of biased attacks on the Barberini family in vol. 1 of *Del Mercvrio* and provides an explanation for the title of his work: "I fell in love with reading the French *Mercvre*. I let myself be blandished by the wish to bring it to Italy, too, and took the greatest care to find the most reliable information" ("Inuaghito della lettura de' Mercurij Oltramontani mi lasciai lusingar dal genio à trasportarne in Italia l'inuentione, impiegando l'vso di tutte le diligenze per rinuenirne le più sicure notitie": Siri, *Del Mercvrio*, vol. 2 (1647), f. a5v). Indeed, the anti-Barberini tone caused various problems that initially prevented Siri from obtaining the required imprimatur; the work was eventually granted the permission to publish, but the actual publisher (Tommaso Baglioni, in Venice, the same who published Galileo's *Sidereal Messenger*) was cautiously changed into Cristoforo della Casa, in Casale. *Le Mercure francoys ov Svitte de l'histoire de la paix* was France's first literary gazette, founded in 1611 by the Paris booksellers Jean and Estienne Richer.

83 For what follows, see Ruffo, "Da Firenze a Lima" (2005). Ruffo also identified a Spanish translation of Siri's biographical sketch of Galileo, by Juan Vasquez de Acuña: *Galileo Galilei filosofo y matematico el mas célebre*, in four pages, published in Lima, Peru, on September 25, 1650. The work is dedicated to Juan de Figueroa, ruler of Lima, and testifies to the success and vast circulation of Siri's work.

he is simply doing so "in imitation of the most praiseworthy authors."[84] As Patrizia Ruffo showed, however, the original manuscript of Vol. 2 of *Del Mercvrio* does not include the short biographies and actually ends just before them. The lives of Galileo and Reni were not part of the original plan, and it also seems that they were not written before October 1646, that is, when volume 2 of Siri's work was being printed.[85] Behind the decision to include such a personal tribute to Galileo was arguably Siri's wish, through it, to pay homage to the Medici family. His meddling with political matters had turned Siri into a persona non grata, and in 1646 he had to leave Venice. At the time he was writing the second book of *Del Mercvrio* he was in a precarious position and was likely trying to keep on good terms with those who had shown benevolence toward him.

In 1610 Galileo had named the newly discovered Jupiter's satellites after the Medici family (*Medicea sidera*), and the Grand Duke reciprocated the homage by appointing Galileo as his Chief Mathematician and Philosopher, allowing him to return to Tuscany after eighteen years at the University of Padua (which was then part of the Republic of Venice). Accordingly, Siri's celebration of Galileo was, in fact, a celebration of the Medici and a way to ingratiate himself in view of possible future patronage. At the same time, Siri had to handle the matter of the trial and condemnation of Galileo with extreme care and had to share the Grand Duke's policy on the issue.

As Paolo Galluzzi has argued, Grand Duke Ferdinando II and his brother Prince Leopoldo encouraged a "strategy of conciliation" with the Church hierarchy:

> In order to continue to exploit the extraordinary celebratory potential of the protection given by Cosimo I and Ferdinando II to Galileo, and of the dedication of the satellites of Jupiter to the Medici dynasty, it was necessary to blur the conflict with the ecclesiastical hierarchy. It seemed opportune to stimulate, on the one hand, the exaltation of the Galilean research tradition as an instrument particularly well suited to display the extraordinarily ordered structure of nature, an evident demonstration of the omnipotence of its Author. On the other hand, every care was taken to erase the intellectual trauma of the trial and condemnation of the Pisan scientist from the general memory.[86]

Accordingly, Siri opens his biographical sketch by recalling Galileo's role as "the most eminent philosopher and mathematician" of the time and contrasting his clarity and his ability "to illuminate and teach the coarsest intellects" with the obscurity of period mathematicians such as Kepler and Viète, who "seem to show off their knowledge by preventing others from understanding them, passing off mathematics as *Sibylline Oracles*."[87] Perhaps more importantl Siri calls attention to the practical aspects of Galileo's science:

---

84 Siri, pp. 120–121.
85 See Ruffo, "Da Firenze a Lima" (2005), pp. 187–188.
86 Galluzzi, "I Sepolcri di Galileo" (1993/1998), pp. 422–423.
87 Siri, pp. 120–121.

The sharpness of his mind, most prolific with speculations, is primarily manifested in the invention of many instruments very useful for civic life. He succeeded in adapting mathematics to philosophy so as to show clearly to us the necessary relationship between them, demonstrating that in order to deeply understand the philosophers' theories it is necessary to appeal to mathematical knowledge.[88]

Siri then focuses on the news about the Dutch spyglass and the invention of the telescope (and its applications). Most notably, alone among the authors of early biographical sketches, Siri goes into some detail about Galileo's telescopic discoveries: the rough surface of the Moon, its librations, the plurality of stars in the Milky Way, "three-bodied" Saturn, sunspots, the phases of Venus, and "the significant change in size of the apparent diameters of Venus and Mars, which had considerable implications, and was required by the theories of the two great astronomers Copernicus and Tycho."[89] Specific references are made to the discovery of the "Medicean Stars" and their possible use for the resolution of the problem of longitude. The *Dialogue* is mentioned, too, as well as—for the first time—Galileo's *Two New Sciences*.

Finally, following the Medici's conciliatory strategy, no mention is made of the trial or the sentence. A historian, Siri was tactfully vague about the end of Galileo's life: "He lived the last eight years of his life outside Florence, partly in some villas close to the city, and partly in Siena." And he ended his life as a pious man:

Galileo endured blindness with a strong and truly philosophical spirit, rising up from that misery by way of relentless speculation; he had already gathered a large amount of material and begun to dictate his ideas. And then he died without pain, after a three-month long illness, ending his days in a most Christian manner [...].[90]

The last paragraph is a description of Galileo's character and multifaceted interests, following the model adopted by Viviani and Gherardini. Most likely, Siri received detailed information about Galileo's achievements and bibliography from the Grand Duke's Secretary of State, Giovan Battista Gondi, who, in turn, might have received a report from Viviani himself, who would later be asked to write Galileo's life by the Grand Duke himself. This remains a mere conjecture, although a number of similar expressions and stylistic features seem to establish a connection between Siri's life and the *Racconto istorico* beyond any doubt.[91]

LORENZO CRASSO AND ISAAC BULLART

Lorenzo Crasso's (1623–1691) biographical entry on Galileo in his famous *Elogii d'hvomini letterati* (1666) is the most baroque portrait of Galileo in this collection,not only for its flowery style but also for the author's choice to supplement it

88 Siri, pp. 000.
89 Siri, pp. 000.
90 Siri, pp. 000.
91 See Ruffo, "Da Firenze a Lima" (2005), pp. 190–192.

with Galileo's epitaph and two encomiastic poems: Francesco Stelluti's celebration of Galileo, first prefaced to *The Assayer*, in 1623; and Giovan Battista Marino's epigram, in a gallery of poetical portraits of eminent people (1620). More than any other work considered here, Crasso's belongs—by its very title—to the literary genre of the *elogium*, intermediary between prose and poetry.

Based on Rossi's opinionated entry in the *Pinacotheca imaginvm illvstrivm*, which was possibly the only published account of Galileo's life readily available at the time, Crasso's short biography betrays a rather surprising ignorance about the subject or, at the very least, levity and carelessness. Indeed, concise as it may be, Crasso's entry is filled with errors. Shortly after the publication of the *Elogii*, Michelangelo Ricci (1619–1682), a pupil of Benedetto Castelli's and a well-known mathematician, pointed them out to Prince Leopoldo:

> Lorenzo Crasso's *Elogii d'hvomini letterati* has just been published. I was very sorry to notice, in the entry on Galileo, that Crasso distinguishes between the Medicean Stars and Jupiter's satellites; that the list of Galileo's published works lacks a substantial number of them; that he describes Galileo as afraid of making public his views on natural philosophy. Also, he states Galileo drew such views from Celio Calcagnini, and Patricius, glossing over Benedetti,[92] who opened the way to Galileo more than anybody else, and was perhaps Galileo's single guide in his philosophizing—as Your Lordship would have noticed, comparing their respective theories, which are so similar. Crasso makes Galileo an illegitimate son, which is totally false; and he also says some other things about Galileo which do not make a whole lot of sense. And whereas as far as Galileo's published works are concerned, Crasso is unworthy of any justification, as it would have been easy for him to see the list in the two volumes printed in Bologna;[93] he deserves to be excused on other issues. Because there is no proper and detailed biography of this

92 Giovanni Battista Benedetti (1530–1590), Italian mathematician and advocate of the Copernican hypothesis. In order to establish priority in what he considered an important discovery, in the prefatory letter to the *Resolvtio* (1553), Benedetti announces to his patron, Abbot Gabriel Guzman, a new (anti-Aristotelian) doctrine on the subject of falling bodies, according to which the (average) speed of free fall is proportional to the weight of a unit volume of the body when weighed in the medium through which it passes; as a consequence, bodies of the same material and shape would fall with the same speed in a given medium, regardless of their size. In the *Demonstratio* (1554), Benedetti purported to give further support to this new doctrine of free fall and to refute the basic Aristotelian propositions that speed of free fall is directly proportional to the gross weight of the body and inversely proportional to the density of the medium. There are two different editions of the *Demonstratio*, although with the same imprint. Whereas in the *Resolvtio* and a first version of the *Demonstratio* the effect of the medium is merely one of buoyancy, in the second version it is implied that there is an additional resistance, namely, the friction encountered by a body as it makes its way through a medium. The implication is that this resistance is proportional not to the volume or weight of the body but to its cross section, or surface. All this is taken up again in Benedetti's *Diversarvm specvlationvm Mathematicarum, & Physicarum Liber* (1585), where the equality of speed of homogeneous bodies is proved only in vacuum. As Israel Drabkin has argued, Galileo's doctrine in *On Motion* accords with that in *Resolvtio* and in the first version of the *Demonstratio* but not with the second, or with the later *Diversarvm specvlationvm Mathematicarum, & Physicarum Liber*; see also the sections from Benedetti's works, as translated in *Mechanics in Sixteenth-Century Italy* (1969), pp. 147–237. Galileo seems to be aware of the problem of resistance apart from buoyancy\, but avoids it by considering only cases where it may be excluded as negligible: see *On Motion*, p. 30 (*OG* 1, p. 266). On the whole issue, see Drabkin, "Two Versions of G. B. Benedetti's *Demonstratio Proportionum Motuum Localium*" (1963).

93 *Opere 1655–1656*, edited by Carlo Manolessi.

great man, by anybody, I would greatly appreciate if this would be an incentive for Viviani to publish his own [...].[94]

As to Galileo's invention of the telescope, although acknowledging his originality and inventiveness, Crasso refers to Giovan Battista Della Porta, who studied the properties of combinations of convex and concave lenses and their magnification power, but these were unable to produce sharp images. Much like Rossi's (without yielding to his sarcasm, though), Crasso's aim—superficial, secondhand, and ill-informed as his account proves to be—seems to be to diminish Galileo's works and discoveries.

When he comes to the most controversial issue, the trial of 1633, Crasso seems to attribute full responsibility to Galileo for his decision to advance opinions "about the rest and motion of the earth and the heavens against what the Roman Church had established."[95] Galileo irresponsibly let his mind travel unconstrained by the accepted wisdom, thereby crossing a boundary he was not supposed to cross. And this was not all: Galileo "went through many troubles, which were assuaged by the advice and help he received from powerful people."[96] Here, Crasso seems to be implying that if Galileo suffered for the consequences of his reckless opinions (what consequences, he does not say), he was spared worse ones; he might have suffered even more had it not been for the good offices of powerful people (who they were, we may only guess). Crasso's remarks sound like a coded message, or an uncomfortable warning.

Like Freher's, Isaac Bullart's (1599–1672) posthumous portrait of Galileo in Book II of the *Académie des Sciences et des Arts* (1682) suffers from the author's failure to complete his research. The tone very much resembles Crasso's, but Rossi was undoubtedly one of the sources. Not only does Bullart repeat the story of Galileo's illegitimate birth, but he does not mention the *Dialogue* or the *Two New Sciences* and remains silent about the trial: Galileo "happily lived to the age of 84 years,"[97] says Bullart repeating Rossi's mistake. Bullart's final summary is too hasty, though, and fails to reproduce Rossi's (or Crasso's) sneaky rhetoric: "he passed away in the year

---

94 Michelangelo Ricci to Prince Leopoldo, 14 November 1666, in Fabroni, *Lettere inedite di uomini illustri* (1773–1775), vol. 2, pp. 142–143: "Sono usciti gli Elogi di uomini letterati di Lorenzo Crasso, e dove parla del Galilei mi ha mosso a compassione il vedere, che egli distingua le stelle Medicee dai pianeti Gioviali: che nel catalogo dell'opere stampate dal Galileo ne lasci fuori una gran parte: che rappresenti il Galileo tanto timido in dar fuori i suoi primi sentimenti circa la Filosofia naturale, i quali vuole che gli cavasse da Celio Calcagnino, e dal Patrizio, tacendo il Benedetti che gli aprì la strada più che ogni altro, e forse fu solo a lui scorta nel suo filosofare, come avrà ben notato V. A. paragonando i concetti dell'uno e dell'altro, che sono tanto conformi. Lo fa bastardo, che è falsissimo, e dice qualche altra cosa di lui poco sussistente; e sebbene quanto ai libri stampati egli non è degno di scusa, perché facil cosa gli era il vederli ne' due tomi stampati in Bologna, merita però scusa in altro, non essendovi di questo grand'uomo la vita bene e copiosamente scritta da nessuno. Goderei se ciò fosse stimolo al Sig. Viviani di pubblicar la sua [...]." As Findlen has observed, Ricci—who was created a cardinal on September 1, 1681, a few months before he died—perceived the urgency of writing an honest and reliable life of Galileo as "a vehicle through which to broker a better understanding of the relationship between science and faith, doing justice to Galileo's merits while also confronting, with the greatest tact and diplomacy and with full recognition for the authority of the institutions that had passed judgment upon Galileo, the implication of his trial": Findlen, "Rethinking 1633" (2012), p. 211.

95 Crasso, pp. 196–197.

96 Crasso, pp. 196–197.

97 Bullart, pp. 240–241.

1642, in a villa on the outskirts of Florence, where he had been confined by a sentence of the Inquisitors of Faith as a punishment for his attempt to equal the uncreated Wisdom, which created the heaven and Earth."[98]

JOACHIM VON SANDRART

Joachim von Sandrart's (1606–1688) short biography of Galileo occupies a very special place among the early ones of Galileo. Sandrart was born in Frankfurt am Main, and traveled extensively in northern Europe and Italy, training first as a printmaker and later earning a reputation as a painter of portraits. After a long and successful career in the foremost artistic circles of England, Italy, and the Netherlands, he returned to Germany and painted for Holy Roman Emperor Ferdinand III, who ennobled him in 1673. In the latter part of his life, Sandrart settled in Nuremberg and turned to publishing projects and teaching. His most important publication is *L'Academia todesca della Architectura, Scultura & Pittura, oder Teutsche Academie der Edlen Bau- Bild- und Mahlerey-Künste*, published in Nuremberg in 1675–1679, in five volumes—the first comprehensive treatise on the history of art written in German. Volume 1 offers theoretical and historical accounts of the arts of architecture, sculpture, and painting, lavishly illustrated with hundreds of prints, many depicting objects in notable collections. It also includes sections on the lives and stylistic characteristics of famous artists from classical antiquity to Sandrart's own time, which earned Sandrart the nickname "The German Vasari." A second volume, published in 1679, expanded on these areas and added a translation of Ovid's *Metamorphoses*. A few years later, Sandrart published an abridged Latin translation of the biographies, the *Academia nobilissimæ artis pictoriæ*, 1683.

Both the German and Latin editions of the *Academia* are collections of biographies of eminent artists, painters, and sculptors, prefaced by short theoretical treatises. In the abridged Latin translation, the most notable exception is the life of Galileo, the only scientist in the collection.[99] Sandrart met Galileo in 1633, in Rome, during the months of the trial. Indeed, whereas Sandrart relies on Gian Vittorio Rossi for information about Galileo's life before 1633—also reporting his ironic remarks on Galileo's generous salaries in both Padua and Pisa—as soon as he gets to about the time when he met with Galileo, Sandrart does not refrain from telling the story from his own, unusual perspective, openly distancing himself from Rossi's harsh remarks.

In the *Dialogue*, says Sandrart, Galileo "disclosed his brilliant theory on the revolution of the earth around the heavens, which are at rest." As a consequence, "he was later summoned to Rome by the Inquisitors and forced to recant." A German writing in a Protestant country, far removed from the constraints imposed by the Catholic Church and with no interest in any attempt at reconciliation, Sandrart is the first, however briefly, to refer to the trial and its consequences in no unclear terms.

---

98 Bullart, pp. 240–241.

99 Another exception may be the inclusion of the life of Athanasius Kircher (1602–1680): see von Sandrart, *Academia nobilissimæ artis pictoriæ* (1683), no. LIX, p. 391 (shortly after Galileo's). Kircher, however, was a German Jesuit who studied at Fulda.

The last part of the account, however, is even more interesting, as Sandrart vividly reports his visits to Galileo in Rome while Galileo was a guest at the residence of the Grand Duke's ambassador:

[...] I was on friendly and kind terms with him, in the *Medicean Palace* in Rome, while he was dealing with the Inquisition. The palace was a true repository of the whole of antiquity, and an extraordinary museum displaying the rarest objects; there, I enjoyed to the highest degree the study of *optics* and *geometry*, and from that eminent scholar and teacher I learned what the world and I plainly ignored. What more [can I say]? By means of his [optical] tube, effortlessly aimed at the Moon, from his bedroom he showed to [my] eye mountains, valleys, forests, lands, lights, shadows, and all sorts of other things. This honor which I received from him led and encouraged me to pay homage to his portrait, by paying, in turn, due tribute.[100]

We can only speculate about Galileo's thoughts and feelings during those days and weeks. Still, Sandrart's account—whose authenticity seems beyond doubt—offers a personal and lively account of Galileo's mood, his availability to scholars and visitors, and his willingness to entertain intellectual conversations. To reciprocate Galileo's kindness, Sandrart painted a portrait of him, from which Bartholomäus II Kilian drew the engraving facing Galileo's short biography in the *Academia*. Unlike most other portraits of Galileo, which may easily be traced to a few specific sources (namely, Francesco Villamena's engraving, in *On Sunspots*, 1613; and Justus Suttermans' painting, in 1635),[101] Sandrart painted Galileo in profile, like an ancient bust.

It seems most fitting that Sandrart's short biography confirms Galileo's interests in visual arts,[102] for Sandrart himself was an artist and scholar with profound interests in the natural world who liked to visit and paint naturalistic locations, such as Mount Vesuvius, the Phlegraen Fields, Stromboli, and Mount Etna.

FAILED PROJECTS AND ELUSIVE BIOGRAPHIES: VOGEL, SALUSBURY, AND BALDIGIANI

On February 2, 1663 *ab Incarnatione* (that is, 1664), the Florentine nobleman Carlo Roberto Dati (1619–1676)—a pupil of Galileo's and one of the founders of the Accademia del Cimento—wrote in a letter to Viviani:

Two days ago Martin Vogel, of Hamburg, visited me here [in Florence]; he is very much interested in natural and mathematical issues, and a very learned young man. I understand he got to know Your Lordship in Rome. He is very devoted

---

100 Von Sandrart, pp. 246–247.

101 See *OGA* 1, D6, pp. 34–35; and D12, pp. 49–51: both portraits are frontal.

102 See, for example, Panofsky, *Galileo as a Critic of the Arts* (1954), and the correspondence between Galileo and painter and architect Lodovico Cardi, also known as Cigoli, in Tognoni, *Il carteggio Cigoli–Galileo* (2009).

to the memory of Galileo, and he is gathering material about all the Lynceans in order to write an accurate history of the Academy [of the Lynxes]. It seems he is hoping to obtain information about Galileo from you, and was very sorry not to find you in Florence.[103]

A member of the *Natio Germanica* at the University of Padua, Martin Vogel (1634–1675) graduated in philosophy and medicine on January 19, 1663, and traveled at length in Italy—as was customary for foreign students, at the time—and particularly visited Rome. There he met with Viviani but was unable to obtain information about Galileo from him. He did manage to gather a number of letters of Galileo's, though, but ended up losing the papers in Florence. He later found two more letters in France, but a number of personal mishaps eventually caused Vogel to abandon the project of a group biography of the Lynceans in 1672.[104] The following year, Vogel resumed it once again, mentioning "the loss of the *Life of Galileo* printed in England" as one of the reasons that prompted his decision.[105] Vogel's untimely death, in 1675, seems to have prevented him from completing or publishing his work, some claims to the contrary notwithstanding.[106]

Vogel's mention of an English biography of Galileo is worth noting. Most likely, he was referring to Thomas Salusbury's (d. 1665–1666) *Life of Galileo*, the first substantial biography of Galileo in any language to appear in print.[107] The work was part of a larger publishing venture, *Mathematical Collections and Translations*, published in two volumes. Each volume comes in two parts, with Salusbury's English translations of Galileo's major works, each with a distinctive title page. Volume I includes the *Dialogue* (*The Systeme of the World: in Four Dialogues. Wherein the Two Grand Systemes Of Ptolomy and Copernicus are largely discoursed of: And the Reasons, both Phylosophical and Physical, as well on the one side as the other, impartially and indefinitely propounded*); the *Letter to Christina* (*The Ancient and Modern Doctrine of Holy Fathers, and Iudicious Divines, concerning The rash citation of the Testimony of Sacred Scripture, in Conclusions meerly Natural, and that may be proved by Sensible Experiments, and Necessary Demonstrations. Written, some years since, to Gratifie The most*

103 "Due giorni sono fu qui da me il Sr Martino Fogelio di Amburgo, giovane curiosissimo delle cose naturali e delle matematiche e grandemente erudito; mi pare dica aver conosciuto V. S. in Roma, è divotissimo della memoria del Sr Galileo. Raccoglie le memorie di tutti i lincei per iscrivere la storia dell'Accademia esattamente, e da lei par che speri le notizie circa il Sr Galileo, onde gli è doluto in estremo non l'aver trovato in Firenze": Carlo dati to Viviani, 2 February 1664, in *Gal.* 254, f. 228*v*.

104 See Favaro, "Serie ventesimaterza di scampoli galileiani: CXLIX. Un mancato biografo di Galileo" (1992), especially the 1664 correspondence between Giovanni Alfonso Borelli and Antonio Magliabechi, on pp. 734–736, and Vogel to Magliabechi, 19 March 1673, ibid., pp. 737–738.

105 "[. . .] il mancamento della Vita del Sig. Galileo stampata in Inghilterra": Vogel to Magliabechi, 19 March 1673, ibid., p. 738.

106 See ibid., pp. 738–739.

107 All my efforts notwithstanding, I have been unable to see the text myself. Most of what follows relies on Nick Wilding's study of Salusbury's biography of Galileo: "The Return of Thomas Salusbury's *Life of Galileo* (1664)" (2008). Wilding saw the only surviving copy of *Mathematical Collections and Translations*, Vol. II, Part 2, at the time it was auctioned. The copy is now in private hands, and inaccessible to scholars. See also Favaro, "Rarità bibliografiche Galileiane: III" (1889) and "Serie ventesimaterza di scampoli galileiani: CXLIX. Un mancato biografo di Galileo" (1914), and Drake, "Introduction" (1967). On Salusbury, see Favaro, "Serie sesta di scampoli galileiani: XLII. Ulteriori notizie intorno alla traduzione inglese di alcune Opere di Galileo" (1891); Drake, "Galileo Gleanings II. A Kind Word for Salusbury" (1958); and Zeitlin, "Salusbury Discovered" (1959).

Serene Christina Lotharinga, Arch-Dutchess of Tvscany); Paolo Foscarini's *Lettera . . . Sopra l'Opinione de' Pittagorici, e del Copernico* (1615); and Benedetto Castelli's *Della misvra dell'acqve correnti* (1628). Volume II, Part 1, includes Galileo's *Two New Sciences* (*Mathematical Discourses and Demonstrations touching Two New Sciences; pertaining to the Mechanicks and Local Motion, with an Appendix of the Centre of Gravity Of some Solids*), *On Mechanics* (*Mechanicks: of the Benefit Derived from the Science of Mechanicks, and from its Instruments*);[108] the *Balance* (*The Ballance; In which, in imitation of Archimedes in the Problem of the Crown, he sheweth how to find the proportion of the Alloy of Mixt-Metals; and how to make the said Instrument*); *On Floating Bodies* (*A Discourse Concerning The Natation of Bodies Vpon, And Submersion In, the Water*); and texts by Descartes, Archimedes, and Tartaglia. Volume II, Part 2, includes three works: Torricelli's treatise on projectiles; Salusbury's *Experiments Statistical, Hydrostatical, and Aerostatical*; and his *Galileus Galileus His Life*.[109]

The one copy of Volume II, Part 2, to survive the Great Fire of London in 1666 was never reprinted and has been lost since the mid-nineteenth century. It reemerged in the early 2000s and was auctioned in London, at Sotheby's, on October 26, 2005, together with the rest of the library of the Earls of Macclesfield, one of the most important British private collections relating to the history of science. The full title of Salusbury's biography of Galileo reads, *The Life Of the Most Excellent Philosopher and Mathematician, Galileus Galileus, A Gentleman of Florence. Extraordinary and Primary Professor of Philosophy and Mathematicks Unto the Most Serene Grandduke Of Tuscanie*; the stated publication date is 1664, in London. The biography of Galileo is divided into five chapters. Here is the table of contents as Salubury himself presented it at the beginning of Volume I, Part 1:

GALILEUS GALILEUS, his LIFE : in Five BOOKS,
BOOK   I.   Containing Five Chapters.
    Chap.   1. His Country.
        2. His Parents and Extraction.
        3. His time of Birth.
        4. His first Education.
        5. His Masters.
  II.   Containing Three Chapters.
    Chap.   1. His judgment in several Learnings.
        2. His Opinion and Doctrine.
        3. His Auditors and Scholars.

108 An unpublished English translation of Galileo's *On Mechanics*, made by Robert Payne in 1636 at the request of William Cavendish, existed prior to Salusbury's version but remained unknown to him. The manuscript is available in London, at the British Library, Harley MS. 6796, ff. 317–330. For his translation, Salusbury likely used the first collected edition of Galileo's works (1655–1656), where *Mechanics* also directly follows *Two New Sciences*.

109 The title page of Vol. I and those of its individual sections bear the date 1661. The title page of Vol. II, Part 1, claims it was published in 1665, but the title pages of its individual sections range from 1663 to 1665. Vol. II, Part 2 (only nine copies of which are known to exist) has the overall date 1664. As Nick Wilding convincingly argued, at the time of the Great Fire of London Vol. II, Part 2, "was still in the process of printing, revision, and perhaps even writing": Wilding, "The Return of Thomas Salusbury's *Life of Galileo* (1664)," p. 248; see also ibid., pp. 250–251. Vol. I was reissued by George Sawbridge in 1667.

III. Containing Four Chapters.
    Chap. 1. His behaviour in Civil Affairs.
           2. His manner of Living.
           3. His morall Virtues.
           4. His misfortunes and troubles.

IV. Containing Four Chapters.
    Chap. 1. His person described.
           2. His Will and Death.
           3. His Inventions.
           4. His Writings.
           5. His Dialogues of the Systeme in particular, containing *Nine Sections.*
              Section 1. Of Astronomy in General; its Definition, Praise, Original.
                  2. Of Astronomers: a Chronological Catalogue of the most famous of them.
                  3. Of the Doctrine of the Earths Mobility, *&c.* its Antiquity, and Progresse from Progresse from *Pythagoras* to the time of *Copernicus.*
                  4. Of the Followers of *Copernicus,* unto the time of *Galileus.*
                  5. Of the severall Systemes amongst Astronomers.
                  6. Of the Allegations against the *Copern.* Systeme, in 77 Arguments taken out of *Ricciolo,* with Answers to them.
                  7. Of the Allegations for the *Copern.* Systeme in 50 Arguments.
                  8. Of the Scriptures Authorities produced against and for the Earths mobility.
                  9. The Conclusion of the whole Chapter.

V. Containing Four Chapters.
    Chap. 1. His Patrons, Friends, and Emulators
           2. Authors judgments of him.
           3. Authors that have writ for, or against him.
           4. A Conclusion in certain Reflections upon his whole Life.[110]

Although a few other notices of Galileo were in print (Girolamo Ghilini's and Leo Allatius's, 1633; Gian Vittorio Rossi's, 1643; and Vittorio Siri's, 1647),[111] the

---

110 Salusbury, *Mathematical Collections and Translations* (1661–1665), Vol. I, Part 1, f. *3v. The actual book and chapter descriptions in the Macclesfield copy present some differences with the project Salusbury laid out in 1661.

111 Indeed, in his biography of Galileo, Salusbury explicitly refers to "that Smooth-Tongu'd Courtier, *Janus Niciis* [*sic*] in his well-known *Pinacotheca Imaginum Illustrium*," whom he calls "a Creature and Dependent upon our Hero's most Mortal Enemies the *Barbarini*": Salusbury, *Mathematical Collections and Translations* (1661–1665), Vol. II, Part 2, f. P1r and f. U2r, quoted in Wilding, "The Return of Thomas Salusbury's *Life of Galileo* (1664)" (2008), p. 258.

main source of information was Viviani. Salusbury felt he could not rely on other sources, and was afraid someone might mislead him on purpose.[112] In Salusbury's own words, reporting his progress to the reader:

> And thus you have the best account I, at present, have been able to give you of the Works of *Galilaeus*; but I hope in a few months to receive those Epistles above named, and the Life of the Author, from the hands of the Great mathematitian [*sic*], and his worthy Disciple, Sig. *Vincenzo Viviani*, which once obtained, I shall ingeniously retract what I have through misinformation erred in.[113]

However, while Viviani sought for, and at times demanded, support and cooperation from all members of the Republic of Letters in Europe in order to gather Galileo's correspondence and related documents, not only did he hesitate to publish the *Racconto istorico* but was frustratingly unwilling to provide other scholars with the substantial help they required to undertake their own biographical projects. Just like Vogel's (although Vogel believed Salusbury had been more successful), Salusbury's requests for access to the documentation he needed were ultimately refused.

All these difficulties notwithstanding, Salusbury succeeded in making the most of the little information he had at his disposal. Even at the very beginning of his venture, in the letter to the Reader with which Volume I of the *Mathematical Collections and Translations* opens, he takes a clear stance:

> [...] his Holiness thereupon conceived an implacable Displeasure against our Author, and thinking no other revenge sufficient, he employed his Apostolical Authority [...] to condemn him and proscribe his Book as Heretical; prostituting the Censure of the Church to his private revenge.[114]

Later on, as Wilding points out, in his *Life of Galileo* Salusbury does not refrain from spelling out his views on Urban VIII and his family, and offers a challenging political interpretation of the real reasons behind the trial, according to which the Barberini sentenced Galileo to attack his patron, the Grand Duke of Tuscany:

> [*I*]*n Anno* MDCXXXIII. *C. Maffeo Barberini* being Pope by the Name of Urban the VIII, he [Galileo] felt what teeth the Inquisition hath. They grounded their proceedings at this time upon that new Piece of his Printed at *Florence* MDCXXXII. touching the two Systemes, Ptolomaick and Copernican [...]. In plain terms, *Urban* VIII. did not love him; but after he had professed to bear him good-will,

---

112 "[...] the variety of Answers to those Letters I writ to *Italy* on purpose to learn the truth herein being such, that I think it very hard to determine of the matter at this distance": Salusbury, *Mathematical Collections and Translations* (1661–1665), Vol. II, Part 2, f. Ff3*v*, quoted in Wilding, "The Return of Thomas Salusbury's *Life of Galileo* (1664)" (2008), p. 258. See also ibid, p. 259.

113 Salusbury, *Mathematical Collections and Translations* (1661–1665), Vol. II, Part 2, f. Nn2*r*, quoted in Wilding, "The Return of Thomas Salusbury's *Life of Galileo* (1664)" (2008), p. 255.

114 Salusbury, *Mathematical Collections and Translations* (1661–1665), Vol. I, Part 1, f. *2*r*. As Finocchiaro remarked, "Salusbury was the first one to formulate publicly the thesis that the root cause of the trial was Pope Urban's personal anger about being caricatured by the character of Simplicio": *Retrying Galileo* (2005), p. 79.

proceeded against him in this unfriendly manner, cantrary [*sic*] to all the Laws of Amity or Generosity. Add to this, that he and his fastidious Nephews, Cardinal *Antonio* and Cardinal *Francisco Barbarini* (who had embroyled all *Italy* in Civil Wars by their mis-government) thought to revenge themselves upon their Natural Lord and Prince, the Great Duke, by the oblique blows which they aimed at him through the sides of his Favourite [. . .] Nor was this all, for the displeasure of the Pope was backt with the Emulation of many Jesuites, contracted upon him upon *Scheiner's* and *Grassio's* [*sic*] score, as also, for that he had too nakedly exposed Truth; whereas those *Institores Imperii* were not willing she should fully discover her self, since they studiously imitate the Aegyptian Priests, and deliver nothing, save to those of their own Tribe, but in Hieroglyphicks. They therefore accuse him of delivering his pernicious Errors (*si Diis placet*) in the Vulgar Italian.[115]

Determined to preserve his monopoly about the writing of Galileo's life,[116] Viviani not only refused to release any information but also tried his best to control what other biographers wrote about his master, and especially about the trial. On May 26, 1678, Father Antonio Baldigiani (1647–1711)—who had joined the Society of Jesus in January 1662 and was well acquainted with the activities and a number of the principal members of the Accademia del Cimento—sent a letter to Viviani, informing him that a new work by the eminent Athanasius Kircher (1602–1680) was about to be published in Amsterdam. One of the last works of the German Jesuit scholar, polymath, and bestselling author, the *Etruria illustrata, seu de admiranda naturae et Artis quae in Etruria suspiciuntur* (also referred to as *Iter Hetruscum*, namely, *Etruscan Journey*), was to be a historical account of the region of ancient Etruria (i.e., Tuscany), "in which are described and explained the origin, sites, nature, politics, major events, sacred and profane monuments, as well as the natural wonders of Etruria, both at the time of the Roman Republic and at recent times, from three perspectives: political, physical, and geographic."[117]

115 Salusbury, *Mathematical Collections and Translations* (1661–1665), Vol. II, Part 2, f. Dd₁*v*, quoted in Wilding, "The Return of Thomas Salusbury's *Life of Galileo* (1664)" (2008), pp. 259–260.

116 In contrast, Viviani welcomed any opportunities to circulate his own work abroad. In the winter of 1660, in Florence, he met Robert Southwell (1635–1702), an Anglo-Irish gentleman with political aspirations and an interest in natural philosophy, and they started a correspondence that would last nearly forty years. Southwell publicized Viviani's work in England, and Viviani urged his employers to recommend Southwell to King Charles II. In the early 1660s, they also hoped for a collaboration between the Royal Society of London (of which Southwell would become president in 1690–1695) and the Accademia del Cimento. In 1662, Southwell told Viviani he would be honored to translate the *Racconto istorico* into Latin, and finally tell the truth about the trial: "[Nicolas Steno, Christiaan Huygens, and Jacob Golius] all celebrate your venture of [publishing] Galileo's *Works*; and I told all three of them I would be honored to translate the *Life of Galileo* into Latin. Golius is particularly happy about that and urges me to tell the whole truth, and the harsh treatment endured by that famous old man, saying that the world is in earnest to know the truth about that remarkable affair" ("Tutti celebrano la sua impresa delle Opere di Galileo; e a tutti tre ho diciarato l'honor che m'arriuerà per la ductione della Vita di Galileo in Lingua Latina. Il Golio principalmente se n'alegra, instigandomi di pigliar sopra di me la diciaratoria di tutta la verita, e del rigido trattamento di quel famoso veccio, dicendo che 'l Mondo brama di saper la uerita di quel gran Caso"): Robert Southwell to Viviani, 2 February 1662, in *Gal.* 161, ff. 362*r*–363*v*: 363*r*. On Viviani and Southwell, see Boschiero, "Robert Southwell and Vincenzio Viviani" (2009).

117 "*Iter Hetruscum*, quo Hetruriæ tum priscæ, tum tempore Reip. Rom. tum posteræ, origo, situs, natura, politica, catastrophæ, monumenta sacra, profana, nec non naturæ admiranda, triplici ratiocinio, politico, physico, geographico describuntur, & explanantur": de Sepibo, *Romani Collegii Societatus* [*sic*]

Baldigiani had been asked to contribute to the editing of Kircher's volume, and he expanded the section on Tuscan science, rewriting or supplementing it with biographical accounts of some of the most eminent Tuscans (*elogij degli huomini letterati*),[118] among whom were Galileo, Evangelista Torricelli, Alessandro Marchetti, and Vincenzo Viviani. The manuscript of the *Iter Hetruscum* was harshly criticized by Domenico Ottolini, one of the censors who examined it and found it replete with all sorts of errors and inaccuracies.[119] Ottolini convincingly argued that it did not meet Jesuit standards of publication, the book never appeared in print, and the manuscript—together with Baldigiani's biographical accounts—was eventually lost. Before that, however, Baldigiani sent the draft of the accounts to Viviani and asked for his opinion.[120]

Viviani—who had already written (but not published) the *Racconto istorico* and kept revising it in light of the new material he was accumulating—was certainly enticed by Baldigiani's letter and eager to read Galileo's (as well as his own) biographical account. On June 14, 1678, Viviani offered a number of comments and corrections but called special attention to one passage of Baldigiani's life of Galileo that particularly bothered him: "*qui si in nonnullis cautior fuisset . . .*," namely, "had he [Galileo] been more cautious in some things. . .".[121] Viviani explicitly asked Baldigiani to delete this remark.

Four days later, in his reply, Baldigiani informed Viviani that Kircher, contrary to Baldigiani's wishes, had deleted the paragraph beginning with the words "*Huic obsurgunt . . .*" ("They reprove him [most likely, Galileo]. . .), and was tempted to cross out the rest of Galileo's biography, had it not been for those who were with him.[122] As to the passage Viviani wished to be deleted, Baldigiani replied on a separate sheet of paper:

> It was not at all easy to convince that Father [Kircher] to cross out that passage [. . .]. Lorenzo Magalotti, who is well acquainted with the spirit of this University and Community,[123] and very well knows what Germans mean by *no* and *nein*, will say that I did as much as Charles did in France,[124] in order to bring him to this point, and obtain whatever I could. I add that we should not judge all countries from Florence, and even less from Rome. I am sure that had the Master of the Sacred Palace[125] (who approves of Kircher's books without even looking at

---

Jesu Musæum celeberrimum (1678), p. 64, no. †39. On Kircher's *Iter Hetruscum* see Rowland, "The Lost *Iter Hetruscum* of Athanasius Kircher (1665–78)" (2009). The title *Etruria illustrata* imitated that of a previous work of Kircher's, *China . . . illustrata*, published in 1667.

118 Baldigiani to Viviani, 26 May 1678, in Favaro, "Miscellanea galileiana inedita" (1882), p. 128.

119 See Siebert, "Kircher and His Critics" (2004), pp. 84–85.

120 See the documents and discussion in Favaro, "Miscellanea galileiana inedita" (1882), pp. 97–156; see also Findlen, "Living in the Shadow of Galileo" (2009), especially pp. 244–251, which also offers a number of translations of relevant documents, upon which some of the translations presented here were based.

121 Viviani to Baldigiani, 14 June 1678, in Favaro, "Miscellanea galileiana inedita" (1882), p. 135.

122 Baldigiani to Viviani, 18 June 1678, ibid, p. 138.

123 Namely, the Roman College and the Society of Jesus.

124 That is, "I did the impossible": the Italian saying refers to Charlemagne's deeds in France, which became the subject matter of a rich epic.

125 The pope's theologian, Raimondo Capizucchi, who was appointed by Clement X in 1673.

INTRODUCTION   XLIX

them) seen that biographical account, he would not have let it go at all. [...] Now, consider that here we must proceed cautiously, in order not to damage others or ourselves.

Furthermore, those words are for Galileo the more honorable ones in the whole account. An entire Congregation[126] declared him a heretic, temerarious and a contradictor of the Scriptures, etc. Who, having signed such a judgement, would then write: *Had he been more cautious in some things?* No Catholic would speak in such a way about someone he considers a heretic. Hence, he who writes like that is stating, in fact, that he does not have such a bad idea of [Galileo], but another of a much lesser relevance, and, as a consequence, he is acquitting him, or approving of him. I add that the words *had he been more cautious in some things* do not say in what he was defective, whether in the theories [he upheld] or, perhaps, in that he might have been somewhat more cautious in advocating them. This is to turn the case from criminal to civil.

He was warned, interrogated, and condemned: what else could one say? Should I not say what happened, but what it should have been? That he was completely innocent, that an entire Congregation was mistaken, that the most holy tribunal was unfair? Who would ever speak in this way, even if he might believe it? And even if he were to speak like that, whom would he persuade? Is it not better to say that he was mortified, and with some reason (since he provided some cause), that he should have behaved with a little more prudence, that he wounded Urban [VIII] and the Barberini, and gave them cause to be justifiably resentful[?]

I, for one, have always spoken in this way, and believed I served him well. And I think that, were he still alive, he would thank me. I think in order to make sure of the key issue, namely, not to be defamed as heretic, or of dubious faith, he would content to be charged of having failed to adopt all the cautions he could have adopted. *Had he been more cautious in some things.* For sure, were one to write like that about Luther, he would be seen as completely approving of Luther, and not as challenging his errors.[127]

---

126 The Congregation of the Holy Office, that is, the Inquisition.

127 "Stentai più d'un poco ad indurre quel Padre a dar di penna a quel passo [...] Il sig. Lorenzo Magalotti che è pratico del genio di questa Università e Comunità, e che sa quello che significa appresso i tedeschi il no et il nain, dirà che io ho fatto quanto Carlo in Francia a ridurlo a questo segno e vincerla come ho potuto. Aggiungo a questo che non bisogna misurar tutti i paesi da Firenze, e molto meno Roma. Son sicuro sicurissimo che se il Ministro del Sacro Palazzo, il quale a Kircher segna i libri senza vedergli, avesse veduto quell'elogio, non ne avrebbe passato né punto né poco [...]. Hor veda come qui bisogna camminar cauti per non rovinar sé e gli altri. Aggiungo a tutto questo che quelle parole sono le più onorevoli al Galileo che sieno in tutto quell'elogio. Una Congregazione intera l'ha dichiarato eretico, temerario, contraddittore alle scritture etc. chi mai avendo firmato un tal giudizio scriverebbe poi: *qui si in nonnullis cautior fuisset?* Niun cattolico parla così di uno ch'ei stimi eretico: dunque chi così scrive vien in atto pratico a protestare di non l'avere in tal cattivo concetto, ma in un altro di assai minor importanza, e in conseguenza vien ad assolverlo, o ad approvarlo. Aggiungo che le parole *qui si in nonnullis* non esprime in che cosa mancasse se nelle dottrine, o nel modo di proporle con qualche cautela maggiore che poteva usare. Questo si chiama trasferire la causa dal criminale al civile. Fu avvisato, esaminato, condannato: che s'aveva a dire? Non dico quel che fu, ma quel che si doveva dire? che fu del tutto innocente, che tutta una Congregazione errò, che il tribunale più santo fu ingiusto: chi mai parlerà in questa forma, quando ancora lo credesse? E quando anche parlasse a quanti lo persuaderebbe? Non è meglio dire che fu mortificato, e con qualche ragione,

Baldigiani, himself a Jesuit, does not refrain from expressing some disagreement with the Inquisitors and wants Viviani to know that, and to acknowledge the magnitude of his achievement.[128] More could not be done and should not be expected: the trial and the sentence of Galileo cannot be completely passed over in silence. Almost half a century after the events, both Viviani and Baldigiani were trying to move on. Both were hoping to come to terms with the Church hierarchy, rehabilitate Galileo, and recover his intellectual and scientific legacy. But whereas in order to do so Viviani sought to present Galileo as a pious Catholic who had simply failed to make it clear that Copernicanism was nothing but a mathematical hypothesis, as he did in the *Racconto istorico*, Baldigiani opted for acknowledging the errors of the past and reinterpreting the events for the greater good.

The two never resolved their differences, nor came to an agreement. Viviani went so far as to ask Baldigiani to remove his own biographical account in its entirety.[129] Eventually, Baldigiani tired of Viviani's incessant requests for amendments, and disclaimed responsibility for the content of the book. Kircher died on November 27, 1680. He had hoped the *Etruria illustrata* could be his last major work to appear in print, but it never did.

Time, perhaps, was not quite ripe, especially in Rome. As Baldigiani would later write to Viviani,

All Rome is up in arms against mathematicians and physic-mathematicians. Extraordinary Congregations have been held, and are being held, by the cardinals of the Holy Office, and before the pope, and there is talk of a general prohibition of all the authors of modern physics. They are making very long lists of their names, and they put at the head of them Galileo, Gassendi, and Descartes, as most harmful for the Republic of Letters and for the integrity of religion.[130]

---

che ne dette qualche occasione, che poteva governarsi con un poco più di prudenza, che ferì Urbano e i Barberini, e dette lor occasione di qualche giusto risentimento. Io per me sempre ho parlato in questa forma, ed ho stimato di fargli un gran servizio, e credo che se e' vivesse me ne ringrazierebbe, credo che per assicurare il punto principale di non essere infamato come eretico, o di dubbia fede, si contenterebbe di soggiacere alla taccia di avere in qualche cosa mancato a tutte quelle cautele che poteva adoperare. *Qui si in nonnullis cautior fuisset*. Certo che se uno così scrivesse di Lutero si stimerebbe perfettissimo Luterano approvatore e non impugnatore de' suoi errori": Baldigiani to Viviani, 18 June 1678, in Favaro, "Miscellanea galileiana inedita" (1882), pp. 143–144.

128 Indeed, at the end of the letter Baldigiani asks Viviani "not to share this paper of mine with others, except for Magalotti, whose opinion I would be very happy to hear now. Just as I would be very happy to hear Your Lordship's, whom I will always acknowledge as my teacher. I also beg you, since the subject matter is as dear to me as it is to you, to send this paper back to me, by attaching it to the first letter you will please me with" ("La prego a non comunicar questa mia carta a nessuno fuori che al Magalotti, del quale volentierissimo sentirò i sentimenti per adesso, siccome quei di VS.ª che sempre riconoscerò per mio maestro. La prego di più per essere in materia sì gelosa a me ed a lei a farmi recapitar questa carta includendola nella sua prima colla quale mi favorirà"): ibid., p. 144.

129 See Viviani to Baldigiani, 23 August 1678, in Favaro, "Miscellanea galileiana inedita" (1882), p. 147.

130 "Tutta Roma sta in arme contra i Matematici e fisico-matematici. Si sono fatte e si fanno Congregazioni Straordinarie de Cardinali del S.° Offizio, e avanti al Papa, e si parla di fare proibizioni generali di tutti gli Autori di Fisiche moderne, e se ne fanno liste lunghissime, e tra essi si mette in capite Galileo, il Gassendo, il Cartesio & come perniciosissimi alla Repubblica letteraria e alla sincerità della Religione.": Baldigiani to Viviani, 25 January 1693, in Favaro, "Miscellanea galileiana inedita" (1882), p. 156.

ACKNOWLEDGMENTS

This book could not have been written without the constant help of fellow scholars and friends, over the course of three years. My profound gratitude to Giuliana Colombo, Maurice Finocchiaro, John Preston, Noel Swerdlow, and especially John Heilbron and Concetta Luna, who carefully revised the texts, offered comments and suggestions, and helped me in countless ways. Special thanks to Richard Adams, Erminia Ardissino, Roberta Corvi, Ciro D'Amato, Franz Daxecker, Stéphane Garcia, Roger Gaskell, Steve Hindle, Michel-Pierre Lerner, Paul Needham, Jay Pasachoff, Giorgio Strano, Federico Tognoni, and James Voelkel for their kind availability and generous support. I am equally indebted to the universities and research institutions that put me in the best possible conditions to pursue my researches: the Chemical Heritage Foundation (now Science History Institute), the University of Pennsylvania, Caltech, and The Huntington Library. I gratefully acknowledge the valuable suggestions of three anonymous referees, who carefully checked my work. Last, but by no means least, I thank Princeton University Press and all those who turned my work into a published book: my editors, Ingrid Gnerlich, Arthur Werneck, and Jessica Yao; my production editor, Karen Carter; and my copyeditor, Barbara Liguori.

EDITORIAL NOTE

All texts are presented in their original language (Italian, Latin, or French) and are here translated into English for the first time.

Bold superscript numbers indicate the original pagination, to facilitate direct references to the sources.

In the original texts, footnotes in numerals record variants (whenever a text is transmitted by two or more testimonies) and corrections to the original texts, using the standard abbreviations for critical apparatuses. Footnotes in letters offer Viviani's annotations (only for texts 1 and 7).

In the English translations, footnotes in letters offer Viviani's annotations. All endnotes, in Arabic numerals, are the editor's own annotations; they provide explanations, references, and additional documents. Whenever additional material is quoted, it appears both in English and in its original language, with corresponding references to the sources.

All translations are by the editor, unless otherwise indicated in the bibliographical references at the end of the book.

The following note provides the source and essential bibliographical information for each text, as well as specific abbreviations used in the footnotes.

1. VINCENZO VIVIANI, *RACCONTO ISTORICO DELLA VITA DEL SIG.ʳ GALILEO GALILEI* (1654)

Viviani's *Racconto istorico*, in the form of a letter to Prince Leopoldo dated April 29, 1654, was left unpublished at the time of the author's death, in 1703. There are two different, holograph manuscripts of the text, both at the National Library in Florence: *Gal.* 11, ff. 73r–118v (= A), and 22r–68r (= B). As Favaro convincingly argues, B was copied from A (itself a copy of a previous manuscript, now lost), and offers a later version of the text.

Viviani's biography was first published in *Fasti consolari dell'Accademia fiorentina*, edited by Salvino Salvini, Florence: Stamperia di S. A. R., by Giovanni Gaetano Tartini and Santi Franchi, 1717, pp. 397–431 (= S), with the editor's introduction (pp. 393–396), and additional material (pp. 432–446), based on a manuscript owned by Abbot Jacopo Panzanini, Viviani's nephew and heir, and now lost. At times, S agrees with A, and at times with B; and at times, S presents independent variants (which are not found in A or B, that is). This also reflects, as Favaro noticed, Viviani's continuous tinkering with the text, which remained *in fieri* throughout the author's life.

The *Racconto istorico* was subsequently included in all editions of Galileo's collected works:

Opere di Galileo Galilei Nobile Fiorentino Primario Filosofo, e Mattematico del Se-
renissimo Gran Duca di Toscana [edited by Tommaso Buonaventuri, in collaboration with Luigi Guido Grandi], Florence: Giovanni Gaetano Tartini
& Santi Franchi, 1718, Vol. I, pp. LX-XC (with Salvini's additional material,
pp. LVII–LX and XCI–CIII);

Opere di Galileo Galilei divise in quattro tomi, In questa nuova Edizione accresciute
di molte cose inedite, [edited by Giuseppe Toaldo], Padua: Stamperia del Seminario & Giovanni Manfrè, 1744, Vol. I, pp. XLIX–LXXVI (with Salvini's
additional material, pp. XLVI–XLVIII and LXXVI–LXXXVIII);

Opere di Galileo Galilei Nobile Fiorentino, Milan: Società Tipografica de' Classici
Italiani, 1808–1811, Vol. I (1808), pp. 9–71 (with Salvini's additional material,
pp. 1–7 and 72–97);

Le Opere di Galileo Galilei: Prima edizione completa condotta sugli autentici ma-
noscritti palatini, Eugenio Albèri editor-in-chief, Florence: Società Editrice
Fiorentina, 1842–1856, Vol. XV (1856), pp. 327–372 (with the editor's preface,
pp. 323–325, and annotations, pp. 381–415);

Le Opere di Galileo Galilei: Edizione Nazionale, Antonio Favaro editor-in-chief,
Florence: Giunti-Barbèra, 1890–1909, Vol. XIX (1907) pp. 599–632 (with the
editor's introduction, pp. 597–598, and annotations).

Other notable editions include the following:

Scelta de' migliori opuscoli. Tanto di quelli che vanno volanti, quanto di quelli che
inseriti ritrovansi negli Atti delle principali Accademie d'Europa, concernenti
le Scienze, e le Arti, che la vita Umana interessano [edited by Fortunato Bartolomeo De Felice], Naples: Giuseppe Raimondi, 1755, Vol. I (the only
one), pp. 269–368 (with the editor's introduction, pp. 267–268, and several
annotations);

Scritti vari di Galileo Galilei, edited by Augusto Conti, Florence: G. Barbèra Editore, 1864, pp. 1–90 (with the editor's annotations, pp. 91–107).

The Proemio (Preface)—here in the appendix at the end of Viviani's *Racconto*
istorico—was first published by Antonio Favaro in "*Inedita Galilaeiana*: Frammenti
tratti dalla Biblioteca Nazionale di Firenze," *Memorie del Reale Istituto Veneto di*
Scienze, Lettere ed Arti, 21, 1879, pp. 433–473: pp. 437–438 (reprinted in Favaro, *Inedita*
Galilaeiana: Frammenti tratti dalla Biblioteca Nazionale di Firenze, Venice: Giuseppe
Antonelli, 1880, pp. 7–8). It was included in *Le Opere di Galileo Galilei: Edizione*
Nazionale, Antonio Favaro editor-in-chief, Florence: Giunti-Barbèra, 1890–1909,
Vol. XIX (1907), pp. 600–601.

The text presented here reproduces Favaro's edition, which is based on *B* (as opposed to all previous editions, which reproduce *S*). Although *S* is the *editio princeps*
of Viviani's text, Favaro opted for following *B*, since Salvini—as was customary of
editors at the time—likely intervened in the original text, correcting and adjusting it
to his own taste and standards. Favaro's edition also records all the variants of *A* and
S; substantial variants are reproduced here, too (in the footnotes to the translation).
Viviani's own footnotes are indicated by letters.

2. Girolamo Ghilini, *Galileo Galilei* (ca. 1633)

Girolamo Ghilini's (1589–1688) *Teatro d'hvomini letterati* first appeared in Milan, in one volume; a second edition was published in Venice, in 1647, in two volumes. Although the publication date of the first edition is not stated on the title page, nor anywhere in the book, it was most likely printed in late 1632 or early 1633. The *terminus post quem* is February 21, 1632, when the *Dialogue* was published in Florence, by Giovanni Battista Landini. In his entry on Galileo, Ghilini explicitly refers to the *Dialogue* as a published work. The *terminus ante quem* is June 22, 1633, when Galileo was pronounced vehemently suspected of heresy by the Holy Office and required to abjure. Ghilini was ordained a priest in 1630, and the work is lavishly dedicated to Pope Urban VIII.[1] Certainly, the *Teatro d'hvomini letterati* was published before Allatius's *Apes Vrbanae* (1633), whose author had to hurriedly replace part of the entry on Galileo to accommodate it to the dramatic turn of events of 1633.

Although the title page of the Milan edition indicates Volume I, no second volume ever appeared in print; Volume I of the Milan edition differs only slightly from Volume I of the second edition (Venice, 1647). The second edition, in six parts, is dedicated to the Venetian patrician Giovan Francesco Loredan (or Loredano, 1607–1661), himself an accomplished author (and founder of the Accademia degli Incogniti, to which Ghilini belonged, too; several works of his were published by Giovanni Guerigli). At the time of Ghilini's death, a third volume (comprising Parts 7–9) had been completed, and a fourth was in progress; both remained unpublished.[2]

The text presented here is based on the second edition: Girolamo Ghilini, *Teatro d'hvomini letterati*, Venice: Giovanni Guerigli, 1647, Vol. I, pp. 68–69 (= H); the few variants of the first edition—Girolamo Ghilini, *Teatro d'hvomini letterati*, Milan: Giovanni Battista Cerri & Carlo Ferrandi, on behalf of Filippo Ghisolfi, [s. d.], Vol. I (and only), pp. 131–133 (= D); reprinted in 1684—are recorded in the apparatus.

3. Leo Allatius, *Galilævs Galilævs* (1633)

Leo Allatius, *Apes Vrbanae, siue De viris illvstribvs, Qui ab Anno MDCXXX. per totum MDCXXXII. Romæ adfuerunt, ac Typis aliquid euulgarunt*, Rome: Lodovico Grignani, 1633, pp. 118–119 (= R2); second edition, Hamburg: Christian Liebzeit, 1711, pp. 162–165.

The text presented here is based on R2. As explained in detail in the introduction (see above, pp. xxxi–xxxiv), the original version of Allatius's work—still available in the Vatican Library, *Vat. lat.* 7075 (= R1)—was written and printed before the 1633 trial, and the entry on Galileo had to be radically revised, shortened, and reset

---

1 In his entry on Ghilini, Filippo Argelati wrote that the first edition was published in 1633: Argelati, *Bibliotheca Scriptorum Mediolanensium* (1745), Vol. I, Part II, no. DCCCL; see especially col. 681 D. Also, in one of the two copies of the first edition at the National Library of Florence (BONAM.139bis), the date 1633 is penciled on the title page, although the entry in the catalogue reports that the book was printed between 1635 and 1677 (the span of years during which the publisher, Filippo Ghisolfi, was active). Dates such as 1635 or 1638, suggested by contemporary scholars, seem unlikely.

2 See Argelati, *Bibliotheca Scriptorum Mediolanensium* (1745), Vol. I, Part II, no. DCCCL, col. 682 C, no. XXII, and Brusoni, *Le glorie de gli Incogniti* (1647), pp. 268–271, especially p. 271.

in consideration of the condemnation of Galileo. Indeed, the *Vat. lat.* 7075 includes the original manuscript version of the text and 38 printed pages (ff. 32r–35v, 37r–40v, 50r–53v, 56r–59v, 69r–70v, and 74r–v), corresponding to two full quires of 8 sheets each, D and F (pp. 49–64 and 81–96), H4–[5] (pp. 119–122), and [I6] (pp. 139–140). Whereas in the first printed version the entry on Galileo occupied pp. 118–120, the final version is limited to pp. 118–119; by contrast, the criticism of Galileo's views are granted more space in the entry on Orazio Grassi (p. 137), and especially on Christoph Scheiner (pp. 68–70). The substantial textual differences between R1 and R2 are recorded in the footnotes.

Together with Rossi's, Siri's, and Crasso's, Allatius' biography of Galileo was later included in *Antiche vite di Galileo scritte da contemporanei ristampate dalle originali e rare edizioni*, Florence: Tipografia Barbèra, 1907, pp. 9–10. This booklet was especially printed and dedicated to Antonio Favaro by his collaborators at the *National Edition* of Galileo's works (Isidoro Del Lungo, Umberto Marchesini, Piero Barbèra, Augusto Alfani, and Arturo Venturi), on the occasion of the marriage of Favaro's son, Giuseppe, to Maria Dolfin, on February 4, 1907.

In the appendix, I offer the Latin text and translation of Allatius's entry on Christoph Scheiner, including several criticisms of Galileo.

4. GIAN VITTORIO ROSSI, *GALILÆVS GALILÆVS* (1643)

Gian Vittorio Rossi, *Pinacotheca imaginvm illvstrivm, doctrinæ vel ingenii laude, virorvm, qui, auctore superstite, diem suum obierunt*, Cologne: Cornelius ab Egmondt [in fact, Amsterdam: Joan Blaeu], 1643, pp. 279–281 (= K). The volume was reprinted in 1645, and two additional volumes were published in 1645 and 1648 (with the imprint Coloniæ Vbiorvm: apud Iodocvm Kalcovivm). In this reprint, all biographies are numbered with Roman numerals: Galileo's is no. CLIII.[3] The whole set of three volumes was later reprinted three times: in Leipzig, by Johann Friedrich Gleditsch (1692) and Thomas Fritsch (1712); and in Wolfenbüttel, by Johann Christoph Meissner (1729). In these reprints (all, in fact, printed in Amsterdam by Willem Jansz Blaeu), the biography of Galileo always appears in Part I, pp. 279–281.

Together with Allatius's, Siri's, and Crasso's, Rossi's biography of Galileo was later included in *Antiche vite di Galileo scritte da contemporanei ristampate dalle originali e rare edizioni*, Florence: Tipografia Barbèra, 1907, pp. 11–12.

5. VITTORIO SIRI, EXCERPT FROM *DEL MERCVRIO OVERO HISTORIA DE' CORRENTI TEMPI* (1647)

Vittorio Siri, *Del Mercvrio Ouero Historia De' correnti tempi*, Casale: Cristoforo della Casa [in fact, Venice: Tommaso Baglioni], 1644–1682, Vol. II, Book III (1647), pp. 1720–1722 (= C).

---

3 In his own biographical entry of Galileo, von Sandrart mentions this number when referring to Rossi's biography: see below, pp. 246–247.

Siri's work is in 15 volumes (in 17 books) and was published between 1644 and 1682. The biography of Galileo does not have a separate title and is not presented as such. It is appended at the end of Book III and is followed by an analogous biographical sketch of the Italian painter Guido Reni (1575–1642), on pp. 1722–1723.

Together with Allatius's, Rossi's, and Crasso's, Siri's biography of Galileo was later included in *Antiche vite di Galileo scritte da contemporanei ristampate dalle originali e rare edizioni*, Florence: Tipografia Barbèra, 1907, pp. 13–15.

6. Vincenzo Galilei, *Alcune notizie intorno alla Vita del Galileo* (ca. 1654)

Alcune notizie intorno alla Vita del Galileo auute da Vinc: Galilei fig.^{lo} et erede del Galileo, in *Gal.* 11, ff. 126r–129r (copy in Viviani's handwriting); "Notizie raccolte da Vincenzio Galilei", in *Le Opere di Galileo Galilei: Edizione Nazionale*, Antonio Favaro editor-in-chief, Florence: Giunti-Barbèra, 1890–1909, Vol. XIX (1907), pp. 594–596.

The text was first published by Favaro in "*Inedita Galilaeiana*: Frammenti tratti dalla Biblioteca Nazionale di Firenze," *Memorie del Reale Istituto Veneto di Scienze, Lettere ed Arti*, 21, 1879, pp. 433–473: pp. 440–443 reprinted in Antonio Favaro, *Inedita Galilaeiana: Frammenti tratti dalla Biblioteca Nazionale di Firenze*, Venice: Giuseppe Antonelli, 1880, pp. 10–13).

7. Niccolò Gherardini, *Vita di Galileo Galilei* (1654)

Gherardini's biography of Galileo is available in full only in *Gal.* 11, ff. 3r–19r, followed by Viviani's annotations (f. 20r); a second, incomplete copy is in Cod. Marucelliano A, LXXI, 6, at the National Library of Florence (= M). The text was first published in Giovanni Targioni Tozzetti, *Notizie degli aggrandimenti delle scienze fisiche accaduti in Toscana nel corso degli anni LX. del secolo XVII.*, Florence: Giuseppe Bouchard, 1780, Vol. II, Part I, no. XII, pp. 62–76 (= T). Targioni Tozzetti drew the text from a different manuscript copy, now lost.

The text presented here reproduces Favaro's "Vita scritta da Niccolò Gherardini," in *Le Opere di Galileo Galilei: Edizione Nazionale*, Antonio Favaro editor-in-chief, Florence: Giunti-Barbèra, 1890–1909, Vol. XIX (1907), pp. 634–646. Favaro's edition is based on *Gal.* 11 (including Viviani's annotations), and records all the variants of *M* and *T*; substantial variants are recorded here, too (in the footnotes to the translation).

8. Vincenzo Viviani, *Lettera al Principe Leopoldo de' Medici intorno all'applicazione del pendolo all'orologio* (1659)

Of this letter (dated August 20, 1659) we have two manuscript copies: one in *Gal.* 85, ff. 39r–50r, at the National Library in Florence, titled *Istoria dell'Oriuolo del Galileo, regolato dal Pendolo* (= G); and one at the Bibliothèque Nationale de France in Paris,

Fonds français, 13039, ff. 147r–155v (= *P*), which Prince Leopoldo de' Medici sent to the French astronomer and mathematician Ismaël Boulliau, with a letter dated October 9, 1659. Both copies are in the handwriting of a professional copyist, and signed by Viviani.

Viviani's letter remained unpublished at the time of Viviani's death, in 1703, and its existence was first revealed in 1774 (*Novelle fiorentine*, 10, p. 150). It was first published in Giovanni Battista Clemente Nelli, *Vita e commercio letterario di Galileo Galilei*, Losanna: [s.n.], 1793, Vol. II, pp. 721–738; Eugenio Albèri included part of it in *Le Opere di Galileo Galilei: Prima edizione completa condotta sugli autentici manoscritti palatini*, Florence: Società Editrice Fiorentina, 1842–1856, Vol. *Supplemento* (1656), pp. 338–342.

The text presented here reproduces the one offered by Favaro: "Lettera al Principe Leopoldo de' Medici intorno all'applicazione del pendolo all'orologio," in *Le Opere di Galileo Galilei: Edizione Nazionale*, Antonio Favaro editor-in-chief, Florence: Giunti-Barbèra, 1890–1909, Vol. XIX (1907), pp. 648–659. Favaro's edition is based on *G*, which offers a later revision of the text, as it presents various modifications by Viviani himself that do not merely correct the copyist's mistakes but revise or add to the text; substantial differences with *P* are recorded, too. Following Favaro, substantial variants of *P* with respect to *G* are recorded here, too (in the footnotes to the translation).

9. LORENZO CRASSO, *GALILEO GALILEI* (1666)

Lorenzo Crasso, *Elogii d'hvomini letterati*, Venice: Sebastiano Combi & Giovanni La Noù, 1666, [Vol. I], pp. 243–251 (= *E*).

Crasso's biography (pp. 243–245) is preceded by an anonymous portrait of Galileo (p. 243) and followed by the epitaph on Galileo's original burial place (p. 246), a poem by Francesco Stelluti (pp. 247–250), an epigram by Giovan Battista Marino (p. 251), and a list of Galileo's main works (p. 251). For the first time, twenty-two years after Galileo's death, his portrait accompanies a biography of him. Illustrated collections of the biographies of eminent figures were to become a distinctive and successful genre, and they would always include an entry on Galileo, of which Crasso's was a sort of archetype.

The epitaph can still be read, under a bust of Galileo, on the east wall of the vestry to the Chapel of the Novitiate (otherwise known as the Medici Chapel), in the Santa Croce Basilica, in Florence. It was also included in *Opere 1655–1656*, Vol. I, p. 22.

Stelluti's poem—an ode of 24 stanzas, each formed by six 7-syllable lines and one 11-syllable line (pattern: ABABBCC)—was originally published in Galileo's *Il saggiatore. Nel quale Con bilancia esquisita e giusta si ponderano le cose contenute nella Libra astronomica e filosofica di Lotario Sarsi Sigensano*, Rome: Giacomo Mascardi, 1623, ff. a2r–a4v; it was reprinted in *Le Opere di Galileo Galilei: Edizione Nazionale*, Antonio Favaro editor-in-chief, Florence: Giunti-Barbèra, 1890–1909, Vol. VI (1896), pp. 207–211. Stillman Drake's beautiful English paraphrase of the poem (in Galileo, *The Assayer*, in *The Controversy on the Comets of 1618: Galileo Galilei, Horatio Grassi, Mario Guiducci, Johann Kepler*, edited by Stillman Drake and Charles D. O'Malley, Philadelphia: University of Pennsylvania Press, 1960, pp. 156–162) is replaced here by an actual (if prosaic) translation.

Marino's epigram—a sonnet (two quatrains + two tercets of 11-syllable lines; pattern: ABBA ABBA, CDC DCD)—was first published in Giambattista Marino, *La Galeria, Distinta in Pittvre, & Sculture*, Venice: Giovanni Battista Ciotti, 1620, p. 192. Other 1620 editions include Ancona: Cesare Scaccioppa; Milan: Giovanni Battista Bidelli; and Naples: Scipione Bonino. Marino's work was later reprinted several times, by different Venetian publishers: Ciotti (1622, 1623, 1626, 1630, and 1635), Cristoforo Tomasini (1647), Francesco Baba (1652 and 1653), Giovanni Pietro Brigonci (1664 and 1667), and Nicolò Pezzana (1674).

Together with Allatius's, Rossi's, and Siri's, Crasso's biography of Galileo was later included in *Antiche vite di Galileo scritte da contemporanei ristampate dalle originali e rare edizioni*, Florence: Tipografia Barbèra, 1907, pp. 17–20.

10. Vincenzo Viviani, *Raggvaglio dell'vltime opere del Galileo* (1674)

Vincenzo Viviani, *Qvinto libro degli Elementi d'Evclide, ovvero Scienza vniversale delle proporzioni spiegata colla dottrina del Galileo, con nuov'ordine distesa, e per la prima volta pubblicata*, Florence: alla Condotta, 1674, pp. 86–88 and 99–106 (= Q). The excerpt is from a section of the book titled "Raggvaglio dell'vltime opere del Galileo" ("Report on Galileo's later works," pp. 86–106); on pp. 89–98, Viviani suspends his account by inserting a long digression on the beauty and usefulness of geometry, addressed to the readers.

11. Isaac Bullart, *Galilée Galilei* (1682)

Isaac Bullart, *Académie des Sciences et des Arts, Contenant les Vies, & les Eloges Historiques des Hommes Illustres, Qui ont excellé en ces Professions depuis environ quatre Siécles parmy diverses Nations de l'Europe: Avec leurs Pourtraits tirez sur des Originaux au Naturel, & plusieurs Inscriptions funebres, exactement recueïllies de leurs Tombeaux*, Paris: Jacques Ignace Bullart, 1682, Vol. II, Book II: *Illustres Philosophes, Mathematiciens, Astrologues & Medecins*, pp. 131–133. In the same year, identical copies of the two volumes (including the biography of Galileo) were published in Amsterdam by Daniel Elzevier, and in Bruxelles by François Foppens (the latter was reprinted in 1695).

For several decades, Bullart (1599–1672) collected portraits of eminent personalities and information about them but never saw the publication of his work. The material he gathered was supplemented by additional documents and eventually published by his son, Jacques Ignace. The *Académie des Sciences et des Arts* was designed as a sort of genealogy of learning, with biographies arranged in chronological order according to a pattern (presented on f. **5r) that reflects the division between arts and sciences, with subdivisions into several disciplinary areas. Volume I includes biographies of politicians, historians, and lawyers (sciences); men of letters and artists (arts); Volume II includes biographies of theologians, philosophers, interdisciplinary scholars and poets (sciences); typographers, geographers, cosmographers, mathematicians, painters, and students of optics and perspective (arts). Within this new encyclopedia, reflecting the divisions and subdivisions between disciplines, images played a major role, presenting portraits of eminent people whose spirit, Bullart says (f. **3*r–v*), is described in the accompanying texts.

The biography of Galileo (pp. 131–132) is preceded by a portrait engraved by Nicolas III De Larmassin (p. 131)—drawn from the portraits of Galileo by Francesco Villamena (in the frontispiece of *On Sunspots*, 1613) and Jacob van der Heyden's (in the Latin translation of the *Dialogue*, published in 1635 by the Elzeviers, the same publisher of Bullart's *Académie des Sciences et des Arts*)—and followed (p. 133) by a French sonnet (14 Alexandrine verses, pattern: ABBA ABAB, CCD EDE) by Guillaume Colletet (1598–1659), whose original manuscript is in Rome, at the Biblioteca Angelica (Cod. 1984, f. 112*r*).

12. JOACHIM VON SANDRART, *GALILÆUS GALILÆI* (1683)

Joachim von Sandrart, *Academia nobilissimæ artis pictoriæ. Sive De veris & genuinis hujusdem proprietatibus, theorematibus, secretis atque requisitis aliis; nimirum de Inventione, Delineatione, Evrythmia & Proportione corporum: de Picturis in albario recente, sive fresco, in tabulis item, atque linteis; de pingendis historiis, imaginibus humanis, iconibusque viventium; de subdialibus & nocturnis; de subactu colorum oleario & aquario, de affectibus & perturbationibus animi exprimendis; de lumine & umbra; de vestibus, deque colorum proprietate, efficacia, usu, origine, natura atque significatione Instructio Fundamentalis, Multarum industria lucubrationum, & plurimorum annorum experientia conquisita*, Nuremberg: Christian Sigmund Froberger, and Frankfurt: Heirs of Michael & Johann Friedrich Endter, and Johan Jacob von Sandrart, 1683, Part II, Book III, Chapter XXVIII, no. LVI, pp. 389–390.

The biography is preceded by a portrait of Galileo, engraved by Bartholomäus II Kilian (1630–1696), from an original drawing by Joachim von Sandrart (who met Galileo at Villa Medici, in Rome, at the time of the 1633 trial). Besides Galileo, represented sideways, the Plate (no. 7 in von Sandrart's book) also includes portraits of the Jesuit scholar Athanasius Kircher (1602–1680), the German painters Johann Heinrich Roos (1631–1685) and Theodor Roos (1638–1698), the Dutch painter Gérard de Lairesse (1641–1711), and Bartholomäus II Kilian himself.

13. PAUL FREHER, *GALILÆUS GALILÆI* (1688)

Paul Freher, *Theatrvm virorum eruditione clarorum*, Vol. II: *In quo vitæ & scripta medicorum & philosophorum, Tam in Germania Superiore & Inferiore, quàm in aliis Europæ Regionibus, Græcia nempè, Hispania, Italia, Gallia, Anglia, Polonia, Hungaria, Bohemia, Dania & Suecia à seculis aliquot, ad hæc usque tempora florentium, secundum annorum emortualium seriem, tanquam variis in scenis repræsentantur*, Nuremberg: Heirs of Andreas Knorz, at the expense of Johann Hofmann, 1688, Part IV, p. 1536 (= F).

Paul Freher's (1611–1682) work is a sort of handbook with biographies of the most eminent European personalities, compiled on the basis of biographical and bibliographical collections, dictionaries, obituaries, and academic eulogies. It is divided into four sections: I. theologians (including popes, cardinals, and professors of theology); II. lawyers and politicians; III. physicians, chemists, botanists, and surgeons; IV. natural philosophers, philologists, historians, mathematicians, and poets. The work was published posthumously, by Karl Joachim Freher (Paul Freher's nephew), in two volumes, with several additions. The biography of Galileo, explicitly drawn from Ghilini's, is accompanied by an engraved plate (no. 81, facing p. 1533), by Johann Azelt, with portraits of Galileo and fifteen other learned men: the Dutch ecclesiastical historian Aubert le Mire (Miraeus, 1573–1640); the German philologist and astronomer Matthias Bernegger (1582–1640), active in Strasbourg, who translated Galileo's *Compass* and *Dialogue* into Latin; the Dutch philosopher Franciscus Mayvart (1585–1640), of Groningen; the Dutch classical scholar and antiquary Johannes van Meurs (Meursius, 1579–1639), professor of history and Greek at Leiden; the Swiss protestant theologian, philosopher, and philologist Ludwig Lucius (1577–1642); the German scholar and poet Johannes Gravius (d. 1644); the Greek-born theologian and classicist Leo Allatius (1586–1669), keeper of the Vatican library; the German historian and philosopher Tobias Andreä (1604–1676), professor of history and Greek at Groningen; the Dutch humanist and philologist Hendrik van der Putten (Erycius Puteanus, 1574–1646), professor of Latin and rhetoric at the Palatine School of Milan; the German theologian and philosopher Jakob Martini (1570–1649); the Dutch polymath Daniel Heins (Heinsius, 1580–1655), professor of poetics, Greek, and political science at Leiden; the Dutch theologian, philologist, and historian Gerhard Johannes Voss (Vossius, 1577–1649), professor of rhetoric, chronology, and Greek at Leiden; the Italian archaeologist, philologist, and librarian Ottavio Ferrari (1607–1682), professor of humanities at Padua; the Swiss-born engraver and publisher Matthäus Merian the Elder (1593–1650), who ran a publishing house in Frankfurt; and the Dutch poet, philologist, and historian Peter Schrijver (Petrus Scriverius, 1576–1660). Galileo's portrait, in particular, was likely based on Francesco Villamena's, in the frontispiece of both *On Sunspots* (1613) and *The Assayer* (1623), as well as of *Opere 1655–1656*; a similar portrait appeared also in the Latin translation of the *Dialogue* (1635).

14. Vincenzo Viviani, *Grati animi monumenta* (1702)

The text of the *Grati animi monumenta* exists in three different versions: in the form of inscriptions on the façade of Viviani's palace in Florence, in manuscript, and in print (in two separate editions).

The first version of the text (1693) can still be read—at least in part—in the form of inscriptions on the façade of Viviani's palace, in Florence; they were painted, not engraved, somewhat hurriedly, and display numerous errors. A few years later, out of concern that they might be destroyed, or go missing, and in order to circulate the text as widely as possible, Viviani decided to publish a revised version of the text.

The original manuscript of this revised version (in the handwriting of a professional copyist, with corrections by Viviani) is at the Wellcome Library of London (Ms. 4949) and was published in Stefano Gattei, "Galileo's Legacy: A Critical Edition and Translation of the Manuscript of Vincenzo Viviani's *Grati Animi Monumenta*," *The British Journal for the History of Science*, 50, 2, 2017, pp. 181–228: pp. 206–213.[4]

The text of the *Grati animi monumenta* (without the title, which may be drawn from the "Monitum lectori") was appended to Vincenzo Viviani, *De locis solidis secunda divinatio geometrica in quinque libros iniuria temporum amissos Aristæi senioris geometræ*, Florence: His Royal Highness's Press at Pietro Antonio Brigonci, 1701, pp. 121–128 of the second half of the book (= L). The *De locis solidis* was actually written in 1646 but was not printed until 1673 (in Florence, by Ippolito Navesi). The book was not immediately circulated, though, but remained with the publisher until the beginning of 1702, when Viviani decided to append new material: an additional eight-page quire (Qq), with the text of the inscriptions; and the "Advice to the reader" ("Monitum lectori", offered here in the appendix) printed on the last page of the previous quire, which had been left blank (p. 120 = f. Pp 4v). The text is followed by the imprimatur (p. 128).[5] The *De locis solidis* also includes three engraved plates by Giovanni Antonio Lorenzini: a portrait of Galileo, presenting the bronze bust cast by Giovanni Battista Caccini at the top of the main entrance of Viviani's palace; the façade of Palazzo dei Cartelloni; and a diagram of the inscriptions.

Later, the text was published in book form: Vincenzo Viviani, *Grati animi monumenta ... Uti fuerunt conscripta Florentiæ in Fronte Ædium A DEO DATARUM Anno Salutis 1693*, Florence: His Royal Highness's Press at Pietro Antonio Brigonci, [1702] (= M). A full eight-page quire, the *Grati animi monumenta* lacks the "Monitum lectori" (as it was an independent publication, an explanation was no longer required)

4 I take this opportunity to correct a few mistakes in my edition of Viviani's manuscript:
   p. 209, line 5: *seculi* should read *seculis*
   p. 211, line 4: after *Inc.* a note should be added in the apparatus: Anno a Salut. Inc. *W*] Anno à Salut. *LM deb.* Anno Salut. *vel* Anno ab Inc.
   p. 211, note 145 should read: læthalis *W*[p-corr]] lætalis *W*[a-corr] lethalis *LM deb.* letalis
   p. 212, line 25: after *Ferd:* a note should be added in the apparatus: Ferd: *W*] FERDIN. *LM*
5 The imprimatur, also included in the manuscript, reads: "Imprimatur | Tommaso della Gherardesca, Vicar General of Florence | On behalf of the Most Reverend Father Inquisitor General of the Florentine Administration, let the Reverend Father Master Antonio Francesco Cioppi, of the Conventual Minors, Consultant of this Holy Office, carefully peruse the present book, titled *Vincentii Viviani ... and report whether it might be granted publication | Written in the Holy Office of Florence, on 9 August 1701 | Friar Lucio Agostino Cecchini, from Bologna, of the Conventual Minors, Vicar General of the Holy Office of Florence | I, Friar Antonio Francesco Cioppi, on behalf of His Most Reverend Father, thoroughly and carefully read, and did not find anything against the law of God and good morals; wherefore I consider this book suitable for publication. | I, Friar Antonio Francesco Cioppi, Consultant of the Holy Office of Florence, wrote this with my own hand | Having carefully considered the report above, let it be printed. | Friar Lucio Agostino Cecchini, from Bologna, of the Conventual Minors, Vicar General of the Holy Office of Florence | Imprimatur, Filippo Buonarroti, Senator Auditor of His Royal Highness" ("Imprimatur | *Thomas de Gherardesca Vic[arius] Gen[eralis] Flor[entiæ]* | De mandato Rev[erendissimi] P[atris] Inquisit[oris] Gen[eralis] Flor[entinae] Adm[inistrationis] R[everendus] P[ater] M[agister] Antonius Franciscus Cioppi Min[or] Convent[ualis] Consultor hujus S[ancti] Officii perlegat attentè præsentem Librum, cui titulus est, *Vincentii Viviani, &c*, & referat an ejusdem possit permitti impressio. Dat[um] in S[ancto] Off[icio] Flor[entiæ] die 9. Aug[usti] 1701. | *Fr[ater] Lucius Augustinus Cecchini de Bon[onia] Min[or] Conv[entualis] Vic[arius] Gen[eralis] S[ancti] Off[icii] Fl[orentiæ]* | Ego Fr[ater] Antonius Franciscus Cioppi ex mandato Patern[itatis] Suæ Reuerendiss[imæ] applicatè perlegi attentè, nec aliquid contra legem Dei, & bonis moribus inveni, quapropter Librum hunc existimo dignum, ut typis mandetur. | *Ego Fr[ater] Antonius Franciscus Cioppi Cons[ultor] S[ancti] Off[icii] Fl[orentiæ] m[anu] p[ropria]* | Attenta supraposita relatione Imprimatur | *Fr[ater] Lucius Augustinus Cecchini de Bon[onia] Min[or] Conv[entualis] Vic[arius] Gen[eralis] S[ancti] Off[icii] Fl[orentiæ]* | Imprimatur Philippus Bonarrota Sen[ator] Reg[iæ] Cels[itudinis] Aud[itor]"): Vincenzo Viviani, *De locis solidis secunda divinatio geometrica* (1701), p. 128).

and offers a distinct title page instead, with a blank verso; also, the lower half of the last page (p. 8 = f. A $4v$) presents an engraved decoration, instead of the text of the imprimatur, as published in the *De locis solidis*.

In fact, *L* and *M* were simultaneous: the setting of types is identical, and only page numbers and quire identifications are different. Also, in the dedication of the *De locis solidis*, Viviani states he is eighty years old ("Vincentius Viviani annum agens octogesimum": f. ✠$5r$). Hence, the book was actually published in 1702.

The history of the *Grati animi monumenta* is more complex than it might seem at first. The title of the work, meaning "testimony of a grateful person," is present only in *M* and is drawn from the "Monitum lectori" in *L*, in which Viviani expresses his gratitude to King Louis XIV of France (who gave Viviani the money to buy the house on whose façade are the inscriptions), to the Grand Dukes of Tuscany, Ferdinand II and Cosimo III de' Medici (Viviani's patrons), and to Galileo. The expression used in the "Monitum lectori," in turn, is reminiscent of the one Viviani uses in the dedication to Louis XIV that opens the *De locis solidis*, which reads "grati animi et obsequii sui monimentum" ("memorial of a grateful person and of his deference": f. ✠$5r$).[6]

Also, there is a progressive shift of emphasis in the structure of the dedications: the *De locis solidis* is dedicated to Louis XIV (no mention of the Grand Dukes of Tuscany, nor of Galileo); in the "Monitum lectori" that introduces *L* in the *De locis solidis*, Viviani expresses his gratitude to Louis XIV, the Medici, and Galileo; lastly, in the title of *M* (containing the dedication), the sequence becomes: Galileo, the Medici, and Louis XIV. When the expression *grati animi monumenta* becomes the title of *M*, the arrangement of Viviani's gratitude, as expressed on the title page of the work, turns from "political" (as in *L*) to "personal," and so Galileo comes first, above all others. Together with the absence of the imprimatur and the date of publication, this indicates that the *Grati animi monumenta* was likely published to be distributed and circulated privately.

Furthermore, Viviani added the "Monitum lectori" in *L* in order to explain why a eulogistic text was appended to a work of geometry. As we read in the "Monitum lectori," Viviani wrote the *De locis solidis* in the house donated by King Louis XIV, the Grand Dukes supported him (Viviani was the Court Mathematician from 1647), and Galileo taught geometry to him. Most likely, these were just excuses, as Viviani was simply trying to publish a polished version of the text of the inscriptions on the façade of his house as soon as possible, and consign it to posterity in what Viviani thought was to be a more enduring form.

The *Grati animi monumenta* was further reprinted by Nelli: Vincenzo Viviani, *Grati animi monumenta . . . Uti fuerunt conscripta Florentiæ in Fronte Ædium A DEO DATARUM Anno Salutis 1693*, edited by Giovanni Battista Clemente Nelli, Florence: Francesco Moücke, 1791. This edition includes the "Monitum lectori" (pp. 3–4), Nelli's preface "To the Learned and Benevolent Reader" ("Erudito et benevolo lectori", pp. 5–6), and three engravings by Giuseppe Calendi: a portrait of Galileo (frontispiece), the façade of Palazzo dei Cartelloni, and the diagram of the inscriptions.

---

6 In the title of Viviani's book, the word *monumenta* is plural, referring to the inscriptions; in the "Monitum lectori," the word is singular, indicating "memorial."

Nelli also included the printed text of the inscriptions in his *Vita e commercio letterario di Galileo Galilei*, Florence: [s.e.], 1793, Vol. II, pp. 857–867 ("Monitum lectori", p. 856); Calendi's engravings of the façade of Palazzo dei Cartelloni and of the diagram of the inscriptions are included, too (Plates VII and VIII, facing pp. 849 and 857, respectively). Calendi's portrait of Galileo is reproduced here as the frontispiece of *Vita e commercio letterario di Galileo Galilei*, Volume I, whereas a second portrait (presenting an aged Galileo) is the frontispiece of Volume II.

Finally, the *Grati animi monumenta* also appeared in *Le Opere di Galileo Galilei: Prima edizione completa condotta sugli autentici manoscritti palatini*, Eugenio Albèri editor-in-chief, Florence: Società Editrice Fiorentina, 1842–1856, Vol. XV (1856), pp. 373–380 (immediately after Viviani's *Racconto istorico*).

The 1701 and 1702 printed versions substantially differ (both in length and detail) from the inscriptions on the façade of Palazzo dei Cartelloni. The differences were first noted by Albèri, who described them as minor. Years later, by contrast, Paolo Galletti (1851–1914)—the owner of the collection of *Galileiana* preserved at the Torre del Gallo, in Florence (to which the manuscript belonged)—called Favaro's attention to the striking differences between the printed version and the actual text of the inscriptions. In 1879, Favaro published (in parallel columns) the two texts in his "*Inedita Galilaeiana*: Frammenti tratti dalla Biblioteca Nazionale di Firenze,", *Memorie del Reale Istituto Veneto di Scienze, Lettere ed Arti*, 21, 1879, pp. 433–473: Appendix II, pp. 467–473 (reprinted in Favaro, *Inedita Galilaeiana: Frammenti tratti dalla Biblioteca Nazionale di Firenze*, Venice: Giuseppe Antonelli, 1880; Appendix II, pp. 37–43).

Both in the *De locis solidis* and in the *Grati animi monumenta*, the text is accompanied by three engravings: a portrait of Galileo, after a 1691 bronze bust cast by Giovan Battista Foggini (1652–1725); the façade of Viviani's palace in Florence; and the detail of the order and location (indicated by letters A-H) of the various parts of the inscriptions on the façade of the palace. All engravings are by Giovanni Antonio Lorenzini (1655–1740).

APPENDIX. MAFFEO BARBERINI, *ADULATIO PERNICIOSA* (1620)

The appendix presents Cardinal Maffeo Barberini's "Adulatio perniciosa," a poem in praise of Galileo and his telescopic discoveries, first published in *Poemata*, Paris: Antoine Estienne, 1620, pp. 46–49 (= *ed. 1620*).

A selection of nine Latin poems by Cardinal Barberini was first included in the 1606 posthumous reprint of Aurelio Orsi's poems (originally published in 1589), edited by Giuliano Castagnacci, which also includes poems by Melchiorre Crescenzi, Claudio Contulio, Giovan Battista Lauri, Vincenzo Palettari, and Marco Antonio Bonciari: Aurelio Orsi *et al.*, *Academicorum Insensator*[um] *Carmina*, Perugia: Accademici Augusti, 1606.[7] It was not until several years later that Barberini published the first collection of his own, in 1620 (31 poems, among which is the *Adulatio perniciosa*).

---

7 The Accademia degli Insensati (*Academia insensatorum*) was a literary society, active in Perugia (which was part of the Papal States, at the time) to the mid-1720s. It was born in 1561, from the merger of a number of previous literary academies (Accademia dei Tranquilli, Accademia degli Scossi, Accademia

Soon after Maffeo Barberini was elected Pope Urban VIII, the collection was reprinted, with minor differences: Urban VIII, *Poemata*, Paris: Antoine Estienne, 1623, pp. 39–42 (= *ed. 1623*). During Urban VIII's pontificate (1623–1644), his poems enjoyed at least 15 further editions and reprints. The 1623 edition was the reference edition for all subsequent ones, six in total, until 1628, all commissioned by the great European powers and all aimed at paying official homage to the new pontiff: Palermo 1624, Cologne 1626, Vienna 1627, Venice 1628, Bologna 1628, and Codogno 1628.

A new, entirely revised edition appeared in 1631 (83 poems); it was edited by the Jesuits of the Roman College and included 35 previously unpublished poems, as well as a description of their various meters. The 1631 edition comes in two versions (with identical texts): a luxury *in quarto* edition (*editio maior*), printed in red and black throughout, with woodcut papal arms on the *verso* of the title page, and an engraved title page and portrait of the author by Claude Mellan, after Gian Lorenzo Bernini: Urban VIII, *Poemata*, Rome: Vatican Printing Office, 1631, with the *Adulatio perniciosa* on pp. 266–270; and *in octavo*, printed in black and white (*editio minor*): Urban VIII, *Poemata*, Rome: The Printing Office of the Reverend Apostolic Chamber, 1631, with the *Adulatio perniciosa* on pp. 217–220 (= *ed. 1631*). This was the model for the Antwerp 1634 edition, and for four additional editions published in Rome by The Printing Office of the Reverend Apostolic Chamber: 1635, 1637, 1638, and 1640. These were probably prompted and edited by Urban VIII himself, and included an increasing number of poems. The 1640 is the fullest edition, including 144 Latin poems and 81 in the vernacular (*Poesie toscane*), and was reprinted in Dillingen in 1641.

In 1642, King Louis XIII of France honored the pope with an elegant edition of both the *Poemata* (107 poems, based on the Antwerp 1634 edition) and the *Poesie toscane* (73 poems, based on the Rome 1635 edition): Urban VIII, *Poemata*, Paris: The Royal Press, 1642, with the *Adulatio perniciosa* on pp. 186–189 (= *ed. 1642*). Whereas along the twenty-one years of the pontificate exaltations of the pope's poetical talent were abundant (and included the production of a substantial number of commentaries, translations, musical adaptations, and imitations), the seemingly boundless appreciation of his literary merits ceased immediately after Urban VIII's death, also owing to the political downfall of the Barberini family during the pontificate of Urban VIII's successor, Pope Innocent X (born Giovanni Battista Pamphilj). The former pope's Latin poems were eventually reprinted in 1726: Urban VIII, *Poemata*, edited by Joseph Brown, Oxford: Clarendon Press, 1726, with the *Adulatio perniciosa* on pp. 179–182 (= *ed. 1726*). Whereas the various editions and reprints differed, at times significantly, in the selection and number of poems, the *Adulatio perniciosa* was always included.

As to the meter, the model for the *Adulatio perniciosa* is one of the meters used by Horace in the *Odes*: 19 Alcaic stanzas, each formed by two Alcaic hendecasyllables (iambic tripody catalectic + dactylic dimeter), one Alcaic enneasyllable (iambic

---

degli Atomi, and Accademia degli Unisoni). Initially, it was formed by local men of letters and noblemen, who met to comment on Francesco Petrarca's poems and to discuss spiritual issues "beyond the constraints of senses" ("al di là della barriera dei sensi," hence the name of the society). At the beginning of the seventeenth century, however, it grew considerably in size and scope, to discuss issues of moral and aesthetic philosophy. Some of the most important Italian poets of the time were among its members, such as Giovanni Battista Guarini (1538–1612) and Torquato Tasso (1544–1595).

pentapody catalectic), and one Alcaic decasyllable (two dactyls + trochaic dipody); the first two lines are divided into two parts by a caesura after the fifth syllable. Structure:

˘_˘_˘ | _˘˘_˘˘ || ˘_˘_˘ | _˘˘_˘˘ || ˘_˘_˘_˘_˘ || _˘˘_˘˘_˘_˘.

Numbered footnotes to the original texts record the critical apparatus. Lettered footnotes to original texts and translations present authorial annotations (such as Viviani's own footnotes to the *Racconto istorico*, or Viviani's annotations on Gherardini's text). All commentaries on the texts by the editor are in the form of numbered endnotes to the translations. The endnotes offer references only to primary sources; a number of relevant secondary sources are referred to in the introduction or else are listed in the "Further Readings" section of the bibliography.

All page references are to works listed in the bibliography. Whenever possible, I made use of available translations; all other translations are the editor's, who took the liberty of modifying existing translations whenever he thought it appropriate.

LIST OF SOURCES AND ABBREVIATIONS

1. Vincenzo Viviani, *Racconto istorico della vita del Sig.ʳ Galileo Galilei* (1654)

Sources = X :	*OG* 19, pp. 599–632.
	Appendix: *OG* 19, pp. 600–601.
A :	Florence, Biblioteca Nazionale Centrale, *Gal.* 11, ff. 73r-118v.
B :	Florence, Biblioteca Nazionale Centrale, *Gal.* 11, ff. 22r-68r.
S :	Salvino Salvini (ed.), *Fasti consolari dell'Accademia fiorentina* (1717), pp. 397–431.

2. Girolamo Ghilini, *Galileo Galilei* (1647)

Source = H :	Girolamo Ghilini, *Teatro d'hvomini letterati* (1647), Vol. I, pp. 68–69.
D :	Girolamo Ghilini, *Teatro d'hvomini letterati* (ca. 1633), Vol. I, pp. 131–133.

3. Leo Allatius, *Galilævs Galilævs* (1633)

Sources = R2 :	Leo Allatius, *Apes Vrbanae* (1633), pp. 118–119.
	Appendix: *Apes Vrbanae* (1633), pp. 68–71.
R1 :	Vatican Library, *Vat. lat.* 7075.

4. Gian Vittorio Rossi, *Galilævs Galilævs* (1643)

Source = K :	Gian Vittorio Rossi, *Pinacotheca imaginvm illvstrivm . . . virorvm* (1643), pp. 279–281.

5. Vittorio Siri, excerpt from *Del Mercvrio Ouero Historia De' correnti tempi* (1647)

Source = C :   Vittorio Siri, *Del Mercvrio* (1647), pp. 1720–1722.

6. Vincenzo Galilei, *Alcune notizie intorno alla Vita del Galileo* (ca. 1654)

Source :   *OG* 19, pp. 594–596.

7. Niccolò Gherardini, *Vita di Galileo Galilei* (1654)

Source = Y :   *OG* 19, pp. 594–596.
N :   Florence, Biblioteca Nazionale Centrale, Cod. Marucelliano *A*, LXXI, 6.
T :   Giovanni Targioni Tozzetti, *Notizie degli aggrandimenti delle scienze fisiche accaduti in Toscana* (1780), Vol. 2, part I, no. XII, pp. 62–76.

8. Vincenzo Viviani, *Lettera al Principe Leopoldo de' Medici intorno all'applicazione del pendolo all'orologio* (1659)

Source = Z :   *OG* 19, pp. 648–659.
G :   Florence, Biblioteca Nazionale Centrale, *Gal.* 85, ff. 39r-50r.
P :   Paris, Bibliothèque Nationale de France, *Fonds français*, 13039, ff. 147r-155v.

9. Lorenzo Crasso, *Galileo Galilei* (1666)

Source = E :   Lorenzo Crasso, *Elogii d'hvomini letterati* (1666), [Vol. I], pp. 243–251.

10. Vincenzo Viviani, *Raggvaglio dell'vltime opere del Galileo* (1674)

Source = Q :   Vincenzo Viviani, *Qvinto libro degli Elementi d'Evclide* (1674), pp. 86–88 and 99–106.

11. Isaac Bullart, *Galilée Galilei* (1682)

Source :   Isaac Bullart, *Académie des Sciences et des Arts* (1682), Vol. II, Book II, pp. 131–133.

12. Joachim von Sandrart, *Galilæus Galilæi* (1683)

Source :	Joachim von Sandrart, *Academia nobilissimæ artis pictoriæ* (1683), Part II, Book III, Chapter XXVIII, no. LVI, pp. 389–390.

13. Paul Freher, *Galilæus Galilæi* (1688)

Source = F :	Paul Freher, *Theatrvm virorum eruditione clarorum* (1688), Vol. II, Part IV, p. 1536.

14. Vincenzo Viviani, *Grati animi monumenta* (1702)

Sources = M :	Vincenzo Viviani, *Grati animi monumenta* (1702). Appendix: Vincenzo Viviani, *De locis solidis* (1701), p. 120.
L :	Vincenzo Viviani, *De locis solidis* (1701), pp. 121–128.
W :	London, Wellcome Library, Ms. 4949.
V :	Vincenzo Viviani's corrections in *W*.

Other abbreviations include those typically used in critical apparatuses: *add. = addidit; ante corr. = ante correctionem; deb. = debuit; del. = delevit; in marg. = in margine; om. = omisit; post corr. = post correctionem; scr. = scripsit; W*$^{s.l.}$ *= W supra lineam.* The apparatus does not record variants in punctuation, capital/small letters, *æ/ae, æ/oe, i/j, u/v,* and accents.

Appendix. Maffeo Barberini, *Adulatio perniciosa* (1620)

Source = ed. 1620 :	Maffeo Barberini, *Poemata*, Paris: Antoine Estienne, 1620, pp. 46–49.
ed. 1623 :	Urban VIII, *Poemata*, Paris: Antoine Estienne, 1623, pp. 39–42.
ed. 1631 :	Urban VIII, *Poemata*, Rome: Vatican Printing Office, 1631, pp. 266–270 = Urban VIII, *Poemata*, Rome: The Printing Office of the Reverend Apostolic Chamber, 1631, pp. 217–220.
ed. 1642 :	Urban VIII, *Poemata*, Paris: The Royal Press, 1642, pp. 186–189.
ed. 1726 :	Urban VIII, *Poemata*, edited by Joseph Brown, Oxford: Clarendon Press, 1726, pp. 179–182.

1

VINCENZO VIVIANI

RACCONTO ISTORICO DELLA VITA DEL SIG.ᴿ GALILEO GALILEI

HISTORICAL ACCOUNT OF THE LIFE OF GALILEO GALILEI

(1654)

Al Ser.^{mo} Principe Leopoldo di Toscana.

Racconto istorico della vita del Sig.^r Galileo Galilei,
Accademico Linceo, Nobil fiorentino,
Primo Filosofo e Matematico dell'Altezze Ser.^{me} di Toscana.

Al Ser.^{mo} Principe Leopoldo di Toscana,
mio Sig.^r et P.ron Col.^{mo}

Ser.^{mo} Principe,

Avendo V. A. S. risoluto di far scriver la vita del gran Galileo di gloriosa memoria, imposemi che, per notizia di chi dall'A. V. S. è destinato per esequire così eroico proponimento, io facesse raccolta di ciò che a me sovvenisse in tal materia, o d'altrove rintracciare io potesse: onde, per obbedire con ogni maggior prontezza a' cenni dell'A. V., reverente le porgo le seguenti memorie, spiegate da me con istorica purità, e con intera fedeltà registrate, avendole estratte per la maggior parte dalla viva voce del medesimo Sig.^r Galileo, dalla lettura delle sue opere, dalle conferenze e discorsi già co' suoi discepoli, dalle attestazioni de' suoi intrinseci e familiari, da pubbliche e private scritture, da più lettere de' suoi amici, e finalmente da varie confermazioni e riscontri che le autenticano per verissime e prive d'ogni eccezzione.

Nacque dunque Galileo Galilei, nobil fiorentino, il giorno 19 di Febbraio del 1563 *ab Incarnatione*, secondo lo stil fiorentino, nella città di Pisa, dov'allora dimoravano i suoi genitori.[1]

Il padre suo fu Vincenzio di Michelangelo Galilei, gentiluomo versatissimo nelle matematiche e principalmente nella musica speculativa, della quale ebbe così eccellente cognizione, che forse tra i teorici moderni di maggior nome non v'è stato sino al presente secolo chi di lui meglio e più eruditamente abbia ⁶⁰⁰scritto, come ne fanno chiarissima fede l'opere sue pubblicate, e principalmente il Dialogo della musica antica e moderna, ch'ei diede alle stampe in Firenze nel 1581. Questi congiunse alla perfezione della teorica l'operativa ancora, toccando a maraviglia varie sorti di strumenti e particolarmente il leuto, in che fu celebratissimo nell'età sua.

Ebbe della Sig.^{ra} Giulia Ammannati sua consorte più figliuoli, et il maggior de' maschi fu Galileo.[2]

---

1 Nacque—genitori *ABX*] Nacque dunque Galileo Galilei, nobil fiorentino, il dì 15 di Febbraio 1564, allo stile romano, in martedì, in Pisa, a ore 22½, altri a ore 3.30 dopo mezzo giorno, e fu quivi nel Duomo battezzato a dì 19 Febbraio detto, in sabato, essendo compari il Sig. Pompeo e Mess. Averardo de' Medici; et il sopraddetto giorno 15 di Febbraio 1564 precede di tre giorni quello nel quale morì in Roma il divino Michelagnolo Buonarroti, che morì alli 18 Febbraio 1564, al Romano *S*

2 Sig.^{ra}—consorte *ABX*] Sig. Giulia Ammannati di Pescia sua consorte, oriunda dall'antica e illustre famiglia degli Ammannati di Pistoia, più figlioli *S*

[599]To the Most Serene Prince Leopoldo of Tuscany

Historical Account of the Life of Galileo
Lyncean Academician, Florentine Nobleman,
Chief Philosopher and Mathematician of the Most Serene Highnesses of Tuscany[1]

To the Most Serene Prince Leopoldo of Tuscany,[2]
Most Honorable Master and Patron

Most Serene Prince,

Your Most Serene Highness decided to have someone write the life of the great Galileo, of glorious memory, and ordered me to gather all that I could recall about it, or anything I could find about it, to supply him whom Your Most Serene Highness would destine to undertake this heroic task. Accordingly, in order to comply with Your Highness's wishes as promptly as possible, I hereby respectfully submit the following recollections. I present them with historical correctness and record them with the best possible accuracy, as I learned the great majority of them from the living voice of Galileo himself, the reading of his works, exchanges and talks with his pupils, statements of his closest friends, public and private documents, numerous letters of his friends, and, lastly, from confirmation and evidence that validate them as very true and indisputable.

Galileo Galilei, Florentine nobleman, was born on 19 February 1563 *ab Incarnatione*,[3] according to the Florentine style, in the city of Pisa, where his parents then resided.[4]

His father was Vincenzo Galilei,[5] son of Michelangelo, a gentleman very well versed in mathematics, and especially theoretical music, of which he had such an excellent knowledge that among the most famous modern theorists perhaps none has written about it better and more learnedly [600]than he has, as his published works most clearly testify, and principally the *Dialogue on Ancient and Modern Music* that he published in Florence in 1581. Vincenzo complemented perfect theory with equally perfect practice, marvelously playing various sorts of musical instruments, and especially the lute, for which he was most celebrated in his lifetime.

Vincenzo had with his wife, Giulia Ammannati,[6] several children, and the oldest of the sons was Galileo.

[601]Cominciò questi ne' prim'anni della sua fanciullezza a dar saggio della vivacità del suo ingegno, poiché nell'ore di spasso esercitavasi per lo più in fabbricarsi di propria mano varii strumenti e machinette, con imitare e porre in piccol modello ciò che vedeva d'artifizioso, come di molini, galere, et anco d'ogni altra macchina ben volgare. In difetto di qualche parte necessaria ad alcuno de' suoi fanciulleschi artifizii suppliva con l'invenzione, servendosi di stecche di balena in vece di molli di ferro, o d'altro in altra parte, secondo gli suggeriva il bisogno, adattando alla macchina nuovi pensieri e scherzi di moti, purché non restasse imperfetta e che vedesse operarla.

Passò alcuni anni della sua gioventù nelli studii d'umanità appresso un maestro in Firenze di vulgar fama, non potendo 'l padre suo, aggravato da numerosa famiglia e constituito in assai scarsa fortuna, dargli comodità migliori, com'averebbe voluto, col mantenerlo fuori in qualche collegio, scorgendolo di tale spirito e di tanta accortezza che ne sperava progresso non ordinario in qualunque professione e' l'avesse indirizzato. Ma il giovane, conoscendo la tenuità del suo stato e volendosi pur sollevare, si propose di supplire alla povertà della sua sorte con la propria assiduità nelli studii; che perciò datosi alla lettura delli autori latini di prima classe, giunse da per sé stesso a quell'erudizione nelle lettere umane, della quale si mostrò poi in ogni privato congresso, ne' circoli e nell'accademie, riccamente adornato, valendosene mirabilmente con ogni qualità di persona, in qualunque materia, morale o scientifica, seria o faceta, che fosse proposta.

In questo tempo si diede ancora ad apprendere la lingua greca, della quale fece acquisto non mediocre, conservandola e servendosene poi opportunamente nelli studii più gravi.

[602]Udì i precetti della logica da un Padre Valombrosano; ma però que' termini dialettici, le tante definizioni e distinzioni, la moltiplicità delli scritti, l'ordine et il progresso della dottrina, tutto riusciva tedioso, di poco frutto e di minor satisfazione al suo esquisito intelletto.

Erano tra tanto i suoi più grati trattenimenti nella musica pratica e nel toccar li tasti e il leuto, nel quale, con l'esempio et insegnamento del padre suo, pervenne a tanta eccellenza, che più volte trovossi a gareggiare co' primi professori di que' tempi in Firenze et in Pisa, essendo in tale strumento ricchissimo d'invenzione, e superando nella gentilezza e grazia del toccarlo il medesimo padre; qual soavità di maniera conservò sempre sino alli ultimi giorni.

601 In the early years of his childhood, Galileo began to display the vivacity of his mind: he spent most of his leisure time constructing various instruments and machines with his own hands, imitating and producing small scale models of man-made objects, such as mills, galleys, and all sorts of other machines of common use. When he lacked some of the parts required for any of his youthful constructions, his imagination supplied them, as in the use of whalebone instead of iron springs, or other things in another part, as need suggested, improving the machine with new ideas and unexpected motions, so that he could finish it and see it operate.

He spent some of his youthful years studying humanities with a teacher of no great reputation[7] in Florence, for his father, having to support a large family with a very small fortune, could not provide him with a better opportunity, as he had wished to. Vincenzo had noticed his son's intelligence and cleverness, and he would have liked to send him to a boarding school, hoping that Galileo would make exceptional progress in any profession to which he would direct him. But the young boy, aware of his meager situation and wanting to improve it, resolved to make up for his poverty by assiduous studies. He committed himself to reading major Latin authors, and all by himself gained the erudition in the humanities that he later demonstrated in private gatherings, as well as in circles and academies, wonderfully exploiting it with all sorts of people, in any subject—moral or scientific, serious or funny—that was proposed.

At the same time, he devoted himself to the study of the Greek language, which he learned quite well andsaved for appropriate use in his more serious studies.

602 Galileo learned the basic principles of logic from a Vallombrosan monk,[8] but the terminology of dialectics, the very many definitions and distinctions, the variety of texts, as well as the order and progress of doctrine—all this seemed to him boring, of little benefit and little satisfaction to his exquisite mind.

Meanwhile, Galileo greatly enjoyed practicing music, fingering the keys and playing the lute. Thanks to the example and teachings of his father, he achieved such a high level of excellence on the lute that he found himself competing with the top masters of those times, in Florence as well as in Pisa. In playing this instrument he displayed the greatest ingenuity and surpassed his very father in his gentle and elegant playing, the sweetness of which he maintained until his last days.[9]

Trattenevasi ancora con gran diletto e con mirabil profitto nel disegnare; in che ebbe così gran genio e talento, ch'egli medesimo poi dir soleva agl'amici, che se in quell'età fosse stato in poter suo l'eleggersi professione, averebbe assolutamente fatto elezione della pittura. Ed in vero fu di poi in lui così naturale e propria l'inclinazione al disegno, et acquistovvi col tempo tale esquisitezza di gusto, che 'l giudizio ch'ei dava delle pitture e disegni veniva preferito a quello de' primi professori da' professori medesimi, come dal Cigoli, dal Bronzino, dal Passignano e dall'Empoli, e da altri famosi pittori de' suoi tempi, amicissimi suoi, i quali bene spesso lo richiedevano del parer suo nell'ordinazione dell'istorie, nella disposizione delle figure, nelle prospettive, nel colorito et in ogn'altra parte concorrente alla perfezione della pittura, riconoscendo nel Galileo intorno a sì nobil arte un gusto così perfetto e grazia sopranaturale, quale in alcun altro, benché professore, non seppero mai ritrovare a gran segno; onde 'l famosissimo Cigoli, reputato dal Galileo il primo pittore de' suoi tempi, attribuiva in gran parte quanto operava di buono alli ottimi documenti del medesimo Galileo, e particolarmente pregiavasi di poter dire che nelle prospettive egli solo gli era stato il maestro.

Trovandosi dunque il Galileo in età di sedici[3] anni in circa con tali virtuosi ornamenti e con gli studii d'umanità, lingua greca e dialettica, deliberò 'l padre suo di mandarlo a studio a Pisa, quantunque con incomodo della sua casa, ma con ferma speranza ch'un giorno l'averebbe sollevata con la professione della medicina, alla quale egl'intendeva ch'e' s'applicasse, come più atta e spedita a potergli somministrar le comodità necessarie; e raccomandatolo ad un parente mercante ch'egli aveva in quella città, quivi inviollo, dove cominciò gli studii di medicina et insieme della vulgata filosofia peripatetica. Ma il Galileo, che dalla natura fu eletto per disvelare al mondo parte di que' segreti che già per tanti secoli restavano sepolti in una densissima oscurità delle menti umane, fatte [603] schiave del parere e de gl'asserti d'un solo, non poté mai, secondo 'l consueto degl'altri, darsele in preda così alla cieca, come che, essendo egli d'ingegno libero, non gli pareva di dover così facilmente assentire a' soli detti et opinioni delli antichi o moderni scrittori, mentre potevasi col discorso e con sensate esperienze appagar sé medesimo. E perciò nelle dispute di conclusioni naturali fu sempre[4] contrario alli più acerrimi difensori d'ogni detto Aristotelico, acquistandosi nome tra quelli di spirito della contradizione, et in premio delle scoperte verità provocandosi l'odio loro; non potendo soffrir che da un giovanetto studente, e che per ancora, secondo un lor detto volgare, non avea fatto il corso delle scienze, quelle dottrine da lor imbevute, si può dir, con il latte gl'avesser ad esser con nuovi modi e con tanta evidenza rigettate e convinte: averando in ciò quel detto di Orazio:

Stimano infamia il confessar da vecchii
Per falso quel che giovini apprendero.

3 sedici *BX*] diciotto *AS*
4 sempre *BX*] spesse volte *AS*

Furthermore, he very much enjoyed drawing, in which he achieved amazing results—and he was so gifted and talented in it that he later used to tell his friends that had it been possible for him, at this age, to choose a profession, no doubt he would have chosen painting. Indeed, his bent for drawing was so natural and innate, and he acquired such an exquisite taste as time went by, that his judgment on paintings and drawings was preferred to those of eminent masters by the masters themselves: by Cigoli, Bronzino, Passignano, and Empoli,[10] and by other famous painters of his time, all of whom were dearest friends of his. They very often asked his opinion when making their designs and arranging their figures, in perspective and in coloring, and in everything that comes together to perfect a painting. In so doing, they acknowledged Galileo's perfect taste in and exceptional gift for their noble art, which they did not find to the same degree in any one else, even a master. Hence the very eminent Cigoli, whom Galileo regarded as the best painter of his time, attributed in large part whatever was good in his work to the excellent teachings of Galileo, and was especially proud to be able to say that, as far as perspective was concerned, Galileo had been his sole master.

When Galileo was about sixteen years old,[11] and being so accomplished and having pursued his studies of humanities, Greek language and dialectics, his father decided to send him to the University of Pisa despite the financial burden. He did hope, however, that one day Galileo would relieve him of the burden by becoming a physician, which Vincenzo wanted him to become, since medicine was the most suitable and fastest way to ease the situation.[12] He entrusted Galileo to a merchant, a relative in that city,[13] where he began his studies of medicine and the accepted Peripatetic philosophy. But Galileo, whom nature had elected to unveil to the world part of those secrets that had been buried in the dense darkness of human minds for so many centuries—[603]enslaved, as they were, by the opinion and statements of a single man—could never accept these teachings blindly, as others were used to do. With his free intellect, Galileo could not see why he should have to accept so easily statements and opinions of ancient or modern authors, when he could satisfy himself by way of argument and sensory experience. Accordingly, he always[14] opposed the most passionate upholders of Aristotle's views in any disputes about natural questions, thereby earning a reputation for a spirit of contradiction. And whenever he discovered a truth, he was rewarded with their hatred, as they could not bear that a young student—someone who had not yet had a proper training in science, as was commonly said—rejected and discarded in new ways and with such evidence the doctrines they had swallowed, as it were, with their mothers' milk. In so doing, they validated what Horace said:

> they hold it a shame to confess in their old age
> that what they learned in youth is false.[15]

Continuò di così per tre o quattr'anni, ne' soliti mesi di studio in Pisa, la medicina e filosofia, secondo l'usato stile de' lettori; ma però in tanto da sé stesso diligentemente vedeva l'opere di Aristotele, di Platone e delli altri filosofi antichi, studiando di ben possedere i lor dogmi et opinioni per esaminarle e satisfare principalmente al proprio intelletto.

In questo mentre con la sagacità del suo ingegno inventò quella semplicissima e regolata misura del tempo per mezzo del pendulo, non prima da alcun altro avvertita, pigliando occasione d'osservarla dal moto d'una lampada, mentre era un giorno nel Duomo di Pisa; e facendone esperienze esattissime, si accertò dell'egualità delle sue vibrazioni, e per allora sovvennegli di adattarla all'uso della medicina per la misura della frequenza de' polsi, con stupore e diletto de' medici di que' tempi e come pure oggi si pratica volgarmente: della quale invenzione si valse poi in varie esperienze e misure di tempi e moti, e fu il primo che l'applicasse alle osservazioni celesti, con incredibile acquisto nell'astronomia e geografia.

Di qui s'accorse che gl'effetti della natura, quantunque apparischin minimi et in niun conto osservabili, non devon mai dal filosofo disprezzarsi, ma tutti egualmente e grandemente stimarsi; essendo perciò solito dire che la natura operava molto col poco, e che le sue operazioni eran tutte in pari grado maravigliose.

Tra tanto non aveva mai rivolto l'occhio alle matematiche, come quelle che, per esser quasi affatto smarrite, principalmente in Italia (benché dall'opera e diligenza del Comandino, e del Maurolico etc.,[5] in gran parte restaurate), per ancora non avendo pigliato vigore, erano più tosto universalmente in disprezzo; e non sapendo comprendere quel che mai in filosofia si potesse dedurre da figure di [604]triangoli e cerchi, si tratteneva senza stimolo d'applicarvisi.

---

5 e del Maurolico etc. BX] *om. AS*

And so Galileo continued in this way for three or four years,[16] during the usual months of study he spent in Pisa, with medicine and philosophy, according to the established way of the professors. In the meantime, however, he carefully perused by himself the works of Aristotle, Plato, and the other ancient philosophers, taking care to learn their doctrines and opinions thoroughly so as to be able to test them, chiefly to satisfy his own mind.[17]

In this period, his sharp mind devised that very simple and regulated measure of time by way of the pendulum, which nobody had previously noticed. One day Galileo happened to observe it while he was in the Duomo, in Pisa, watching the motion of a lamp, and by making very exact experiments ascertained the equality of its vibrations.[18] It occurred to him to adapt it to medical use, for the measure of pulse beats, to the amazement and pleasure of the physicians of the time; this very practice is still commonly used nowadays.[19] Galileo later made use of this discovery in several experiments and measures of times and motions, and he was the first to apply it to celestial observations, from which astronomy and geography benefited in astonishing ways.

As a consequence, he realized that nature's effects, however minimal and negligible they might appear, are never to be ignored by philosophers; by contrast, all must be equally and carefully taken into account. He used to say that nature achieved much with very little,[20] and that its operations were all equally astonishing.

In all this, he had never paid any attention to mathematics. Almost completely abandoned, especially in Italy (although by the work and diligence of Commandino, Maurolico, and others in great part restored[21]), not only was mathematics still weak at the time but it was generally despised. Without any incentives, failing to understand that something could be deduced in philosophy from drawings of [604]triangles and circles, Galileo refrained from applying himself to it.

Ma il gran talento e diletto insieme ch'egli aveva, come dissi, nella pittura, prospettiva e musica, et il sentire affermare frequentemente dal padre che tali pratiche avevan l'origin loro dalla geometria, gli mossero desiderio di gustarla, e più volte pregò il padre che volesse introdurvelo; ma questi, per non distorlo dal principale studio di medicina, differiva di compiacerlo, dicendogli che quando avesse terminato i suoi studii in Pisa, poteva applicarvisi a suo talento. Non per ciò si quietava il Galileo; ma vivendo allora un tal Mess. Ostilio Ricci di Fermo, matematico de' SS. paggi di quell'Altezza di Toscana e dipoi lettore delle matematiche nello Studio di Firenze, il quale, come familiarissimo di suo padre, giornalmente frequentava la sua casa, a questo s'accostò, pregandolo instantemente a dichiarargli qualche proposizione d'Euclide, ma però senza saputa del padre. Parve al Ricci di dover saziar così virtuosa brama del giovane, ma volle ben conferirla al Sig.ʳ Vincenzio suo padre, esortandolo a permetter che il Galileo ricevesse questa satisfazione. Cedé il padre all'instanze dell'amico, ma ben gli proibì il palesar questo suo assenso al figliuolo, acciò con più timore continuasse lo studio di medicina. Cominciò dunque il Ricci ad introdurre il Galileo (che già aveva compliti diciannove[6] anni) nelle solite esplicazioni delle definizioni, assiomi e postulati del primo libro delli Elementi; ma questi sentendo preporsi principii tanto chiari et indubitati, e considerando le domande d'Euclide così oneste e concedibili, fece immediatamente concetto che se la fabbrica della geometria veniva alzata sopra tali fondamenti, non poteva esser che fortissima e stabilissima. Ma non sì tosto gustò la maniera del dimostrare, e vedde aperta l'unica strada di pervenire alla cognizione del vero, che si pentì di non essersi molto prima incamminato per quella. Proseguendo 'l Ricci le sue lezzioni, s'accorse il padre che Galileo trascurava la medicina e che più si affezionava alla geometria; e temendo che egli col tempo non abbandonasse quella, che gli poteva arrecar maggior utile e comodità nell'angustie della sua fortuna, lo riprese più volte (fingendo non saperne la cagione), ma sempre in vano, poiché tanto più quegli s'invaghiva della matematica, e dalla medicina totalmente si distraeva; ond'il padre operò che 'l Ricci di quando in quando tralasciasse le sue lezzioni, e finalmente ch'allegando scuse d'impedimenti desistesse affatto dall'opera. Ma accortosi di ciò il Galileo, già che il Ricci non gli aveva per ancora esplicato il primo libro delli Elementi, volle far prova se per sé stesso poteva intenderlo sino alla fine, con desiderio di arrivare almeno alla 47, tanto famosa; e vedendo che gli sortì d'apprendere il tutto felicemente, fattosi d'animo, si propose di voler scorrer qualch'altro libro: e così, ma furtivamente dal padre, andava studiando, con tener gl'Ippocrati e Galeni appresso l'Euclide, per poter con essi prontamente occultarlo quando 'l padre gli fosse sopraggiunto.

6 diciannove BX] i venti due AS

However, his great talent for painting, perspective, and music, as I said, and the pleasure he drew from them, together with his father's often-repeated statement that they were all grounded in geometry, encouraged Galileo to give it a try. He often begged his father to teach him mathematics, but Vincenzo refused, in order not to divert him from his studies of medicine. He told Galileo that he could devote himself to mathematics, as he wished to do, once he had completed his studies in Pisa; but this was not enough to quiet Galileo. Ostilio Ricci, from Fermo, who later became lecturer of mathematics at the University of Florence, was then a mathematician at the court of the Grand Duke of Tuscany.[22] Ricci was very close to Galileo's father and visited his house daily. Galileo approached him and repeatedly asked him for the explanation of some of Euclid's propositions, but unbeknown to his father. Ricci wanted to quench Galileo's youthful and meritorious thirst but decided to tell Vincenzo about it, urging him to comply with Galileo's wish. Vincenzo yielded to his friend's requests but forbade him to let his son know it, so that he would more anxiously pursue his studies of medicine. Thus Ricci began to introduce Galileo (who had already turned nineteen[23]) to the standard explanations of definitions, axioms, and postulates of the first book of the *Elements*—but Galileo, presented with such clear and undisputable principles, and deeming Euclid's postulates so reasonable and admissible, thought immediately that if the edifice of geometry was erected upon these foundations, it could not be other than extremely solid and stable. But as soon as he appreciated the manner of demonstrating a statement, and saw, before his eyes, the only way to reach knowledge of the truth, he resented not having started along that path long before. As Ricci continued his lessons Galileo's father realized that his son was increasingly ignoring medicine as he was growing increasingly fond of geometry. Thus fearing, as time went by, that Galileo would abandon the former, from which he could have benefited more, from the point of view of utility and opportunity, given the tight financial circumstances, Vincenzo repeatedly scolded Galileo (pretending not to know the reason [for his lack of interest in medicine]), but always in vain, for the more Galileo grew fond of mathematics, the more he turned away from medicine. And so his father had Ricci suspend his lessons from time to time, and eventually stop them, offering, as an excuse, that he could no longer do it. Ricci had not yet explained the first book of the *Elements* to him, but once Galileo realized what had happened, he decided to see whether he could understand it in its entirety, hoping to get at least to Proposition 47, which is so famous.[24] And having managed to understand everything perfectly, he steeled himself for working through some other books. Thus, unknown to his father, Galileo studied on, keeping Hippocratic and Galenic books[25] side by side with the Euclid so as to be able to hide it in case his father came by.

Finalmente sentendosi traportar dal diletto et acquisto che parevagli d'aver [605]conseguito in poco tempo da[7] tale studio, nel ben discorrere argumentare e concludere, assai più che dalle logiche e filosofie di tutto il tempo passato, giunto al sesto libro d'Euclide, si risolse di far sentire al padre il profitto che per sé stesso aveva fatto nella geometria, pregandolo insieme a non voler deviarlo donde sentivasi traportare dalla propria inclinazione. Udillo 'l padre, e conoscendo dalla di lui perspicacità nell'intendere e maravigliosa abilità nell'inventare varii problemi ch'egli stesso gli proponeva, che 'l giovane era nato per le matematiche, si risolse in fine di compiacerlo.

Tralasciando dunque il Galileo lo studio di medicina, in breve tempo scorse gl'Elementi d'Euclide e l'opere de' geometri di prima classe; et arrivando all'Equiponderanti e al trattato *De his quae vehuntur in aqua* d'Archimede, sovvennegli un nuovo modo esattissimo di poter scoprire il furto di quell'orefice nella corona d'oro di Hierone:[a] et allora scrisse la fabbrica e uso di quella sua bilancetta, per la quale s'ha cognizione delle gravità in specie di diverse materie e della mistione o lega de' metalli, con molt'altre curiosità appresso; quali, benché poi dal Galileo non sieno state fatte pubbliche con le stampe, parte però furono conferite da lui a quei che se gli facevano amici, e parte vanno intorno in private scritture: onde non è gran fatto s'alcuno l'ha publicate per sue o se ne è valso, mascherandole, come di propria invenzione.

Con questi et altri suoi ingegnosi trovati, e con la sua libera maniera di filosofare e discorrere, cominciò ad acquistar fama d'elevatissimo spirito; e conferendo alcune delle sue speculazioni meccaniche e geometriche[8] con il Sig.[r] Guidubaldo de' Marchesi dal Monte, gran matematico di quei tempi, che a Pesaro dimorava, acquistò seco per lettere strettissima amicizia, et ad instanza di lui s'applicò alla contemplazione del centro di gravità de' solidi, per supplire a quel che ne aveva già scritto il Comandino; e ne' ventiuno anni[9] di sua età, con due anni soli di studio di geometria, inventò quello ch'in tal materia si vede scritto nell'Appendice impressa alla fine de' suoi Dialogi delle due Nuove Scienze della meccanica e del moto locale, con gran satisfazione e maraviglia del medesimo Sig.[r] Guidubaldo, il quale per così acute invenzioni l'esaltò a segno appresso il Ser.[mo] Gran Duca Ferdinando Primo e l'Eccel.[mo] Principe Don Giovanni de' Medici, ch'in breve divenne a loro gratissimo e familiare: che perciò vacando nel 1589 la cattedra delle matematiche in Pisa, di proprio moto della medesima Ser.[ma] Altezza ne fu provvisto, correndo egli l'anno vigesimo sesto dell'età sua.

a nel 1586 trovò questa bilancia.

7 poco tempo da *BX*] pochi mesi di *AS*

8 *post* geometriche *scr.* (nell'invenzion delle quali havea acutezza, e facilità sopraordinarie) *A* (nell'invenzion delle quali havea, come s'è detto, acutezza, e facilità sopraordinarie) *S*

9 ventiuno anni *BX*] ventiquattr'anni *A* ventun'anno *S*

Eventually, carried away by the pleasure and profit in discoursing, arguing, and demonstrating he thought he had [605]gained from that study in a short period of time[26]—much more than from all the logics and philosophies of past times—he decided when he reached the sixth book of Euclid to let his father know the progress he had made in geometry by himself, at the same time begging him not to divert him from the direction his own inclination drove him to. Vincenzo listened to him, and knowing from his perspicacity in understanding and wonderful skill in solving various problems which he proposed to Galileo, that the young man was born to do mathematics, he at last decided to comply with his wish.

Thus Galileo abandoned his studies of medicine and in a short span of time went through Euclid's *Elements* and the works of the most eminent geometers. Having reached Archimedes's *On the Equilibrium of Planes* and *On Floating Bodies*,[27] a new and most exact method for discovering the goldsmith's fraud about Hiero's crown[28] struck him,[a] and so he described the construction and use of his little balance by which we can ascertain the specific gravity of different kinds of matter, and the mixture, or alloy, of metals, as well as many other curiosities. Although Galileo never published them, he told his friends about some of them, and circulated others privately. Unsurprisingly, someone published them, or took advantage of them, disguising them as his own discovery.

Thanks to these and other ingenious inventions of his, and to his free way of philosophizing and conversing, Galileo became known as a man with an extraordinary mind. Exchanging some speculations on mechanics and geometry[29] with Guidobaldo, of the Marquises del Monte, a prominent mathematician of the time living in Pesaro,[30] he became very close to him through correspondence. On Guidobaldo's instance he devoted himself to the study of the center of gravity of solids to expand on what Commandino had written about it. And at twenty-one,[31] having studied geometry for only two years, he invented what we now find on the subject in the appendix to his *Dialogues Concerning Two New Sciences*, pertaining to mechanics and local motion, to the satisfaction and amazement of Guidobaldo.[32] For such ingenious inventions, Guidobaldo praised him greatly to the Most Serene Grand Duke Ferdinand I[33] and the Most Excellent Prince Don Giovanni de' Medici.[34] Galileo soon earned their admiration and friendship and became close to them, and when the chair of mathematics became available in Pisa, in 1589, the Most Serene Highness, on his own initiative, offered it to Galileo, who was then 26 years old.

a He devised this balance in 1586.

[606]In questo tempo, parendogli d'apprendere ch'all'investigazione delli effetti naturali necessariamente si richiedesse una vera cognizione della natura del moto, stante quel filosofico e vulgato assioma *Ignorato motu ignoratur natura*, tutto si diede alla contemplazione di quello: et allora, con gran sconcerto di tutti i filosofi, furono da esso convinte di falsità, per mezzo d'esperienze e con salde dimostrazioni e discorsi, moltissime conclusioni dell'istesso Aristotele intorno alla materia del moto, sin a quel tempo state tenute per chiarissime et indubitabili; come, tra l'altre, che le velocità de' mobili dell'istessa materia, disegualmente gravi, movendosi per un istesso mezzo, non conservano altrimenti la proporzione delle gravità loro, assegnatagli[10] da Aristotele, anzi che si muovon tutti con pari velocità, dimostrando ciò con replicate esperienze, fatte dall'altezza del Campanile di Pisa con l'intervento delli altri lettori e filosofi e di tutta la scolaresca; e che né meno le velocità di un istesso mobile per diversi mezzi ritengono la proporzion reciproca delle resistenze o densità de' medesimi mezzi, inferendolo da manifestissimi assurdi ch'in conseguenza ne seguirebbero contro al senso medesimo.[11]

Sostenne perciò questa cattedra con tanta fama e reputazione appresso gl'intendenti di mente ben affetta e sincera, che molti filosofastri suoi emuli, fomentati da invidia, se gli eccitarono contro; e servendosi di strumento per atterrarlo del giudizio dato da esso sopra una tal macchina, d'invenzione d'un eminente soggetto, proposta per votar la darsina di Livorno, alla quale il Galileo con fondamenti meccanici e con libertà filosofica aveva fatto pronostico di mal evento (come in effetto seguì), seppero con maligne impressioni provocargli l'odio di quel gran personaggio: ond'egli, rivolgendo l'animo suo all'offerte che più volte gl'erano state fatte della cattedra di Padova, che per morte di Giuseppe Moleti stette gran tempo vacante, per consiglio e con l'indirizzo del Sig.ʳ Marchese Guidubaldo s'elesse, con buona grazia del Ser.ᵐᵒ Gran Duca, di mutar clima, avanti che i suoi avversarii avessero a godere del suo precipizio. E così dopo tre anni di lettura in Pisa, ne' 26 di Settembre del 1592, ottenne dalla Ser.ᵐᵃ Republica di Venezia la lettura delle matematiche in Padova per sei anni: nel qual tempo inventò varie macchine in servizio della medesima Republica, con suo grandissimo onore e utile insieme, come dimostrano gl'amplissimi privilegi ottenuti da quella; et a contemplazione de' suoi scolari scrisse allora varii trattati, tra' quali uno di fortificazione, secondo l'uso di quei tempi, uno di gnomonica e prospettiva pratica, un compendio di sfera, et un trattato di meccaniche,[b] che va attorno manoscritto, e che poi nel 1634, tradotto in lingua franzese, fu stampato [607]in Parigi dal Padre Marino Mersennio, e ultimamente nel 1649 publicato in Ravenna dal Cav.ʳ Luca Danesi: trovandosi di tutti questi trattati, e di molti altri, più copie sparse per l'Italia, Germania, Francia, Inghilterra et altrove, trasportatevi da' suoi medesimi discepoli, la maggior parte senza l'inscrizione del suo nome, come fatiche delle quali ei non faceva gran conto, essendo di esse tanto liberale donatore quanto fecondo compositore.[12]

b nel 1593 scrisse le Meccaniche e altre cose.

---

10 loro, assegnatagli *ABX*] loro assolute, assegnata loro *S*

11 *post* medesimo *scr.* che tutto si vede poi diffusamente trattato da lui nelli ultimi Dialogi delle due Nuove Scienze *A* che tutto si vede poi diffusamente trattato da lui nelli suddetti Dialoghi delle Nuove Scienze *S*

12 *post* compositore *scr.* Ben è vero che questa sua natural liberalità in comunicare i suoi scritti, le proprie invenzioni et i suoi nuovi pensieri indifferentemente a ciascuno, gli fu spesso contracambiata da altretanta ingratitudine e sfacciataggine, non essendo mancati o chi con disprezzo tentasse avvilirle o chi se ne facesse onore, come di parti de' propri ingegni *AS*

[606]At that time, Galileo seems to have learned that inquiring into natural effects necessarily requires a true knowledge of the nature of motion (given that philosophical and commonly used axiom, *Ignorato motu ignoratur natura*[35]) and devoted himself fully to the study of it. In so doing, to the utmost bewilderment of all philosophers, Galileo—by way of experiences and rigorous demonstrations and arguments[36]—showed that very many of Aristotle's statements about motion, which were regarded as perspicacious and beyond doubt, were false. Among them was the proposition that the speeds of movable objects made of the same kind of matter but differing in weight, and moving through the same medium, are in proportion to their weights, as Aristotle taught;[37] rather, they all move with the same speed.[38] He showed this with repeated experiences from the top of the Tower of Pisa, in the presence of the other lecturers and philosophers, as well as of all their students.[39] He also showed it false that the speed of the same movable object through different media is inversely proportional to the resistance or density of the same media, inferring this from the most obvious absurdities that would follow in consequence, contrary to experience itself.[40]

Galileo held this chair with such fame and repute among knowledgeable people with sympathetic and honest minds that many philosophasters—would-be emulators, instigated by envy—roused themselves against him. To bring him down, they used his critical remarks about a machine devised by an eminent person[41] for emptying the wet dock of Leghorn. By appealing to arguments grounded on mechanics and philosophical liberty, Galileo had predicted that the machine was doomed to fail (as, in fact, it did), and with nasty innuendoes they managed to turn that eminent person against Galileo. Consequently, he accepted the offer that time and again he had received for the chair in Padua, which had lain vacant since the death of Giuseppe Moleti.[42] Thus, following the recommendation of Marquis Guidobaldo del Monte, and with the Most Serene Grand Duke's approval, he decided to move before his adversaries could benefit from his downfall. And so, after lecturing for three years in Pisa, on September 26, 1592, Galileo received the lectureship of mathematics in Padua from the Most Serene Republic of Venice.[43] His tenure lasted six years, during which Galileo devised several machines in support of the Republic, earning both honor and profit from them, as the very generous privileges he obtained from the Republic clearly show.[44] Galileo wrote a number of treatises for his students, among which was one on fortification, as was customary [for mathematicians] in those days, one on gnomonics and practical perspective, a synopsis on the sphere, and a treatise on mechanics[b] that circulated in manuscript form and was later translated into French in 1634, printed [607]in Paris by Father Marin Mersenne, and eventually published in Ravenna by Cavalier Luca Danesi.[45] There are several copies of all these treatises, as well as of many others, scattered around Italy, Germany, France, England, and elsewhere, taken there by Galileo's close disciples. Most of them do not bear the author's name; he did not worry about such things, as he was as much a generous donor as he was a prolific writer.[46]

b In 1593 he wrote *On Mechanics* and other works.

In questi medesimi tempi ritrovò i termometri, cioè quelli strumenti di vetro, con acqua et aria, per distinguer le mutazioni di caldo e freddo e la varietà de' temperamenti de' luoghi; la qual maravigliosa invenzione dal sublime ingegno del gran Ferdinando Secondo, nostro Ser.ᵐᵒ Padron regnante, è stata modernamente ampliata et arricchita con nuovi effetti di molte vaghe curiosità e sottigliezze, quali, coperte con ingegnose apparenze, sono da quelli che ne ignorano le cagioni stimate prestigiose.

Circa l'anno 1597 inventò il suo mirabile compasso geometrico e militare, cominciando sin da quel tempo a fabbricarne gli strumenti et insegnarne l'uso in voce et in scritto a' suoi discepoli, esplicandolo a molti principi e gran signori di diverse nazioni, tra' quali furono l'Ill.ᵐᵒ et Eccel.ᵐᵒ Sig.ʳ Gio. Federigo Principe d'Olsazia, et appresso il Ser.ᵐᵒ Arciduca D. Ferdinando d'Austria, dopo l'Ill.ᵐᵒ et Eccel.ᵐᵒ Sig.ʳ Filippo Langravio d'Assia, Conte di Nidda, et il Ser.ᵐᵒ di Mantova, et altri infiniti, che lungo sarebbe il registrargli qui tutti.

Proseguendo il Galileo le sue private e pubbliche lezzioni con applauso sempre maggiore, li 29 di Ottobre del 1599 fu ricondotto alla medesima lettura per altri sei anni, con augumento di provvisione.

In questo mentre, dimostrandosi con strana e portentosa maraviglia del cielo, nella costellazione del Serpentario, la nuova stella del 1604, fu dal Galileo con tre lunghe e dottissime lezzioni pubblicamente discorso sopra così alta materia; nelle quali intese provare che la nuova stella era fuori della regione elementare et in luogo altissimo sopra tutti i pianeti, contro l'opinione della scuola peripatetica e principalmente del filosofo Cremonino, che allora procurava di sostenere il contrario e di mantenere il cielo del suo Aristotele inalterabile et esente da qualunque accidentaria mutazione.

In questi medesimi tempi fece studio et osservazione particolare sopra la virtù della calamita, e con varie e replicate esperienze trovò modo sicuro di armarne qualunque pezzo, che sostenesse di ferro ottanta e cento volte più che disarmato; alla qual perfezione non si è mai pervenuto da alcun altro a gran segno.

⁶⁰⁸Aveva, come s'è detto, sol per utile e diletto de' suoi discepoli, scritto varii trattati et inventato molti strumenti, tra' quali uno era il sopradetto compasso, non però con pensiero d'esporlo al publico: ma presentendo che altri s'apparecchiava per appropriarsene l'invenzione, scrisse in fretta una general descrizione de' suoi usi, riserbandosi ad altra occasione a darne fuori una più ampia dichiarazione insieme con la sua fabbrica; e nel Giugno del 1606 la diede alle stampe in Padova, con titolo delle *Operazioni del Compasso Geometrico e Militare*, dedicato al Ser.ᵐᵒ D. Cosimo, allora Principe di Toscana e suo discepolo.[13] Quest'opera fu dopo tradotta in latino da Mattia Berneggero tedesco, e stampata in Argentina nel 1612 insieme con la fabbrica del compasso e con alcune annotazioni, e ristampatavi ancora nel 1635, sì come più volte in Padova et altrove.

Ne' 5 d'Agosto del 1606 fu ricondotto dalla medesima Republica lettor matematico per altri sei anni, con nuovo augumento di provvisione, ch'era poi maggiore della solita darsi a qualunque de' suoi antecessori.

13 suo discepolo *BX*] poi padre di V. A. *AS*

During this same time, Galileo invented thermometers, that is, those glass instruments filled with air and water that are used to measure the changes of heat and cold, and the variation of temperatures in different places.[47] With his sublime mind, the great Ferdinando II, our Most Serene Sovereign,[48] developed and improved on this amazing invention with curious and subtle new effects, which, cleverly hidden under ingenious appearances, are deemed prestigious by those who do not know their causes.

Around 1597, Galileo devised his admirable geometric and military compass[49] and then produced the instrument and taught his pupils how to use it, both verbally and in writing, explaining it to many princes and noblemen from different countries. Among them were the Most Eminent and Excellent John Frederick, Prince of Alsace; the Most Serene Archduke Ferdinand of Austria; the Most Eminent and Excellent Philip, Landgrave of Hesse and Count of Nidda; and the Most Serene <Duke> of Mantua; and many more, whose names it would be too long to list here.[50]

Galileo continued lecturing privately and publicly, with ever more success, and on October 29, 1599, his lectureship was renewed for six more years, with an increased salary.[51]

During that period, a new star appeared with unusual and extraordinary magnificence in the sky, in the constellation of the Serpent, in 1604.[52] Galileo gave three extensive and very learned public lectures about this highly important matter, with which he meant to show that the new star stood outside the elementary region,[53] far above all the planets. This clashed with the view of the Peripatetic school, and especially with that of the philosopher Cremonini, who at that time taught the view opposite Galileo's, and strived to preserve the heavens of his beloved Aristotle incorruptible and exempt from any accidental change.[54]

In these same years, Galileo studied and observed in detail the properties of magnets, and with various repeated experiments found a sure way to fit an armature to any such pieces so that they could hold up quantities of iron 80 or 100 times larger than they could without armatures—a level no one else has ever reached with such perfection.[55]

[608]As already stated, Galileo wrote several treatises and invented many instruments for the sole benefit and delight of his pupils. Among them was the compass mentioned earlier, which was not meant to be publicly displayed. However, sensing that someone else was about to seize it as his own discovery, Galileo hastened to write a general description of its uses, intending to publish a more detailed description later on that would explain its construction. In June 1606 he published it in Padua, under the title *Operations of the Geometric and Military Compass*, and dedicated it to the Most Serene Cosimo, Prince of Tuscany, who was one of his pupils.[56] This work was later translated into Latin by Matthias Bernegger, a German, and printed in Strasbourg in 1612, together with the <instructions for the> construction of the compass and some additional annotations; this was reprinted in 1635, as it was in Padua and elsewhere.[57]

On August 5, 1606, the Republic of Venice renewed Galileo's lectureship in mathematics for six more years, with a further increase of salary, which was already higher than his predecessors had usually been paid.[58]

Nel 1607 trovandosi il Galileo fieramente offeso e provocato da un certo Baldassar Capra milanese, che si era allora temerariamente appropriata l'invenzione del suddetto compasso col tradurlo in latino e stamparlo nell'istessa città di Padova in faccia del medesimo autore, con titolo di *Usus et fabrica circini cuiusdam proportionis*,[14] fu questi necessitato a publicare una sua *Difesa* in volgare, per evidente dimostrazione di furto così detestabile e vergognoso; difendendosi insieme dalle calunnie et imposture del medesimo Capra, il quale in una sua *Considerazione astronomica circa la stella nuova del 1604*, stampata già più di due anni avanti, l'aveva acerbamente lacerato, mosso da invidia per l'universale applauso che avevano ricevuto le tre suddette lezzioni del Galileo, fatte sopra la nuova stella. Ma il Capra per mezzo di queste sue abominevoli azzioni ne riportò il dovuto premio d'una perpetua ignominia, poiché dalli Eccel.ᵐⁱ SS. Reformatori dello Studio di Padova, dopo essersi, con rigoroso processo formato contro di quello, assicurati a pieno di tanta temerità, fu comandato supprimersi tutte le copie stampate del libro di detto Capra e proibitone la publicazione, et all'incontro conceduto al Galileo d'esporre alla luce la suddetta Difesa, per ricatto della propria reputazione et oppressione di quella del medesimo Capra.

Non fu già valevole tal Difesa a reprimere l'audacia o la troppa confidenza di alcuni altri d'altre nazioni, i quali, allettati o traportati dalla novità e vaghezza dell'invenzione o dalla mirabil copia e facilità de' suoi usi, non esponessero alle stampe, come interamente lor proprio, questo ingegnoso compasso del Galileo, [609]publicandolo, o con diverse inscrizioni in altra forma ridotto o con nuove linee et ad altri usi ampliato, senza pur far menzione del principale autore di tal strumento; l'operazioni del quale, dove non erano pervenute stampate, si trovavano già molto prima in ogni provincia d'Europa manuscritte, e divulgate da quelli istessi forestieri a' quali in Padova il medesimo Galileo le aveva prodigamente, con altri suoi scritti, comunicate. Ma l'ardire di questi o l'ingratitudine, oltre al farsi palese dalla suddetta Difesa, vien dannata dalla medesima azione, et autenticata dalla gloriosa fama del Galileo, che per l'altre opere e invenzioni di assai maggior maraviglia si è poi saputo acquistare sopra quelli che pochi altri et assai deboli parti col proprio ingegno hanno saputo produrre.

---

14 *post* proportionis *scr.* osando in oltre di tacciare d'impudentissimo usurpatore l'istesso Sig.ʳ Galileo *A* (osando inoltre chiamarlo sfacciatissimo usurpatore) *B*

In 1607, Galileo was hurt and annoyed by one Baldassarre Capra, from Milan, who had recklessly appropriated the invention of the geometric and military compass by translating <Galileo's text> into Latin and publishing it in Padua, right in Galileo's face, under the title *Use and Construction of Proportional Compasses.*[59] Galileo felt compelled to publish a *Difesa* (*Defense*) of himself in the vernacular, so as to provide clear evidence of the dreadful and shameful plagiarism. He also defended himself from the libels and lies of Capra, who had harshly insulted him in his *Astronomical Thoughts about the New Star of 1604,*[60] printed more than two years earlier, prompted by his envy of the universal approval with which Galileo's three lectures on the new star had been received. Thanks to his repugnant actions, however, Capra was deservedly awarded eternal infamy. Once the Most Excellent Wardens of the University of Padua had ascertained his crime by a rigorous formal procedure against him, they ordered that all printed copies of his book were to be destroyed, and its publication forbidden; and they granted Galileo the right to make his *Difesa* public, to redeem his reputation and destroy Capra's.[61]

Galileo's *Difesa* was not enough to curb the audacity and excess of confidence of some other people in other nations[62] who, enticed or carried away by the novelty and originality of the invention, or by the amazing variety and facility of its uses, published [descriptions of] Galileo's ingenious compass as if it were their own. [609]The counterfeits bore different inscriptions, had different shapes, or else presented new lines and were enhanced to serve more purposes. No mention was made of the principal author of the instrument. Its operations, where they were not described in print, had been known in all provinces of Europe via manuscripts circulated by those very foreigners to whom Galileo had generously communicated them, together with other works of his. But the audacity of these people, or their ingratitude, became plain through his *Difesa* and was condemned by the action itself. Furthermore, Galileo's claims were made undeniable by the glorious fame he earned for his other and much more marvelous works and inventions, over those who proved able to invent only a few, very weak products.

Intorno all'Aprile o al Maggio del 1609 si sparse voce in Venezia, dove allora trovavasi il Galileo, che da un tale Olandese fusse stato presentato al Sig.^r Conte Maurizio di Nassau un certo occhiale, co 'l quale gli oggetti lontani apparivano come se fusser vicini; né più oltre fu detto. Con questa sola relazione, tornando subito il Sig.^r Galileo a Padova, si pose a specularne la fabbrica, quale immediatamente ritrovò la seguente notte: poiché il giorno appresso, componendo lo strumento nel modo che se lo aveva immaginato, non ostante l'imperfezione de' vetri che poté avere, ne vidde l'effetto desiderato, e subito ne diede conto a Venezia a' suoi amici; e fabbricandosene altro di maggior bontà, sei giorni dopo lo portò quivi, dove sopra le maggiori altezze della città fece vedere et osservare gl'oggetti in varie lontananze a' primi Senatori di quella Republica, con lor infinita maraviglia; e riducendo lo strumento continuamente a maggior perfezione, si risolse finalmente, con la solita prodigalità nel comunicare le sue invenzioni, di far libero dono di questa ancora al Ser.^mo Principe o Doge Leonardo Donati et insieme a tutto 'l Senato Veneto, presentando con lo strumento una scrittura nella quale ei dichiarava la fabbrica, gl'usi e le maravigliose conseguenze che in terra e in mare da quello trar si potevano.

In gradimento di così nobil regalo fu immediatamente, con generosa dimostrazione della Ser.^ma Republica, ne' 25 d'Agosto del 1609 ricondotto il Sig.^r Galileo a vita sua alla medesima lettura, con più che triplicato stipendio del maggiore che fusse solito assegnarsi a' lettori di matematica.

Considerando fratanto il Sig.^r Galileo che la facultà del suo nuovo strumento era sol d'appressare et aggrandire in apparenza quelli oggetti i quali senz'altro artifizio, quando possibil fusse accostarglisi, con eguale o maggior distinzione si scorgerebbero, pensò ancora al modo di perfezionar assai più la nostra vista con fargli perfettamente discernere quelle minuzie le quali, benché situate in qualunque breve distanza dall'occhio, gli si rendono impercettibili; et allora inventò i microscopii d'un convesso e di un concavo, et insieme d'uno e di più convessi, applicandogli a scrupolosa osservazione de' minimi componenti delle materie e della mirabile struttura delle parti e membra delli insetti, nella piccolezza de' quali [610]fece con maraviglia vedere la grandezza di Dio e le miracolose operazioni della natura. In tanto, non perdonando né a fatiche né a spese, studiava nella perfezione del primo strumento, detto il telescopio o volgarmente l'occhiale del Galileo; e conseguitala a gran segno, lasciando di rimirar gl'oggetti terreni, si rivolse a contemplazioni più nobili.

About April or May 1609, word spread, in Venice, where Galileo then was, that some Dutchman had offered Maurice, Earl of Nassau,[63] a spyglass by means of which objects far away looked as if they were close by; nothing more was known. With just this piece of information, Galileo went back immediately to Padua and began to think about the construction of this spyglass, which he immediately devised the following night. The next day, putting together the instrument as he had imagined it, Galileo immediately observed the desired effect, notwithstanding the imperfection of the lenses he managed to find. He immediately reported his success to his friends in Venice and constructed a better instrument, which he brought to Venice six days later;[64] and from the most elevated places in the city he let the highest-ranking senators of the Republic see and observe objects at various distances, to their utter amazement.[65] Later, improving the instrument more and more, Galileo eventually decided, with his usual generosity in communicating his inventions, to present the instrument as a gift to the Most Serene Prince, or Doge Leonardo Donati,[66] and with him to the entire Venetian Senate, accompanying it with a written description of the construction and uses, as well as the amazing applications of the new instrument on both land and sea.[67]

In appreciation of this noble present, on August 25, 1609, Galileo's lectureship was renewed for life in a generous act by the Most Serene Republic, and his salary was raised to more than three times the highest that had ever been paid to a lecturer of mathematics.[68]

Meanwhile, considering that the function of his new instrument was only to bring near and apparently enlarge objects that, without any artifice, if it were possible to approach them, would appear with equal or higher distinction, Galileo thought of a way to greatly improve our eyesight so that it could perfectly discern minute details that are invisible to the eye, no matter how close they are to it. And so he invented microscopes with a concave lens and a convex one, or else with sets of convex lenses. Galileo applied them to the meticulous scrutiny of the minutest components of different materials or to the most amazing structure of parts and limbs of insects, in the smallness of which [610]he wonderfully showed the greatness of God and the miraculous operations of nature.[69] In the meantime, without saving himself any effort or expense, Galileo kept working to improve the first instrument, dubbed a "telescope," or, commonly, Galileo's spyglass. And having obtained a high level of perfection, he stopped gazing at earthly objects and turned to contemplating nobler ones.[70]

E prima, riguardando il corpo lunare, lo scoperse di superficie ineguale, ripieno di cavità e prominenze a guisa della terra. Trovò che la Via Lattea e le nebulose altro non erano ch'una congerie di stelle fisse, che per la loro immensa distanza, per la lor piccolezza rispetto all'altre, si rendevano impercettibili alla nuda e semplice vista. Vidde sparse per lo cielo altre innumerabili stelle fisse, state incognite all'antichità: e rivolgendosi a Giove con altro migliore strumento, ch'egli s'era nuovamente preparato, l'osservò corteggiato da quattro stelle, che gli s'aggirano intorno per orbi determinati e distinti, con regolati periodi ne' lor moti; e consecrandogli all'immortalità della Ser.ma Casa di V. A., gli diede nome di Stelle o Pianeti Medicei: e tutto questo scoperse in pochi giorni del mese di Gennaio del 1610 secondo lo stile romano, continuando tali osservazioni per tutto 'l Febbraio susseguente; quali tutte manifestò poi al mondo per mezzo del suo Nuncio Sidereo, che nel principio di Marzo pubblicò con le stampe in Venezia, dedicandolo all'augustissimo nome del Ser.mo Don Cosimo, Gran Duca di Toscana.

Queste inaspettate novità publicate dal Nunzio Sidereo, che immediatamente fu ristampato in Germania et in Francia, diedero gran materia di discorsi a' filosofi et astronomi di que' tempi, molti de' quali su 'l principio ebbero gran repugnanza in prestargli fede,[15] e molti temerariamente si sollevarono, altri con scritture private et altri più incauti sin con le stampe,[c] stimando quelle vanità e delirii o finti avvisi del Sig.r Galileo, o pure false apparenze et illusioni de' cristalli; ma in breve gl'uni e gl'altri necessariamente cedettero alle confermazioni de' più savii, all'esperienze et al senso medesimo. Non mancarono già de' così pervicaci et ostinati, e fra questi de' constituiti in grado di publici lettori,[d] tenuti per altro in gran stima, i quali, temendo di commetter sacrilegio contro la deità del loro Aristotele, non vollero cimentarsi all'osservazioni, né pur una volta [611]accostar l'occhio al telescopio; e vivendo in questa lor bestialissima ostinazione, vollero, più tosto che al lor maestro, usar infedeltà alla natura medesima.

Proseguendo col telescopio l'osservazioni celesti, nel principio di Luglio del 1610 scoperse Saturno tricorporeo, dandone avviso ad alcuni matematici di Italia e di Germania et a' suoi amici più cari[e] per mezzo di cifre e caratteri trasposti, che doppo ordinati dal medesimo Sig.r Galileo, a richiesta dell'Imperatore Ridolfo Secondo, dicevano:

Altissimum Planetam tergeminum observavi.

Vidde ancora nella faccia del sole alcuna delle macchie, ma per allora non volle publicare quest'altra novità, che poteva tanto più concitargli l'invidia o persecuzione di molti ostinati Peripatetici (conferendola solo ad alcuno de' suoi più confidenti di Padova e di Venezia e di altrove[f]), per prima assicurarsene con replicate osservazioni, e poter intanto formar concetto della essenza loro e con qualche probabilità almeno pronunciarne la sua oppinione.

c Martino Orchio, Francesco Sizii et altri.

d Dottor Cremonino, lettor in Padova.

e Al P. D. Benedetto Castelli (Brescia). A Lodovico Cigoli, pittore; Al Padre Grembergero, gesuita; Al P. Clavio, gesuita; Al Sig.r Luca Valerio (Roma). A Mons.r Gualdo; A Mons.r Pignoria (Padova). A Mons.r Giuliano de' Medici; A Gio. Kepplero, Matematico dell'Imperatore (in Praga).

f A Mons.r Gualdo, a Mons.r Pignoria, D. Benedetto Castelli, Fra Paolo e Fra Fulgenzio serviti, Sig.r Filippo Contarini, Sig.r Sebastiano Venieri.

15 *in marg. scr.* Dottor Anton Magini *A scr. et del.* Dottor Gio. Anton Magini Matematico di Bologna *B*

First, observing the lunar body, Galileo discovered that it had an uneven surface, full of cavities and elevations, just like the earth.[71] He found that the Milky Way and nebulae were nothing but congeries of fixed stars, which, owing to their enormous distance or their small size with respect to others, cannot be seen by naked and plain sight.[72] He saw countless fixed stars scattered around the sky, unknown to ancient observers.[73] And turning to Jupiter with another and better instrument, which he had constructed from scratch, Galileo observed it courted by four stars, which revolve around it on well-defined and distinct orbs, in exact periods.[74] He dedicated them to Your Highness's immortal Most Serene Family, naming them Medicean Stars, or Planets.[75] All this Galileo discovered in the span of few days in January 1610, according to the Roman style. He pursued the same observations throughout the following February and presented them all to the world by means of his *Sidereal Messenger*, published in Venice at the beginning of March, dedicating it to the most august name of the Most Serene Cosimo, Grand Duke of Tuscany.[76]

The unexpected novelties published in *The Sidereal Messenger*, which was immediately reprinted in Germany and France,[77] offered many subjects for philosophers and astronomers of those times to discuss. At first, many of them manifestly distrusted him,[78] and many boldly protested, some with privately circulated writings, others less cautiously in print,[c][79] deeming Galileo's observations unreliable and nonsensical, fake announcements, or else false appearances and illusions of the lenses. But soon both groups necessarily yielded to the confirmations provided by wiser people, to their observations, and to sense experience itself. Some [however] were so obstinate and stubborn—and among them were some public university professors,[d][80] otherwise held in great esteem—that, afraid of outraging their divine Aristotle, they did not want to make observations, and not even once [611]put their eye near the telescope. Persisting in this most insane stubbornness, they decided to be disloyal to nature rather than to their master.

Galileo continued his observations of the heavens with the telescope, and at the beginning of July 1610 he discovered three-bodied Saturn,[81] announcing it to some mathematicians in Italy and Germany and to his dearest friends[e][82] by way of a coded anagram. At the request of Emperor Rudolf II,[83] Galileo reordered the letters, to read:

Altissimum Planetam tergeminum observavi.[84]

Moreover, he saw a few spots on the surface of the Sun but decided against publishing this further novelty at that moment, as it might have aroused the envy or opposition of many die-hard Peripatetics[85] (he merely conveyed the discovery to some of his closest friends in Padua, Venice, and elsewhere[f][86]). Before that, Galileo confirmed them by renewed observations, and took time to devise an explanation of their nature and state a well-grounded opinion about them.

c Martin Horký, Francesco Sizzi, and others.

d Dr. Cremonini, professor in Padua.

e To Father Don Benedetto Castelli, in Brescia. To Ludovico Cigoli, painter; Father Grienberger, SJ; Father Clavius, SJ; and Mr Luca Valerio, in Rome. To Monsignor Gualdo and Monsignor Pignoria, in Padua. To Monsignor Giuliano de' Medici and Johannes Kepler, Imperial Mathematician, in Prague.

f To Monsignor Gualdo; Monsignor Pignoria; Don Benedetto Castelli; Father Paolo and Father Fulgenzio, OSM; Mr. Filippo Contarini; and Mr. Sebastiano Venier.

L'avviso di tante e non più udite maraviglie, scoperte in cielo dal Sig.ʳ Galileo nella città di Padova, eccitò nelli animi d'ogni nazione veementissimo desiderio di accertarsene col senso stesso. Ma nel Ser.ᵐᵒ D. Cosimo de' Medici non cedé punto a questa comune curiosità la sua regia munificenza, poi che volle con propria lettera de' 10 Luglio 1610 richiamarlo di Padova al suo servizio con titolo di Primario e Sopraordinario Matematico dello Studio di Pisa, senz'obligo di leggervi o risedervi, e di Primario Filosofo e Matematico della sua Ser.ᵐᵃ Altezza, assegnandogli a vita amplissimo stipendio, proporzionato alla somma generosità di un tanto Principe.

Licenziatosi adunque il Sig.ʳ Galileo dal servizio della Ser.ᵐᵃ Republica, verso la fine d'Agosto[16] se ne venne a Firenze, dove da quelle Ser.ᵐᵉ Altezze, da' litterati ⁶¹²e dalla nobiltà fiorentina, fu accolto et abbracciato con affetti di ammirazione; e subito si diede a far vedere i nuovi lumi e le nuove maraviglie del cielo, con stupore e diletto universalissimo.

Quivi, del mese di Novembre, nel continuare l'osservazioni che fin d'Ottobre[17] aveva cominciate intorno alla stella di Venere, che parevagli andare crescendo in mole, l'osservò finalmente mutar figure come la luna, propalando quest'altra ammirabile novità tra gl'astronomi e matematici d'Europa con tal anagramma:

Haec immatura a me iam frustra leguntur o, y;

il quale, ad instanza pure del medesimo Imperatore e di molti curiosi filosofi, fu risoluto e deciferato dal Sig.ʳ Galileo nel vero senso così:

Cynthiae figuras aemulatur mater amorum.

Intorno alla fine di Marzo del 1611, desiderato il Sig.ʳ Galileo et aspettato da tutta Roma, quivi si condusse, e nell'Aprile susseguente fece vedere i nuovi spettacoli del cielo a molti SS. Prelati e Cardinali; e particolarmente nel Giardino Quirinale, presente il Sig.ʳ Card.ˡᵉ Bandini e i Mons.ʳⁱ Dini, Corsini, Cavalcanti, Strozzi, Agucchia, et altri Signori, dimostrò le macchie solari: e questo fu sei mesi prima delle più antiche osservazioni fatte da un tal finto Apelle,[g] il quale poi vanamente pretese l'anteriorità di questo discoprimento, poi che le sue prime osservazioni non furon fatte prima che del mese d'Ottobre susseguente.[18]

Quivi inoltre, nel mese d'Aprile 1611, gli sortì di incontrare con assai precisione i tempi de' periodici movimenti de' Pianeti Medicei, predicendo per molte notti future le loro costituzioni, e facendole osservare a molti tali quali egli le haveva pronosticate.[19]

---

g P. Cristoforo Scheiner, gesuita.

---

16 L'avviso—d'Agosto *BX*] Circa la fine d'Agosto (sollecitato dal suddetto suo Principe a sbrigarsi di Padova) *AS*

17 d'Ottobre *ABX*] nel mese di Settembre *S*

18 *post* susseguente *scr.* quando per altro è noto che il Galileo l'aveva scoperta qualche mese avanti ch'ei tornasse di Padova, cioè un anno prima nel 1610 *S*

19 Quivi—pronosticate *BX*] *om. AS*

The announcement of so many unheard of marvels discovered in the heavens by Galileo in Padua caused the men of every nation to intensely desire to ascertain them through sensory experience. But the Most Serene Cosimo de' Medici in his royal munificence would not be outdone in this widespread curiosity. By letter, on July 10, 1610, he called Galileo back from Padua to his service as Chief and Extraordinary Mathematician at the University of Pisa, without any obligation either to teach or reside there, and as Chief Philosopher and Mathematician of the Most Serene Highness, bestowing on Galileo a lifelong and very substantial stipend, proportional to the generosity of that Prince.[87]

Thus, Galileo resigned from his service to the Most Serene Republic <of Venice>[88] and went back to Florence about the end of August,[89] where their Most Serene Highnesses, intellectuals, [612]and noblemen welcomed him and showered him with affectionate signs of admiration; and Galileo showed the new lights and the new marvels of the heavens to everybody's amazement and enjoyment.

Here, in November, he continued the observations of the star Venus he had started in October:[90] it looked to him as if Venus was growing in size, and he eventually saw it change shape like the Moon.[91] Galileo announced this new extraordinary novelty to Europe's astronomers and mathematicians by way of the following anagram:

Haec immatura a me iam frustra leguntur o, y;

which, upon the request of the emperor himself and of many inquiring philosophers, Galileo resolved and unscrambled into its actual meaning:

Cynthiae figuras aemulatur mater amorum.[92]

Around the end of March 1611, Galileo went to Rome,[93] where the whole city was waiting for him. In the following month of April, he showed the new heavenly marvels to many bishops and cardinals; in particular, he showed the sunspots to Cardinal Bandini and Monsignors Dini, Corsini, Cavalcanti, Strozzi, Agucchi, and others in the gardens of the Quirinale.[94] All this preceded by six months the earliest observations by a certain Apelles,[g] [95] who later vainly insisted on the priority of his discovery, whereas his earliest observations were not made before the end of the following month of October.[96]

Furthermore, in that same place, during the month of April 1611 Galileo succeeded in finding the times of the periodic movements of the Medicean Planets, predicting their positions for several nights ahead, and having many people observe them as he had anticipated.[97]

---

g Father Christoph Scheiner, SJ.

Avendo dunque egli solo veduto il primo nel cielo tante e così gran maraviglie, state occulte all'antichità, era ben dovere ch'egli in avvenire con nome di Linceo dovesse chiamarsi; onde allora fu quivi ascritto nella famosissima Accademia de' Lincei, poco avanti instituita dal Sig.ʳ Principe Federigo Cesi, Marchese di Monticelli.

Sopragiungendo l'estate, se ne tornò a Firenze, dove ne' varii congressi de' letterati, che frequentemente si facevano d'avanti al Ser.ᵐᵒ G. Duca Cosimo, fu una volta introdotto discorso sopra il galleggiar in acqua e il sommergersi de' corpi, e tenuto da alcuni che la figura fosse a parte di questo effetto, ma dal Sig.ʳ Galileo sostenuto il contrario; ond'egli, per commessione della medesima Altezza, [613]scrisse quell'erudito *Discorso sopra le cose che stanno in acqua e che in quella si muovono*, dedicato al suddetto Serenissimo e stampato in Firenze nell'Agosto del 1612: nell'ingresso del qual trattato diede publicamente[20] notizia delle novità delle macchie solari; e poco doppo ristampandosi il medesimo Discorso con alcune addizzioni, nella prima di esse inserì il parer suo circa il luogo, essenza e moto di dette macchie, avvisando inoltre d'aver per mezzo di quelle osservato il primo un moto e revoluzione del corpo solare in sé stesso nel tempo di circa un mese lunare; accidente, benché nuovo in astronomia, eterno nondimeno in natura, a cui perciò il Sig.ʳ Galileo referiva, come a men remoto principio, le cagioni d'effetti e conseguenze maravigliose.

In occasione delle dispute che nacquero in proposito del galleggiare, soleva dire il Sig.ʳ Galileo, non vi esser più sottile né più industriosa maestra dell'ignoranza, poiché per mezzo di quella gl'era sortito di ritrovare molte ingegnose conclusioni e con nuove et esatte esperienze confermarle per satisfare all'ignoranza delli avversarii, alle quali per appagare il proprio intelletto non si sarebbe applicate.

Contro la dottrina di tal Discorso si sollevò tutta la turba peripatetica, et immediatamente si veddero piene le stamperie di opposizioni[h] et apologie, alle quali fu poi nel 1615 abondantemente risposto dal P. D. Benedetto Castelli, matematico allora di Pisa e già discepolo del Sig.ʳ Galileo, a fine di sottrarre il suo maestro da occuparsi in così frivole controversie.[21]

Stava bene il Sig.ʳ Galileo tutto intento a' celesti spettacoli, quando però non veniva interrotto da indisposizioni o malattie che spesso l'assalivano, cagionate da lunghe e continuate vigilie et incomodi che pativa nell'osservare; e trovandosi poco lontano da Firenze nella villa delle Selve del Sig.ʳ Filippo Salviati, amico suo nobilissimo[22] e d'eminentissimo ingegno, quivi fece scrupolosissime osservazioni intorno alle macchie solari: et avendo ricevuto lettera dal Sig.ʳ Marco Velsero, Duumviro d'Augusta, accompagnata con tre del suddetto Apelle sopra l'istesso argumento, ne i 4 di Maggio del 1612 rispose a quella con varie con [614]siderazioni sopra le lettere del

h Lodovico delle Colombe. Vincenzio di Grazia. Giorgio Coresi, lettore in Pisa. Dottor... Palmerini.[23]

20 diede pubblicamente *BX*] manifestò i tempi de' periodici movimenti de' Pianeti Medicei, che prossimamente aveva investigati l'Aprile del 1611, mentre era in Roma, dando ancora *AS*

21 *post* controversie *scr*. ripiene di perverse malignità non meno che di crassissima ignoranza *AS*

22 nobilissimo *BX*] parzialissimo *AS*

23 Giorgio—Palmerini *BX*] Giorgio Coresi, Dottor Papazzoni, lettori in Pisa *A* Giorgio Coresio Lettore in Pisa. Dottor Tommaso Palmerini *S*

Having been the first to observe in the heavens so many and such great marvels unknown in antiquity, it was only proper for Galileo to be named a Lynx for the future. Accordingly, he was called to the most famous Academy of the Lynxes, which Prince Federico Cesi, Marquis of Monticelli, had established a short time before.[98]

As the summer approached, Galileo headed back to Florence.[99] Here, during the many meetings of intellectuals that often took place in the presence of the Most Serene Grand Duke Cosimo, the subject of floating and immersing bodies in water was introduced.[100] Some claimed that the shape of bodies was responsible for their behavior, but Galileo argued against this. Thus, His Highness charged him [613]with writing the learned *Discourse on Bodies in Water*, dedicated to the Most Serene Prince and printed in Florence in August 1612.[101] At the beginning of that *Discourse* Galileo publicly announced his news about sunspots;[102] shortly afterward, on the occasion of the reprinting of the *Discourse*, he included some additions. In the first of these Galileo expressed his opinion about the place, essence, and motion of these spots, and announced that, thanks to the spots, he had observed for the first time a motion and revolution of the body of the sun around itself, lasting approximately one lunar month. This circumstance was new in astronomy, but nonetheless perennial in nature, and for this reason Galileo ascribed to it, as if to a less remote principle, the cause of marvelous effects and consequences.[103]

As to the controversies that sprouted about floating bodies, Galileo used to say that there is no subtler or more productive teacher than ignorance, as it allowed him to discover many ingenious results and confirm them with new and exact experiments, so as to satisfy his opponents' ignorance, and he would not have devoted himself to them had it been to satisfy his own intellect.

The whole of the Peripatetic mob rose against the theses of the *Discourse*, and print shops were immediately flooded with confutations[h] [104] and vindications, to which Father Benedetto Castelli, who was then a mathematician in Pisa and had been a pupil of Galileo's, extensively responded in 1615, so as to save his master from such futile controversies.[105]

Galileo wholly devoted himself to observing heavenly wonders when not interrupted by ailments and diseases that often assailed him; they were caused by long and extended vigils, and by the discomforts he suffered in the course of observations. While he was at a short distance from Florence, in Filippo Salviati's (his noblest[106] and sharpest-minded friend) Villa delle Selve, he made extremely meticulous observations of sunspots;[107] he also received a letter from Markus Welser, one of the chief men of Augsburg,[108] accompanied by three additional letters by the aforementioned Apelles on the same subject. On May 4, 1612, Galileo replied to Welser with various

h Ludovico delle Colombe; Vincenzo Di Grazia; Giorgio Coresio, professor in Pisa; Doctor . . . Palmerini.

medesimo Apelle, replicando ancora con altra de' 14 d'Agosto susseguente; e ricevendo dal Sig.ʳ Velsero altre speculazioni e discorsi d'Apelle, scrisse la terza lettera del primo di Dicembre prossimo, sempre confermandosi con nuove e più accurate ragioni ne' suoi concetti: e di qui nacque l'*Istoria e Demostrazioni delle Macchie Solari e loro accidenti*, che nel 1613 fu publicata in Roma dall'Accademia de' Lincei insieme con le suddette lettere e disquisizioni del finto Apelle, dedicandola al medesimo Sig.ʳ Filippo Salviati, nella villa del quale aveva il Sig.ʳ Galileo osservato e scritto sopra queste apparenze; vedendosi in detta istoria ciò che di vero, o di probabile almeno, è stato detto fin ora sopra argumento così difficile e dubbio.

Ma non contento d'aver, con le sue peregrine speculazioni e con tanti nobili scoprimenti, introdotto nuovi raggi di chiarissima luce nelli umani intelletti, illustrando e restaurando insieme la filosofia et astronomia, non prima investigò ne' Pianeti Medicei alcuni lor varii accidenti, che pensò di valersene ancora per universal benefizio delli uomini nella nautica e geografia, sciogliendo perciò quell'ammirando problema per il quale in tutte l'età passate si sono in vano affaticati gl'astronomi e matematici di maggior fama, che è di poter in ogn'ora della notte, in qualunque luogo di mare o terra, graduare le longitudini. Scorgeva bene ch'al conseguimento di ciò si richiedeva un'esatta cognizione de' periodi e moti di quelle stelle, a fine di fabbricarne le tavole e calcular l'efemeridi per predire le loro constituzioni, congiunzioni, eclissi, occultazioni et altri particolari accidenti, da lui solo osservati, e che quella non si poteva ottenere se non dal tempo, con moltissime e puntuali osservazioni: però sin che non gli sortì conseguirla, si astenne di proporre il suo ammirabil trovato; e quantunque in meno di quindici mesi dal primo discoprimento de' Pianeti Medicei arrivasse ad investigare i lor movimenti con notabile aggiustatezza per le future predizioni, volle però con altre più esquisite osservazioni, e più distanti di tempo, emendargli.

Dell'anno adunque 1615 in circa, trovandosi il Sig.ʳ Galileo d'aver conseguito quanto in teorica e in pratica si richiedeva per la sua parte all'effettuazione di così nobile impresa, conferì il tutto al Ser.ᵐᵒ G. Duca Cosimo, suo Signore: il quale, molto ben conoscendo la grandezza del problema e la massima utilità che dall'uso di esso poteva trarsi, volle egli stesso, per mezzo del proprio residente in Madrid, muoverne trattato con la Maestà Cattolica del Re di Spagna, il quale già prometteva grandissimi onori e grossissime recognizioni a chi avesse trovato modo sicuro di navigar per la longitudine con l'istessa o simil facilità che si cammina per latitudine. E desiderando S. A. che tal invenzione, come pro **615**porzionata alla grandezza di quella Corona, fosse con pronta resoluzione abbracciata, compiacevasi che il Sig.ʳ Galileo, per facilitare i mezzi per condurla a buon fine, conferisse a S. Maestà un altro suo nuovo trovato, pur di grandissimo uso et acquisto nella navigazione, da S. A. stimatissimo e custodito con segretezza; et era l'invenzione d'un altro differente occhiale, col quale potevasi dalla cima dell'albero o del calcese d'una galera riconoscer da lontano la qualità, numero e forze de' vasselli nemici, assai prima dell'inimico medesimo, con egual prestezza e facilità che con l'occhio libero, guardandosi in un tempo stesso con amendue gl'occhi, e potendosi di più aver notizia della loro lontananza dalla propria galera, et in modo occultar lo strumento sì che altri non ne apprenda la fabbrica.

614remarks on Apelles' letters, followed by another letter on August 14. Having received from Welser more speculations and considerations by Apelles, Galileo wrote a third letter on the following December 1, always offering new and more accurate reasons in support of his own theses. From this sprang the *History and Demonstrations Concerning Sunspots and Their Properties*, which was published in Rome by the Academy of the Lynxes in 1613, together with the letters just mentioned and the considerations of the fake Apelles. It was dedicated to Filippo Salviati, in whose villa Galileo had observed and written about these phenomena.[109] In the *History*, we may see what truth—or probable truth, at least—has been expressed so far about such a difficult and obscure subject.

However, Galileo did not content himself with having, through his extraordinary speculations and with so many famous discoveries, introduced new rays of the clearest light into human minds, exalting and renewing philosophy and astronomy together. Rather, he investigated some of the many features of the Medicean Planets, which he thought of using again for the general benefit of mankind in navigation and geography, solving the remarkable problem that the most eminent astronomers and mathematicians of all previous ages had vainly attempted to solve: the problem of determining longitudes at any hour in the night, in any place on the earth or at sea.[110] Galileo realized full well that to do that required a precise knowledge of the periods and motions of those stars, so as to compile tables and compute ephemerides in order to predict their positions, conjunctions, eclipses, concealments, and other specific properties he alone had observed. He realized that this knowledge could be gathered only over time, with very many and careful observations. Accordingly, Galileo refrained from proposing his admirable discovery until he managed to achieve the desired result; and even though he came to investigate the motions of the Medicean Planets with the precision required for future predictions in less than fifteen months after he first observed them, he nonetheless decided to revise them with observations more recent, more precise, and more distant in time.

In 1615, roughly, Galileo accomplished what was both theoretically and practically required to implement his noble undertaking. He reported everything to his Lord, the Most Serene Grand Duke Cosimo, who was well aware of the difficulty of the problem and of the extreme utility its solution would be. Through his ambassador in Madrid, the Grand Duke decided to come to an agreement with His Catholic Majesty the King of Spain,[111] who had already promised the highest honors and the greatest rewards to anyone who found the safest way to sail east-west as easily as north-south. Furthermore, His Highness wished that that discovery—commensurate with **615**the grandeur of his own Crown—be embraced promptly. And in order to facilitate its accomplishment, the Grand Duke permitted Galileo to present His Majesty with yet another discovery of his, which was extremely useful and convenient for navigation, and which His Highness highly valued and kept carefully hidden. It was a new kind of spyglass, by means of which one could discern the quality, number, and power of the enemy's ships from the top of a tree or of a galley's mast, well before the enemy could see the observer. This could be done with the speed and ease of the naked eye, using both eyes at the same time. The observer could also simultaneously assess the enemy's distance from his own galley and in such a way as to hide the instrument so that no one would know its construction.[112]

Ma come per lo più accader suole delle nobili e grandi imprese, che quanto sono di maggior conseguenze, tanto maggiori s'incontrano le difficoltà nel trattarle e concluderle, dopo molti anni di negoziato non fu possibile indurre, per varii accidenti, i ministri di quella Corona all'esperienza del cercato artifizio, non ostante ch'il Sig.ʳ Galileo si fosse offerto di trasferirsi personalmente in Lisbona o Siviglia o dove fosse occorso, con provedimento di quanto all'esecuzione di tal impresa si richiedesse, e con larga offerta di instruire ancora i medesimi marinari e quelli che dovevano in nave operare, e di conferire liberamente a chi fosse piaciuto a S. Maestà tutto ciò che s'appartenesse alla proposta invenzione. Svanì dunque il trattato con Spagna, restando però a S. A. S. et al Sig.ʳ Galileo l'intenzion di promuoverlo altra volta in congiunture migliori.

In tanto le tre comete che apparvero nel 1618, et in specie quella che si vedde nel segno di Scorpione, che fu la più conspicua e di più lunga durata, aveva tenuto in continuo esercizio i primi ingegni d'Europa; tra' quali il Sig.ʳ Galileo, con tutto che per una lunga e pericolosa malattia, ch'ebbe in quel tempo, poco potesse osservarla, a richiesta però del Ser.ᵐᵒ Leopoldo Arciduca d'Austria, che trovandosi allora in Firenze volle onorarlo con la propria persona visitandolo sino al letto, vi fece intorno particolar reflessione, conferendo alli amici i suoi sentimenti sopra questa materia: onde il Sig.ʳ Mario Guiducci, uno de' suoi parzialissimi, compilando intorno a ciò l'opinioni delli antichi filosofi e moderni astronomi e le probabili conietture che sovvennero al Sig.ʳ Galileo, scrisse quel dottissimo *Discorso delle Comete* che fu impresso in Firenze nel 1619, dove reprovando tra l'altre[24] alcune opinioni del Matematico del Collegio Romano,[i] poco avanti promulgate in una disputa astronomica sopra le dette comete, diede con esso occasione a tutte le controversie che nacquero in tal proposito, e di più a tutte le male sodisfazioni che il Sig.ʳ Galileo da quell'ora sino alli ultimi giorni [616] con eterna persecuzione ricevé in ogni sua azione e discorso.

---

i Padre Orazio Grassi savonese, gesuita.

---

24 *post* l'altre *scr.* come filosofo A come filosofo libero S

But, as it often happens with noble and great enterprises—the greater their consequences, the greater the difficulties faced in working on them and completing them—after several years of negotiation, due to various events, it proved impossible to have the \<Spanish\> Crown's ministers test the instrument they required, despite Galileo's own offer to move personally to Lisbon or Seville, or wherever it was necessary, bringing whatever was necessary for the accomplishment of that enterprise. He also generously offered to instruct the sailors and those who were going to operate on the ships, as well as to reveal to whomever His Majesty appointed everything related to the proposed invention. The negotiation with Spain came to nothing, then, although His Most Serene Highness and Galileo were determined to take it up again under better circumstances.

Meanwhile, three comets that appeared in 1618, and particularly one observed in the Scorpio star sign, which was the brightest and lasted longer, kept the most eminent minds of Europe busy.[113] Among these was Galileo, despite a long and dangerous illness that prevented him from observing it at length. However, at the request of the Most Serene Leopold, Archduke of Austria,[114] who happened to be in Florence at the time and did Galileo the honor of paying him a visit while he was in bed, Galileo considered the matter deeply and conveyed his thoughts about it to his friends.[115] Mario Guiducci, one of his closest friends, gathered opinions of ancient philosophers and modern astronomers on the topic, including Galileo's probable conjectures, and wrote a most learned *Discourse on the Comets*, printed in Florence in 1619.[116] In it, Guiducci criticized, among others,[117] a few opinions of the Mathematician of the Roman College,[i] who had previously expressed his views in an astronomical disputation on the comets. In so doing, Guiducci brought on all the controversies that arose about the comets, as well as the fierce persecution that from then until his final days [616]Galileo suffered for every action or discourse of his.

---

i Father Orazio Grassi, SJ, from Savona.

Poi che il suddetto Matematico, offendendosi fuor del dovere e contro l'obligo di filosofo che le sue proposizioni non fossero ammesse senz'altro esame per infallibili e vere, o pure anche invidiando alla novità de' concetti così dottamente spiegati nel sopradetto Discorso delle Comete, indi a poco publicò una certa sua *Libra astronomica e filosofica*, mascherata con finto nome di Lotario Sarsio Sigensano, nella quale trattando con termini poco discreti il Sig.^r Mario Guiducci e con molesti punture il Sig.^r Galileo, necessitò questo a rispondere col suo *Saggiatore*, scritto in forma di lettera al Sig.^r D. Virginio Cesarini, stampato in Roma nel 1623 dalli Accademici Lincei e dedicato al Sommo Pontefice Urbano Ottavo; per la qual opera chiaramente si scorge, quanto si deva alle persecuzioni delli emuli del Sig.^r Galileo, ch'in certo modo sono stati autori di grandissimi acquisti in filosofia, destando in quello concetti altissimi e peregrine speculazioni, delle quali per altro saremmo forse restati privi.

Ben è vero, all'incontro, che le calunnie e contradizioni de' suoi nemici et oppositori, che poi lo tennero quasi sempre angustiato, lo resero ancora assai ritenuto nel perfezionare e dar fuori l'opere sue principali di più maravigliosa dottrina. Che però non prima che dell'anno 1632 publicò il *Dialogo de' due Massimi Sistemi Tolemaico e Copernicano*, per il soggetto del quale, sin dal principio che andò lettore a Padova, aveva di continuo osservato e filosofato, indottovi particolarmente dal concetto che gli sovvenne per salvare con i supposti moti diurno et annuo, attribuiti alla terra, il flusso e reflusso del mare, mentre era in Venezia; dove insieme col Sig.^r Gio. Francesco Sagredo, signore principalissimo di quella Republica, di acutissimo ingegno, e con altri nobili suoi aderenti trovandosi frequentemente a congresso, furono, oltre alle nuove speculazioni promosse dal Sig.^r Galileo intorno alli effetti e proporzioni de' moti naturali, severamente discussi i gran problemi della constituzione dell'universo e delle reciprocazioni del mare: intorno al quale accidente egli poi nel 1616, che si trovò in Roma, scrisse ad instanza dell'Emin.^mo Card.^le Orsino un assai lungo Discorso, che andava in volta privatamente, diretto al medesimo Sig.^r Cardinale. Ma presentendo che della dottrina di questo suo trattato, fondata sopra l'assunto del moto della terra, si trovava alcuno che si faceva autore, si risolse di inserirla nella detta opera del Sistema, portando insieme, indeterminatamente per l'una parte e per l'altra, quelle considerazioni che, avanti e dopo i suoi nuovi scoprimenti nel cielo, gl'erano sovvenute in comprobazione dell'opinione Copernicana e le altre solite addursi in difesa della posizione Tolemaica, quali tutte, ad instanza di gran personaggi egli aveva raccolte, et ad imitazione di Platone spiegate in dialogo, introducendo quivi a parlare il suddetto Sig.^r Sagredo et il Sig.^r Filippo Salviati, soggetti di vivacissimo spirito, d'ingegno libero e suoi carissimi confidenti.

The aforementioned mathematician, taking great offense, violated the philosopher's duty not to regard his own statements as true and infallible without further examination, or else being jealous of the novelty of the ideas expounded so learnedly in the aforementioned *Discourse on the Comets*, published shortly afterward an *Astronomical and Philosophical Balance* under the false name Lotario Sarsi Sigensano.[118] In it, the author did not treat Mario Guiducci very kindly, and attacked Galileo viciously. Thus, Galileo felt obliged to respond with the *Assayer*, written in the form of a letter to Virginio Cesarini. It was printed by the Academy of the Lynxes in Rome, in 1623, and dedicated to Pope Urban VIII.[119] From Galileo's reply we may gather the extent to which we are indebted to these jealous competitors. We owe to them very important advances in philosophy, because they inspired Galileo with very bright ideas and extraordinary speculations, of which we might have been deprived had it not been for them.

On the other hand, however, the libels and attacks of his enemies and opponents, which always tormented Galileo, made him extremely reluctant to finish and publish his major works, containing his most startling theories. Hence it was not until 1632 that Galileo published the *Dialogue of the Two Chief World Systems, Ptolemaic and Copernican*, on which he had been working, both by observing and philosophizing, ever since he was appointed lecturer in Padua. He drew his inspiration from an idea he had in Venice for explaining the high and low tide by attributing it to the alleged diurnal and annual motions of the earth.[120] When Galileo was in Venice with Giovanni Francesco Sagredo, a man of sharpest mind and a most eminent citizen of that Republic, as well as with other noblemen close to him, with whom he had frequent exchanges, they earnestly debated, as well as Galileo's new thoughts about the effects and ratios of natural motions, the noteworthy problems of the structure of the universe and the alternating tidal movements. When he was in Rome, in 1616, upon the Most Eminent Cardinal Orsini's request, Galileo wrote a very long discourse on this topic, which he dedicated to the cardinal and circulated privately.[121] However, anticipating that someone might claim the content of this treatise, based on the assumption of the earth's movement, as his own, he decided to include it in his <*Two Chief World*> *Systems*. Galileo offered, equally, on one side and the other, the arguments he had devised, before and after his discoveries of the heavenly novelties, in support of the Copernican opinion and those that were customarily advanced in support of the Ptolemaic standpoint. He gathered the arguments together at the request of very important people and expounded on them in the form of a dialogue, following Plato's example, thereby introducing Sagredo and Filippo Salviati, two persons with the brightest minds, free spirits, and his dearest confidants.[122]

[617]Ma essendosi già il Sig.r Galileo per l'altre sue ammirabili speculazioni con immortal fama sin al cielo inalzato, e con tante novità acquistatosi tra gl'uomini del divino, permesse l'Eterna Providenza ch'ei dimostrasse l'umanità sua con l'errare, mentre nella discussione de' due sistemi si dimostrò più aderente all'ipotesi Copernicana,[25] già dannata da S. Chiesa come repugnante alla Divina Scrittura. Fu perciò il Sig.r Galileo, dopo la publicazione de' suoi Dialogi, chiamato a Roma dalla Congregazione del S. Offizio: dove giunto intorno alli 10 di Febbraio 1632 *ab Incarnatione*, dalla somma clemenza di quel Tribunale e del Sovrano Pontefice Urbano Ottavo, che già per altro lo conosceva troppo benemerito alla republica de' letterati,[26] fu arrestato nel delizioso palazzo della Trinità de' Monti appresso l'ambasciador di Toscana, et in breve (essendogli dimostrato il suo errore) retrattò, come vero catolico, questa sua opinione; ma in pena gli fu proibito il suo Dialogo,[27] e dopo cinque mesi licenziato di Roma (in tempo che la città di Firenze era infetta di peste), gli fu destinata per arresto,[28] con generosa pietà, l'abitazione del più caro et stimato amico ch'avesse nella città di Siena, che fu Mons.r Arcivescovo Piccolomini: della qual gentilissima conversazione egli godé con tanta quiete e satisfazione dell'animo, che quivi ripigliando i suoi studii trovò e dimostrò gran parte delle conclusioni meccaniche sopra la materia delle resistenze de' solidi, con altre speculazioni; e dopo cinque mesi in circa, cessata affatto la pestilenza nella sua patria, verso il principio di Dicembre del 1633 da S. S.tà gli fu permutata la strettezza di quella casa nella libertà della campagna, da esso tanto gradita: onde tornò alla sua villa d'Arcetri, nella quale egli già abitava più del tempo,[29] come situata in buon'aria e assai comoda alla città di Firenze, e perciò facilmente frequentata dalle visite delli amici e domestici, che sempre gli furono di particolar sollievo e consolazione.

25 mentre—Copernicana *BX*] nell'aderire ne' suoi Dialogi all'opinione de' Pitagorici e del Copernico sopra la stabilità del sole e mobilità della terra *A* mentre nella discussione de i due Sistemi si dimostrò forse più aderente all'Ipotesi Copernicana *S*

26 *in marg. scr.* e fin cantatone le sue lodi con publici componimenti *B*

27 *post* Dialogo *scr.* imponendogli perpetuo silenzio sopra questa materia *A*

28 *in marg.* et assegnatogli per carcere la propria casa per tempo a beneplacito di S. S.tà; et dopo ottenendo di partirsi di Roma (per non si esporre a manifesto pericolo della vita, per la peste che ancora teneva infetta la città di Firenze), gli fu prescritta con generosa pietà *B*

29 onde—tempo *BX*] onde si ritirò nella villa di Bellosguardo e dopo in quella d'Arcetri, nelle quali anche per propria elezione gustava di abitar più del tempo *A* onde se ne tornò alla sua villa di Bellosguardo e dopo in quella d'Arcetri, nelle quali aper propria elezione gustava prima d'abitar più del tempo *S*

[617]The immortal fame of Galileo's amazing speculations rose to the heavens, and his many innovations earned him a divine reputation among men. Consequently, Eternal Providence allowed him to show his humanity by erring. While arguing for and against the two systems, Galileo showed himself closer[123] to the Copernican hypothesis,[124] which the Holy Church had condemned as contradicting the Divine Scriptures. Therefore, after the publication of his *Dialogue*, the Congregation of the Holy Office summoned Galileo to Rome. He arrived around 10 February 1632, *ab Incarnatione*, and by the greatest mercy of that Tribunal and of the Sovereign Pope Urban VIII (who otherwise knew he was most meritorious in the Republic of Letters[125]) was confined in the very beautiful palace in Trinità dei Monti, the residence of the Tuscan ambassador.[126] And having been shown his mistake, Galileo promptly retracted his opinion, as a true Catholic.[127] As a punishment, however, his *Dialogue* was banned.[128] After five months (during which the plague infected the city of Florence) Galileo was dismissed from Rome, and by an act of generous compassion was sent, for his house arrest, to the residence of Archbishop Monsignor Piccolomini, his dearest and most esteemed friend in Siena.[129] Galileo enjoyed such tranquility and satisfaction in their conversations that he resumed his studies and found and proved most of his mechanical propositions on the resistance of solids and other conjectures.[130] After approximately five months, the plague having completely come to an end in Galileo's homeland, at the beginning of December 1633 His Holiness turned the confinement to the archbishop's house into the freedom of the countryside, which Galileo greatly enjoyed. Thus, he went back to his villa in Arcetri, in which he had spent much of his time,[131] as it had the benefit of good air and was conveniently located close to Florence.[132] Friends and members of his family often visited him there, and their visits were always of great comfort and solace to him.

[618]Non fu già possibile che quest'opera del Mondano Sistema non capitasse in paesi oltramontani: e perciò indi a poco in Germania fu tradotta e publicata in latino dal suddetto Mattia Berneggero, e da altri nelle lingue franzesi, inglesi e tedesche; et appresso fu stampato in Olanda, con la versione latina fatta da un tal Sig.[r] Elia Deodati, famosissimo iurisconsulto di Parigi e grandissimo litterato, un tal Discorso scritto già in volgare dal Sig.[r] Galileo circa l'anno 1615, in forma di lettera indirizzata a Madama Ser.[ma] Cristina di Lorena, nel tempo in che si trattava in Roma di dichiarare come erronea l'opinione Copernicana e di proibire il libro dell'istesso Copernico: nel qual Discorso intese il Galileo avvertire, quanto fosse pericoloso il valersi de' luoghi della Sacra Scrittura per l'esplicazione di quelli effetti et conclusioni naturali che poi si possino convincer di falsità con sensate esperienze o con necessarie dimostrazioni. Per l'avviso delle quali traduzioni e nuove publicazioni de' suoi scritti restò il Sig.[r] Galileo grandemente mortificato, prevedendo l'impossibilità di mai più supprimergli, con molti altri ch'egli diceva trovarsi già sparsi per l'Italia e fuori manuscritti, attenenti pure all'istessa materia, fatti da lui in varie occasioni nel corso di quel tempo in che era vissuto nell'opinione d'Aristarco[30] e del Copernico; la quale ultimamente, per l'autorità della romana censura, egli aveva catolicamente abbandonata.

Per così salutifero benefizio che l'infinita Providenza si compiacque di conferirgli in rimuoverlo d'error così grave, non volle il Sig.[r] Galileo dimostrarsele ingrato con restar di promuover l'altre invenzioni di altissime conseguenze. Che perciò nel 1636 si risolse di far[31] libera offerta alli Ill.[mi] et Potentissimi Stati Generali delle Provincie Unite d'Olanda del suo ammirabil trovato per l'uso delle longitudini, col patrocinio del Sig.[r] Ugon Grozio, ambasciador residente in Parigi per la Maestà della Regina di Svezia, e con l'ardentissimo impiego del suddetto Sig.[r] Elia Deodati, per le cui mani passò poi tutto il negoziato. Fu dalli [619]Stati avidamente abbracciata sì generosa offerta, e nel progresso del trattato fu gradita con lor umanissima lettera, accompagnata con superba collana d'oro, della quale il Sig.[r] Galileo non volle per allora adornarsi, supplicando gli Stati a compiacersi che il lor regalo si trattenesse in altre mani sin che l'intrapreso negozio fosse ridotto a suo fine, per non dar materia a' maligni suoi emuli di spacciarlo come espilator de' tesori di gran Signori per mezzo di vane oblazioni e presuntuosi concetti.

30 d'Aristarco *ABX*] di Pittagora *S*

31 conseguenze—far *BX*] conseguenze, e col tacere le sue nuove speculazioni che gli rimanevano di publicare, ma con atti di generosità e gratitudine non si saziava di esaltarla, propalando le di lei maraviglie e grandezze. Con tal gratissima resoluzione nel 1636 volle far *A* conseguenze, o col tacere le nuove speculazioni, che gli rimanevano di publicare, anzi con atti di generosità, e di gratitudine, non si saziava d'esaltarla, propalando le di lei maraviglie e grandezze. Con tal gratissima resoluzione nel 1636. fece *S*

618 It proved impossible to prevent this work on the *System of the World* from crossing the mountains to other countries. Thus, it was translated into Latin by Matthias Bernegger, and published in Germany shortly afterward; still others translated it into French, English, and German.[133] Later, in Holland, an eminent lawyer from Paris and prominent intellectual, Elia Diodati, published his Latin translation of the letter [*Discorso*] Galileo had written in the vernacular to the Most Serene Lady Christina of Lorraine in 1615,[134] when it was being debated in Rome whether to declare the Copernican hypothesis mistaken and to ban Copernicus's book. Through this letter Galileo wanted to warn against the dangers of making use of passages from the Scriptures to explain natural effects and propositions that could later be shown false by way of sensory experiences and necessary demonstrations.[135] When he received news of these translations and new publications of his works, Galileo was greatly mortified,[136] foreseeing the impossibility of ever suppressing them or other works of his on the same subjects, which he said were being circulated in manuscript form in Italy and abroad—pieces written on different occasions when he was convinced of the truth of Aristarchus's[137] and Copernicus's opinion, an opinion that he had abandoned shortly before, as a Catholic, owing to the authority of the Roman censorship.

Galileo did not wish to prove ungrateful for the beneficial gift infinite Providence kindly decided to give him by releasing him from such a terrible mistake. Thus, he did not refrain from promoting his other, most fruitful inventions.[138] Accordingly, in 1636 he decided to offer his marvelous discovery that allowed for the correct computation of longitudes to the Most Eminent and Powerful States General of the United Provinces of Holland, with the support of Hugo Grotius, Her Majesty the Queen of Sweden's ambassador in Paris, and with the most energetic help of Elia Diodati, through whose hands the whole negotiation then passed.[139] The States **619**warmly welcomed such a generous offer, and as negotiations continued Galileo received a most friendly letter, accompanied by a magnificent golden chain.[140] Galileo did not want to accept it then, however, and requested the States to agree to place their gift in somebody else's hands until the negotiations that had just started proved successful. He did not wish to give his calumniators an opportunity to slander his name by claiming he robbed grand noblemen of their treasure by empty offers and presumptuous ideas.

Gli destinarono ancora, in evento di felice successo, grossissima recognizione. Havevano già deputato per l'esamina et esperienza della proposta quattro Commessarii, principalissimi matematici, esperti in nautica, geografia et astronomia, [j] a' quali poi il Sig.[r] Galileo conferì liberamente ogni suo pensiero e secreto concernente alla speculativa e pratica del suo trovato, et in oltre ogni suo immaginato artifizio per ridurre, quando fosse occorso, a maggior facilità e sicurezza l'uso del telescopio nelle mediocri agitazioni della nave per l'osservazioni delle Stelle Medicee. Fu da quei Commessarii esaminata e con ammirazione approvata così utile et ingegnosa proposizione. Fu eletto da' medesimi Stati il Sig.[r] Martino Ortensio, uno de' quattro Commessarii, per transferirsi d'Olanda in Toscana e abboccarsi col Sig.[r] Galileo, per estrarre ancor di più dalla sua voce tutti quei documenti et instruzioni più particolari circa la teorica e pratica dell'invenzione. Insomma, nella continuazione per più di cinque[32] anni di questo trattato, non fu per l'una parte o per l'altra pretermessa diligenza e resoluzione per venire alla conclusione di tanta impresa. Ma a tanto non concorrendo per ancora il Divino volere, ben si compiacque che il nostro Galileo fosse riconosciuto per primo e solo ritrovatore di questa così bramata invenzione, sì come di tutte le celesti novità e maraviglie, e che per ciò si rendesse immortale e be [620]nemerito insieme alla terra, al mare, et quasi dico al cielo stesso; ma volle con varii accidenti impedire l'esecuzione dell'impresa, differendola ad altri tempi, con reprimer intanto il fastoso orgoglio degli uomini, che averebbero per tal mezzo con egual sicurezza passeggiato l'incognite vie dell'oceano come le più cognite della terra.

j Presidente eletto dalli Stati per l'esame della invenzione: Sig.[r] Lorenzo Realio, Governatore generale delle Indie Orientali. Deputati o Commessarii: Sig.[r] Martino Ortensio, Matematico d'Amsterdam; Sig.[r] Guglielmo Blavio, geografo; Sig.[r] . . . Golio, professore di matematica in Leida;[33] Sig.[r] Isaac Beecchmanno, professore di matematica e Riformatore della Scuola Dodracena.

32 cinque *BSX*] sei *A*
33 Sig.[r] . . . Golio—Leida *BX*] *om. A* Sig.[r] Giacomo Golio Professore di Matematica in Leida *S*

They promised him even greater rewards in case of success. They had already appointed four commissioners to examine and test Galileo's proposal; to these most eminent mathematicians, experts in navigation, geography, and astronomy[j] [141] Galileo freely imparted every single thought or secret of his about the theory and practice of his discovery, adding all the ideas he had considered for facilitating the use of the telescope to observe the Medicean Stars during moderate rolling of the ship, as well as to make its use safer. The commissioners examined the useful and ingenious proposal, and endorsed it. The States chose one of the four commissioners, Hortensius, to go from Holland to Tuscany to consult Galileo and hear from him all those details and specific instructions about the theory and practice of his invention.[142] All in all, throughout the more than five-year-long[143] negotiations, neither party failed in diligence or determination to draw the enterprise to a close. The Divine Will did not concur, though. It was pleased that our Galileo be acknowledged as the first and only inventor of this longed-for device, and as the discoverer of all the heavenly novelties and marvels, and that for these he become most worthy and immortal, [620]together with the earth, the sea and, I daresay, the heavens themselves. But it preferred to prevent the enterprise from succeeding by various accidents, by deferring it to other times, and meanwhile restraining the pompous pride of men who, with its help, would have crossed the unknown paths of the ocean with the ease with which they traverse the byways of earth.

j President elected by the States to examine the invention: Laurens Reael, Governor-General of the East Indies. Members or commissioners: Hortensius, mathematician from Amsterdam; William Blaeu, geographer; . . . Golius, professor of mathematics in Leiden; Isaac Beeckman, professor of mathematics and rector of the [Latin] School of Dordrecht.

Per lo che, avendo il Sig.ʳ Galileo per lo spazio di ventisette anni sofferto grandissimi incomodi e fatiche per rettificare i moti de' satelliti di Giove, i quali finalmente con somma aggiustatezza egli aveva conseguiti per l'uso delle longitudini; e di più avendo per esattissime osservazioni pochi anni avanti, e prima d'ogn'altro, avvertito col telescopio un nuovo moto o titubazione nel corpo lunare per mezzo delle sue macchie; non permettendo la medesima Providenza Divina che un sol Galileo disvelasse tutti i segreti che forse per esercizio de' futuri viventi ella tiene ascosi nel cielo; nel maggior calore di questo trattato, nell'età di settanta quattro anni in circa, lo visitò con molestissima flussione ne gl'occhi, e dopo alcuni mesi di travagliosa infermità lo privò affatto di quelli, che soli, e dentro minor tempo d'un anno, avevan scoperto, osservato et insegnato vedere nell'universo assai più che non era stato permesso a tutte insieme le viste umane in tutti i secoli trascorsi. Per questo compassionevole accidente fu egli necessitato a consegnar nelle mani del P. D. Vincenzio Renieri suo discepolo, che fu poi Matematico di Pisa, tutti i proprii scritti, osservazioni e calculi intorno a' detti Pianeti, acciò quegli, supplendo alla sua cecità, ne fabbricasse le tavole e l'efemeridi, per donarle poi alli Stati e comunicarle al Sig.ʳ Ortensio, che qua dovea comparire. Ma nello spazio di breve tempo vennero avvisi non solo della morte di questo, ma ancora delli altri tre Commissarii deputati a tal maneggio, a pieno instrutti et assicurati della verità della proposta e della certezza e modo del praticarla. Et finalmente, quando dal Sig.ʳ Ughenio, primo Consigliere e Segretario del Sig.ʳ Principe d'Oranges, e dal Sig.ʳ Borelio, Consigliere e Pensionario della città d'Amsterdam, personaggi di chiarissima fama e litteratura, si procurava incessantemente di riassumere e perfezionare il negoziato con i medesimi Stati;[34] e che il Sig.ʳ Galileo aveva deliberato, con lor consenso, d'inviar colà il P. D. Vincenzio Renieri, come informatissimo d'ogni secreto, con le tavole et efemeridi de' Pianeti Medicei, per conferire il tutto et instruirne chiunque a lor fosse piaciuto; quando, dico, da questi, che già apprendevano la proposta per infallibile e di sicurissimo evento, ciò si trattava con ogni maggior fervore; mancò la vita all'autore di sì grand'invenzione, come dico appresso: e qui si troncò totalmente ogni trattato con gli Stati d'Olanda. Non però qui s'estinse la maligna influenza, ostinatasi ad opprimere con tanti modi, o più tosto a differire, la conclusione d'opera così egregia; poiché nel 1648, quando il sud[621]detto P. Renieri aveva ormai in ordine di publicare (come l'Altezze Lor Ser.ᵐᵉ asseriscono d'aver veduto)[35] l'efemeridi con le tavole e canoni per calcolare in ogni tempo le future constituzioni de' Pianeti Medicei, elaborate sugli studii e precetti conferitigli dal Sig.ʳ Galileo e conseguiti da esso nelle vigilie di tanti anni, fu il detto Padre sopragiunto d'improvisa e quasi repentina malattia, per la quale si morì; et in questo accidente fu, non si sa da chi, spogliato il suo studio delle suddette opere già perfezionate e quasi di tutti gli scritti et osservazioni, tanto delle consegnategli dal Sig.ʳ Galileo che delle proprie, sopra questa materia: perdita tanto più deplorabile, quanto che si richiede per resarcirla assai maggior tempo di quel che fu di bisogno al Sig.ʳ Galileo, perspicacissimo osservatore, per ottenere una perfetta cognizione de' periodi e moti di quei Pianeti.

34 dal Sig.ʳ Ughenio—Stati BSX] da i medesimi Stati si reassumeva il trattato, con aver fatte nuove elezioni di Commissari A

35 (come—veduto) BSX] *om. A in marg. add. et del.* come asserisce d'aver veduto ancora l'Alt.ª Vostra insieme con il Ser.ᵐᵒ Card.ᵉ Gio. Carlo, suo fratello B

Galileo endured the greatest troubles and pains throughout twenty-seven years to straighten out the motions of Jupiter's satellites, which he eventually did with very great precision, so as to use them for the computation of longitudes. Furthermore, he noticed a few years earlier, before anybody else, by very precise telescopic observations, a new motion or libration of the body of the Moon, thanks to its spots.[144] The same Divine Providence did not allow a single Galileo to unveil all the secrets which it keeps hidden in the heavens, possibly to engage other men in the future. Thus, at the most important moment of these negotiations, when he was about seventy-four years old, Divine Providence visited Galileo with a very annoying fluxion in the eyes, and after a few months of tormented infirmity, deprived him completely of those <eyes>[145] that alone, and within less than a year, had discovered, observed, and taught others to see in the universe much more than had been granted all human sight together, in all previous centuries. Because of this pitiful accident Galileo had to hand all his writings, observations, and computations about the Medici stars to Father Vincenzo Renieri, a pupil of his, who would later become professor of mathematics in Pisa, so that he could compensate for Galileo's blindness and compile tables and ephemerides to be conveyed to the States by Hortensius, who was to come to Florence.[146] However, in a brief span of time, news came not only of Hortensius's passing, but of the passing of all three other commissioners appointed for the negotiations[147] who had been fully informed about and tested the validity of Galileo's proposal and the way in which it could be made properly to work. Finally, when Huygens, chief councilor and secretary to the Prince of Orange, and Boreel, councilor and pensionary of the city of Amsterdam, both most eminent and learned personalities,[148] continually attempted to resume negotiations with the States, and bring them to a positive end;[149] and when Galileo decided, with Huygens's and Boreel's agreement, to send to Holland Father Vincenzo Renieri, who was very well acquainted with all the secrets, and brought with him the tables and ephemerides of the Medicean planets, so as to convey everything and instruct anybody appointed—when, I say, Huygens and Boreel had ascertained the validity and the sure success of Galileo's idea, and they were most ardently working on it, then the author of this great invention passed away, as I will say later, thereby totally breaking off negotiations with the States of Holland. This, however, was not the end of the malicious influence that obstinately stopped in many ways, or rather postponed, the accomplishment of such an excellent enterprise, for in 1648, when [621]Father Renieri was ready to publish (as Their Most Serene Highnesses[150] maintain they saw) the ephemerides, together with the tables and canons to compute the positions of the Medicean Planets at any future moments, which he had compiled according to the theories and instructions Galileo had imparted to him, and which Galileo had obtained over sleepless nights through many years, the said Father was overtaken by a sudden and unexpected disease, from which he died. When this happened, somebody—nobody knows who—stripped his room of these works, which had been completed, and of nearly all the writings and observations on this subject, both those handed down to him by Galileo and his own. This loss was most lamentable, as in order to make up for it much more time would be required than Galileo took to obtain a perfect knowledge of the periods and motions of those planets, since he was a very skilful observer.

Ma differiscasi pure per qualsivoglia accidente la pratica di così nobil trovato, et altri si affatichi di rintracciare con i proprii sudori i movimenti di quelle Stelle, o pur altri, adornandosi delle fatiche del primo discopritore, tenti farsene l'autore per estrarne premii et onori; ché sì come per graduare le longitudini il mezzo de' compagni di Giove è l'unico e solo in natura, e perciò questo solo sarà un giorno praticato da tutti gl'osservatori di terra e mare, così il primato e la gloria dell'invenzione sarà sempre del nostro gran Galileo, autenticata da regni interi e dalle republiche più famose d'Europa, et a lui solo sarà perpetuamente dovuta la correzzione delle carte marine e geografiche e l'esattissima descrizione di tutto 'l globo terrestre.

Aveva già il Sig.^{r} Galileo risoluto di mai più esporre alle stampe alcuna delle sue fatiche, per non provocarsi di nuovo quelli emuli che per sua mala sorte in tutte l'altre opere sue egli aveva sperimentati; ma ben, per dimostrar gratitudine alla natura, voleva comunicar manuscritte quelle che gli restavano a varii personaggi a lui ben affetti et intelligenti delle materie in esse trattate. E perciò avendo eletto in primo luogo il Sig.^{r} Conte di Noailles, principalissimo signor della Francia, quando questi nel 1636 ritornava dall'ambasciata di Roma, gli presentò una copia de' suoi Dialogi o pur *Discorsi e Demostrazioni matematiche intorno a due nuove scienze della meccanica e del moto locale*; i fondamenti del quale, insieme con moltissime conclusioni, acquistò sin nel tempo che era in Padova et in Venezia, conferendole a' suoi amici,[k] che si trovarono a varie [622]esperienze ch'egli di continuo faceva intorno all'esamine di molti curiosi problemi e proposizioni naturali. Accettò il Sig.^{r} Conte come gioia inestimabile l'esemplare manuscritto del Sig.^{r} Galileo; ma giunto a Parigi, non volendo defraudare il mondo di tanto tesoro, ne fece pervenir copia in mano alli Elsevirii di Leida, i quali subito ne intrapresero l'impressione, che restò terminata nel 1638.

k Sig.^{r} Filippo Salviati. Sig.^{r} Gio. Francesco Sagredo. Sig.^{r} Daniello Antonini, nobil udinese. Sig.^{r} Paolo Aproino, nobil trivisano. Fra Paolo servita, Teologo della Republica, et altri, etc.

But let the use of this superb discovery be deferred, for whatever reasons; and let others strain themselves to recover, by their own efforts, the motions of those stars; or else let still others attempt to present themselves as their authors, embellishing themselves with the efforts of the first discoverer, in order to gain awards and honors—for just as appealing to Jupiter's companions is the one and only means, in nature, to measure longitudes, and thus will one day be used by all observers, on earth as well as on the sea,[151] so the primacy and the glory of the invention will always be our great Galileo's, validated by entire kingdoms and by Europe's most famous republics, and to him alone shall we owe, forever, the correction of nautical charts and geographic maps, and the most precise description of the whole of the terrestrial globe.

Galileo had decided against publishing any of his works ever again, lest he arouse the jealousies he had experienced, to his misfortune, through all his other works. However, in order to show his gratitude toward nature, he wanted to convey his extant manuscripts to various people who were fondly attached to him and could understand the subjects treated in them. Accordingly, when his first choice, the Count of Noailles, a most eminent person in France, was returning from his embassy in Rome, Galileo entrusted him with a copy of his Dialogues, that is, *Discourses and Mathematical Demonstrations concerning Two New Sciences pertaining to Mechanics and Local Motions,*[152] whose groundwork, together with very many proofs, he had laid down in Padua and Venice. There he showed them to his friends,[k][153] who attended the many [622]experiments he kept performing in his examination of several interesting natural problems and propositions.[154] The count received Galileo's manuscript as a priceless gem; back in Paris, however, and unwilling to defraud the world of this treasure, he saw to it that a copy reached the Elsevier's, in Leiden. They immediately undertook the printing, which was completedin 1638.[155]

k Filippo Salviati; Giovan Francesco Sagredo, Daniello Antonini, nobleman from Udine; Paolo Aproino, a nobleman from Treviso; and Paolo Sarpi, OSM, theologian of the Republic; and others.

Poco dopo questa inaspettata pubblicazione, concedendomisi l'ingresso nella villa d'Arcetri, dove allor dimorava il Sig.ʳ Galileo, acciò quivi io potesse godere de' sapientissimi suoi colloquii e preziosi ammaestramenti, e contentandosi questi che nello studio delle matematiche, alle quali poco avanti mi ero applicato, io ricorresse alla viva sua voce per la soluzione di quei dubbii e difficoltà che per natural fiacchezza del mio ingegno[36] bene spesso incontravo, accadde che nella lettura de' Dialogi sopradetti, arrivando al trattato de' moti locali, dubitai, come pur ad altri era occorso, non già della verità del principio sopra 'l quale è fondata l'intera scienza del moto accelerato, ma della necessità di supporlo come noto; onde io, ricercandolo di più evidenti confermazioni di quel supposto, fui cagione ch'egli nelle vigilie della notte, che allora con gran discapito della vita gli erano familiarissime, ne ritrovò la dimostrazione geometrica,[37] dipendente da dottrina da esso pur dimostrata contro ad una conclusione di Pappo (qual si vede nel suo trattato di Meccaniche, stampato dal suddetto P. Mersennio), et a me subito la conferì, sì come ad altri suoi amici ch'eran soliti visitarlo: et alcuni mesi dopo, compiacendosi di tenermi poi di continuo appresso la sua disciplina, per guidarmi, benché cieco come egli era di corpo, d'intelletto però lucidissimo, per il sentiero di quelli studii ch'egli intendeva ch'io proseguisse, imposemi ch'io facesse il disteso di quel teorema, per la difficoltà che gli arrecava la sua cecità nell'esplicarsi dove occorreva usar figure e caratteri; e di questo ne mandò più copie per l'Italia et in Francia alli amici suoi. Per una simil occasione di dubitare mi aveva ancora esplicato una sua considerazione o dimostrazione sopra la 5ᵃ e 7ᵃ definizione del quinto libro d'Euclide, dettandola a me dopo in dialogo per inserirla in detto suo libro appresso la prima proposizione del moto equabile, quando si fosse ristampato; et è quell'istessa dimostrazione che, a richiesta di V. A. S., fu poi distesa dal Sig.ʳ Evangelista Torricelli, che l'aveva sentita dal medesimo Sig.ʳ Galileo.[38]

⁶²³Negli 11 di Marzo del 1639 avendo V. A. S. con filosofica curiosità ricercato per lettera il Sig.ʳ Galileo del parer suo circa il libro *De lapide Bononiensi* del filosofo Liceti, e particolarmente sopra la dottrina del capitolo 50, dove l'autore oppone alla di lui oppinione sopra il candore o luce secondaria della luna, risposele tra pochi giorni, come è noto all'A. V., con dottissima lettera dell'ultimo dell'istesso mese, che cadde nel 1640, procurando per essa di mantener saldi i proprii pensieri con ragioni e conietture vivissime e sottilissime; alla qual lettera poi replicò il suddetto Liceti con assai grosso volume, che egli publicò nel 1642 insieme con detta lettera.

36 *post* ingegno *scr.* e per la novità della materia, di natura fisica e però non interamente geometrica *S*
37 geometrica *ABX*] geometrica meccanica *S*
38 *post* Galileo *scr.* nel tempo che dimorò appresso di lui *S*

I was granted entrance to the villa in Arcetri, where Galileo was living, shortly after this unexpected publication, to enjoy his most learned talks and precious teachings.[156] He was pleased that in my studies of mathematics, to which I had earlier devoted myself,[157] I turned to him for the solution of those doubts and difficulties into which I often ran due to the natural weakness of my mind.[158] It so happened that when I got to the section on local motions in the *Discourses*, I doubted—as others had done, too—not the truth of the principle upon which the whole science of accelerated motion is founded but the necessity of assuming it as known.[159] I asked him for more solid confirmations of that assumption, and therefore prompted him, through sleepless nights—something to which he was only too accustomed, to the detriment of his life—to find its geometric[160] demonstration, which he drew from a thesis he had proven against a proposition by Pappus (as we see it in his treatise *On Mechanics*, printed by Father Mersenne).[161] He told me immediately, as he did other friends of his who used to visit him. A few months later, he kindly decided to keep me constantly with him to teach me, so as to lead me—for he was physically blind, but had the brightest mind—along the paths of the studies he wanted me to pursue, and required me to write down that theorem, given the difficulty he had, because of his blindness, to express himself when he had to appeal to diagrams and figures; he then circulated several copies of this to his friends in Italy and France. I expressed my doubts in a similar vein once again, and he explained to me a consideration, or demonstration, of his of the 5th and 7th definition of Euclid's fifth book,[162] later to dictate it to me in person, so that it could be inserted in the aforementioned book of his, after the first proposition on equable motion, whenever it was reprinted. This very demonstration was later expounded, at Your Most Serene Highness's request, by Evangelista Torricelli, who had heard it from Galileo himself.[163]

[623]On March 11, 1639, prompted by philosophical curiosity, Your Most Serene Highness wrote Galileo a letter asking for his opinion about the book *De lapide Bononiensi*, by philosopher Liceti, and especially about the thesis expounded in chapter 50, where the author combats Galileo's opinion about the Moon's whiteness or secondary light.[164] Within a few days, as Your Highness knows, Galileo replied with a most learned letter dated March 31, 1640. With it, Galileo stuck to his views, offering the deepest and surest reasons and conjectures in their support. To this letter Liceti replied with a very thick volume, which he published in 1642, also including Galileo's letter.[165]

Nel tempo di trenta mesi ch'io vissi di continuo appresso di lui sino alli ultimi giorni della sua vita,[39] essendo egli spessissimo travagliato da acerbissimi dolori nelle membra, che gli toglievano il sonno e 'l riposo, da un perpetuo bruciore nelle palpebre, che gl'era di insopportabil molestia, e dall'altre indisposizioni che seco portava la grave età, defatigata da tanti studii e vigilie de' tempi addietro, non poté mai[40] applicare a disporre in carta l'altre opere che gli restavano già risolute e digerite nella sua mente, ma per ancora non distese, come pur desiderava di fare. Aveva egli concetto (già che i Dialoghi delle due Nuove Scienze erano fatti pubblici) di formare due Giornate da aggiugnersi all'altre quattro; e nella prima intendeva inserire, oltre alle due suddette dimostrazioni, molte nuove considerazioni e pensieri sopra varii luoghi delle Giornate già impresse, portando insieme la soluzione di gran numero di problemi naturali di Aristotele e di altri suoi detti et oppinioni, con discoprirvi manifeste fallacie, et in specie nel trattato *De incessu animalium*; e finalmente nell'ultima Giornata promuovere un'altra nuova scienza, trattando con progresso geometrico della mirabil forza della percossa, dove egli stesso diceva d'aver scoperto e poter dimostrare acutissime e recondite conclusioni, che superavano di gran lunga tutte l'altre sue speculazioni già publicate. Ma nell'applicazione a così vasti disegni, sopragiunto da lentissima febbre e da palpitazione di quore, dopo due mesi di malattia che a poco a poco gli consumava gli spiriti, il mercoledì dell'8 di Gennaio del 1641 *ab Incarnatione*, a hore quattro di notte, in età di settantasette anni, mesi dieci e giorni venti, con filosofica e cristiana constanza rese l'anima al suo Creatore, inviandosi questa, per quanto creder ne giova, a godere e rimirar più d'appresso quelle eterne et immutabili maraviglie, che per mezzo di fragile artifizio con tanta avidità et impazienza ella aveva procurato di avvicinare agl'occhi di noi mortali.

[624]D'inestimabil pregiudizio all'università de' litterati e al mondo tutto fu questa perdita inconsolabile, che ci privò non solo della miniera fecondissima del discorso d'un tanto filosofo, che già per inviolabil decreto di natura dovea mancare, ma più dell'oro purissimo delle speculazioni, estratto già e conservato nella sua lucidissima mente, forsi senza speranza di mai più recuperarlo per opera di alcun altro. Di queste rimasero solo appresso il figliuolo e nipoti alcuni pochi fragmenti per introdursi nella contemplazione della forza della percossa, con la suddetta dimostrazione del principio della scienza del moto accelerato, e con l'altra della 5ª e 7ª definizione del quinto libro d'Euclide.

Il corpo suo fu condotto dalla villa d'Arcetri in Firenze, e per commessione del nostro Ser.mo Gran Duca fatto separatamente custodire nel tempio di S. Croce, dove è l'antica sepoltura della nobil famiglia de' Galilei, con pensiero d'ereggergli augusto e suntuoso deposito in luogo più conspicuo di detta chiesa, e così, non meno ch'in vita, generosamente onorar dopo morte l'immortal fama del secondo fiorentino Amerigo, non già discopritore di poca terra, ma d'innumerabili globi e nuovi lumi celesti, dimostrati sotto i felicissimi auspicii della Ser.ma Casa di V. A.

39 alli—vita BX] alli ultimi giorni della sua vita, e posso ancor dire de' miei studi A all'ultimo respiro della sua vita, che per altri sinistri accidenti, occupazioni e impieghi sopravvenutimi posso dir l'ultimo delli studii miei più giocondi e più quieti S
40 non poté mai BSX] poco poté A

Over the thirty months during which I lived with him, until his very last days,[166] very severe pains very often afflicted his limbs, kept him awake, and prevented him from resting. A persistent burning sensation in his eyelids caused unbearable torment, and other ailments brought by advanced age, by the continuous studies and sleepless nights of the previous years, prevented[167] him from devoting himself to writing down other works that were already finished and polished in his mind but were not yet on paper. After the *Dialogues on Two New Sciences* had been published, he thought of writing two more Days to add to the other four.[168] In the first, besides the two demonstrations mentioned above, he planned to include many new reflections and thoughts about a number of passages of the Days already printed, bringing together the solution of a large number of Aristotle's natural problems, and other statements and opinions of his, calling attention to plain fallacies (especially in the treatise *On the Progression of Animals*).[169] And eventually, in the final Day, he intended to promote another new science, by analyzing with ordered geometric demonstrations the most noteworthy force of percussion,[170] about which he claimed he had discovered and could prove very ingenious and hidden proofs, which by far surpassed all other speculations of his previously published. However, while devoting himself to these extended plans, he was struck by a long-lasting fever and palpitations. After two months of illness that progressively eroded his life, on Wednesday, January 8, 1641, *ab Incarnatione*,[171] at 4 a.m., aged 77 years, 10 months, and 20 days, with philosophical and Christian perseverance he returned his soul to his Creator. Thus, his soul—if believing in this could be of any solace—went to enjoy and inspect more closely those eternal and unalterable wonders which by way of a fragile instrument it eagerly and impatiently succeeded in drawing closer to our mortal eyes.

[624]This devastating loss momentously affected the community of scholars and the whole world, as it deprived us not only of the very rich mine of discourse of so great a philosopher, which, by the inviolable decree of nature, had to perish, but also of the purest gold of his speculations,[172] extracted and stored in his splendid mind, with no hope, perhaps, of anybody else recovering it. Of these, a few fragments remained with Galileo's son and grandsons concerning the force of percussion, with the demonstration of the principle of the science of accelerated motion, as well as with the demonstration of the 5th and 7th definition of Euclid's fifth book.

Galileo's body[173] was brought from Arcetri to Florence, and by commission of our Most Serene Grand Duke separately enshrined in the church of Santa Croce, where the ancient tomb of Galileo's noble family is located.[174] The plan was to erect an august and sumptuous monument to him in a more prominent location in that church, in order that, no less than in his lifetime, it would be possible to lavish honor, after his death, on the immortal fame of the second Florentine Amerigo:[175] not a discoverer of some bit of land, though, but of countless globes and new heavenly lights, unveiled under the most fortunate auspices of the Most Serene House of Your Highness.

Fu il Sig.^r Galileo di gioviale e giocondo aspetto, massime in sua vecchiezza, di corporatura quadrata, di giusta statura, di complessione per natura sanguigna, flemmatica et assai forte, ma per fatiche e travagli, sì dell'animo come del corpo, accidentalmente debilitata, onde spesso riducevasi in stato di languidezza. Fu esposto a molti mali accidenti et affetti ipocondriaci e più volte assalito da gravi e pericolose malattie, cagionate in gran parte da' continui disagi e vigilie nell'osservazioni celesti, per le quali bene spesso impiegava le notti intere. Fu travagliato per più di 48 anni della sua età, sino all'ultimo della vita, da acutissimi dolori e punture, che acerbamente lo molestavano nelle mutazioni de' tempi in diversi luoghi della persona, originate in lui dall'essersi ritrovato, insieme con due nobili amici suoi, ne' caldi ardentissimi d'una estate in una villa del contado di Padova, dove postisi a riposo in una stanza assai fresca, per fuggir l'ore più noiose del giorno, e quivi addormentatisi tutti, fu inavvertentemente da un servo aperta una finestra, per la quale solevasi, sol per delizia, sprigionare un perpetuo vento artifizioso, generato da moti e cadute d'acque che quivi appresso scorrevano. Questo vento, per esser fresco e umido di soverchio, trovando i corpi loro assai alleggeriti di vestimenti, nel tempo di due ore che riposarono, intro⁶²⁵dusse pian piano in loro così mala qualità per le membra, che svegliandosi, chi con torpedine e rigori per la vita e chi con dolori intensissimi nella testa e con altri accidenti, tutti caddero in gravissime infermità, per le quali uno de' compagni in pochi giorni se ne morì, l'altro perdé l'udito e non visse gran tempo, et il Sig.^r Galileo ne cavò la sopradetta indisposizione, della quale mai poté liberarsi.

Non provò maggior sollievo nelle passioni dell'animo, né miglior preservativo della sanità, che nel godere dell'aria aperta; e perciò, dal suo ritorno di Padova, abitò quasi sempre lontano dalli strepiti della città di Firenze, per le ville d'amici o in alcune ville vicine di Bellosguardo o d'Arcetri: dove con tanto maggior satisfazione ei dimorava, quanto che gli pareva che la città in certo modo fosse la prigione delli ingegni speculativi, e che la libertà della campagna fosse il libro della natura, sempre aperto a chi con gl'occhi dell'intelletto gustava di leggerlo e di studiarlo; dicendo che i caratteri[41] con che era scritto erano le proposizioni, figure e conclusioni geometriche, per il cui solo mezzo potevasi penetrare alcuno delli infiniti misterii dell'istessa natura. Era perciò provvisto di pochissimi libri, ma questi de' migliori e di prima classe: lodava ben sì il vedere quanto in filosofia e geometria era stato scritto di buono, per dilucidare e svegliar la mente a simili e più alte speculazioni; ma ben diceva che le principali porte per introdursi nel ricchissimo erario della natural filosofia erano l'osservazioni e l'esperienze, che, per mezzo delle chiavi de' sensi, da i più nobili e curiosi intelletti si potevano aprire.

41 i caratteri *X*] i caratteri e l'alfabeto *AS* i caratteri dell'alfabeto *B*

Galileo was jovial and lighthearted, especially in his old age; he was of solid build, fair height, and a naturally hot-tempered complexion, phlegmatic, and very strong. This strength was weakened accidentally, though, by struggles and tribulations, of the spirit as well as of the body, which often reduced it to lethargy. Galileo was subject to several misfortunes and hypochondriac feelings, and was affected by serious and dangerous maladies over and over again, largely caused by repeated discomforts and sleepless nights while observing the heavens, to which he often devoted entire nights. For over forty-eight years, until the last year of his life, Galileo was tormented by very severe pains and pangs, which affected various parts of his body at every change of season. These originated when Galileo and two noblemen friends of his happened to be in a villa in the country, near Padua, during a steaming hot summer; they went to rest in a very cool room to escape the most oppressive hours of the day, and all fell asleep. A servant inadvertently opened a window, through which an unnatural and constant wind, caused by the stirring and falling of waters that flowed nearby, usually blew. This wind was exceedingly fresh and humid, and blew on their barely dressed bodies for two hours, while they were sleeping, thereby slowly [625]affecting their limbs. When they awoke, they felt numbness and stiffness in their bodies, or had very acute headaches, or other maladies. All fell very gravely ill: one of Galileo's two friends died a few days later; the other lost his hearing and did not live long. Galileo survived with the ailment from which he never fully recovered.[176]

He never found greater relief from the passions of the soul, nor a better way to safeguard his health, than by enjoying fresh air. Accordingly, after his return from Padua, Galileo lived most of the time away from the uproar of Florence, either in his friends' villas or in others near Bellosguardo or Arcetri. He lived there with the greater satisfaction the more it seemed to him that the city was a sort of prison for speculative minds, and that the liberty of the countryside was the book of nature, always open to those who enjoyed reading and studying it with their minds' eyes. He said that the letters[177] with which it was written were geometric propositions, diagrams, and proofs—the only means by which one could penetrate some of the infinite number of nature's innermost mysteries.[178] Therefore, he had very few books, but the best and finest. He praised the good that had been written in philosophy and geometry, to enlighten and awaken the mind to similar or higher speculations—but he insisted that the main gates that gave access to the treasures of natural philosophy were observations and experiments, which the noblest and most inquiring minds could unlock with keys provided by the senses.

Quantunque le piacesse la quiete e la solitudine della villa, amò però sempre d'avere il commercio di virtuosi e d'amici, da' quali era giornalmente visitato e con delizie e regali sempre onorato. Con questi piacevagli trovarsi spesso a conviti, e, con tutto fosse parchissimo e moderato, volentieri si rallegrava; e particolarmente premeva nell'esquisitezza e varietà de' vini d'ogni paese, de' quali era tenuto continuamente provvisto dall'istessa cantina del Ser.ᵐᵒ Gran Duca e d'altrove: e tale era il diletto ch'egli aveva nella delicatezza de' vini e dell'uve, e nel modo di custodire le viti, ch'egli stesso di propria mano le potava e legava nelli orti delle sue ville, con osservazione, diligenza et industria più che ordinaria; et in ogni tempo si dilettò grandemente dell'agricoltura, che gli serviva insieme di passatempo e di occasione di filosofare intorno al nutrirsi e al vegetar delle piante, sopra la virtù prolifica de' semi, e sopra l'altre ammirabili operazioni del Divino Artefice.

Ebbe assai più in odio l'avarizia che la prodigalità. Non rispiarmò a spesa alcuna in far varie prove e osservazioni per conseguir notizie di nuove et ammirabili conseguenze. Spese liberalmente in sollevar i depressi, in ricevere et ⁶²⁶onorare forestieri, in somministrar le comodità necessarie a poveri, eccellenti in qualch'arte o professione, mantenendogli in casa propria finché gli provvedesse di convenevol trattenimento. E tra quei ch'egli accolse, tralasciando di nominar molti giovani fiamminghi, tedeschi e d'altrove, professori di pittura o scultura e di altro nobil esercizio, o esperti nelle matematiche o in altro genere di scienza, farò solo particolar menzione di quegli che fu l'ultimo in tempo, e in qualità forse il primo, e che già discepolo del P. D. Benedetto Castelli, ormai fatto maestro, fu dal medesimo Padre inviato e raccomandato al Sig.ʳ Galileo, affinché questi gustasse d'aver appresso di sé un geometra eminentissimo, e quegli, allora in disgrazia della fortuna, godesse della compagnia e protezione d'un Galileo. Parlo del Sig.ʳ Evangelista Torricelli, giovane d'integerrimi costumi e di dolcissima conversazione, accolto in casa, accarezzato e provvisionato dal Sig.ʳ Galileo, con scambievol diletto di dottissime conferenze. Ma la congiunzione in terra di due lumi sì grandi ben esser quasi momentanea dovea, mentre tali son le celesti. Con questi non visse il Sig.ʳ Galileo più che tre mesi; morì ben consolato di veder comparso al mondo, e per suo mezzo approssimato a' benigni influssi della Ser.ᵐᵃ Casa di V. A., così riguardevol soggetto. Et il Padre Castelli conseguì ancora l'intento: giaché, mancato il Sig.ʳ Galileo, essendo, a persuasione del Sig.ʳ Senatore Andrea Arrighetti, anch'esso discepolo del Sig.ʳ Galileo, trattenuto in Firenze il Sig.ʳ Torricelli, fu questo da V. A. S. (con l'ereditario instinto di protegere e sollevare i possessori d'ogni scienza e per la particolar affezzione e natural talento alle matematiche) favorito appresso il Ser.ᵐᵒ nostro G. Duca, e da questo onorato col glorioso titolo di suo Filosofo et Matematico, e con regia liberalità invitato a pubblicar quella parte dell'opere sue che l'ànno reso immortale, et altra prepararne di maraviglia maggiore, che, prevenuto da invidiosa e immatura morte, lasciò imperfetta, ma, postuma e bramata sin d'oltre a' monti, spera tra poco⁴² la luce.

42 tra poco *ABX*] una volta *S*

However much he enjoyed quiet and solitude in his villa, Galileo loved to have exchanges with virtuous persons and friends, who visited him daily and honored him with delights and gifts. He enjoyed having meals often with them, and although he was very frugal and measured, he was easily cheerful. He particularly dwelt on the quality and variety of wines from all countries, with which he was constantly supplied from the wine cellar of the Most Serene Grand Duke,[179] and from elsewhere. Galileo took such delight in the delicacy of the wines and grapes, and the cultivation of the vines, that he pruned and grafted them himself in the gardens of his villas, with extraordinary attention, diligence, and industriousness. He always took great pleasure from agriculture, which appealed to him both as pastime and opportunity to philosophize about the nourishment and growth of plants, the prolific virtues of seeds, and other marvelous operations of the Divine Maker.

Galileo hated stinginess much more than prodigality. He did not refrain from spending anything required to perform tests and observations to gather information that might have new and remarkable consequences. He spent money liberally to relieve unfortunates, to welcome and [626]honor foreigners, to provide what poor people who excelled in some art or profession required, supporting them in his own house as long as he could do so satisfactorily. He welcomed many young people from Flanders, Germany, and elsewhere: students of painting or sculpture, or of other noble occupations, or else experts in mathematics or in other kinds of sciences. I shall omit their names but specifically mention here the last one from a chronological point of view, and possibly the first in quality. When a former pupil of Father Benedetto Castelli's became a teacher he was sent and recommended by that priest to Galileo, so that the latter might have a most eminent geometer with him, and the former—who was then down on his luck—could enjoy the company and protection of Galileo. I am referring to Evangelista Torricelli, a most upright young man and very pleasant in conversation, whom Galileo welcomed to his house, cherished, and supported, taking great pleasure from their most learned conversations. But the coupling, on earth, of these two great luminaries was fated to be very short-lived, just like the conjunctions of heavenly bodies. Galileo lived with him no more than three months, but he died with the comfort of seeing a person so worthy of esteem coming into the world, and with his help approaching the propitious influence of the Most Serene House of Your Highness. And Father Castelli accomplished what he wished: after Galileo died, Torricelli remained in Florence on the advice of Senator Andrea Arrighetti (himself a pupil of Galileo's),[180] and Your Most Serene Highness (owing to your inherited instinct for protecting and supporting students of any sciences, and your particular affection and talent for mathematics) brought him to the attention of our Most Serene Grand Duke, who honored him with the prestigious position of his Philosopher and Mathematician. Also, with royal liberality he encouraged Torricelli to publish the part of his work that made him immortal, and to prepare others even more astonishing, which he left unfinished, prevented by envious and premature death;[181] however, this second part, much awaited beyond the Alps, will see light soon.[182]

Non fu il Sig.ʳ Galileo ambizioso delli onori del volgo, ma ben di quella gloria che dal volgo differenziar lo poteva. La modestia gli fu sempre compagna; in lui mai si conobbe vanagloria o iattanza. Nelle sue avversità fu constantissimo, e soffrì coraggiosamente le persecuzioni delli emuli. Muovevasi facilmente all'ira, ma più facilmente si placava. Fu nelle conversazioni universalmente amabilissimo, poiché discorrendo sul serio era ricchissimo di sentenze e concetti gravi, e ne' discorsi piacevoli l'arguzie et i sali non gli mancavano. L'eloquenza [627] poi et espressiva ch'egli ebbe nell'esplicare l'altrui dottrine o le proprie speculazioni, troppo si manifesta ne' suoi scritti e componimenti per impareggiabile e, per così dire, sopraumana. [43]

Fu dotato dalla natura d'esquisita memoria; e gustando in estremo la poesia, aveva a mente, tra gl'autori latini, gran parte di Vergilio, d'Ovidio, Orazio e di Seneca, e tra i toscani quasi tutto 'l Petrarca, tutte le rime del Berni, e poco meno che tutto il poema di Lodovico Ariosto, che fu sempre il suo autor favorito e celebrato sopra gl'altri poeti, avendogli intorno fatte particolari osservazioni e paralleli col Tasso sopra moltissimi luoghi. Questa fatica gli fu domandata più volte con grandissima instanza da amico suo, mentre era in Pisa, e credo fusse il Sig.ʳ Iacopo Mazzoni, al quale finalmente la diede, ma poi non poté mai recuperarla, dolendosi alcuna volta con sentimento della perdita di tale studio, nel quale egli stesso diceva aver avuto qualche compiacenza et diletto. Parlava dell'Ariosto con varie sentenze di stima e d'ammirazione; et essendo ricercato del suo parere sopra i due poemi dell'Ariosto e del Tasso, sfuggiva prima le comparazioni, come odiose, ma poi, necessitato a rispondere, diceva che gli pareva più bello il Tasso, ma che gli piaceva più l'Ariosto, soggiugnendo che quel diceva parole, e questi cose. E quand'altri gli celebrava la chiarezza et evidenza nell'opere sue, rispondeva con modestia, che se tal parte in quelle si ritrovava, la riconosceva totalmente dalle replicate letture di quel poema, scorgendo in esso una prerogativa solo propria del buono, cioè che quante volte lo rileggeva, sempre maggiori vi scopriva le maraviglie e le perfezioni; confermando ciò con due versi di Dante, ridotti a suo senso:

Io non lo lessi tante volte ancora,
Ch'io non trovasse in lui nuova bellezza.

Compose varie poesie in stil grave et in burlesco, molto stimate da' professori.

Intese mirabilmente la teorica della musica, e ne diede evidente saggio nella prima Giornata delli ultimi Dialogi sopradetti.

Oltre al diletto ch'egli aveva nella pittura, ebbe ancora perfetto gusto nell'opere di scultura et architettura e in tutte l'arti subalternate al disegno.

---

43 *in marg. scr. et del.* e per ciò scrisse nella lingua materna *B*

Galileo did not crave for common honors but yearned for the glory that could set him apart from common people. He was always discrete; he never knew vainglory or boasting. In hardship, he showed great constancy and boldly endured the torment of his opponents. He was quick to anger but easier to calm down. In conversation, he was always very good-natured. In serious exchanges he had plenty of momentous opinions and ideas, and in pleasant ones he never lacked jokes and wit. His persuasiveness [627] and incisiveness in explaining others' views or his own speculations shine in his work and writings as incomparable and, if you will, superhuman.[183]

Nature endowed Galileo with a very fine memory; he very much enjoyed poetry, and knew by heart, among Latin authors, most of Virgil, Ovid, Horace, and Seneca; and among Tuscan ones, nearly all of Petrarch's work, all the poems by Berni, and nearly all of Ludovico Ariosto's entire poem. Ariosto was always his favorite author: he celebrated him above all other poets, wrote about him, and drew parallels with Tasso on very many occasions.[184] When Galileo was still in Pisa, a friend of his (I believe it was Jacopo Mazzoni[185]) repeatedly and persistently asked him for this work on Ariosto; Galileo eventually gave it to him, and never managed to retrieve it. He complained about the loss of this work occasionally, as he had enjoyed doing it and was proud of what he had done. Galileo spoke of Ariosto using many expressions of respect and admiration, and when he was asked for an opinion about Ariosto's and Tasso's poems, at first he avoided any comparisons, deeming them invidious, but then, when pressed, said that Tasso seemed to him more pleasing, but he liked Ariosto better, adding that the former spoke words, the latter, things. And whenever someone praised the clarity and intelligibility of his works, Galileo humbly replied that whichever of those qualities was to be found there he owed to repeated reading of Ariosto's poem. Galileo saw in it a feature only good works have: no matter how many times he read it, he always found marvels and virtues in it. And he used to say that by recalling two lines by Dante, which he adjusted to the meaning he wished to give them:

Never have I read it so many times,
that I could not find new beauty in it.[186]

Galileo wrote both grave and merry poems, which were held in high esteem by leading literary scholars.

He understood musical theory amazingly well, as he clearly showed in the first Day of his later Dialogues, to which I referred above.[187]

In addition to his great enjoyment of painting, Galileo had an excellent taste for sculpture and architecture, and for all arts subordinate to drawing.

Rinovò nella patria, e si può dire nell'Italia, le matematiche e la vera filosofia; e questo non solo con le pubbliche e private lezzioni nelle città di Pisa, Padova, Venezia, Roma e Firenze, ma ancora con le continue dispute che ne' congressi avanti di lui si facevano, instruendo particolarmente moltissimi curiosi [628]ingegni e gran numero di gentiluomini, con lor notabili acquisti. Et in vero il Sig.ʳ Galileo ebbe dalla natura così maravigliosa abilità d'erudire, che gli stessi scolari facevan in breve tempo conoscer la grandezza del lor maestro.[1]

Alle publiche sue lezzioni di matematica interveniva così gran numero d'uditori, che vive ancor oggi in Padova la memoria, autenticata da soggetto di singolarissima fama e dottrina, stato già quivi scolare del Sig.ʳ Galileo, che egli fu necessitato (e tali son le parole di Mons.ʳ Vescovo Barisone) d'uscire della scuola destinata alla sua lettura et andare a leggere nella scuola grande delli artisti, capace di mille persone, e non bastando questa, andare nella scuola grande de' legisti, maggiore il doppio, e che spesse volte questa ancora era pienissima; al qual concorso et applauso niun altro lettore in quello Studio (ancorché di professione diversa dalla sua, e perciò dall'universale più abbracciata) è mai giunto a gran via. Accrescevasi questo grido dal talento sopranaturale ch'egl'ebbe nell'esaltar le facultà matematiche sopra tutte l'altre scienze, dimo[629]strando con assai ricca et maestosa maniera le più belle e curiose conclusioni che trar si possino dalla geometria, esplicandole con maravigliosa facilità, con utile e diletto insieme delli ascoltanti. E per chiara confermazione di ciò si consideri la qualità de' personaggi che in Padova gli voller esser discepoli; e tralasciando tanti Principi e gran Signori italiani, franzesi, fiaminghi, boemi, transilvani, inglesi, scozzesi e d'ogn'altra nazione, sovviemmi aver inteso ch'il gran Gustavo re di Svezia, che fu poi fulmine della guerra, nel viaggio che da giovane fece incognito per l'Italia, giunto a Padova vi si fermò con la sua comitiva per molti mesi, trattenutovi principalmente dalle nuove e peregrine speculazioni e curiosissimi problemi che giornalmente venivano promossi e risoluti dal Sig.ʳ Galileo nelle publiche lezzioni e ne' particolari congressi, con ammirazione de' circostanti; e volle nell'istessa casa di lui (con l'interesse d'esercitarsi insieme nelle vaghezze della lingua toscana) sentire l'esplicazione della sfera, le fortificazioni, la prospettiva e l'uso di alcuni strumenti geometrici e militari,[45] con applicazione et assiduità di vero discepolo, discoprendogli in fine con amplissimi doni quella regia maestà ch'egli s'era proposto di occultare.

---

1 Nota di alcuni gentiluomini fiorentini che furono discepoli[44] del Sig.ʳ Galileo. 1. Mons.ʳ Nerli, Arcivescovo di Firenze; 2. Mons.ʳ Piccolomini, Arcivescovo di Siena; 3. Mons.ʳ Rinuccini, Arcivescovo di Fermo; 4. Mons.ʳ Giuliano de' Medici, Arcivescovo di Pisa; 5. Mons.ʳ Marzimedici, Arcivescovo di Firenze; 6. Mons.ʳ Giovanni Ciampoli, Segretario de' Brevi di Papa Urbano Ottavo (3–6: defunti.); 7. Sig.ʳ Senator Filippo Pandolfini; 8. Sig.ʳ Senator Andrea Arrighetti; 9. Sig.ʳ Cav.ʳ Tommaso Rinuccini; 10. Sig.ʳ Pier Francesco Rinuccini, Residente a Milano; 11. Sig.ʳ Mario Guiducci; 12. Sig.ʳ Niccolò Arrighetti; 13. Sig.ʳ Braccio Manetti; 14. Sig.ʳ Canonico Niccolò Cini; 15. Sig.ʳ Conte Piero de' Bardi; 16. Sig.ʳ Filippo Salviati; 17. Sig.ʳ Iacopo Soldani; 18. Sig.ʳ Iacopo Giraldi; 19. Sig.ʳ Michelangelo Buonarruoti; 20. Sig.ʳ Alessandro Sertini (11–20: defunti), et altri.

---

44 discepoli BX] scolari A scolari e seguaci S
45 l'uso—militari BSX] l'uso del suo nuovo Compasso e di altri militari strumenti A

In his homeland, and we may well say in the whole of Italy, he renewed mathematics and true philosophy. And Galileo did this not only through his public and private lectures in Pisa, Padua, Venice, Rome, and Florence but also in the continuous debates that arose before him in assemblies, teaching in particular many inquiring [628]minds and a large number of gentlemen, with great benefit to them. In fact, Galileo was so naturally and beautifully gifted with a talent for teaching that his pupils themselves quickly testified to their master's greatness.[1 188]

So many auditors attended his public lectures in mathematics that their memory is still alive in Padua. It is attested to by one of most eminent reputation and learning who became Galileo's pupil that he felt compelled (these are Monsignor Bishop Barisone's own words[189]) to move from his own classroom at the university and lecture in the bigger classrooms assigned to students of arts, which could hold a thousand people. And when this was no longer enough, he moved to the big hall for lawyers, which was twice as large; and yet this, too, was often filled. No other lecturer of that university (however different his discipline was from Galileo's, and therefore more widely followed) ever achieved this success and approval. His fame grew from his extraordinary talent to extol mathematics over all other sciences, [629]proving the most beautiful and bizarre statements in geometry in a most graceful and elegant way, explaining them with a most amazing aptitude to the simultaneous benefit and pleasure of the audience. To clearly confirm this, suffice it to mention the quality of the people who wanted to become his disciples in Padua: many princes and noblemen from Italy, France, Flanders, Bohemia, Transylvania, England, Scotland, and all other nations. Besides them, I understood that the great Gustaf, King of Sweden, who was later a great warrior, during his incognito travel to Italy went to Padua and stayed there with his retinue for several months; he was detained mainly by the new and extraordinary speculations, and most intriguing problems, that Galileo proposed and resolved on a daily basis, either in public lectures or in private assemblies, to the awe of those who attended. The king visited Galileo in his own house (so as also to practice the beautiful Tuscan language) to be taught spherical trigonometry, fortification, perspective, and the use of some geometric and military instruments.[190] He set himself to studying regularly, as a true disciple, eventually revealing to Galileo, by way of most generous gifts, the royal majesty he had proposed to conceal from everybody else.[191]

---

1 Some Florentine gentlemen who were Galileo's disciples: 1. Monsignor Nerli, Archbishop of Florence; 2. Monsignor Piccolomini, Archbishop of Siena; 3. Monsignor Rinuccini, Archbishop of Fermo; 4. Monsignor Giuliano de' Medici, Archbishop of Pisa; 5. Monsignor Marzimedici, Archbishop of Florence; 6. Monsignor Giovanni Ciampoli, Secretary of Briefs for Pope Urban VIII (4–6: deceased); 7. Senator Filippo Pandolfini; 8. Senator Andrea Arrighetti; 9. Knight Tommaso Rinuccini; 10. Pier Francesco Rinuccini, living in Milan; 11. Mario Guiducci; 12. Niccolò Arrighetti; 13. Braccio Manetti; 14. Canon Niccolò Cini; 15. Earl Piero de' Bardi; 16. Filippo Salviati; 17. Jacopo Soldani; 18. Jacopo Giraldi; 19. Michelangelo Buonarroti; 20. Alessandro Sertini (11–20: deceased); and others.

Fuori di Padova poi, nel tempo delle vacanze di Studio, e prima nell'estate del 1605, il Ser.^{mo} D. Cosimo, allora Principe di Toscana, volle pur sentire l'esplicazione del suo Compasso, continuando poi il Sig.^r Galileo per molti anni in quella stagione ad instruire nelle matematiche il medesimo Serenissimo, mentre già era Gran Duca, e con l'Altezza Sua gl'altri Ser.^{mi} Principi D. Francesco e D. Lorenzo.

Tra i professori di matematica suoi discepoli, ne usciron cinque famosi lettori publici di Roma, Pisa e Bologna.^m A questi soleva dire ch'eglino con maggior ragione dovevano render grazie a Dio et alla natura, che gl'avesse dotati d'un privilegio sol conceduto a quei della lor professione, che era di potere con sicurezza giudicar del talento et abilità di quelli uomini i quali, applicati alla geometria, si facevano loro uditori; poi che la pietra lavagna, sopra la quale si disegnano le figure geometriche, era la pietra del paragone delli ingegni, e quelli che non riuscivano a tal cimento si potevano licenziare non solo come inetti al filosofare, ma com'inabili ancora a qualunque maneggio o esercizio nella vita civile.

⁶³⁰Quanto queste virtuose doti et eminenti prerogative, ch'in eccesso risplenderono nel Sig.^r Galileo, fossero in ogni tempo conosciute et ammirate dal mondo con evidenti dimostrazioni di stima, scorgesi dalli amplissimi onori di richieste e regali fattigli in varie occasioni da i più insigni litterati d'Europa, da i Ser.^{mi} Duchi di Parma, Baviera, Mantova e Modena, da i Ser.^{mi} Arciduchi d'Austria Leopoldo e Carlo,⁴⁶ da tanti Ill.^{mi} et Emin.^{mi} Prelati e Cardinali, dalle Ser.^{me} e Potentiss.^{me} Republiche di Venezia e d'Olanda, dalli invittissimi Re Vladislao di Pollonia e Gustavo di Svezia, dalla Maestà Catolica del Re di Spagna e dalli Augustissimi Imperadori Ridolfo, Mattia e Ferdinando, e da tanti altri Signori, Principi e Potentati. Scorgesi dalle lettere con le quali molti di questi a lui ricorrevano come ad oracolo, ricercandolo del parer suo intorno alle novità de' celesti discoprimenti e loro conseguenze, sopra varii effetti naturali e sopra conclusioni o dubbii filosofici, astronomici o geometrici: che se così fosse facile il far raccolta delle sue ingegnose risposte come si può dell'altrui proposte, certo è che si accumulerebbe un tesoro di inestimabil valore, per la novità delle dottrine e per la sodezza di quei concetti di che ell'eran sempre feconde.⁴⁷ Scorgesi in fine dalla stima e venerazione in che fu tenuto dal mondo tutto, poi che niun litterato di qualche fama, niun signore o principe forestiero, passò per Padova o per Firenze, che non procurasse di visitarlo in città o nella villa, dove egli fosse; et allora stimavano d'aver bene spesi i lor lunghi viaggi, quando, tornando alle patrie loro, potevano dire d'aver conosciuto un tant'uomo e avuto seco discorso: e a imitazione di quei nobili che fin dall'ultime regioni d'Europa si portavano a Roma sol per vedere il famoso Livio, quando per altro le grandezze di quella Republica trionfante non ve gli averebber condotti, quanti gran personaggi e signori da remote provincie a posta intrapresero per l'Italia il cammino per veder un sol Galileo!

m 1. D. Benedetto Castelli, in Pisa e Roma; 2. Sig.^r Niccolò Aggiunti, in Pisa; 3. Sig.^r Dino Peri, in Pisa; 4. D. Vincenzio Renieri, in Pisa; 5. Fra Buonaventura Cavalieri, in Bologna.

46 Leopoldo e Carlo *ABX*] Ferdinando, Leopoldo e Carlo *S*

47 *post* feconde *scr. in marg.* Scorgesi dall'applauso universale con che sono state ricevute l'opere sue, i suoi dogmi e dottrine, e dalle più famose accademie e nelle più celebri catedre lette, insegnate et abbracciate *B*

During the university holidays, away from Padua, and even earlier, in the summer of 1605, the Most Serene Duke Cosimo, who was then Prince of Tuscany, wished to have Galileo's compass explained to him. Subsequently, for many years, Galileo taught mathematics to Cosimo during the holidays, even after he became Grand Duke,[192] and together with him he taught the other Most Serene Princes Francesco and Lorenzo.[193]

Among the professors of mathematics who were his pupils, five became well known public lecturers in Rome, Pisa, and Bologna.[m] [194] Galileo used to tell them that, more than others, they had to thank God and nature, for they had a privilege granted only to those practising their profession, namely, to be able to assess with certainty the talent and skills of pupils who had devoted themselves to geometry. The blackboard slate upon which they drew geometric diagrams was the touchstone of their mind: those who failed in that endeavor could well be dismissed as not only incapable of philosophizing but also ill-suited to any business or office in civil life.

[630]The extent to which these virtuous skills and eminent traits, which abundantly shone in Galileo, were always appreciated and admired in the world by a clear display of its consideration we gather from the honor of requests[195] and gifts presented to him in various circumstances by the most eminent intellectuals of Europe, by the Most Serene Dukes of Parma, Bavaria, Mantua, and Modena; by the Most Serene Archdukes of Austria, Leopold and Karl;[196] by many most illustrious and eminent bishops and cardinals;[197] by the Most Serene and Powerful Republics of Venice and Holland;[198] by the Most Undefeated Kings Władisław of Poland and Gustaf of Sweden;[199] by His Catholic Majesty the King of Spain; by the Most August Emperors Rudolf, Matthias, and Ferdinand;[200] and by many other noblemen, princes, and powers. We gather this from the letters with which many of them appealed to Galileo as if to an oracle, asking for his opinion about novel heavenly discoveries and their consequences, about various natural effects, as well as about demonstrations and doubts, whether philosophical, astronomical, or geometric. If we could collect his ingenious responses as easily as we can the others' requests, we might certainly gather a priceless treasure, for the novelty of his views and the solidity of the ideas that could always be drawn from them.[201] Lastly, we may see it from the esteem and veneration in which he was held by the entire world, since no intellectual of any reputation, nor foreign nobleman or prince, ever passed through Padua or Florence without paying him a visit, either in the city or at his villa, wherever he was.[202] And when they returned to their homelands, they thought their long travels were well worth it, for they could claim they met and talked with so great a man. Just like those who traveled to Rome from the farthest regions of Europe to see the famous Livy, as if they had no other pressing reasons to visit that triumphant Republic,[203] how many great personalities and noblemen undertook the journey to Italy from the remotest provinces in order to see one Galileo?

m 1. Benedetto Castelli, in Pisa and Rome; 2. Niccolò Aggiunti, in Pisa; 3. Dino Peri, in Pisa; 4. Vincenzo Renieri, in Pisa; 5. Father Buonaventura Cavalieri, in Bologna.

Ma non potendo registrar qui tutti i segni di benevolenza e di stima con i quali fu questo sempre gradito et ammirato da' grandi, epilogando tutte le di [631]lui glorie in quest'unica e singolare, sovvenga all'A. V. che trovandosi egli nell'anno 1638 aggravato[48] da malattia nella sua abitazione di Firenze, l'istesso Ser.mo Gran Duca di Toscana oggi felicemente regnante, insieme con V. A. S., lo visitò sino al letto, porgendogli di propria mano soavissimi ristorativi, con dimorarvi sopra due ore; gustando, come sapientissimo Principe, di coltivar le sue nobili e curiose speculazioni con la conferenza e discorso del suo primario Filosofo. Esempio in vero di singolare affezzione verso un proprio vassallo, per il quale non men risplende un'eminente virtù in chi conferisce, che in chi riceve, onore sì glorioso.

Di simili visite fu ancor prima e dopo, come è ben noto all'A. V. S., più e più volte onorato dal medesimo Ser.mo Gran Duca[n] e da lor altri Ser.mi Principi, che, a posta movendosi di Firenze o dalla Villa Imperiale, si trasferivano in Arcetri, per godere della sapientissima erudizione di quel buon Vecchio, o per consolarlo nell'angustie dell'animo e nella sua compassionevole cecità.

Dicalo l'A. V. S., che più frequentemente delli altri si compiacque onorarlo con la maestà della sua presenza, in tempo in che ella, mirabilmente avanzandosi nelle scienze matematiche, dilettavasi comunicar seco quei pensieri che nello studio dell'opere di lui le sovvenivano, dando allora materia al gran Galileo di far quel giudizio ch'in oggi, vivendo, goderebbe vedere a pieno verificato; mentre egli a me più volte con stupore affermava di non aver mai incontrato, fra tanti suoi uditori, chi più di V. A. gli avesse dimostrato prontezza d'ingegno e maturità di discorso, da sperarne maravigliosi progressi non tanto nelle matematiche quanto nelle filosofiche discipline, e conseguentemente, secondo la di lui regola sopradetta, nelli affari importanti.[49]

Questo per ora è sovvenuto alla sterilità della mia memoria intorno a soggetto così fecondo, e tanto ho potuto raccogliere d'altrove, in tempo assai scarso, [632]delle più antiche notizie, e privo della maggior parte delli amici più vecchi di quel grand'uomo, che mi potevano somministrare maggior numero di virtuosi detti e memorabili azioni che risplenderono nel corso della sua vita. Compiacciasi non di meno l'A. V. S.ma di gradire per ora questa dovuta dimostrazione d'obbedienza et ossequio, con il quale io mi rassegno

Di casa, li 29 Aprile 1654.
Di V. A. Ser.ma

Umiliss.mo e Devotiss.mo Servo Oblig.mo
Vincenzio Viviani.

n Detto eroico di S. A., originato da queste visite: *Sempre ch'io avrò un Galileo, farò così.*

48 trovandosi—aggravato *ABX*] nelli 8 di Settembre del 1638, aggravato egli *S*
49 nelli affari importanti *BX*] ne i governi politici *AS*

Here I cannot record all evidence of the benevolence and respect with which Galileo was always welcomed and admired by great people. I shall summarize all of [631]his glories into this single and remarkable one: as Your Most Serene Highness might remember, in 1638[204] Galileo was lying ill in his house in Florence, and the Most Serene Grand Duke of Tuscany, who is still reigning today, together with Your Highness, paid him a visit, going to his bedside and also giving him most soothing restoratives. You stayed in his house for two hours, and as a most learned Prince you enjoyed nurturing your noble and inquiring thoughts through exchanges and conversation with your Chief Philosopher. This was a most remarkable display of affection toward one's servant, in which the virtue of him who bestows such an honor shines no less than that of the one who receives it.

As Your Most Serene Highness knows very well, the same Most Serene Grand Duke[n] honored Galileo with these visits very many times, before that one and after it. And so did other Most Serene Princes, who traveled purposely from Florence or the Royal Villa to Arcetri in order to enjoy the most learned scholarship of that good Old Man, or else to comfort him in the anxiety of his soul and his pitiful blindness.[205]

Let Your Most Serene Highness, who more frequently than others wanted to honor Galileo with the majesty of your presence, say that while you were making great progress in the mathematical sciences, you enjoyed reporting the thoughts that came to your mind when studying the works of the great Galileo; that gave him the grounds to express the judgment that, were he alive today, he would have been delighted to see so completely fulfilled. He told me many times that, to his amazement, he had never met, among those who attended his lectures, anyone who displayed a quickness of mind and maturity of argumentation at a level higher than Your Highness did. And he hoped for astonishing progress not so much in mathematics but, rather, in philosophy—and, consequently, according to that rule of his, in truly important matters.[206]

This much, so far, has the sterility of my memory allowed me to recollect on such a fertile subject. And this much was I able to gather from other sources, in a very short time, [632]about the events that took place before I met Galileo. I did so without the help of the older friends of that great man, who could have provided me with a larger number of the virtuous sayings and memorable actions that blazed forth in the course of his life. At least for the time being, may Your Most Serene Highness be pleased by this devout demonstration of my obedience and deference, with which I remain

From home, 29 April 1654.   The Most Humble and Devout, Obliged Servant
Of Your Most Serene Highness   Vincenzo Viviani

---

n After these visits, Your Highness bravely used to say: *As long as I have a Galileo, I will behave like that.*

[600]*Proemio.*

Lo scriver la vita de gl'huomini illustri non sarebbe per avventura necessario, se non dove le loro azioni, appoggiandosi a fondamenti caduchi, hanno bisogno del sostentamento degli scrittori.

Vive alcuno nelle voci del volgo e nelle menti degli uomini; ma se dalla penna dello scrittore non venga sostenuto, tosto se ne spegnerà la memoria.

Fondano altri la loro eternità ne' marmi, bronzi, obelischi e sontuosi edifizi, che sembrano immortali, e pure hanno lor morte, se da gli scrittori non gli fia prorogato la vita.

Viverà il Colombo con i suoi scoprimenti terrestri, assicurati da tal fondamento che ingombra quasi la metà della terra: ma se la navigazione per qualche mondano accidente fosse impedita, o per altra cagione interrotto il commercio, in pochi anni, obliato il fondamento, ne languirebbe la ricordanza, se però da diligente scrittore non ne fosse stato compilato la storia.

Grande e maravigliosa si può sicuramente dire l'accortezza e la fortuna del Sig.ʳ Galileo, che, aiutata dal suo divino intelletto, ha incontrato fondamento celeste. Onde con i suoi ammirabili discoprimenti, senza tema d'esser impediti o consumati dal tempo o nascosi alla vista e cognizione de' mortali, e spiegati con la sua singolar sapienza, s'è procacciato fama gloriosa e durabile quanto durerà l'universo.

Non era adunque d'huopo al Sig.ʳ Galileo ch'altri scrivesse la vita di lui per eternare la sua fama, fondata sopra la permanenza del cielo. Ma perché la generosa curiosità della maggior parte delli amatori delle buone lettere, invaghiti dell'eccellenza, chiarezza e novità della sua dottrina e de' suoi maravigliosi componimenti, si son mostrati ansiosi d'avanzarsi nella di lui cognizione e delle sue qualità, vita e costumi, hanno cagionato che io, forse con presun[601]tuoso ardire, mi sia messo a rappresentarne parte: non già ch'io creda d'accrescer un momento alla sua gloria immortale, ma per sodisfare, per quanto potrò, alla sete di quelli ch'hanno assaporato la sua unica e mirabile filosofia.

Ben è vero ch'io mi troverei grandemente ingannato se, scrivendo la di lui vita, presumessi d'inalzare le sue lodi; per che, se bene di sì eminente soggetto non si può dire senza lode, tuttavia lasciando questa cura a più sublime eloquenza, m'ingegnerò di spiegare semplicemente quel tanto che delle sue azioni fin ad ora ho saputo raccogliere.

Nacque dunque Galileo Galilei, nobil fiorentino, il 19 Febbraio del 1563 *ab Incarnatione*, nella città di Pisa, dove allora per domestici affari dimoravano i suoi genitori.

Il padre fu Vincenzio di Michel Angelo Galilei, gentiluomo versatissimo nelle matematiche e principalmente nella musica speculativa, della quale hebbe così eccellente cognizione, che forsi tra i teorici moderni di maggior nome non v'è stato fino al presente secolo chi di lui

[600]*Preface*

As I see it, writing the biography of eminent people would not be necessary unless their actions were based on shaky grounds and needed the support of biographers.

Some such men live in the words of common people, as well as in the minds of men—but their memory will soon fade away if a biographer's quill does not support it.

Others entrust their eternity to marbles, bronzes, obelisks, and lavish palaces, which look immortal—and still they perish if writers do not prolong their life.

Columbus will live: his discoveries of new lands are grounded on foundations that cover nearly half of the earth.[207] But if navigation, for some earthly reason, were hindered, or commerce ceased for some other motives, once the foundations were forgotten, the memory would die out in a few years if no diligent writer had drawn up their history.

We can rightfully say Galileo's discernment and fate were great and astonishing: assisted by his divine intellect, they set their foundations in the heavens. Thanks to his amazing discoveries—which have no risk of being diminished or worn out by time, or hidden from the sight and knowledge of mortals, and are explained by his truly unique wisdom—he has earned glorious and eternal fame, for as long as the universe lasts.

Thus, Galileo did not need anyone to write his biography in order to immortalize his fame, grounded as it is in the everlasting heavens. However, owing to their noble curiosity, the vast majority of amateurs of good literature, attracted by the excellence, clarity, and novelty of his theories and of his extraordinary works, wanted to know more about him, his qualities, his life and habits, and so caused me, in my presumptuous [601]audacity, to write about some of them. I do not believe that I will add anything to his immortal fame, but I shall try my best to quench the thirst of those who savored his unique and extraordinary philosophy.

If by writing his biography I presumed to sing his praises, I would gravely err, for although it is impossible to speak of such an eminent subject without praise, I shall leave it to others to do with more sublime eloquence. I will strive to explain plainly as many of his deeds as I have been able to gather.

Galileo Galilei, Florentine nobleman, was born on 19 February 1563 *ab Incarnatione*, in the city of Pisa, where his parents resided, looking after family matters.

His father was Vincenzo Galilei, son of Michelangelo, a gentleman very well versed in mathematics, and especially theoretical music, of which he had such an excellent knowledge that among the most famous modern theorists, up to the present century, perhaps none

Notes

1. These first few lines are in *B*, f. 24*r*. On f. 24*v*, we read: "To the Most Serene Prince Leopoldo of Tuscany. Various information for the story of the life of Galileo Galilei, Florentine nobleman, Lyncean Academician, Chief Philosopher and Mathematician of the Most Serene Highnesses of Tuscany, gathered by Vincenzo Viviani at the request of the Most Serene Prince Leopoldo" ("Al Ser.^mo Principe Leopoldo di Toscana. Notizie varie per la descrizione della vita del Sig.^r Galileo Galilei, Nobil fiorentino, Accademico Linceo, Primo Filosofo e Matematico dell'Altezze Ser.^me di Toscana, raccolte ad instanza del Ser.^mo Principe Leopoldo da Vincenzio Viviani").

2. Leopoldo de' Medici (1617–1675), son of Grand Duke Cosimo II and Maria Magdalena of Austria, and brother of Ferdinando II de' Medici (Grand Duke of Tuscany from 1621 to 1670). Named a cardinal in 1667 by Pope Clement IX, he was a scholar and a patron of the arts and sciences: in 1641, he was named a member of the Accademia della Crusca, and in 1657, together with his brother Ferdinando, he founded the Accademia del Cimento, to promote observation of nature through the Galilean method. Members of the latter Academy (whose motto was "provando e riprovando," namely, "trying and trying again") included Viviani, Giovanni Alfonso Borelli, Francesco Redi, Nicolas Steno, Carlo Dati, Candido Del Buono, and Alessandro Marsili. When the meetings ended, Lorenzo Magalotti, the Academy's secretary, published *Saggi di natvrali esperienze* (1667), a report and appraisal of the Academy's activity: experiments first concerned air pressure and the effects of vacuum, the freezing of liquids, the properties of heat, sound and light propagation, magnetic and electric phenomena; later on, researches focused on mathematics, acoustics, thermodynamics, hydrostatics, celestial mechanics, human and natural physiology, and optics.

3. According to the style *ab Incarnatione* (short for *ab Incarnatione Domini*), in use in Tuscany and elsewhere in Europe before the Gregorian reform of the calendar (implemented in 1582), the year started or ended on March 25, the day of the festivity of the Incarnation of Christ (anniversary of the Annunciation to the Virgin Mary and conception of Jesus). The style had two variants: according to the use in Florence (as well as Siena, Lucca, and Prato), the year began on March 25; according to the use in Pisa, by contrast, the year ended on March 25. The reason for the difference is that there was no year 0: the Annunciation took place on March 25 of the year 1 BC; as Jesus was born on December 25, the following March 25, according to the Pisan calendar, marked the end of year 1, and March 26 the first day of year AD 2; according to the Florentine calendar, by contrast, the first March 25 after Jesus's birthday marked the beginning of year AD 1. Viviani was writing in Florence, and adopted the Florentine style: his February 19, 1563 (i.e., 1564, *stylo Pisano*) is to be understood as February 19, 1564, of the Gregorian calendar (*stylo Romano*). Viviani conflates the date of Galileo's birth (on February 15: see *OG* 19, pp. 23–24) with that of his baptism (on February 19, in the Pisa cathedral: see ibid., p. 25): as Galileo himself noted when preparing his own horoscope, he was born on February 15, 1564 (i.e., 1563 *ab Incarnatione stylo Florentino*; 1564 *stylo Pisano*). There, Galileo wrote his birth date as "1564. 15. febr. h. 22. 30" (*Gal.* 81, 7*r*, line 1): in the Florentine system, the new day began at sunset (of the previous day, according to our calendar); so he says he was born 22 hours and a half after the sunset (at 5:30 p.m.) of February 14, that is, at 4 p.m.; see ibid., line 2: "15. febr. h. 4. p.m." See also next note.

4. Instead of this paragraph, *S* has: "Galileo Galilei, Florentine nobleman, was born in Pisa, on Tuesday, 15 February 1564, Roman style, at 4 p.m. (according to others, at 3.30 p.m.); there, in the cathedral, he was baptized on Saturday, February 19, with Pompeo and Averardo de' Medici as his godfathers. That day, 15 February 1564, precedes by three days that in which the divine Michelangelo Buonarroti died in Rome, on 18 February 1564, Roman style." On Averardo de' Medici (1518–1601) and Pompeo de' Medici (16th century) see *OG* 19, p. 25, and *OGA* 1, p. 233.

5. Vincenzo Galilei (1520–1591) was an accomplished musician, a musical theorist, and a lute player; he was one of the most eminent and active members of the Florentine Camerata (also known as "Camerata de' Bardi"), a group of humanists, musicians, and intellectuals who gathered under the patronage of Count Giovanni de' Bardi (1534–1612), to discuss trends in the arts, especially music and drama. Vincenzo clashed with the classic tradition, advocated by his master Gioseffo Zarlino (1517–1590), the most important musical theorist of the sixteenth century. He especially focused on consonances and dissonances, thus laying the foundations for baroque music. Among his works were *Fronimo* (1568), *Dialogo della musica antica, et della moderna* (1581), and *Discorso intorno all'opere di messer Gioseffo Zarlino da Chioggia* (1589). Members of the Camerata included Giulio Caccini, Jacopo Peri, and Piero Strozzi (musicians), Girolamo Mei, and Ottavio Rinuccini (literary men).

6. *S* has: "[Vincenzo] had from his wife, Giulia Ammannati di Pescia, from the ancient and eminent Ammannati family, from Pistoia, several children." Giulia Venturi degli Ammannati (1538–1620) was from Villa Basilica, near Pescia: see *OG* 19, pp. 17–20. She married Vincenzo on July 5, 1562 (see *OG* 19, pp. 17–18), and moved in with him in Pisa. Besides Galileo, their firstborn, they had six children: Benedetto, Virginia (15713–1623), Anna, Michelangelo (1575–1631), Livia (b. 1578), and Lena. At the death of Vincenzo, it was up to the firstborn, Galileo, to support the whole family.

7. After the family moved to Florence, Vincenzo's hometown, in 1574, Galileo began to study Latin, Greek, and other basic subjects under a teacher of whom it is known only that he charged 5 lire a month: see Muzio Tedaldi's letters to Vincenzo Galilei (January 13 and February 9, 1574) in *OG* 10, nos. 1*–2*, pp. 17–18. On the basis of manuscripts in his possession, Nelli identified the teacher as Jacopo Borghini, from Dicomano: see Nelli, *Vita e commercio letterario di Galileo Galilei* (1793), vol. 1, pp. 26–27.

8. Vallombrosa is a Benedictine monastery (elevated to abbey in 1713) in Reggello, some 30 kilometers southeast of Florence. It was founded in 1038 by the nobleman Giovanni Gualberto (canonized in 1193) and became the house of the Congregation of Vallombrosans, or the Vallumbrosan Order. Galileo entered the monastery at the age of 11, as a novice, and received part of his education there. The library was excellent, and the monks were distinguished for their learning in theology, as well as rhetoric, mathematics, cosmology, and astrology; Galileo's lifelong astrological interests may well have begun here. Among his boyhood friends was Orazio Morandi, who later became abbot general of the order (1617–1621). Galileo always remained in friendly contact with Morandi, who also supported him in the quarrel with Francesco Sizzi over the discovery of Jupiter's satellites; see Sizzi, *ΔIANOIA astronomica, optica, physica* (1611). Morandi turned the monastery into a center for the meeting of intellectuals and the study of astrology. Although the Catholic Church was officially hostile to astrology, Pope Urban VIII had a personal interest in astrological prognostications and tolerated Morandi until he started to foretell the death of the pontiff, thus stirring the ambitions of many cardinals of the Roman Curia, in a difficult moment for the pope. Galileo was Morandi's guest in Rome, in 1630, while he was seeking the imprimatur for the *Dialogue*, and shortly after Galileo's departure Morandi was arrested. He died in prison later that year; on April 1, 1631, Urban VIII issued the bull *Contra astrologos iudiciarios*, by which astrologers were sentenced to death.

9. Galileo competed with his younger brother Michelangelo, who excelled in playing the lute and was known as a composer and lutenist both in Italy and abroad, especially in Bavaria, Poland, and Lithuania.

10. Lodovico Cardi, known as "il Cigoli" (from his birthplace, a suburb of San Miniato, in the province of Pisa, 1559–1613), was a painter and architect of the late Mannerist and early Baroque period, active in Florence and Rome. He was taught in perspective and mathematics by the same Ostilio Ricci who was the early teacher of Galileo. The latter's senior by five years, Cigoli remained devoted to him and corresponded with him extensively. While in Rome, in 1610, after the publication of the *Sidereal Messenger*, Cigoli made careful, independent observations of the sunspots, and in his last work (the *Assumption of the Virgin* in the dome of the papal chapel in Santa Maria Maggiore) paid tribute to Galileo by representing the Moon under the Virgin's feet as revealed by Galileo's telescope. Agnolo di Cosimo, known as "Bronzino" (possibly from his bronze-colored hair, 1503–1572), was a Florentine Mannerist painter who was active in Florence and worked mainly for the Grand Duke Cosimo I de' Medici. Domenico Cresti (or Crespi), known as "il Passignano" (from the name of the district of Tavarnelle Val di Pesa, near Florence, where he was born, 1559–1638), was a painter of the late-Renaissance or Counter-Mannerism style; he was educated by the Vallombrosan monks. Jacopo Chimenti, known as "l'Empoli" (from his hometown, 1551–1640), was a Florentine Reformist painter who worked mainly in Empoli, his native city.

11. *A* and *S* have: "eighteen."

12. Galileo's father's great-grandfather's brother, Galileo Galilei (1370–before 1451), was a highly respected physician and professor of medicine at the University of Florence. He was buried in the Santa Croce Basilica, in Florence: the inscription on the gravestone (which can still be seen on the floor of the central nave, not far from Galileo's funeral monument) reads: "In his lifetime, Galileo Galilei, once Bonaiuti, buried hereunder, achieved the highest level in Philosophy and Medicine, and taught them; in his capacity of highest civil officer, he showed the greatest care for the state. In the venerable memory of his well-lived life, his son Benedetto affectionately brought about this grave, for his own father, for himself and for his descendants" ("TEMPORIBVS HIC SVIS PHILOSOPHIE / ATQVE MEDICINE CVLMEN FVIT ET MAGISTER / GALILEVS DE GALILEIS OLIM BONAIVTIS QVI / ETIAM SVMMO IN MAGISTRATV MIRO / QVODAM MODO REM PVBLICAM DILEXIT / CVIVS SANCTE MEMORIE BENE ACTE / VITE PIE BENEDICTVS FILIVS HVNC TVMV / LVM PATRI SIBI SVISQVE POSTERIS EDIDIT").

13. Muzio Tedaldi (ca. 1520–1591), Vincenzo's good friend, took care of his family during Vincenzo's frequent absences from Pisa, until the whole family moved to Florence, in 1574. Tedaldi worked as customhouse officer in Pisa (*Doganerius duanæ civitatis Pisanæ*), and was the godfather of Galileo's younger brother, Michelangelo (at his christening, on December 21, 1575). In 1578, Tedaldi married Bartolomea Ammannati, Galileo's cousin on his mother's side.

14. *A* and *S* have: "often."

15. Horace, *Epistulae* II 1, 84–85: "because they hold it a shame to yield to their juniors, and / to confess in their old age that what they learned when they were beardless should be made away with" ("[. . .] quia turpe putant parere minoribus, et quae / imberbes didicere senes perdenda fateri").

16. Galileo remained in Pisa until 1585, as can be gathered from the list of his father's expenses to support him there, which ends in May 1585: see *OG* 19, p. 35.

17. Antonio Favaro edited Galileo's early writings and published them under the collective title *Iuvenilia*, in *OG* 1.

18. This is the origin of the story about Galileo's apparent discovery of the (approximate) isochronism of the pendulum's motion. In two manuscripts related to the letter to Prince Leopoldo on the application of a pendulum to clocks (see pp. 000), Viviani referred Galileo's observation of the swinging lamp to two different years, 1580 (*Gal.* 227, f. 60r) and 1583 (*Gal.* 228, f. 6v). See, for example, *Gal.* 227, f. 60r: "That of the pendulum is one of Galileo's oldest observations and discoveries about nature. It dates back, roughly, to 1580, when he was a student in Pisa: one day, while he was in the city's cathedral, he happened to notice a swinging lamp, hanging from a very long rope. As a child, he had also practiced music (under the guidance of the great Vincenzo, his father, who later wrote a very learned dialogue on ancient and modern music), and had his mind imprinted with the equality of times that governs music: accordingly, pondering that motion, it was easy for him to see it as isochronous, both in the large swings at the beginning of its motion, and in the shorter ones at its end, when approaching rest" ("Questa del pendolo si è una delle più antiche osservazioni e scoperte in natura del Galileo, e fu circa all'anno 1580 quando egli era studente a Pisa, nel trovarsi egli un giorno in quel Duomo, dove si abbatté di vedere lasciata in moto una lampada, pendente da una lunghissima corda; e come quello che da giovanetto s'era anche esercitato nella musica (sotto la disciplina del quel gran Vincenzio, suo padre, che sì dottamente poi scrisse in dialogo della musica antica e moderna), e che perciò aveva impressa nell'animo l'egualità de' tempi co' quali essa si regola, riflettendo a quel moto, gli fu facile il giudicarlo in mente sua equitemporaneo, sì nell'andate larghe al principio del moto, come nelle strette nel fine verso la quieta"). As far as we know, Galileo first mentioned the discovery of the isochronism of pendulums in his letter to Guidobaldo del Monte, on 29 November 1602: see *OG* 10, no. 88, pp. 97–100. See also the discussions in the *Dialogue*, pp. 230–231 and 449–450 (*OG* 7, pp. 256–257 and 474–476); and the *Two New Sciences*, pp. 87–88, 97–99, and 226–227 (*OG* 8, pp. 128–129, 139–141, and 277–278).

19. Viviani is here referring to the *pulsilogium*, or pulse meter, a device used to measure the pulse rate: in fact, this invention is due to the Venetian physiologist and physician Santorio Santorio (1561–1636), a pioneer of the application of physical measurement devices to medicine. He graduated in Padua in 1582 and worked for several years in Croatia; he returned to Padua in 1599, and became a close friend of Galileo, Paolo Sarpi, and Giovanni Francesco Sagredo. See Santorio, *Methodi vitandorvm errorum omnium qui in arte Medica contingunt* (1603), Book V, ch. 7, f. 109r–v.

20. See *Dialogue*, p. 117, in the margin: "Nature does not act by means of many things when she can act by means of few" ("La natura non opera con molte cose quello che può operar con poche": *OG* 7, p. 143); an analogous expression is also in the body of the text.

21. Born in Urbino, Federico Commandino (1509–1575) studied at the universities of Padua and Ferrara, where he received his doctorate in medicine. He translated the works of ancient mathematicians (Euclid, Aristarchus, Apollonius, Hero, Ptolemy, Pappus, and Eutocius) and was responsible for the publication of the works of Archimedes (*Archimedis opera nonnulla*, 1558), thus playing a key role in the renaissance of mathematics in Europe in the sixteenth century. In 1565 Commandino published the *Liber de centro gravitatis solidorum*, on the center of gravity of solids. Among his pupils was Guidobaldo del Monte, a good friend and supporter of Galileo's. Commandino frequently corresponded with Francesco Maurolico (1494–1575), of Messina, one of the most eminent mathematicians of his time. Maurolico made important contributions to the fields of geometry, optics, mechanics, music, and astronomy, and edited works of Greek authors, including Archimedes, Apollonius, Theodosius, and Serenus. His correspondents included Christopher Clavius (1538–1612), a prominent German Jesuit mathematician from the Roman College, who supported Galileo in the early stages of his career.

22. Ostilio Ricci (1540–1603), of Fermo, in the Marches, was a mathematician of the school of Niccolò Tartaglia (ca.1499–1557). He taught at the Academy of the Arts and Drawing, founded by Giorgio Vasari in Florence, in 1550. Ricci was the court mathematician of the Grand Duke Francesco I in Florence, in 1580, when he met Galileo, whom he taught Euclid and Archimedes.

23. *A* and *S* have: "twenty-two."

24. It is the so-called Pythagoras's theorem: "In right-angled triangles the square of the side subtending the right angle is equal to the sum of the squares of the sides containing the right angle" (*Elements*, Book I, proposition 47).

25. The works of Hippocrates and Galen, as well as *The Canon of Medicine*, compiled by Avicenna (980–1037) and largely based on Galen's works, were the standard texts for students of medicine in European universities in the sixteenth century.

26. *A* and *S* have: "in a few months."

27. The 1543 edition of Archimedes's works, edited by Tartaglia and published in Venice by Venturino Ruffinelli, offered William of Moerbeke's Latin translations of *On the Equilibrium of Planes* (*De centris gravium vel de planis æquerepentibus*) and Book I of *On Floating Bodies* (*De insidentibus aquae*); there was no indication that the latter was only Book I of Archimedes's work, though, nor that a second book existed. Moerbeke's translation of Book I, together with Book II (from the legacy of Tartaglia), were later reprinted by Curzio

Troiano Navò in Venice, in 1565. That same year Federico Commandino published in Bologna his own revision of Moerbeke's translation of Archimedes's *On Floating Bodies: De iis quae vehuntur in aqua* (1565).

28. According to Vitruvius (*De architectura*, IX 9–12), Hiero II of Syracuse suspected he was being cheated by a goldsmith to whom he had supplied the gold to make a votive golden crown, and asked Archimedes to check whether all the gold had been used, as agreed. In complying with the tyrant's order, Archimedes discovered the principle of displacement needed to measure the density of the crown, thereby revealing the fraud of the goldsmith, who had replaced some of the gold with silver. In the 1580s Galileo offered a solution to the problem of Hiero's crown different from that commonly related by the Archimedean tradition: he devised an instrument (similar to the Westphal balance) to determine the specific gravity of metals, and described it in an early work, the *Balance*: see *OG* 1, pp. 215–220. The *editio princeps* of the *Balance* is included in Odierna, *Archimede redivivo con la stadera del momento* (1644), pp. 1–8.

29. *A* and *S* add: "(which he devised [*S*: "as said"] with extraordinary ingenuity and ease)."

30. Guidobaldo del Monte, or Guidubaldo Bourbon del Monte (1545–1607), was a prominent Italian mathematician and astronomer. Born in Pesaro, he inherited the title of Marchese (Marquis) from his father, Ranieri, whom the Duke of Urbino had honored with nobility for his role as a soldier and also as the author of two books on military architecture. Guidobaldo enrolled in the University of Padua in 1564, but crucial were his later studies of mathematics under Francesco Commandino. One of the most eminent experts of mathematics and mechanics of the sixteenth century, he published, among other works, the *Mechanicorvm liber* (1577), the *Planisphaeriorvm vniversalivm theorica* (1579), and the most influential *Perspectivae libri sex* (1600). In the late 1580s, Galileo submitted to Guidobaldo a few theorems on the center of gravity of solids (as recalled by Salviati in the *Two New Sciences*, pp. 259–260; *OG* 8, p. 313), earning his admiration and friendship. Del Monte held Galileo in high esteem (see, for example, Del Monte to Galileo, 28 May 1588, in *OG* 10, no. 17, p. 34), and together with his brother Francesco Maria (who was created a cardinal in 1588) supported him for a chair in mathematics at Pisa and Florence. In 1589, after unsuccessful attempts at Pisa and Florence (see Galileo to Del Monte, 16 July 1588, in *OG* 10, no. 19, p. 36; and Del Monte to Galileo, 22 July 1588, ibid., no. 20, p. 37), Guidobaldo was instrumental to the appointment of Galileo to the chair of mathematics at the University of Pisa (see Del Monte to Galileo, 3 August 1589, in *OG* 10, no. 27, p. 41). Galileo did not like the position, or at least the salary that came with the job, so Del Monte helped him get his next appointment at Padua (see Del Monte to Galileo, 11 February 1592, no. 35, p. 47; and 10 January 1593, in *OG* 10, no. 45, p. 54). Guidobaldo always remained critical of Galileo's discovery of the isochronism of the pendulum.

31. *A* has: "twenty-four."

32. Galileo's *Theoremata circa centrum gravitatis solidorum* (in *OG* 1, pp. 187–208), supplementing Commandino's *Liber de centro gravitatis solidorum* (1565), dating to the mid-1580s, but were published only in 1638, as an appendix to *Two New Sciences*, pp. 261–280 (*OG* 1, pp. 187–208). See also Galileo's recollections in his letter to Élie Diodati, on 12 December 1636 (in *OG* 16, no. 3398*, p. 524).

33. Ferdinando I de' Medici (1549–1609), the fifth son of Grand Duke Cosimo I, was Grand Duke of Tuscany from 1587 to 1609, having succeeded his older brother Francesco I. Also a cardinal, he left the office in 1589, when he married Christina of Lorraine.

34. Giovanni I de' Medici (1567–1621), natural son of the Grand Duke of Tuscany, Cosimo I. A military commander and diplomat (he was Florentine ambassador in Madrid, and Commander-in-Chief of the Republic of Venice in 1616–1617), he was also a poet, a painter, and an architect, and a member of the Florentine Academy; also, he played a major role as broker for the circulation of art objects and political information on behalf of the Medici family. A pupil of Bernardo Buontalenti's, he collaborated with Matteo Nigetti in the design of the Principi Chapel in the church of San Lorenzo, in Florence. In 1589, he supported Galileo's appointment to the chair of mathematics at the University of Pisa, but later (as Viviani himself recalls shortly afterward, however implicitly, since the *Racconto istorico* is addressed to Prince Leopoldo, a member of the Medici family) clashed with him over the construction of a hydraulic machine Giovanni had devised to empty the wet dock of Leghorn; eventually, this contributed to Galileo's resigning his chair and taking up an appointment at the University of Padua. In 1611, after Galileo had returned to Florence, Giovanni played a role in the controversy on floating bodies, siding with Lodovico delle Colombe against Galileo. Delle Colombe's *Discorso apologetico* (1612)—in which Galileo is criticized for his attack on Aristotle, and for his claim (following Archimedes) that bodies' floating or sinking depends on the difference between their specific gravity and that of the medium, rather than on their shape—is dedicated to Giovanni de' Medici. For the whole controversy, also involving Vincenzo Di Grazia and Benedetto Castelli, see *OG* 4, pp. 311–789.

35. "Ignorance of motion is ignorance of nature": it is one of the axioms of Aristotle's physics. See *Physica*, III 1, 200 b 12–15: "Nature is a principle of motion and change, and it is the subject of our inquiry. We must therefore see that we understand what motion is; for, if it were unknown, nature would be unknown, too" ("Ἐπεὶ δ᾽ ἡ φύσις μέν ἐστιν ἀρχὴ κινήσεως καὶ μεταβολῆς, ἡ δὲ μέθοδος ἡμῖν περὶ φύσεώς ἐστι, δεῖ μὴ λανθάνειν τί ἐστι κίνησις· ἀναγκαῖον γὰρ ἀγνοουμένης αὐτῆς ἀγνοεῖσθαι καὶ τὴν φύσιν").

36. See, for example, Galileo's *Letter to Castelli*, p. 50: "whatever sensory experience places before our eyes or necessary demonstrations prove to us concerning natural effects should not in any way be called into question on account of scriptural passages whose words appear to have a different meaning, since not every statement of the Scripture is bound to obligations as severely as each effect of nature" ("quello de gli effetti naturali che o la sensata esperienza ci pone innanzi a gli occhi o le necessarie dimostrazioni ci concludono, non debba in conto alcuno esser revocato in dubbio per luoghi della Scrittura ch'avesser nelle parole diverso sembiante, poi che non ogni detto della Scrittura è legato a obblighi così severi com'ogni effetto di natura": *OG* 5, p. 283). See also the *Letter to Christina*, p. 93 (*OG* 5, p. 317).

37. See Aristotle, *Physica*, IV 8, 215 a 24–216 a 20; see, in particular, 216 a 13–16: "We see that bodies which have a greater impulse either of weight or of lightness, if they are alike in other respects, move faster over an equal space, and in the ratio which their magnitudes bear to each other" ("ὁρῶμεν γὰρ τὰ μείζω ῥοπὴν ἔχοντα ἢ βάρους ἢ κουφότητος, ἐὰν τἆλλα ὁμοίως ἔχῃ [τοῖς σχήμασι], θᾶττον φερόμενα τὸ ἴσον χωρίον, καὶ κατὰ λόγον ὃν ἔχουσι τὰ μεγέθη πρὸς ἄλληλα"). See also *De caelo*, I 7, 275 a 32-b 2.

38. See *On Motion*, pp. 26–38, and particularly pp. 31–32: "It is therefore clear that, if we know the ratio of the speeds of those bodies that differ only in weight, but not in size, we also know the ratios of those that differ in every other way. And so, in order to find this ratio and to show, in opposition to Aristotle's view, that also for bodies of different material this ratio is not equal to the ratio of their weights, we shall prove certain propositions. On these depends the outcome not only of this investigation but also of the investigation of the ratio of the speeds of the same body moving in different media" ("Patet igitur quomodo, data proportione motuum eorum mobilium quae differunt tantum gravitate et non mole, dentur etiam proportiones eorum, quae quovis alio modo differant. Ut igitur proportionem hanc inveniamus et, contra Aristotelis sententiam, ostendamus, nullo pacto mobilia, etiam si diversae speciei, proportionem suarum gravitatum servare, ea demonstrabimus ex quibus non solum huius quaestionis, verum etiam et quaestionis de proportione motuum eiusdem mobilis in diversis mediis, exitus pendet [. . .]": *OG* 1, p. 267; see also pp. 262–273).

39. Viviani's account is the only source for this story, undoubtedly one of the most famous and controversial anecdotes in the history of science. In fact, Galileo's *On Motion* offers a number of references to experiments performed from a "tower," or "high tower" (*turris*, or *alta turris*): see *On Motion*, pp. 27, 38, 87, 97, 101, 107, and 127 (*OG* 1, pp. 263, 273, 317, 326, 329, 334, and 406, respectively). None of these passages explicitly refers to the Leaning Tower of Pisa. However, according to Viviani, the experiment was performed when Galileo was a professor at the University of Pisa, between 1589 and 1592; and although the dating of the annotations that make up Galileo's *On Motion* is uncertain, scholars generally agree this work was written ca. 1590. Moreover, in the seventeenth century, other professors in Pisa did perform experiments from the top of the Leaning Tower: Giorgio Coresio, a professor of Greek, in 1612; Vincenzo Renieri, a follower of Galileo's and a friend of Viviani's, professor of mathematics, in 1641; and Carlo Rinaldini, a professor of philosophy and collaborator of Viviani's (see *OG* 4, p. 242; *OG* 18, pp. 305–306; and *ODG* 1, p. 341, respectively). Perhaps Galileo did drop things off the Leaning Tower of Pisa (or from another tower in the town, or else a high building) to test this theory, too. The story might have some truth in it, then, although Viviani might have simply embellished an actual experiment by presenting it in the form of a realistic anecdote, thereby conforming to the genre and style to which the *Racconto istorico* belonged.

40. In both *A* and *S*, the sentence continues after a colon: "all this is treated at length in the last [*S*: in the above-mentioned] *Dialogues on the New Sciences*". See the *Two New Sciences*, pp. 66–67: "SALVIATI: Then if we had two movables whose natural speeds were unequal, it is evident that were we to connect the slower to the faster, the latter would be partly retarded by the slower, and this would be partly speeded up by the faster. Do you not agree with me in this opinion? SIMPLICIO: It seems to me that this would undoubtedly follow. SALVIATI: But if this is so, and if it is also true that a larger stone is moved with eight degrees of speed, for example, and a smaller one with four [degrees], the joining both together, their composite will be moved with a speed less than eight degrees. But the two stones joined together make a larger stone than that first one which was moved with eight degrees of speed; therefore this greater stone is moved less swiftly than the lesser one. But this is contrary to your assumption. So you see how, from the supposition that the heavier body is moved more swiftly than the less heavy, I conclude that the heavier moves less swiftly" ("SALVIATI: Quando dunque noi avessimo due mobili, le naturali velocità de i quali fussero ineguali, è manifesto che se noi congiugnessimo il più tardo col più veloce, questo dal più tardo sarebbe in parte ritardato, ed il tardo in parte velocitato dall'altro più veloce. Non concorrete voi meco in quest'opinione? SIMPLICIO: Parmi che così debba indubitabilmente seguire. SALVIATI: Ma se questo è, ed è insieme vero che una pietra grande si muova, per esempio, con otto gradi di velocità, ed una minore con quattro, adunque, congiugnendole amendue insieme, il composto di loro si moverà con velocità minore di otto gradi: ma le due pietre, congiunte insieme, fanno una pietra maggiore che quella prima, che si moveva con otto gradi di velocità: adunque questa maggiore si muove men velocemente che la minore; che è contro alla vostra supposizione. Vedete dunque come dal suppor che 'l mobile più grave si muova più velocemente del men grave, io vi concludo, il più grave muoversi men velocemente": *OG* 8, pp. 107–108).

41. Most likely, Giovanni de' Medici: see note 34. See also Gherardini, pp. 000.

42. Giuseppe Moleti, or Moleto (1531–1588), was born in Messina, Sicily, where he studied medicine and mathematics, the latter as a pupil of the mathematician Francesco Maurolico. He moved to Venice in about 1556, where he made the acquaintance of the Paduan bibliophile and polymath Gian Vincenzo Pinelli (1535–1601), who became a lifelong friend and correspondent. In 1570, he took the position of mathematics tutor of Vincenzo, the son of Guglielmo Gonzaga, Duke of Mantua, a position he held for seven years. By May of 1577, Moleti had left Mantua to take up the chair of mathematics at the University of Padua left vacant by the death of Pietro Catena, in 1576. A renowned mathematician, Moleti was consulted by Pope Gregory XIII for the reform of the calendar: he published the *Tabvlae Gregorianae Motuum Octauæ Sphæræ ac Luminarium* (1580), which included *De corrigendo Ecclesiastico Calendario Libri duo*. At Padua, he taught mainly geometry, astronomy, and optics, but he also lectured on mechanics, and, after Catena, he was the second ever to lecture on the pseudo-Aristotelian *Mechanical Problems*; Galileo was to follow suit. In his unfinished *Dialogue on Mechanics* (1576, published in Laird, *The Unfinished Mechanics of Giuseppe Moletti* (2000)), in two Days, he describes an experiment in which two balls, "one of 20 pounds of lead and the other of 1 pound, also of lead," are let fall from "a tall tower," reporting that "both arrive at one and the same time, even if the test were done not once but many times. But what is more, a ball of wood, either larger or smaller than one of lead, let fall from the same height at the same time as the lead ball, would descend and touch the earth or ground at the same moment in time" (Laird, *The Unfinished Mechanics of Giuseppe Moletti* (2000), p. 147; "[…] se dalla cima d'un'alta torre noi lascieremo venir giù due palle, l'una di piombo di 20 libre et l'altra parimente di piombo di una libra", "[…] vengono tutti in uno stesso tempo, et di ciò se n'è fatta la pruova non una volta ma molte. Ma che è di più, che una balla di legno, ò più ò men grande d'una di piombo, lasciata venir giù d'una stessa alteza nello stesso tempo con quella di piombo, descendono ò toccano la terra ò 'l suolo nello stesso momento di tempo": ibid., pp. 146, 148). Galileo sent Moleti his work on the centers of gravity of solids (see *OG* 1, pp. 182–183), a copy of which survives at the Biblioteca Ambrosiana, in Milan, as part of Pinelli's collection, who acquired Moleti's papers at the time of his death.

43. See *OG* 19, pp. 111–112; his initial yearly salary was 180 florins.

44. See, for example, the privilege he was granted for a machine that could lift water: *OG* 19, pp. 126–129.

45. Galileo's *Breve instruzione all'Architettura militare* (in *OG* 2, pp. 15–75) and *Trattato di fortificazione* (ibid., pp. 77–146); *Trattato della sfera ovvero cosmografia* (ibid., pp. 211–255); the works on gnomonics and optics—two subjects Galileo taught in Padua for several years—are now lost. Galileo's *On Mechanics*, which he used for his lectures at Padua, was translated into French by Mersenne as *Les Méchaniques de Galilée* (Paris: Henri Guénon, 1634; critical edition by Bernard Rochot, Paris: Vrin, 1966). A French theologian, philosopher, mathematician, and music theorist, Marin Mersenne (1588–1648) was at the heart of an extended network of scholars in the first half of the sixteenth century. This included René Descartes, Gilles Personne de Roberval, Nicolas-Claude Fabri de Peiresc, and Constantijn Huygens, to mention but a few. Mersenne also translated the *Two New Sciences* (*Les novvelles pensées de Galilée*, Paris: Henri Guénon, 1639, and Paris: Pierre Rocolet, 1639; critical edition by Pierre Costabel and Michel-Pierre Lerner, Paris: Vrin, 1973, 2 vols.). Luca Danesi, or Danese (1598–1672), from Ravenna, studied humanities and graduated *in utroque iure*, after which he devoted himself to mathematics and architecture. Danesi's edition of Galileo's work on mechanics (*Della scienza mecanica e dell'vtilità Che si traggono da gl'Istromenti di quella. Opera Cauata da manoscritti dell'Eccellentissimo Matematico Galileo Galilei dal Cavallier Lvca Danesi da Ravenna*, Ravenna: Stamperia Camerale, 1649) is, in fact, the first edition of Galileo's work: Danesi, however, did not know Galileo's original manuscript, but a different one, which he supplemented with new material, so much that the work he eventually published was later included in Danesi's own collected works: see Danesi, *Opere* (1670), part IV, pp. 1–67.

46. In both *A* and *S*, the text continues with these words: "In fact, this innate generosity in sharing his own works, inventions, and novel ideas with anybody, unconcerned, was often reciprocated with as much ingratitude and insolence: indeed, there were people who attempted to disdainfully diminish them, or else to pride themselves with them, as if they were the products of their own wit."

47. According to Viviani, Galileo's invention would date to 1593–1597; from Benedetto Castelli to Ferdinando Cesarini, 20 September 1638 (*OG* 17, no. 3786*, p. 377), we gather that the invention should be traced back to 1603, at the latest. However, from Galileo to Cesare Marsili, 25 April 1626 (*OG* 13, no. 1776, p. 320), we conclude that Galileo invented thermometers around 1606.

48. Ferdinando II de' Medici (1610–1670), Grand Duke of Tuscany since 1621, was the most generous patron of both Viviani and Galileo (also helping Galileo during the last years of his life in Arcetri), as well as of other scientists such as Evangelista Torricelli (1608–1647). In 1642, he founded the Sperimentale Accademia Medicea, and supported the Accademia del Cimento (founded in 1657 by Leopoldo de' Medici, Ferdinando's brother), whose members included, besides Viviani and Torricelli, Giovanni Alfonso Borelli (1608–1679) and Nicolas Steno (1638–1686).

49. Galileo's *Compass*; see also *OG* 2, pp. 345–361.

50. John Frederick of Holstein-Gottorp (1579–1634), the Lutheran Administrator of the Prince-Bishopric of Bremen, of Lübeck, and of Verden; Ferdinand II (1578–1637), a member of the House of

Habsburg, Archduke of Austria and Holy Roman Emperor from 1619; Philip III (1581–1643), Landgrave of Hesse-Butzbach from 1609; Vincenzo I Gonzaga (1562–1612), Duke of Mantua and Monferrato. Among Galileo's pupils were also Federico Baldissera Bartolomeo Cornaro, or Cornèr (1579–1653), who was elevated to cardinal by Pope Urban VIII in 1626; and Guido Bentivoglio (1577–1644). A close friend of Pope Urban VIII's, Bentivoglio was made a cardinal in 1621, and was one of the inquisitors who pronounced and signed Galileo's sentence in 1633: see *OG* 19, pp. 402 and 406. See also the *Compass*, p. 41 (*OG* 2, p. 370).

51. Galileo's salary was raised to 320 florins a year: see *OG* 19, p. 113.

52. Supernova 1604, also known as Kepler's nova, occurred in the Milky Way, in the constellation Ophiucus (Serpentarius). Visible to the naked eye, at its peak it was brighter than any other star in the night sky and was visible during the day for over three weeks. The first recorded observation was in Padua, on October 10, 1604 (see *OG* 2, p. 293); Kepler began observing it on October 17, in Prague: see Kepler, *Gründtlicher Bericht Von einem ungewohnlichen Newen Stern* (1604) and *De stella nova* (1606) (it was the second nova to be observed in a generation, after Tycho Brahe observed one in the constellation of Cassiopeia, in 1572). In Italy, the observation of the new star sparked a large controversy, involving Galileo, Baldassarre Capra (1580–1626), and Lodovico delle Colombe, among others. Galileo came to know about the new star from Giacomo Alvise Cornaro (1539–1608), who had been informed by Capra (see *OG* 2, pp. 520 and 294); as Viviani writes, Galileo gave three crowded lectures on it in early December 1604 (of which we have only fragments: see *OG* 2, pp. 277–284; see also Galileo to Onofrio Castelli, January 1605, in *OG* 10, no. 113, pp. 134–135). In early February 1605, the Milanese physician Baldassarre Capra attacked him in *Consideratione astronomica Circa la noua, & portentosa Stella* (1604), trying to discredit Galileo. A few weeks later, the *Dialogo in perpvosito De La Stella Nvova* (1605), in rustic Paduan dialect, was published under the name Cecco di Ronchitti (a second edition appeared in Verona, later that year) but actually was written by Girolamo Spinelli, a friend of Galileo's, in support of the latter's views. Lodovico delle Colombe stated his belief that Galileo was hiding behind the print name Cecco di Ronchitti and published the *Discorso* (1606), in which he attempted to show the validity of Aristotle's theory and explain the new star in accordance with the "true" Peripatetic philosophy. Delle Colombe's *Discorso* was attacked in print, too, in a book by one Alimbero Mauri (behind whose name some scholars suspect was hiding Galileo himself), *Considerazioni sopra alcvni lvoghi del Discorso di Ludouico delle Colombe* (1606), to which delle Colombe replied in *Risposte piacevoli, e cvriose alle considerazioni di certa Maschera saccente nominata Alimberto Mauri* (1608).

53. That is, below the sphere of the Moon, where matter is made of the four elements: earth, water, air, and fire.

54. Cesare Cremonini (1550–1631) was appointed at the University of Padua in 1590, to replace the eminent philosopher Jacopo Zabarella (1533–1589). A staunch Aristotelian scholar, in the tradition of Alexander of Aphrodisias and Averroes, he was regarded as one of the most important philosophers of his time. See note 80.

55. Galileo's studied the property of the lodestone between 1600 and 1609, following the publication of William Gilbert's *De magnete*, in 1600. He had various exchanges about it with Paolo Sarpi (see Sarpi to Galileo, 2 September 1602, in *OG* 10, no. 83*, p. 91) and Giovanni Francesco Sagredo: see Sagredo to Galileo, 8 August 1602 (ibid., no. 80*, p. 89) and 28 October 1609 (ibid., no. 246, p. 262). In 1607–1608, Galileo exchanged several letters on the subject with the Grand Duke's secretary Belisario Vinta (1542–1613) and with Curzio Picchena (1553–1626), and presented Ferdinando I with a lodestone fit with an armature (see Benedetto Castelli's report, in *Gal.* 111, f. 203*v*). See also *Dialogue*, pp. 399–410 (*OG* 7, pp. 425–436).

56. Instead of the latter sentence, *A* and *S* have: "who later became the father of Your Highness." It is a reference to Galileo's *Compass*, dedicated to Cosimo II de' Medici (1590–1621), Grand Duke of Tuscany from 1609. He was tutored by Galileo between 1605 and 1608, and the two became friends until Cosimo's sudden death, at thirty years of age; see note 192.

57. Matthias Bernegger (1582–1640) was a German philologist and astronomer. A friend and correspondent of Johannes Kepler's (1571–1630) and Wilhelm Schickard's (1592–1635), he translated Galileo's *Compass* into Latin as *De proportionum instrumento a se invento, quod meritò Compendium dixeris uniuersæ Geometriæ, Tractatus*, Strasbourg: Karl Kieffer and Johann Carolus, 1612 (reprinted in 1613); new edition, *Tractatvs De proportionvm instrvmento, Quod merito Compendium vniuersæ Geometriæ dixeris*, Strasbourg: David Hautt, 1635. Later on, upon the request of Élie Diodati (1576–1661), Bernegger translated into Latin Galileo's *Dialogue*: see note 133.

58. Galileo's yearly salary was further raised to 520 florins: see *OG* 19, p. 114.

59. In both *A* and *B*, the text continues with this sentence: "also venturing to accuse Galileo himself of being a very insolent usurper."

60. In fact, this work was actually written by Capra's teacher, Simon Mayr, and published under only Capra's name. See *The Assayer*, pp. 163–168 (*OG* 6, pp. 213–217). See also note 62.

61. As Galileo himself recalled (*Difesa*, pp. 530 and 536; see also *OG* 10, pp. 173–176), he had shown Capra the compass in the house of Giacomo Alvise Cornaro. After Galileo published the *Compass*, in 1606, Capra published *Vsvs et fabrica circini cvivsdam proportionis* (1607), a blatant plagiarism of Galileo's work, in

which he claimed the authorship for the invention of the new instrument. Galileo sued him, and thanks to the testimony of Paolo Sarpi (who claimed he had received a copy of the instrument from Galileo in 1597: see *Difesa*, p. 544), won the lawsuit. Furthermore, Capra proved unable to explain the functioning of the instrument and how to use it, thereby making his plagiarism evident. On May 4, 1607, he was sentenced to destroy all copies of the book that were by him or the publisher (see *OG* 19, p. 224), but some of them had already been sold (in Italy and abroad), and so Galileo decided to vindicate himself and give the incident the widest audience possible, by publishing the *Difesa* (1607). As a consequence, in 1620 Capra was not admitted to the Medical School in Milan: see *OG* 2, pp. 625–630.

62. Galileo believed that Capra's teacher, the German astronomer Simon Mayr, or Marius (1573–1625)—who would later claim priority over Galileo on the discovery of Jupiter's moons, in the *Mundus Iovialis* (1614), in which he named them "sidera Brandeburgica": see Mayr 1614, f. B1r–v—had an important role in motivating, if not actually writing, Capra's *Consideratione astronomica* (1605). See *The Assayer*, pp. 164–165 (*OG* 6, pp. 214–215).

63. Maurice of Orange (1567–1625), stadtholder and commander of the armed forces of the United Provinces of Holland; before he became Prince of Orange upon the death of his eldest half-brother Philip William, in 1618, he was known as Maurice of Nassau.

64. See *The Assayer*, pp. 211–212 (*OG* 6, pp. 257–258); and Galileo to Benedetto Landucci, 29 August 1609, in *OG* 10, no. 231, pp. 253–254.

65. According to the chronicle of Antonio Priuli (in *OG* 19, p. 587), these were Senators Zaccaria Contarini, Lodovico Falier, Zaccaria Sagredo, Piero Contarini and Lorenzo Soranzo, Sebastiano Venier and Ventura Cavalli (actually, Cavanis).

66. Leonardo Donà (also Donato, or Donati, 1536–1612) was the 90th Doge of Venice, reigning from January 10, 1606, until his death. His (as well as his predecessor Marino Grimani's, 1532–1605) determination to limiting the power of the papacy within the Republic of Venice led him to conflict with Cardinal Camillo Borghese (1552–1621), the future Pope Paul V, elected in 1605. In late 1605, the Republic charged two priests as common criminals and had them face secular courts, thus denying their clerical immunity. The pope protested, but Donà (who had just been elected Doge) rejected the protest, at the urging of Paolo Sarpi. Paul V issued a papal interdict on the Republic of Venice, thus excommunicating the entire Venetian population. Donà ordered all Roman Catholic clergy to ignore the interdict and continue to perform the Mass, on pain of immediate expulsion from the Republic. All the Venetian clergy ignored the interdict, except for the Jesuits, who left (or were forced to leave) the Republic. Both Donà and Sarpi were excommunicated, and although the interdict was lifted in 1607, the Jesuits were not allowed back until 1655. On October 5, 1607, at the instigation of the pope, Sarpi was attacked by assassins and seriously wounded with several stiletto thrusts, but he eventually recovered.

67. In September 1608, Hans Lipperhey (1570–1619), a German-born spectacle maker, presented Count Maurice of Nassau a spyglass; the news arrived in Venice in mid-July 1609. Galileo's own account in *The Assayer*, which was likely Viviani's source, is worth reading in full: "[in Venice, where I happened to be at the time, news came that a Hollander had presented to Count Maurice [of Nassau] a glass by means of which distant things might be seen as perfectly as if they were quite close. That was all. Upon hearing this news, I returned to Padua, where I then resided, and set myself to thinking about the problem. The first night after my return, I solved it, and the following day I constructed the instrument and sent word of this to the same friends in Venice with whom I had been discussing the subject the previous day. Immediately afterward, I applied myself to the construction of another and better one, which I took to Venice six days later; there it was seen with great admiration by nearly all the principal gentlemen of that republic for more than a month on end, to my considerable fatigue. Finally, at the suggestion of one of my friendly patrons, I presented it to the ruler in a full meeting of the Council. How greatly it was esteemed by him, and with how much admiration it was received, is testified by ducal letters still in my possession which reveal the munificence of that serene ruler in recompense for the invention presented to him, reappointing and confirming me for life to my professorship at the University of Padua at double my former salary, which was then more than triple that of some of my predecessors" (pp. 211–212; "[. . .] in Vinezia, dove allora mi ritrovavo, giunsero nuove che al Sig. Conte Maurizio era stato presentato da un Olandese un occhiale, col quale le cose lontane si vedevano così perfettamente come se fussero state molto vicine; né più fu aggiunto. Su questa relazione io tornai a Padova, dove allora stanziavo, e mi posi a pensar sopra tal problema, e la prima notte dopo il mio ritorno lo ritrovai, ed il giorno seguente fabbricai lo strumento, e ne diedi conto a Vinezia a i medesimi amici co' quali il giorno precedente ero stato a ragionamento sopra questa materia. M'applicai poi subito a fabbricarne un altro più perfetto, il quale sei giorni dopo condussi a Vinezia, dove con gran meraviglia fu veduto quasi da tutti i principali gentiluomini di quella republica, ma con mia grandissima fatica, per più d'un mese continuo. Finalmente, per consiglio d'alcun mio affezzionato padrone, lo presentai al Principe in pieno Collegio, dal quale quanto ei fusse stimato e ricevuto con ammirazione, testificano le lettere ducali, che ancora sono appresso di me, contenenti la magnificenza di quel Serenissimo Principe in ricondurmi, per ricompensa della presentata invenzione, e confermarmi in vita nella mia lettura nello

Studio di Padova, con dupplicato stipendio di quello che avevo per addietro, ch'era poi più che triplicato di quello di qualsivoglia altro mio antecessore": *OG* 6, pp. 257–258). See also the *Sidereal Messenger*, pp. 39–40 (*OG* 3.1, pp. 60–61), as well as Galileo to Leonardo Donati, 24 August 1609, and to Galileo's brother-in-law, Benedetto Landucci, 29 August 1609 (*OG* 10, nos. 228 and 231, pp. 250–251 and 253–254).

68. Galileo's yearly salary was eventually raised to 1,000 florins (with no possibilities of further raises), for life: see *OG* 19, p. 116.

69. The description of Galileo's microscope, built by combining a convex lens and a concave one, was first offered by Galileo's pupil and admirer John Wedderburn, in his *Confvtatio* (1610), pp. 163–164: "a few days ago I heard the author himself [Galileo] telling the Most Excellent Cremonini, noblest philosopher, various things greatly deserving to be known: among others, how he distinguishes perfectly with his spyglass [*perspicillum*] the organs of motion and those of the senses in the most minute animals; and particularly in a certain insect that has each eye covered by a thick membrane, which, however, is pierced by seven slits like the visor of a fully-armored warrior, leaving the way open to the species emanating from visible objects" ("Audiveram paucis ante diebus auctorem ipsum Excellentissimo D. Cremonino purpurato philosopho varia narrantem, scitu dignissima, et inter cetera, quomodo ille minimorum animantium organa, motus et sensus ex perspicillo ad unguem distinguat; in particulari autem de quodam insecto, quod utrumque habet oculum membrana crassiuscula vestitum, quae tamen, septem foraminibus ad instar larvae ferreae militis cataphracti terebrata, viam praebet speciebus visibilium"). Galileo's first, very brief reference to the microscope is to be found in *The Assayer* (1623), p. 245: "Let him [Sarsi] take any material, whether stone or wood or metal, and holding it up to the Sun gaze at it most intently. [. . .] and if looking at this, he will make use of a telescope [*telescopio*] set for looking at very near objects, he will see what I mean very much more distinctly and without any need of these bodies being resolved into dew or moist vapors" ("Prenda egli qualsivoglia materia, o sia pietra o sia legno o sia metallo, e tenendola al Sole, attentissimamente la rimiri, ch'egli vi vederà tutti i colori compartiti in minutissime particelle; e s'ei si servirà, per riguardargli, d'un telescopio accommodato per veder gli oggetti vicinissimi, assai più distintamente vederà quant'io dico, senza verun bisogno che quei corpi si risolvano in rugiada o in vapori umidi": *OG* 6, p. 290). On 11 May 1624, in a letter to Federico Cesi, Giovanni Faber (Johann Schmidt, 1574–1629) wrote: "I spent yesterday evening with our Galileo, who lives near the Madalena. He gave a most beautiful microscope [*ochialino*] to Cardinal Zollern for the Duke of Bavaria. I saw a fly Galileo himself showed me: I was amazed, and told Galileo he is another Creator, as he makes things appear that hitherto nobody knew had been created" ("Sono stato hier sera col Sig.ʳ Galilei nostro, che habita vicino alla Madalena. Ha dato un bellissimo ochialino al Sig. Card. di Zoller per il Duca di Baviera. Io ho visto una mosca che il Sig.ʳ Galileo stesso mi ha fatto vedere: sono restato attonito, et ho detto al Sig.ʳ Galileo che esso è un altro Creatore, atteso che fa apparire cose che finhora non si sapeva che fossero state create": *OG* 13, no. 1631*, pp. 177–178). Later that year, on September 23, Galileo provided Cesi with the most detailed description of the microscope to be found in his papers: "I am sending Your Excellency a microscope [*occhialino*] to see the tiniest things close up, which I hope will afford you much pleasure and entertainment, as has been the case with myself. I delayed in sending it because I had not yet brought it to perfection, having had problems in finding a way of grinding the lenses perfectly. The object is attached to the mobile circle, which is in the base, and is moved about so as to see all of it, since what is seen at one look is only a small part. And since the distance between the lens and the object must be very precise, in order to look at objects in relief, it must be possible to bring the glass closer or farther away, according to whether one is looking at this or that part. And so the foot of the tube [*cannoncino*] has been made mobile, to be guided as we wish. Also, it must be used in very calm, bright air, and better in sunlight itself, so that the object is well lit. I have considered a great number of tiny animals with infinite admiration. Among them the flea is most horrible; the mosquito and the moth are very beautiful. And with great satisfaction I have seen how it is that flies and other tiny creatures can walk attached to mirrors, and even upside down. But Your Excellency will have occasion to observe thousands and thousands of details, and I beg you to notify me of the most curious of them. To sum up: we might contemplate infinitely the grandeur of nature, and how subtly she works, and with what indescribable diligence" ("Invio a V. E. un occhialino per veder da vicino le cose minime, del quale spero che ella sia per prendersi gusto e trattenimento non piccolo, ché così accade a me. Ho tardato a mandarlo, perché non l'ho prima ridotto a perfezzione, havendo hauto difficoltà in trovare il modo di lavorare i cristalli perfettamente. L'oggetto si attacca sul cerchio mobile, che è nella base, e si va movendo per vederlo tutto, atteso che quello che si vede in un'occhiata è piccola parte. E perché la distanza tra la lente e l'oggetto vuol esser puntualissima, nel guardar gl'oggetti che hanno rilievo bisogna potere avvicinare e discostare il vetro, secondo che si guarda questa o quella parte; e però il cannoncino si è fatto mobile nel suo piede, o guida che dir la vogliamo. Devesi ancora usarlo all'aria molto serena e lucida, e meglio è al sole medesimo, ricercandosi che l'oggetto sia illuminato assai. Io ho contemplati moltissimi animalucci con infinita ammirazione: tra i quali la pulce è orribilissima, la zanzara e la tignuola son bellissimi; e con gran contento ho veduto come faccino le mosche et altri animalucci a camminare attaccati a' specchi, et anco di sotto in su. Ma V. E. haverà campo larghissimo di osservar mille e mille particolari, de i quali la

prego a darmi avviso delle cose più curiose. In somma ci è da contemplare infinitamente la grandezza della natura, e quanto sottilmente ella lavora, e con quanta indicibil diligenza": ibid., no. 1665, pp. 208–209). On April 13, 1625, Faber eventually suggested to Cesi (who had proposed the name *telescopio* for Galileo's spyglass) that the new instrument—until then variably referred to as *perspicillum, telescopio, occhialino,* or *cannoncino*—be named *microscopio* (see ibid., no. 1719**, p. 264).

70. See the *Sidereal Messenger*, pp. 39–40: "And first I prepared a lead tube in whose ends I fitted two glasses, both plane on one side while the other side of one was spherically convex and of the other concave. Then, applying my eye to the concave glass, I saw objects satisfactorily large and close. Indeed, they appeared three times closer and nine times larger than when observed with natural vision only. Afterward I made another more perfect one for myself that showed objects more than sixty times larger. Finally, sparing no labor or expense, I progressed so far that I constructed for myself an instrument so excellent that things seen through it appear about a thousand times larger and more than thirty times closer than when observed with the natural faculty only. It would be entirely superfluous to enumerate how many and how great the advantages of this instrument are on land and at sea. But having dismissed earthly things, I applied myself to explorations of the heavens" ("[. . .] ac tubum primo plumbeum mihi paravi, in cuius extremitatibus vitrea duo Perspicilla, ambo ex altera parte plana, ex altera vero unum sphaerice convexum, alterum vero cavum aptavi; oculum deinde ad cavum admovens obiecta satis magna et propinqua intuitus sum; triplo enim viciniora, nonuplo vero maiora apparebant, quam dum sola naturali acie spectarentur. Alium postmodum exactiorem mihi elaboravi, qui obiecta plusquam sexagesies maiora repraesentabat. Tandem, labori nullo nullisque sumptibus parcens, eo a me deventum est, ut Organum mihi construxerim adeo excellens, ut res per ipsum visae millies fere maiores appareant, ac plusquam in terdecupla ratione viciniores, quam si naturali tantum facultate spectentur. Huius Instrumenti quot quantaque sint commoda, tam in re terrestri quam in maritima, omnino supervacaneum foret enumerare. Sed, missis terrenis, ad Caelestium speculationes me contuli": *OG* 3.1, pp. 60–61).

71. *Sidereal Messenger*, pp. 37–38: "It is most beautiful and pleasing to the eye to look upon the lunar body, distant from us about 60 terrestrial diameters, from so near as if it were distant by only two of these measures, so that the diameter of the same Moon appears as if it were 30 times, the surface 900 times, and the solid body about 27,000 times larger than when observed only with the naked eye. Anyone will then understand with the certainty of the senses that the Moon is by no means endowed with a smooth and polished surface but is rough and uneven and, just as the face of the earth itself, crowded everywhere with vast prominences, deep chasms, and convolutions" ("Pulcherrimum atque visu iucundissimum est, lunare corpus, per sex denas fere terrestres semidiametros a nobis remotum, tam ex propinquo intueri, ac si per duas tantum easdem dimensiones distaret; adeo ut eiusdem Lunae diameter vicibus quasi terdenis, superficies vero noningentis, solidum autem corpus vicibus proxime viginti septem millibus, maius appareat, quam dum libera tantum oculorum acie spectatur: ex quo deinde sensata certitudine quispiam intelligat, Lunam superficie leni et perpolita nequaquam esse indutam, sed aspera et inaequali; ac, veluti ipsiusmet Telluris facies, ingentibus tumoribus, profundis lacunis atque anfractibus undiquaque confertam existere": *OG* 3.1, pp. 59–60). See also pp. 41–59 (*OG* 3.1, pp. 62–75).

72. *Sidereal Messenger*, p. 38: "Moreover, it seems of no small importance to have put an end to the debate about the Galaxy or Milky Way and to have made manifest its essence to the senses as well as the intellect; and it will be pleasing and most glorious to demonstrate clearly that the substance of those stars called nebulous up to now by all astronomers is very different from what has hitherto been thought" ("Altercationes insuper de Galaxia, seu de Lacteo circulo, substulisse, eiusque essentiam sensui, nedum intellectui, manifestasse, parvi momenti existimandum minime videtur; insuperque substantiam Stellarum, quas Nebulosas hucusque Astronomorum quilibet appellavit, digito demonstrare, longeque aliam esse quam creditum hactenus est, iocundum erit atque perpulcrum": *OG* 3.1, p. 60). See also pp. 64–65 (*OG* 3.1, pp. 78–79).

73. *Sidereal Messenger*, p. 37: "Certainly it is a great thing to add to the countless multitude of fixed stars visible hitherto by natural means and expose to our eyes innumerable others never seen before, which exceed tenfold the number of old and known ones" ("Magnum sane est, supra numerosam inerrantium Stellarum multitudinem, quae naturali facultate in hunc usque diem conspici potuerunt, alias innumeras superaddere oculisque palam exponere, antehac conspectas nunquam, et quae veteres ac notas plusquam supra decuplam multiplicitatem superent": *OG* 3.1, p. 59). See also pp. 59–64 (*OG* 3.1, pp. 75–78).

74. *Sidereal Messenger*, p. 38: "But what greatly exceeds all admiration, and what especially impelled us to give notice to all astronomers and philosophers, is this, that we have discovered four wandering stars, known or observed by no one before us. These, like Venus and Mercury around the Sun, have their periods around a certain star notable among the number of known ones, and now precede, now follow, him, never digressing from him beyond certain limits. All these things were discovered and observed a few days ago by means of a glass contrived by me after I had been inspired by divine grace" ("Verum, quod omnem admirationem longe superat, quodve admonitos faciendos cunctos Astronomos atque Philosophos nos apprime impulit, illud est, quod scilicet quatuor Erraticas Stellas, nemini eorum qui ante nos cognitas aut

observatas, adinvenimus, quae circa Stellam quandam insignem e numero cognitarum, instar Veneris atque Mercurii circa Solem, suas habent periodos, eamque modo praeeunt, modo subsequuntur, nunquam extra certos limites ab illa digredientes. Quae omnia ope Perspicilli a me excogitati, divina prius illuminante gratia, paucis abhinc diebus, reperta atque observata fuerunt": *OG* 3.1, p. 60). See also pp. 66–85 (*OG* 3.1, pp. 79–94). The observation of Jupiter's satellites provided the key evidence in support of the Copernican hypothesis: "We have moreover an excellent and splendid argument for taking away the scruples of those who while tolerating with equanimity the revolution of the planets around the Sun in the Copernican system are so disturbed by the attendance of one Moon around the earth while the two together complete the annual orb around the Sun that they conclude that this constitution of the universe must be overthrown as impossible. For here we have only one planet revolving around another while both run through a great circle around the Sun; but our vision offers us four stars wandering around Jupiter like the Moon around the earth while all together with Jupiter traverse a great circle around the Sun in the space of twelve years" (ibid., pp. 86–87; "Eximium praeterea praeclarumque habemus argumentum pro scrupulo ab illis demendo, qui in Systemate Copernicano conversionem Planetarum circa Solem aequo animo ferentes, adeo perturbantur ab unius Lunae circa Terram latione, interea dum ambo annuum orbem circa Solem absolvunt, ut hanc universi constitutionem, tanquam impossibilem, evertendam esse arbitrentur: nunc enim, nedum Planetam unum circa alium convertibilem habemus, dum ambo magnum circa Solem perlustrant orbem, verum quatuor circa Iovem, instar Lunae circa Tellurem, sensus nobis vagantes offert Stellas, dum omnes simul cum Iove, annorum spatio, magnum circa Solem permeant orbem": *OG* 3.1, p. 95).

75. As we read in the title page of the *Sidereal Messenger*: "[. . .] four planets flying around the star of Jupiter at unequal intervals and periods with wonderful swiftness, which, unknown by anyone until this day, the author first detected recently and decided to name *Medicean Stars*" (p. 28; "Apprime verò in Qvatvor Planetis Circa Iovis Stellam disparibus interuallis, atque periodis, celeritate mirabili circumuolutis; quos, nemini in hanc vsque diem cognitos, nouissimè Author depræhendit primus; atque Medicea Sidera nvncvpandos decrevit": *OG* 3.1, p. 53). See also ibid., p. 33 (*OG* 3.1, p. 56). On the choice of the name for Jupiter's satellites, see the exchange between Galileo and Belisario Vinta, the Grand Duke's secretary, on February 13 and 26, 1610 (in *OG* 10, no. 265, pp. 283–284; and no. 266*, pp. 284–285, respectively). Galileo referred to Jupiter's satellites as both "planets" (Latin *planetae*, Greek πλανῆται, namely, "erring object") and "stars" (Latin *stellae*, Greek ἀστέρες). Both terms were based on Aristotelian cosmology, hence the distinction between "erring stars" and "fixed stars."

76. Galileo made his first telescopic observations (of the Moon) on November 30, 1609, and recorded the last observation (Jupiter's satellites) to be included in the book on March 2, 1610. He began writing the book on January 16, and signed the dedication to Grand Duke Cosimo II de' Medici on March 12. The printing of the *Sidereal Messenger* was completed the next day; the print run was of 550 copies. In both *A* and *S* we further read: "and father of Your Highness, who, as a sign of royal gratitude, called him back from Padua to His service, with a personal letter dated 10 July 1610. He appointed Galileo Chief and Superordinary Mathematician of the University of Pisa, without any obligation to teach or to reside there, and Chief Philosopher and Mathematician of Your Most Serene Highness, bestowing on him a very generous salary, commensurate with the very great generosity of such a Prince" ("e padre di V. A., il quale in segno di regia gratitudine con propria lettera de' X Luglio del 1610 lo richiamò di Padova al suo servizio, con titolo di Primario e Sopraordinario Matematico dello Studio di Pisa, senz'obligo di leggervi o risedervi, e di Primario Filosofo e Matematico della sua Ser.^ma Altezza, assegnandogli amplissimo stipendio, proporzionato alla somma generosità d'un tanto Principe"). See Cosimo II to Galileo, 10 July 1610, in *OG* 10, no. 359*, pp. 400–401.

77. The *Sidereal Messenger* was reprinted in Frankfurt, by Zacharias Palthen, in 1610. There was no reprint in France: the first edition by a French scholar was that of a friend and correspondent of Galileo's, Pierre Gassendi (1592–1655), who published the *Sidereal Messenger* as an appendix (with a separate title page and independent page numbers) to the second edition of his *Institutio Astronomica* (1653^2, pp. 1–50).

78. Among them, Giovanni Antonio Magini (1555–1617)—as we read in the margin of *A* (and *B*, though the note was later deleted)—professor of astronomy at the University of Bologna, and the Jesuit Christopher Clavius (Christoph Clau, or Schlüssel, 1537–1612), professor of mathematics at the Roman College, and an early supporter of Galileo. As reported by Martin Hasdale, Magini wrote to the Prince-Elector-Archbishop of Cologne (Ernest of Bavaria, who received a copy of the telescope from Galileo, and lent it to Kepler) these words: "As to Galileo's book and instrument, I believe it is a fraud: for, when I looked at the solar eclipse with obscured lenses I myself had made, they made me see three Suns. Likewise, I believe, happened to Galileo, who must have been deceived by the glare of the Moon. Many others rebut Galileo's opinion, among them Dr Papazzoni [. . .]. It seems to me ridiculous that these four new planets go around planet [Jupiter], as Galileo maintains" ("Quanto al libro et stromento del Gallilei, io credo che sia un inganno, perchè quando con occhiali colorati, fatti da me, guardavo l'ecclipsi solare, mi faceva vedere 3 soli; così anco credo che sia avvenuto al Gallilei, quale si deve essere ingannato dal reflesso della luna. Sono molti altri che oppugnano questa openione del Gallilei, et tra gli altri il Dottore Papazzone [. . .].

Ma per tornare al proposito, mi pare una cosa ridicolosa di quei 4 nuovi pianetti, che presuppone il Gallilei che vadino intorno al pianeta . . .": Hasdale to Galileo, 28 April 1610, in *OG* 10, no. 303, p. 345). According to Cigoli, the "Clavisi" (that is, Clavius and his friends), "do not believe a single thing; and Clavius, among them, told a friend of mine that he laughed at the four stars and that it will be necessary to make a spyglass that creates them and then shows them. Let Galileo keep his opinion, he added, and he will keep his own" ("questi Clavisi [. . .] non credono nulla; et il Clavio fra gli altri, capo di tutti, disse a un mio amico, delle quattro stelle, che se ne rideva, et che bisognierà fare uno ochiale che le faccia e poi le mostri, et che il Galileo tengha la sua oppinione et egli terrà la sua": Cigoli to Galileo, 1 October 1610, ibid., no. 403, p. 442). A couple of weeks later, however, after had he made his own telescopic observations, Clavius changed his mind and wrote to Galileo that "Your Lordship truly deserves great praise, for you were the first to observe this [i.e., Jupiter's satellites]. And early on we observed very many stars in the Pleiades, Cancer, Orion, and the Milky Way, which cannot be observed without the instrument [i.e., the telescope]" ("Veramente V. S. merita gran lode, essendo il primo che habbi osservato questo. Già molto prima havevamo vedute moltissime stelle nelle Pleiadi, Cancro, Orione et Via Lactea, che senza l'instromento non si veggono": Clavius to Galileo, ibid., no. 437, p. 484; there follows the report of Clavius' own observations).

79. Martin Horký (ca. 1584–ca. 1646), a Bohemian mathematician and a pupil of Magini's, published a libel against Galileo's work, *Brevissima peregrinatio contra Nvncivm siderevm* (1610), in which he roundly dismissed Galileo's discoveries regarding the surface of the Moon and the moons of Jupiter as either flaws in Galileo's spyglass, or a cynical attempt on Galileo's part to earn money. A guest in Magini's house in Bologna, Horký was kicked out as soon as the astronomy professor learned about the book (see the exchange of letters between Giovanni Antonio Roffeni, Galileo, Magini, and Antonio Santini in June 1610, in *OG* 10, nos. 334, 335, 337, 338**, and 344, pp. 376–379 and 384–385); Magini also sent a letter to Galileo, making it clear that he was not aware of Horký's work (Magini to Galileo, 23 October 1610, ibid., no. 414*, p. 450). Horký sought Kepler's help but received only a harsh reprimand (see Kepler to Horký, 9 August 1610, ibid., no. 376, p. 419; and Kepler to Galileo, 9 August 1610, ibid., no. 374, p. 414), after which Horký regretted his mistake: see Kepler to Galileo, 25 October 1610, ibid., no. 419, pp. 457–458; and Martin Hasdale to Galileo, 19 December 1610, ibid., no. 439, p. 491). Horký's weak arguments were refuted by John Wedderburn, a pupil of Galileo's in Padua, in the *Confvtatio* (1610); and by Giovanni Antonio Roffeni, a friend of Magini's, in the *Epistola apologetica contra cæcam peregrinationem Cuiusdam furiosi Martini, cognomine Horkij* (1611). However, the *Brevissima peregrinatio* was but the first attack on Galileo's discoveries. Francesco Sizzi, or Sizi (d. 1618) wrote the *ΔIANOIA astronomica, optica, physica* (1611) and, like Horký, Sizzi appealed mainly to astrological arguments against the existence of Jupiter's satellites. Lodovico delle Colombe also circulated a manuscript, *Contro il moto della terra*, written in late 1610 or early 1611, and published in *OG* 3.1, pp. 253–290. Here is how Daniello Antonini, a former pupil of Galileo's in Padua, described the situation to Galileo: "I have not seen the work against Your Lordship [Sizzi's *ΔIANOIA astronomica, optica, physica*]! I looked for a copy here in Brussels, and I haven't found it; I sent someone to get a copy in Antwerp, and I also wrote to a few selected mathematicians to have their opinions. But I assume it is a Cremoninade [see note 80]. Oh, how solid are Plutarch's remarks [in *De facie in orbe lunae*] compared with Your Lordship's arguments! How can it be that there are such gullible men around and—what is worse—that they are regarded as the learned ones? What on earth could be done to have them acknowledge the truth if having them look at it with their own eyes is not enough? On the one hand, I laugh at them; but, on the other, I get angry—and I am tempted to say, as that good devout man said: if I were the Lord, I could not stand that such dumb people live. But I believe that this almighty Lord let them be so that they may serve as jesters for mother nature" ("Non ho veduto ancora l'opra scritta contro V. S.; ho cercato qui in Brusselles, et non l'ho trovata, onde ho mandato in Anversa per haverla, et anco scritto a certi pochi mathematici per haver i loro pareri: ma m'immagino che sarà una Cremoninata. O come camina bene la osservation di Plutarco contro V. S.! Possibile che si ritrovino al mondo huomini così goffi, et quel ch'è peggio, che sian quelli stimati li saputi? Che cosa si potrebbe far al mondo per farli confessar la verità, se il fargliela veder con gl'occhi proprii non basta? D'una parte me ne rido, dal'altra mi vien colera et voglia quasi di dire, come disse quel buon religioso: Se io fussi Meser D. Dio, non soportarei che vivesse tal razza d'huomini irragionevoli. Ma credo che questo Meser D. Dio, che regna, lasci costoro acciò servano per bufoni alla madre natura" (Antonini to Galileo, in *OG* 11, no. 544, 24 June 1611, p. 129).

80. A renowned professor of natural philosophy at the University of Padua, Cesare Cremonini (1550–1631) was a good friend of Galileo's. A man of independent judgment, Cremonini was an enemy of the Jesuits, who accused him of materialism and relayed their complaints to Rome, and a perennial target of the Inquisition, which prosecuted him for his literal interpretations of Aristotle and for holding the Averroist heresy of "double truth" (on May 17, 1611, the Tribunal of the Holy Inquisition considered whether to appoint Galileo in a trial against him: see *OG* 19, p. 275). Cremonini always refused to retract and remained unscathed because Padua was under tolerant Venetian rule, which kept him out of reach of a full trial. We know of his attitude toward the new telescopic discoveries from two letters Paolo Gualdo sent to Galileo, on May 6 and July 29, 1611 (see *OG* 11, no. 526, p. 100; and no. 564, p. 165). In the latter, Gualdo reports an

exchange with Cremonini: "This is what displeased Galileo, that you did not want to see [what Galileo showed him through the telescope]", said Gualdo; to which Cremonini retorted: "I believe that nobody else but him saw [what Galileo claimed to be seeing]; moreover, looking through those glasses makes my mind dizzy. Enough, I am done with that"; Gualdo responded: "Your Lordship swore allegiance to the word of the Master, and is right to follow divine antiquity" ("Questo è quello, dico, c'ha dispiacciuto al S.ʳ Galilei, ch'ella non habbia voluto vederle. Rispose: Credo che altri che lui non l'habbia veduto; e poi quel mirare per quegli occhiali m'imbalordiscon la testa: basta, non ne voglio saper altro. Io risposi: V. S. *iuravit in verba Magistri*; e fa bene a seguitare la santa antichità": ibid., p. 165). Cremonini's attitude, however, was much less dogmatic than it is usually portrayed, as seeing what Galileo claimed to see through the telescope was far more difficult than we would think today. Not only were the quality and optical power of the lenses he used quite low, but the instrument itself did not allow the whole face of the Moon to be seen, for example, but only part of it. Furthermore, seeing what Galileo claimed to see required abandoning an age-old and well-established tradition, in times when deceptive tricks or visual gimmicks (by way of systems of mirrors or lenses) were far from uncommon, and were also carefully studied by various people, including Giovanni Battista Della Porta (1535–1615), in Naples, and Athanasius Kircher (1602–1680), in Rome.

81. Galileo first observed the "handled" shape of Saturn on July 25, 1610, shortly before leaving Padua to head back to Florence. See Galileo's letter to Belisario Vinta on July 30, 1610: "I discovered another most extravagant wonder [. . .]: the star of Saturn is not a single one but is a compound of three stars: they touch one another and do not move or change with respect to one another; they are in a row, along the line of the Zodiac, and the central star is approximately three times as big as the two lateral ones; they are arranged like this ⊂○⊃" ("ho scoperto un'altra stravagantissima meraviglia [. . .]. Questo è, che la stella di Saturno non è una sola, ma un composto di 3, le quali quasi si toccano, né mai tra di loro si muovono o mutano; et sono poste in fila secondo la lunghezza del zodiaco, essendo quella di mezzo circa 3 volte maggiore delle altre 2 laterali: et stanno situate in questa forma ⊂○⊃": *OG* 10, no. 370, p. 410). Astronomers struggled to see what Galileo claimed to have observed about Saturn, and it was not until 1655 that Christiaan Huygens, thanks to a much better telescope than Galileo's, first observed a ring around Saturn and published his discovery in the *Systema Satvrnivm* (1659).

82. Benedetto Castelli (1578–1643), Galileo's friend and collaborator, from Brescia; Lodovico Cardi, "il Cigoli" (1559–1613); Christian Grienberger (1561–1636), Austrian Jesuit astronomer, member of the Roman College; Christopher Clavius (Christoph Clau, 1538–1612), Jesuit German mathematician and astronomer, who substantially contributed to the reform of the calendar and was arguably one of the most respected astronomers of his time; Luca Valerio (1553–1618), a mathematician and correspondent of Galileo's, who developed ways to find volumes and centers of gravity of solids using the methods of Archimedes; Paolo Gualdo (1553–1621), a friend of Galileo's and biographer of Gian Vincenzo Pinelli; Lorenzo Pignoria (1571–1631), a learned man from the Pinelli circle; Giuliano de' Medici (1574–1636), ambassador of the Grand Duke of Tuscany to the Habsburgs' court in Prague; Johannes Kepler (1571–1630), Imperial Mathematician at the court of Rudolf II, in Prague, and author of two important works in support of his telescopic discoveries, the *Dissertatio* (1610, dedicated to Giuliano de' Medici) and the *Narratio* (1611).

83. Rudolf II (1552–1612), Holy Roman Emperor from 1576. Patron of the arts and the sciences, and great collector of curiosities, he supported Tycho Brahe (1546–1601) in Prague from 1599 to his death, and appointed Kepler as Imperial Mathematician in 1601. Tycho's *Astronomiæ instauratae mechanica* (1598), as well as Kepler's *Ad Vitellionem Paralipomena, Quibus Astronomiæ Pars Optica traditur* (1604), and *Astronomia Nova ΑΙΤΙΟΛΟΓΗΤΟΣ: sev Physica coelestis, tradita commentariis de motibvs stellæ Martis* (1609), are dedicated to Rudolf II.

84. Before publication, Galileo wanted to preserve his priority over the discovery of three-bodied Saturn and prevent others from claiming the discovery as their own. In order to do so, he circulated the announcement in the form of an anagram: *Smaismrmilmepoetaleumibunenugttauiras* (see Kepler, *Dioptrice* (1611), p. 344). Kepler struggled to solve it, as Giuliano de' Medici reported to Galileo on 23 August 1610 (in *OG* 10, no. 384*, p. 426; see also Martin Hasdale to Galileo, 17 August 1610, ibid., no. 378, pp. 420–421) and eventually came up with a tentative solution, *Salve umbistineum geminatum Martia proles*, that is, "Hail, doubly embossed offspring of Mars" (Kepler, *Narratio* (1611), p. 319; and *Dioptrice* (1611), p. 344). Since Galileo had previously discovered the four moons of Jupiter, Kepler conjectured that Galileo might have found new moons around Mars, too. That was not the case, though, as Galileo revealed to Giuliano de' Medici on November 13, 1610: *Altissimum planetam tergeminum observavi*, "I observed the highest planet [i.e., Saturn] as three-bodied" (*OG* 10, no. 427, p. 474). Saturn is said to be the "highest" planet because it was believed to be the farthermost planet from the Sun; Uranus, Neptune, and Pluto were discovered in 1781 (by William Herschel), 1846 (by Urbain Le Verrier), and 1930 (by Clyde W. Tombaugh), respectively.

85. In his letter to Galileo of January 6, 1612 (accompanying a copy of Scheiner's *Tres epistolae de macvlis solaribvs*), from Augsburg, Markus Welser refers to Scheiner's discovery as to something Galileo had known for some time already: "'The kingdom of heaven has suffered a violent assault, and the most violent take it by force' [*Matthew* 11, 12 KJV]. Your Lordship was the first to scale the walls, and earned the mural crown.

Now others are following you, their courage increased as they realize it would be an open display of cowardice not to pursue further so successful and honored achievement an undertaking, now that you have broken the ice once. Take a look at what this friend of mine ventured to do; and, as I believe, if it does not seem to be something entirely new to you, I nevertheless hope it will be to your liking, seeing how on this side of the mountains [the Alps] as well there is no lack of men who follow in your footsteps" ("*Regnum caelorum vim patitur, et violenti rapiunt illud.* V. S. è stato il primo alla scalata, et ne ha riportato la corona murale. Hora le vanno dietro altri, con tanto maggior coraggio, quanto più conoscono che sarebbe viltà espressa non secondare sì felice et honorata impresa, poiché lei ha rotto il giaccio [*sic*] una volta. Veda ciò che si è arrischiato questo mio amico; et se a lei non riuscirà cosa totalmente nova, come credo, spero però che le sarà di gusto, vedendo che ancora da questa banda de' monti non manca chi vada dietro alle sue pedate": *OG* 11, no. 637, p. 357). In his first rejoinder to Scheiner (May 4, 1612) Galileo states he observed sunspots 18 months earlier (*On Sunspots*, p. 90; OG 5, p. 95), that is, in November or December 1610—as Galileo would later recall in his letters to Maffeo Barberini (*OG* 11, no. 684*, 2 June 1612, p. 305) and Giuliano de' Medici (ibid., no. 706, 23 June 1612, p. 335). See also *Dialogue*, pp. 345–358 (*OG* 7, pp. 372–375), and Fulgenzio Micanzio's letter to Galileo, 27 September 1631 (*OG* 14, no. 2210, p. 299).

86. Paolo Sarpi (1552–1623) was a Venetian Servite monk and polymath, theological advisor to the Venetian Senate; he played a major role in the Venetian Republic's defiance of the papal interdict (1605–1607) and wrote an anti-Roman *Historia del Concilio tridentino* (1619); a friend, patron, and advisor of Galileo's, Sarpi was also a renowned experimental scientist and a proponent of the Copernican system. Fulgenzio Micanzio (1570–1654), a Servite monk, a friend and supporter of Galileo's, was a disciple of Sarpi's, eventually becoming his biographer and successor as theologian to the Venetian Republic in 1623. Filippo Contarini (1573–1610) was a pupil of Galileo's in Padua. Sebastiano Venier (1572–1640), a learned man, was ambassador for the Republic of Venice, and reformer of the University of Padua.

87. See Galileo's letter to the Grand Duke's secretary, Belisario Vinta, 7 May 1610, in which he agrees on the salary the two of them had discussed earlier and asks to be appointed as Mathematician and Philosopher, motivating his request with the fact that he had "spent more years studying philosophy than months studying pure mathematics" ("professando io di havere studiato più anni in filosofia, che mesi in matematica pura": *OG* 10, no. 307, p. 353). Vinta communicated the acceptance of Galileo's conditions and the Grand Duke's consent on May 22 and June 26 (see ibid., no. 311, pp. 355–356, and no. 342*, pp. 383–384); a full description of the employment followed on June 5, 1610: "These Highnesses ordered to offer Your Lordship the title of Chief Mathematician of the University of Pisa and Philosopher of the Most Serene Grand Duke, with no obligation to teach at the University of Pisa nor to live in the city of Pisa, and with a salary of 1,000 scudi, Florentine currency, so as to provide you with all the means to pursue your studies and finish your works" ("Hanno queste AA. deliberato di dar titolo a V. S. di Matematico primario dello Studio di Pisa et di Filosofo del Ser.ᵐᵒ Gran Duca, senz'obligo di leggere et di risedere né nello Studio né nella città di Pisa, et con lo stipendio di mille scudi l'anno, moneta fiorentina, et con esser per darle ogni commodità di seguitare i suoi studii et di finir le sue compositioni": Vinta's letter to Galileo, ibid., no. 327, p. 369). Grand Duke Cosimo II personally wrote to Galileo on July 10, 1610: "The eminence of your doctrine and of your valiant abilities, accompanied by your unique talent in mathematics and philosophy, as well as the most obedient affection, devotion, and service you have always shown to us, caused us to wish to have you here with us. And you, in turn, have always let us know that, once back in your homeland, you would have been immensely pleased and delighted to come and be in our service, on a regular basis, not only as Chief Mathematician of our University in Pisa, but specifically as our personal Chief Mathematician and Philosopher. Accordingly, we decided to have you here with us, and chose and appointed you as Chief Mathematician of the aforementioned university, and as our personal Chief Mathematician and Philosopher" ("L'eminenza della vostra dottrina et della valorosa vostra sufficienza, accompagnata da singular bontà nelle matematiche et nella filosofia, et l'ossequentissima affezzione, vassallaggio, et servitù che ci havete dimostrata sempre, ci hanno fatto desiderare di havervi appresso di noi; et voi a rincontro ci havete fatto sempre dire che, ripatriandovi, havereste ricevuto per sodisfazione et grazia grandissima di poter venire a servirci del continuo, non solo di Primario Matematico del nostro Studio di Pisa, ma di proprio Primario Matematico et Filosofo della nostra persona: onde, essendoci risoluti di havervi qua, vi habbiamo eletto et deputato per Primario Matematico del suddetto nostro Studio, et per proprio nostro Primario Matematico et Filosofo" (*OG* 10, no. 359*, p. 400).

88. In both *A* and *S*, this sentence and the preceding paragraph are replaced by: "About the end of August (urged by the above-mentioned Prince to release himself from Padua)".

89. On June 15, 1610, Galileo formally renounced his appointment at the University of Padua (see *OG* 19, p. 125) and arrived in Florence on September 9 (Galileo to Cosimo II, October 1610, ibid., no. 401*, p. 439).

90. *S* has: "in September."

91. From two letters Galileo sent to Christopher Clavius and Benedetto Castelli on 30 December 1610, we gather that Galileo began to observe the phases of Venus in October 1610 (see *OG* 10, nos. 446–447, pp. 500 and 503): "[. . .] as soon as I started to observe Venus as it began to be visible in the evening sky,

I saw that its shape was circular, though very small. Later, when I continued my observations, it was grow-ing in magnitude significantly, always maintaining its circular shape. Approaching maximum elongation, however, Venus began to lose its circular shape on the other side from the sun, and within a few days it acquired a semicircular shape. This shape it maintained for several days, that is, until it began to move toward the Sun, slowly abandoning the tangent. It now begins to assume a notable corniculate shape, and will continue to decrease as long as it remains visible in the evening sky. When the time comes, we will see it in the morning sky, with its tiniest horns on the other side from the sun. At about maximum elongation, they will form a half-circle, which they will maintain unchanged for many days" ("come nel principio della sua apparizione vespertina la cominciai ad osservare et la veddi di figura rotonda, ma piccolissima: continuando poi le osservazioni, venne crescendo in mole notabilmente, et pur mantenendosi circolare, sin che, avvicinandosi alla maxima digressione, cominciò a diminuir dalla rotondità nella parte aversa al sole, et in pochi giorni si ridusse alla figura semicircolare; nella qual figura si è mantenuta un pezzo, ciò è sino che ha cominciato a ritirarsi verso il sole, allontanandosi pian piano dalla tangente: hora comincia a farsi notabilmente cornicolata, et così anderà assottigliandosi sin che si vedrà vespertina; et a suo tempo la vedremo mattutina, con le sue cornicelle sottilissime et averse al sole, le quali intorno alla massima digressione faranno mezzo cerchio, il quale manterranno inalterato per molti giorni": ibid., p. 500).

92. After he observed the phases of Venus, Galileo circulated another anagram to establish his priority: *Haec immatura a me iam frustra leguntur o y*, "These unripe ones are being uselessly gathered by me, o y" (Galileo to Giuliano de' Medici, 11 December 1610, in *OG* 10, no. 435, p. 483). Once again, Kepler went crazy in his effort to solve it, offering several possible solutions in a letter to Galileo on January 9, 1611 (*OG* 11, no. 455, pp. 15–16). His best attempt was *Nam Iovem gyrari macula hem rufa testatur*, "Behold! A reddish spot testifies that Jupiter is indeed rotating." Eventually, Galileo provided the solution: *Cynthiae figuras aemulatur mater amorum*, "The mother of love [i.e., Venus] imitates the faces of Cynthia [i.e., the Moon]", that is, Venus has phases, too (letter to Giuliano de' Medici, 1 January 1611, in *OG* 11, no. 451, pp. 11–12). See also *On Sunspots*, pp. 375–376: "the new triform Venus, imitator of the Moon" ("la nuova triforme Venere, emula della Luna": *OG* 5, p. 81). To call attention to their similarities, Venus is here described with an epithet usually attributed to the Moon, given its different appearances: Artemis/Diana (waxing crescent moon), Selene/Luna (full moon), and Hecate/Persephone (waning crescent moon); see, for example, Horace, *Carmina*, III 22, 4 (*diva triformis*), and Vergil, *Aeneis*, IV 511 (*tergemina Hecate*). The discovery of the phases of Venus was of fundamental importance, as it implied that Venus orbited the Sun, thus dealing a fatal blow to the Aristotelian-Ptolemaic model of the universe (which Christoph Scheiner apparently failed to realize, as Galileo pointed out in *On Sunspots*, pp. 92–94; *OG* 5, pp. 98–100) and at the same time offering important evidence in favor of the Copernican theory: "from this amazing evidence we have sensible and certain demonstrations of two great questions, which so far have been debated by the greatest minds of the world. The one being that planets are all dark, by their own nature (for the same thing happens with Mercury as with Venus); the other, that Venus necessarily revolves around the Sun, as Mercury and all the other planets do, too. This was indeed believed by the Pythagoreans, Copernicus, Kepler, and me—but was not sensibly demonstrated, as it is now for Venus and Mercury. Kepler and the other Copernicans will have reasons to be proud of their having believed and philosophized correctly, although we have been, and will be again in the future, regarded by all philosophers *in libris* [that is, who merely rely on what they read in books] as incompetent and almost foolish" ("dalla quale mirabile esperienza haviamo sensata et certa dimostrazione di due gran questioni, state sin qui dubbie tra' maggiori ingegni del mondo. L'una è, che i pianeti tutti sono di loro natura tenebrosi (accadendo anco a Mercurio l'istesso che a Venere): l'altra, che Venere necessariamente si volge intorno al sole, come anco Mercurio et tutti li altri pianeti, cosa ben creduta da i Pittagorici, Copernico, Keplero et me, ma non sensatamente provata, come hora in Venere et in Mercurio. Haveranno dunque il Sig. Keplero et gli altri Copernicani da gloriarsi di havere creduto et filosofato bene, se bene ci è toccato, et ci è per toccare ancora, ad esser reputati dall'universalità de i filosofi *in libris* per poco intendenti et poco meno che stolti": Galileo to Giuliano de' Medici, 1 January 1611, in *OG* 11, no. 451, p. 12).

93. Galileo left for Rome on March 3, 1611 (see *OG* 19, p. 200), and arrived on March 29 (Giovanni Niccolini to Grand Duke Cosimo II, 30 March 1611, in *OG* 11, no. 504*, p. 78): at a few stops along the way, as well as in Rome, he continued observing Jupiter's satellites (see *OG* 3.2, pp. 442–443).

94. Cardinal Ottavio Bandini (1558–1629); Pietro Dini (ca. 1570–1625), a nephew of Bandini's, whom he succeeded as Archbishop of Fermo, in The Marches, in 1621; Archbishop Ottavio Corsini (1588–1641); Orazio Cavalcanti (b. 1562); Giulio Strozzi (1583–1660); Giovanni Battista Agucchi (1570–1632), who had been introduced to Galileo in Rome, by mathematician Luca Valerio, and was appointed Nuncio in Venice by Urban VIII, in 1623. See also *On Sunspots*, pp. 375–376 (*OG* 5, pp. 81–82), and Dini to Galileo, 2 May 1615 (in *OG* 12, no. 1115, p. 175). On April 19, 1611, Cardinal Roberto Bellarmino wrote to the mathematicians of the Roman College, saying he had himself made amazing observations of the Moon and Venus, and asking their opinion about Galileo's telescopic discoveries; a few days later, on April 24, Jesuits Christopher Clavius, Christopher Grienberger, Odo van Maelcote, and Giovanni Paolo Lembo responded, cautiously

confirming Galileo's telescopic observations but avoiding commitment to any interpretation of them: see *OG* 11, no. 515, pp. 87–88, and no. 520, pp. 92–93. See also Galileo to Belisario Vinta, 1 April 1611, ibid., no. 505, pp. 79–80; Galileo to Virginio Orsini, 8 April 1611, ibid., no. 510**, pp. 82–83; and Galileo to Filippo Salviati, 22 April 1611, ibid., no. 517, pp. 89–90.

95. Christoph Scheiner (1573–1650), who entered the Jesuit order in 1595 and studied mathematics and metaphysics at the University of Ingolstadt. Early in his career he became an expert on the mathematics of sundials and also invented a pantograph for copying and enlarging drawings. Upon hearing about Galileo's telescopic discoveries, in 1610, he set out to obtain good telescopes and turned his attention to the Sun, where he discovered sunspots in March or April 1611. His publications about them sparked a controversy with Galileo over the nature of sunspots and the priority over their discovery. In *Tres epistolae de macvlis solaribvs* (1612)—dedicated to Markus Welser and signed "Apelles latens post tabulam" ("Apelles, hiding behind the painting," with reference to a well-known story told by Plinius in the *Naturalis historia*, XXXV 84–85; see also Valerius Maximus, *Facta et dicta memorabilia*, VIII 12, 3)—Scheiner attempted to rescue the perfection of the Sun, and by implication of the heavens in general, by postulating that sunspots were caused by satellites orbiting the Sun, projecting their shadows onto the Sun's disk.

96. *S* continues: "when it is otherwise well known that Galileo had discovered them [i.e., sunspots] a few months before he returned from Padua, that is, one year earlier, in 1610."

97. At first, Galileo was unable to determine the periods of Jupiter's satellites: see the *Sidereal Messenger*, p. 66 (*OG* 3.1, p. 80). He succeeded in obtaining a first approximation of their periods in April 1611, after long and continued observations (see *OG* 3.2, pp. 442–444), as we gather from what Galileo himself states in *On Floating Bodies*, pp. 18–20 (*OG* 4, pp. 63–64). He announced the discovery to a few friends, including Daniello Antonini (see Antonini to Galileo, 24 June 1611, in *OG* 11, no. 544, p. 129), in order to state his priority, but kept working on them until he eventually communicated the important result to Giuliano de' Medici on June 23, 1612 (see ibid., no. 706, p. 335).

98. Galileo was inducted to the Accademia dei Lincei (Academy of the Lynxes, or Lincean Academy) on April 25, 1611: see *OG* 19, p. 265. The Academy was founded on August 17, 1603, by naturalist Federico Cesi, (1585–1630), Marquis of Monticelli, in association with mathematician Francesco Stelluti (1577–1653), of Fabriano; polymath Anastasio De Filiis (1577–1608), of Terni; and the Dutch physician Johannes van Heeck (1574–1616). The Academy aimed at understanding nature, at both microscopic and macroscopic levels, through a method of research based upon observation and experiment. The name Lincei came from Giovanni Battista Della Porta's *Magia naturalis* (1589, in twenty books), on whose title page is an engraving of a lynx, with the words "aspicit et inspicit" ("looks and examines"). The lynx was the emblem of an inquiry aimed at both the visible aspects of nature and the more secret and hidden ones, as tradition ascribed to the lynx a most acute eyesight, capable of seeing the exterior and interior of things. The motto Cesi chose for the Academy was "minima cura si maxima vis," "Take care of small things if you want to achieve the greatest results." The life of the Academy—which published Galileo's *On Sunspots* (1613) and *The Assayer* (1623), and also supported Galileo in his clashes with prominent scholars and the Church hierarchy—was cut short by Cesi's sudden death. After an attempt to reestablish it in 1801, the Academy resumed its activities in 1847, and was eventually refounded in 1874.

99. Galileo left Rome on June 4, 1611 (Piero Guicciardini to Belisario Vinta, 4 June 1611, in *OG* 11, no. 538*, p. 121) and arrived in Florence on June 13 (Belisario Vinta to Piero Guicciardini, 13 June 1611, ibid., no. 540*, p. 125).

100. See *On Floating Bodies*, pp. 20–22 (*OG* 4, pp. 64–65). Cosimo enjoyed gathering scientists and literary men at his court and attending their debates, which he encouraged and instigated. Years later, his son Leopoldo—himself a scholar and patron of arts and sciences—would found the Accademia del Cimento: see note 2.

101. *On Floating Bodies* (1612); see also *OG* 4, pp. 17–56.

102. *A* and *S* have: "revealed the times of the periodic movements of the Medicean Planets, which he had investigated about April 1611, while in Rome, also giving news of the newly discovered [sunspots]".

103. See *On Floating Bodies*, pp. 19–20: "I add to these things the observation of some dark spots that are seen in the Sun's body, which, changing their position in that, offer a strong argument either that the Sun revolves or perhaps that other stars similar to Venus and Mercury circulate around the Sun, sometimes invisible by reason of their small elongations, smaller even than that of Mercury, and visible only when they come between the Sun and our eyes—or perhaps they indicate both things. Certainty on such matters must not be scorned or hidden. *Continued observations have finally assured me that such spots are materials contiguous to the Sun's body, there continually produced in numbers and then dissolved, some in shorter and some in longer times, being carried around by rotation of the Sun itself, which completes its period in about a lunar month—a great event, and even greater for its consequences*" ("Aggiungo a queste cose l'osservazione d'alcune macchiette oscure, che si scorgono nel corpo solare: le quali, mutando positura in quello, porgono grand'argomento, o che 'l Sole si rivolga in sé stesso, o che forse altre stelle, nella guisa di Venere e di Mercurio, se gli volgano intorno, invisibili in altri tempi per le piccole digressioni e minori di quella di Mercurio, e solo visibili

quando s'interpongono tra 'l Sole e l'occhio nostro, o pur danno segno che sia vero e questo e quello; la certezza delle quali cose non debbe disprezzarsi o trascurarsi. *Ànnomi finalmente le continuate osservazioni accertato, tali macchie esser materie contigue alla superficie del corpo solare, e quivi continuamente prodursene molte, e poi dissolversi, altre in più brevi ed altre in più lunghi tempi, ed esser dalla conversione del Sole in sé stesso, che in un mese lunare in circa finisce il suo periodo, portate in giro; accidente per sé grandissimo, e maggiore per le sue conseguenze*": *OG* 4, p. 64). The second edition of *On Floating Bodies* appeared a few months after the first (in the same year, 1612), with an identical title page (only the words "second edition" were added); the lines here reproduced in italics were added in the second edition. From the observation of sunspots, Galileo concluded that the Sun rotates around its axis, dragging the planets with it: see *On Sunspots*, pp. 90 and 109 (*OG* 5, pp. 96 and 117).

104. Lodovico delle Colombe, *Discorso apologetico d'intorno al Discorso di Galileo Galilei* (1612); Vincenzo Di Grazia, *Considerazioni sopra 'l Discorso di Galileo Galilei* (1613); Giorgio Coresio, *Operetta intorno al galleggiare de corpi solidi* (1612). At the end of his footnote, Viviani first referred to Flaminio Papazzoni, whose name was then replaced by that of Tommaso Palmerini: one of them, according to Viviani, is the author of the *Considerazioni sopra il Discorso del Sig. Galileo Galilei Intorno alle cose, che stanno in su l'Acqua, o che in quella si muouono*, anonymously published by one "Accademico Incognito" ("unknown academician"), in Pisa, in 1612. The identity of the anonymous author must have posed a challenge to Galileo's own contemporaries, as Giovanni Battista Agucchi identified him with philosopher Flaminio Papazzoni, from Bologna (Agucchi to Galileo, 1 September 1612, in *OG* 11, no. 755*, p. 390). In his extensive biography of Galileo, Giovanni Battista Clemente Nelli identified him with Tommaso Palmerini, a Peripatetic philosopher from Pisa (Nelli, *Vita e commercio epistolare di Galileo Galilei* (1793), vol. 1, p. 314); by contrast, Antonio Favaro (*OG* 4, p. 6) identified the author with Arturo Pannocchieschi d'Elci, General Commissioner of the University of Pisa (1608–1614), who signed the dedicatory letter of the book, claiming he translated the text from Latin (ibid., p. 147). This conclusion was disputed by Stillman Drake, who agreed with Agucchi, but still remains the most plausible hypothesis for the identification of the "unknown academician." In Viviani's footnote, instead of the ellipsis *S* has: "Tommaso"; *A* does not mention Palmerini.

105. *A* and *S* continued with these words: "full of wicked malice as much as crassest ignorance." Benedetto Castelli replied to both delle Colombe's and Di Grazia's criticisms in the *Risposta alle opposizioni del S. Lodovico delle Colombe, e del S. Vincenzio di Grazia* (1615). The work was published anonymously but opens with a dedication to Enea Piccolomini signed by Castelli. Galileo's pupil also exposed Coresio's mistakes in the *Errori dei più manifesti commessi da Messer Giorgio Coresio, nella sua Operetta del galleggiare della figura*, which remained unpublished (and is now available in *OG* 4, pp. 247–286). Galileo significantly contributed to both these works. See also *OG* 4, pp. 443–447.

106. *A* and *S* have: "closest."

107. Filippo Salviati (1582–1614) was one of Galileo's closest friends. He is Galileo's spokeperson in the *Dialogue*, with the double role of countering Simplicio's Aristotelian positions and at the same time correcting Sagredo's ingenuousness, thereby explaining away the obvious difficulties in the Copernican theory. Salviati often hosted Galileo at his Villa delle Selve, in Lastra a Signa, on the outskirts of Florence, where Galileo could rest and work in tranquility: see note 152. The villa provided the setting for much of Galileo's work on sunspots; there he made observations for three months, from February 12 to May 3 (see *OG* 5, pp. 253–254).

108. Markus Welser (1558–1614), a member of a noble family from Augsburg, studied in Padua, Paris, and Rome, where he became a member of the Lincean Academy. He returned to Augsburg in 1584, and started a trading company with his brother; most important, he was the node of a wide network of scholars and scientists throughout Europe, and his extensive patronage activity made him one of the key figures of European *Späthumanismus*. In 1594, he founded a publishing house, "Ad insigne pinus," which remained active for twenty-five years. Besides Galileo and Christoph Scheiner, he corresponded with Isaac Casaubon, Joseph Justus Scaliger, and Joachim Camerarius the Younger.

109. Welser's first letter to Galileo (dated January 6, 1612) accompanied a copy of Scheiner's *Tres epistolae de macvlis solaribvs* (1612), which included three letters by Scheiner to Welser on sunspots (dated November 12, December 19, and December 26, 1611). Galileo responded on May 5, from Villa Le Selve. Welser wrote a second letter on June 1, to which Galileo responded on August 14, from Florence. In his letters, Galileo argued that sunspots are on or near the surface of the Sun, that they change their shapes, and that they are often seen to originate and disappear on the solar disk, thereby claiming the Sun is not perfect. In the meanwhile, Scheiner wrote two further letters to Welser (on January 16 and April 14, 1612), and after reading Galileo's first letter (5 May 1612) to Welser, he wrote yet another letter to Welser (on July 25), repeating that sunspots were shadows cast by satellites orbiting the Sun, and arguing that Jupiter had more satellites than those discovered by Galileo. This second series of letters of Scheiner's was published by Welser as *De macvlis solaribvs et stellis circa Iouem errantibus, accvratior disqvisitio* (1612). Again, Scheiner signed with the pseudonym "Apelles latens post tabulam," adding "vel, si mavis, Ulysses sub Aiacis clypeo" ("or, if you wish, Ulysses [hiding] under Ajax's shield"; the source of this image is Homer, *Ilias*, XVII

123–137: having killed Patroclus and stripped his body of the armor, Hector wants to drag him away, cut off his head, and give the corpse to the dogs of Troy, "but Ajax came up, carrying his shield that was like a city wall" ("Αἴας δ᾽ ἐγγύθεν ἦλθε φέρων σάκος ἠΰτε πύργον," ibid., 128); Hector gave ground backward, and "Ajax covered the son of Menoetius round about with his broad shield, and stood like a lion over his whelps, one that huntsmen have encountered in the forest as he leads his young" ("Αἴας δ᾽ ἀμφὶ Μενοιτιάδη σάκος εὐρὺ καλύψας / ἑστήκει ὥς τίς τε λέων περὶ οἷσι τέκεσσιν, / ᾧ ῥά τε νήπι᾽ ἄγοντι συναντήσωνται ἐν ὕλη / ἄνδρες ἐπακτῆρες")). Welser wrote two additional letters to Galileo (on September 28 and October 5), to which Galileo replied with his third letter on sunspots, dated December 1, from Villa Le Selve. The exchange between Welser and Galileo was eventually published in *Istoria e dimostrazioni matematiche intorno alle macchie solari e loro accidenti* (*On Sunspots*, 1612). In this work, Galileo politely but firmly refuted Scheiner's arguments. Later on, Scheiner published *Ocvlvs hoc est: fvndamentvm opticvm* (1619), on the optics of the eye, building on Kepler's works, and *Rosa Vrsina* (1630), which became the standard work on sunspots for more than a century. On the engraved frontispieces of these two works, Scheiner restated his priority over the discovery of sunspots. His last work (*Prodromvs pro Sole mobili et terra stabili* (1651)), criticizing Galileo's *Dialogue*, was published posthumously, although it appears to have been written earlier, in 1633 (see Scheiner to Gassendi, 23 February 1633, in *OG* 15, no. 2418*, p. 47).

110. Once he had discovered Jupiter's satellites, Galileo proposed that navigators use their positions and eclipses to compute longitudes. He carefully calculated the satellites' periods and refined his tables of their motions to a high degree of reliability, so as to determine the longitude of the place of observation at any hour of the night (a ready method of finding longitudes at sea had long been an object of search by all the maritime powers of Europe). In July 1612, Belisario Vinta started a correspondence on this subject with the representative of the Spanish Court, Orso d'Elci; the correspondence went on for a few months (Galileo wrote a letter on 7 September, too), but stopped in October. The negotiations were resumed in June 1616 and protracted until December 1620 (also including Galileo's design for the *celatone*, in 1617: see also note 112), with several exchanges between Orso d'Elci, Curzio Picchena, and Giuliano de' Medici, but once again led nowhere. The issue was finally taken up again in May 1630, but failed once again; any plans were eventually given up in October 1632, after several exchanges between Esaù Del Borgo and Andrea Cioli.

111. Philip III of Spain (1578–1621): a member of the House of Habsburg, he was King of Spain and Portugal, and reigned from 1598 until his death.

112. Here Viviani, as many other scholars after him (see, for example, Nelli, *Vita e commercio letterario di Galileo Galilei* (1793), vol. 1, p. 281), ascribes to Galileo the invention of binocular spyglasses, identifying them with the *celatone*. These scholars possibly relied on Galileo's letters to Curzio Picchena, 22 March 1617 (in *OG* 12, no. 1251, pp. 311–312); to Orso d'Elci, in June 1617 (ibid.., no. 1260, pp. 321–328); and to Leopold V, Archduke of Austria, on 23 May 1618 (ibid., no. 1324, p. 390). Galileo had been commissioned with the design of new devices to facilitate the use of the telescope on the galleys of the Grand Duke, and in 1617 he presented the *celatone* to solve the problems caused by oscillations, which prevented the efficient use of the telescope on ships. The *celatone*, however, took the form of a headgear with a telescope taking the place of one eyehole, the other being free. Galileo himself made this very clear in his letter to Laurens Reael, on June 6, 1637 (see *OG* 17, no. 3496, p. 99). In fact, the inventor of binoculars was the cartographer Ottavio Pisani, who communicated it to Galileo himself in a letter dated September 15, 1613: "I am working on a book on general optics, and I have a lot to say about the assembly of this spyglass and the symmetry of the lenses, what its length must be, and how to construct it. Yet, I do not make this spyglass so as to be used with one eye, but with two eyes, and I turn both eyes into one direction" ("Ego paro librum de tota prospectiva, et habeo multa circa construxionem huius pespicilli [*sic*] et symmetriam vitrorum, quanta debet esse longitudo, quis modus formandi. Verum ego non facio hunc pespicillum [*sic*] uno oculo apponendum, sed duobus oculis, et ambos oculos volvo in unum": *OG* 11, no. 924*, p. 565). See also his other letter to Galileo, dated July 18, 1614: "I made one of those spyglasses that Your Lordship, almost like a novel and celestial Amerigo [Vespucci], turned to the heavens—I made, I say, a telescope *with two eyes*, just like others have one; the body is short, and shaped like an oval" ("Io ho fatto uno di quelli occhiali che V. S., quasi nuovo et celeste Americo, have rivolto al cielo; ho fatto, dico, uno telescopio *a due occhi*, come li altri sono ad uno: il corpo è poco, e di figura ovale": *OG* 12, no. 1030*, pp. 86–87). In 1617, Galileo taught Castelli how to make binoculars so that he could test the instrument and inform Galileo of any possible defects: see Castelli to Galileo, 24 May 1613 (in *OG* 12, no. 1257**, pp. 319–320). Pisani also corresponded with Kepler about the binoculars: see Pisani to Kepler, 7 October 1613 (in *KGW* 17, no. 665, p. 76), and Kepler to Pisani, 16 December 1613 (ibid., no. 675, p. 96). Kepler told Pisani he had thought about the binoculars, too, and worked on it two years earlier; eventually, however, he gave up the instrument. See also Pisani's undated reply to Kepler (ibid., no. 676, pp. 98–99) and his letter on 4 January 1618 (ibid., no. 778, pp. 244–246).

113. Three comets appeared in the sky, beginning with August 1618; the third was particularly bright and remained visible almost until the end of January 1619. During the whole time that the comet was visible, Galileo was confined by illness to his bed and had to rely on reports from friends: see *The Assayer*, p. 176 (*OG* 6, p. 225), Paolo Gualdo to Galileo, 30 November 1618, and Virginio Cesarini to Galileo, 1 December

1618 (in *OG* 12, no. 1355*, p. 421, and no. 1357, pp. 422–423). The following March, the Jesuits published *De tribvs cometis* (1619), in which (unlike Aristotle, who believed them to take place below the sphere of the Moon) comets were described as celestial objects, traveling well beyond the lunar sphere; indeed, their parallax was barely measurable. Although anonymous, its author was soon recognized in Orazio Grassi, professor of mathematics at the Roman College (and designer of the church of Sant'Ignazio, in Rome). Grassi relied on Tycho Brahe's analysis of the comet of 1577.

114. Leopold V, Archduke of Further Austria (1586–1632).

115. See Galileo to Leopold, 23 May 1618 (in *OG* 12, no. 1324, p. 389); see also Castelli to Galileo, 1 March 1618 (ibid., no. 1307*, p. 374), and Cesi to Galileo, 20 April 1618 (ibid., no. 1315, p. 383).

116. Mario Guiducci (1585–1646) studied law at the University of Pisa and was awarded a doctorate *in utroque iure* in 1610. After the publication of Grassi's *De tribvs cometis* (1619), validating Tycho's geo-heliocentric model of the universe, Galileo intervened to defend the Copernican worldview. Together with Guiducci, he wrote the *Discorso delle comete* (1619), which appeared only under the name of Guiducci. In it, comets are (erroneously) explained not as celestial objects but as optical effects produced by sunlight on vapors originated from the earth. Grassi replied to Guiducci in the *Libra astronomica ac philosophica* (1619), pointing to Galileo as the true author of the *Discorso delle comete* and openly criticizing the Copernican theory. Galileo would eventually reply in person in 1623, with *The Assayer*. In 1620, Guiducci published a letter to Tarquinio Galluzzi, in which he claimed that Galileo did not write the *Discorso* but that he (Guiducci) popularized Galileo's ideas.

117. *A* has: "as a natural philosopher"; and *S*: "as a free natural philosopher." These key words were transcribed into *B*, too, but were eventually deleted: Viviani resolutely tried to tone down his account of Galileo's life as best as he could.

118. Lotario Sarsi Sigensano is the anagram of Oratio Grassi Salonensi, i.e., from Savona, where Grassi was born in 1583.

119. Virginio Cesarini (1595–1624): a friend of Cardinals Roberto Bellarmino's and Maffeo Barberini's, as well as of Federico Cesi's, he was enrolled in the Academy of the Lynxes in 1618. Here he met with Giovanni Ciampoli and Galileo, who had a profound influence on Cesarini's scientific and philosophical views (see Cesarini to Galileo, 1 October 1610, in *OG* 12, no. 1349, pp. 413–415). Cesarini was appointed the Papal Chamberlain by Gregory XV and served under Urban VIII (Maffeo Barberini), who also appointed him Master of Chamber. After the publication of Grassi's *Libra*, Cesarini urged Galileo to publish a rejoinder (see Cesarini to Galileo, 7 May 1622, in *OG* 13, no. 1523, p. 89). This took the form of a long letter to Cesarini and was eventually published in 1623 as *The Assayer*, printed by the Academy of the Lynxes, and dedicated to the newly elected pope, Urban VIII. In the *Libra* (Latin for "balance"), Grassi aimed at "weighing" different views about the nature of comets; with *The Assayer*, Galileo sarcastically offered a better assessment of the arguments: the assayer (*saggiatore*) was a balance used by jewelers to weigh precious metals and was therefore much more accurate than the *libra*, commonly used in markets.

120. While working on the book, in the early 1620s, Galileo referred to it as *Discorso sopra 'l flusso e reflusso*, that is, *Discourse on the Ebb and Flow* (see Galileo to Cesare Marsili, 7 December 1624, in *OG* 13, no. 1688*, p. 235); later, he referred to it as *Dialogi intorno al flusso e reflusso* (*Dialogues about the Ebb and Flow*: see Galileo to Élie Diodati, 29 October 1629, in *OG* 14, no. 1962*, p. 49). In order to be granted the imprimatur, however, he was required to change its title, as approval of it would look like approval of Galileo's theory of tides using the motion of the earth as proof: see Niccolò Riccardi's 24 May 1631 letter to the Inquisitor, in *OG* 19, p. 327. The *Dialogue* is presented as a series of discussions, over a span of four days, between two philosophers and a layman: Salviati (after the astronomer and Galileo's friend Filippo Salviati, a member of the Academy of the Lynxes), who argues for the Copernican position and offers some of Galileo's views directly; Sagredo (after the Venetian mathematician and one of Galileo's closest friends Giovanni Francesco Sagredo, 1571–1620), an intelligent and initially neutral layman; and Simplicio (after Simplicius of Cilicia, author of extensive commentaries on Aristotle's philosophy; but the word also means "simpleton," in Italian), a dedicated and loyal follower of Ptolemy's and Aristotle's. During the fourth and last day of their exchanges, tides become the central issue, and Salviati purports to explain the ebb and flow of the sea by appealing to the movements of the earth, according to the Copernican hypothesis.

121. See Galileo's *Discorso del flusso e reflusso del mare*, written in Rome, in the form of a letter to Cardinal Alessandro Orsini, on 8 January 1616 (in *OG* 5, pp. 377–395; new edition in *OGA* 3, pp. 233–254). A patron of Galileo's, Alessandro Orsini (1592–1626) was brought up at the court of Ferdinando I, Grand Duke of Tuscany, and was created a cardinal by Paul V in 1615. In 1616, after a letter from the Grand Duke, Cardinal Orsini protected Galileo, speaking in his support to the pope: see Galileo's letters to the successor of Belisario Vinta as the Grand Duke's Secretary of State, Curzio Picchena, on February 6 and 20, 1616 (in *OG* 12, nos. 1174, pp. 231–232, and 1182, p. 238); see also Piero Guicciardini to Cosimo II, 4 March 1616, ibid., no. 1185, p. 242.

122. As Tommaso Campanella put it, "Everyone plays his own role perfectly. Simplicio seems the amusement of this philosophical comedy, who demonstrates simultaneously the stupidity of his sect, its

way of arguing, its fickleness, and stubbornness, and whatever comes with all this. Assuredly, there is no need to envy Plato. Salviati is a great Socrates who helps in giving birth more than giving birth himself; and Sagredo is a free spirit: without being corrupted by the schools, he judges them all with a great deal of wisdom" ("Ognun fa la parte sua mirabilmente; e Simplicio par il trastullo di questa comedia filosofica, ch'insieme mostra la sciocchezza della sua setta, il parlare, e l'instabilità, e l'ostinatione, e quanto ci va. Certo che non havemo a invidiar Platone. Salviati è un gran Socrate, che fa parturire più che non parturisce, et Sagredo un libero ingegno, che senza esser adulterato nelle scole giudica di tutte con molta sagacità": Campanella to Galileo, 5 August 1632, in *OG* 14, no. 2284, p. 366). Most remarkably, Viviani's briefly refers to Galileo's *magnum opus* as a dialogue between two people, Filippo Salviati and Giovanni Francesco Sagredo, avoiding even a mention of the third character, Simplicius (in whose words Pope Urban VIII read a mockery of his own arguments).

123. *S* has: "he possibly showed himself closer."

124. *A* has: "since in his *Dialogues* he advocated the opinion of the Pythagoreans and of Copernicus on the stillness of the Sun and the mobility of the earth."

125. *B* has, in the margin: "and even praised Galileo publicly, with poems." In 1620, Maffeo Barberini published the *Adulatio perniciosa*, a poem in praise of Galileo and his telescopic discoveries: see the appendix to this book.

126. Francesco Niccolini (1584–1650), the Grand Duke's ambassador in Rome from 1621 to 1643. Together with his wife Caterina Riccardi, he closely assisted Galileo, before, during, and after the trial, as is documented by the letters by the two of them to Galileo and his daughter Maria Celeste, as well as by Niccolini's many reports to the Grand Duke's secretary, Andrea Cioli, in 1630–1634.

127. Galileo set up for his journey to Rome on January 20, 1633 (see Andrea Cioli to Francesco Niccolini, 21 January 1633, in *OG* 15, no. 2390*, p. 29); because of the plague, he had to spend a quarantine period near Acquapendente, on the border between the Grand Duchy of Tuscany and the Papal States. He arrived at Rome in the evening of February 13 and took up lodging at the residence of the Tuscan ambassador, Francesco Niccolini, at Villa Medici (see Niccolini to Cioli, 14 February 1633, ibid., no. 2408, p. 40). Galileo first appeared before the Commissary of the Holy Office on April 12 and was lodged in the chambers of the prosecutor of the Tribunal (Niccolini to Cioli, 16 April 1633, ibid., no. 2471, p. 94); it was not until April 30 that he was granted permission to return to the Tuscan embassy, as a substitute prison (*OG* 19, p. 344). On the morning of June 21 Galileo was requested to present himself to the Holy Office; the next day, Galileo had to put on the penitential garb and was led to the Dominican convent of S. Maria sopra Minerva (in the center of Rome, very close to the Pantheon). There he was ordered to kneel down while the sentence of condemnation was being read. He was pronounced, sentenced, and declared "vehemently suspected of heresy, namely, of having held and believed a doctrine which is false and contrary to the divine and Holy Scripture: that the sun is the center of the world and does not move from east to west, and the earth moves and is not the center of the world, and that one may hold and defend as probable an opinion after it has been declared and defined contrary to Holy Scripture. Consequently, you have incurred all the censures and penalties proclaimed and issued against such delinquents by the sacred canons and other general and particular statutes" ("Diciamo, pronunciamo, sententiamo e dichiaramo che tu, Galileo sudetto, per le cose dedotte in processo e da te confessate come sopra, ti sei reso a questo S. Off.° vehementemente sospetto d'heresia, cioè d'haver tenuto e creduto dottrina falsa e contraria alle Sacre e divine Scritture, ch'il sole sia centro della terra e che non si muova da oriente ad occidente, e che la terra si muova e non sia centro del mondo, e che si possa tener e difendere per probabile un'opinione dopo esser stata dichiarata e diffinita per contraria alla Sacra Scrittura; e conseguentemente sei incorso in tutte le censure e pene dai sacri canoni et altre constitutioni generali e particolari contro simili delinquenti imposte e promulgate": *OG* 19, p. 405). The *Dialogue* was prohibited, and Galileo was forced to abjure:

> I, Galileo, son of the late Vincenzio Galilei, Florentine, aged seventy years, arraigned personally for judgment, kneel before you Most Eminent and Most Reverend Cardinals Inquisitors-General against heretical depravity in the entire Christendom. Having before my eyes and touching with my hands the Holy Gospels, I swear that I have always believed, I believe now, and with God's help I will believe in the future all that the Holy Catholic and Apostolic Church holds, preaches, and teaches. However, whereas, after having been judicially instructed with injunction by this Holy Office, to abandon altogether the false opinion that the sun is the center of the world and does not move, and the earth is not the center of the world and moves, and not to hold, defend or teach this false doctrine in any way whatever, either orally or in writing; and after it had been notified to me that this doctrine is contrary to the Holy Scripture—I wrote and published a book in which I discuss this very doctrine, which had been already condemned, and offer very effective arguments in its favor, without refuting them in any way. Therefore, I have been judged vehemently suspected of heresy, namely, of having held and believed that the sun is the center of the world and motionless, and the earth is not the center and moves.

Therefore, desiring to remove from the minds of Your Eminences, and of every faithful Christian, this vehement suspicion, rightly conceived against me, with sincere heart and unfeigned faith I abjure, curse, and detest the aforementioned errors and heresies, and in general each and every other error, heresy, and sect contrary to the Holy Church. And I swear that in the future I will never again say or assert, orally or in writing, anything which might cause a similar suspicion about me. By contrast, if I should come to know any heretic, or anyone suspected of heresy, I will denounce him to this Holy Office, or else to the Inquisitor or Ordinary of the place where I happen to be.

Moreover, I swear and promise to comply with and observe entirely all the penances that have been or will be imposed upon me by this Holy Office. Should I fail to keep any of these promises and oaths—God forbid—I submit myself to all the penalties and punishments proclaimed and issued against such delinquents by the sacred canons and other general and particular statutes. So help me God and these Holy Gospels of His, which I touch with my hands.

I, the said Galileo Galilei, have abjured, sworn, promised, and bound myself as above; and in witness of the truth, I have signed with my own hand the present document of my abjuration, and recited it word for word in Rome, at the convent of the Minerva, this twenty-second day of June 1633.

Io Galileo, fig.^{lo} del q. Vinc.^{o} Galileo di Fiorenza, dell'età mia d'anni 70, constituto personalmente in giuditio, et inginocchiato avanti di voi Emin.^{mi} et Rev.^{mi} Cardinali, in tutta la Republica Christiana contro l'heretica pravità generali Inquisitori; havendo davanti gl'occhi miei li sacrosanti Vangeli, quali tocco con le proprie mani, giuro che sempre ho creduto, credo adesso, e con l'aiuto di Dio crederò per l'avvenire, tutto quello che tiene, predica et insegna la S.^{a} Cattolica et Apostolica Chiesa. Ma perché da questo S. Off.,^{o} per haver io, dopo d'essermi stato con precetto dall'istesso giuridicamente intimato che omninamente dovessi lasciar la falsa opinione che il sole sia centro del mondo e che non si muova e che la terra non sia centro del mondo e che si muova, e che non potessi tenere, difendere né insegnare in qualsivoglia modo, né in voce né in scritto, la detta falsa dottrina, e dopo d'essermi notificato che detta dottrina è contraria alla Sacra Scrittura, scritto e dato alle stampe un libro nel quale tratto l'istessa dottrina già dannata et apporto ragioni con molta efficacia a favor di essa, senza apportar alcuna solutione, sono stato giudicato vehementemente sospetto d'heresia, cioè d'haver tenuto e creduto che il sole sia centro del mondo et imobile e che la terra non sia centro e che si muova;

Pertanto, volendo io levar dalla mente delle Eminenze V.^{re} e d'ogni fedel Christiano questa vehemente sospitione, giustamente di me conceputa, con cuor sincero e fede non finta abiuro, maledico e detesto li sudetti errori et heresie, e generalmente ogni et qualunque altro errore, heresia e setta contraria alla Sta Chiesa; e giuro che per l'avvenire non dirò mai più né asserirò, in voce o in scritto, cose tali per le quali si possa haver di me simil sospitione; ma se conoscerò alcun heretico o che sia sospetto d'heresia, lo denontiarò a questo S. Offitio, o vero all'Inquisitore o Ordinario del luogo dove mi trovarò.

Giuro anco e prometto d'adempire et osservare intieramente tutte le penitenze che mi sono state o mi saranno da questo S. Off.^{o} imposte; e contravenendo ad alcuna delle dette mie promesse e giuramenti, il che Dio non voglia, mi sottometto a tutte le pene e castighi che sono da' sacri canoni et altre constitutioni generali e particolari contro simili delinquenti imposte e promulgate. Così Dio m'aiuti e questi suoi santi Vangeli, che tocco con le proprie mani.

Io Galileo Galilei sodetto ho abiurato, giurato, promesso e mi sono obligato come sopra; et in fede del vero, di mia propria mano ho sottoscritta la presente cedola di mia abiuratione et recitatala di parola in parola, in Roma, nel convento della Minerva, questo dì 22 Giugno 1633. (*OG* 19, pp. 406–407)

The *Dialogue* was listed in the 1634 edition of the *Index librorum prohibitorum*, and it was not until 1710 that it was reprinted (in Naples, even though the title page declares Florence as the place of publication; no mention of the publisher), with some additional material: Galileo's *Letter to Christina*, Paolo Antonio Foscarini's *Lettera* (1615), an excerpt from the Introduction of Kepler's *Astronomia nova* (1609), an excerpt from Diego de Zuñiga's *In Iob Commentaria* (1591), and both Galileo's sentence and abjuration (both from Giovanni Battista Riccioli, *Almagestvm Novvm* (1651)). The *Dialogue* was not included in *Opere 1718* but was included in *Opere 1744*, vol. 4: this reproduces the 1632 text (with the exception of the marginal postils stating that the earth truly moves, which are either removed or corrected), prefaced by the condemnation of Galileo, his abjuration, and an introductory essay by Father Augustin Calmet. The general prohibition of works on heliocentrism was retired in 1758, and works advocating the Copernican theory were permitted in print in 1822. The *Dialogue* remained in the *Index* of prohibited books until 1835.

128. Here, *A* continues as follows: "imposing eternal silence on this matter." These words appear in *B*, too, but were crossed out. In the margin, *B* also has this annotation: "and he was sentenced to house arrest at the will of His Holiness; and after he was allowed to leave Rome (in order not to risk his life, as the plague was still infecting the city of Florence), by generous mercy he was required to . . ." It was meant to replace part of the text that follows, but never did; Viviani remained undecided about the phrasing. In the early 1690s Viviani asked the Jesuit scholar Antonio Baldigiani (whose moral authority and intellectual abilities earned him the respect of colleagues inside the Vatican and at the Roman College) to help

lift the ban on the *Dialogue*, reiterating the same arguments offered in the *Racconto istorico*: "It is not an underground publication: the book was granted permission by the Master of the Sacred Palace [Niccolò Riccardi], appointed by Pope Urban [VIII], and later printed in Florence with the permission of other superiors. It was written by a Catholic and pious author, and dedicated to the most religious Grand Duke of Tuscany, and dealt not with doctrines that were deliberately against faith but with merely philosophical, mathematical, astronomical, and physical issues that were written, put forth, and debated freely by most eminent personalities such as Gassendi, Riccioli, Tacquet, and a hundred more. Ultimately, it is a work in which if the author showed himself perhaps more inclined toward the Copernican than the Ptolemaic hypothesis, against the injunction he had received, he later changed his mind, acknowledged his errors, was absolved, lived and died most devoutly, with all assistance from the Church. I know that, as I was present, together with two parish priests. We assisted him during his illness until his death, which followed with total remission to God, full awareness through the last moment, and the edification all those who attended, among whom were his son and daughter-in-law, all the family and Torricelli" ("Qui si tratta di un Libro non stampato alla macchia, ma licenziato prima in Roma dal Maestro del Sacro Palazzo consegnatogli da un papa Urbano, e di poi con permissione di altri superiori stampato in Firenze, composto da un autor cattolico e pio: dedicato ad un Granduca di Toscana religiosissimo, e trattante non di dottrine ex professo contro la fede, ma di materie mere filosofiche, matematiche, astronomiche e fisiche, scritte, poi ventilate e discusse con libertà da persone Excellentissime come fu il Gassendi, il Riccioli, il Tacquet e cento altri, e finalmente d'un'opera nella quale se l'autore si dimostrò forse più propenso all'ipotesi Copernicana che alla Tolemaica, contro i precetti che ne aveva avuti, egli poi si disdisse, confessò l'error suo, ne fu assoluto, visse e morì santissimamente con tutti gli aiuti della Chiesa, come io lo so, che fui presente con due sacerdoti curati che l'assistimmo nella malattia alla di lui morte seguìta con totale remissione in Dio, cognizione fino all'ultimo et edificazione dei circostanti, fra quali intervenne anche il figliuolo e la sua moglie, tutta la lor famiglia ed il Torricelli"): Viviani to Baldigiani, undated, in Favaro, "Miscellanea galileiana inedita" (1882), p. 154.

129. On July 2, 1633, Galileo was granted permission to leave Rome and move to the residence of the Archbishop of Siena, Ascanio Piccolomini (1590–1671), a former pupil of Bonaventura Cavalieri's. Galileo left Rome on July 6, and arrived in Siena three days later (see Niccolini to Galileo, 2 July 1633, in *OG* 15, no. 2564*, p. 168; Niccolini to Cioli, ibid., no. 2576, p. 174; and *OG* 19, p. 364).

130. Galileo resumed his work on *Two New Sciences* during the time he spent at Piccolomini's residence, in Siena, as is confirmed by Mario Guiducci's letter to Galileo, 22 October 1633 (*OG* 15, no. 2755*, p. 309).

131. *A* and *S* have, instead: "Thus he retired [*S*: he went back] to his villa in Bellosguardo, and later in that in Arcetri, in which he enjoyed spending most of his time, even by his own choice."

132. On December 3, 1633, Niccolini informed Galileo he had been granted permission to leave Siena and return to his villa in Arcetri (see *OG* 15, no. 2802, p. 344; see also *OG* 19, p. 389), where he arrived by December 17, when Cavalieri sent him a letter with a geometric theorem they once discussed (see *OG* 15, no. 2820, p. 356).

133. As soon as the *Dialogue* was published, Fulgenzio Micanzio (a learned Servite friar, and close associate of Paolo Sarpi's) predicted it would be translated into several languages: "The deed is done: you wrote one of the most extraordinary work ever produced by a philosophical spirit. Banning it will not diminish the glory of its author; it will be read in spite of evil resentments, and, as Your Lordship will see, it will be translated into other languages" ("Il colpo è fatto: ella ha fatta un'opera delle più singolari che sia uscita da ingegno filosofico: il vietarli il corso non diminuirà la gloria dell'authore: si leggerà a dispetto dell'invidia maligna, e vedrà V. S. che si trasportarà in altre lingue": Micanzio to Galileo, 14 August 1632, in *OG* 14, no. 2286, p. 372). Later on, Micanzio assured Galileo that the *Dialogue* would not be forgotten: "Ever since your book was published, I know I predicted it would have been translated and published in all languages: neither Your Lordship nor any power will ever be able to prevent that; so do not worry. A friend of yours, who is now enjoying heaven [= Paolo Sarpi] wrote a *History of the Council of Trent*. Rome banned it [. . .]. I have that work in Italian, Latin English, French: let Your Lordship know for sure that the same will be for your *Dialogue*: and who cares if someone grumbles about it. But were you to be harmed a hair for this, we would have to conclude that your champions lack both in good sense and in reputation, and your persecutors in honor, courage, and piety" ("Sino da principio che venne alla stampa il suo libro, so d'haverle predetto che saria tradotto e stampato in tutte le lingue: né V. S. né alcun potere lo può impedire; perciò non se ne affanni. Un suo amico, che gode nel Cielo, scrisse un'Historia del Concilio Tridentino. Roma lo prohibì [. . .]. Io l'ho in Italiano, Latino, Inglese, Francese: creda pur certo V. S. che l'istesso ha da essere de' suoi Dialoghi; e sbattasi chi vuole. Ma se per questo a V. S. fosse torto un pelo, converrebbe ben conchiudere che non fosse in chi la debbe difendere né senso né riputatione, come ne' suoi persecutori né honore né anima né religione": Micanzio to Galileo, 5 August 1634, in *OG* 16, no. 2972*, p. 120). Matthias Bernegger's Latin translation of the *Dialogue* appeared as *Systema cosmicvm, In quo Qvatvor Dialogis De Duobus Maximis Mundi Systematibus, Ptolemaico & Copernicano, Vtriusque rationibus Philosophicis ac Naturalibus indefinite propositis, disseritur*, Strasburg: Bonaventura & Abraham Elzevier (publishers) and David Hautt (printer), 1635; this was reprinted several times, notably in Lyon (1641) and Leiden (1700).

Thomas Salusbury's English translation was published in 1661, with independent title page, in Salusbury's *Mathematical Collections and Translations*, vol. I, Part 1, pp. 1–424: *The Systeme of the World: In Four Dialogues. Wherein the Two Grand Systemes Of Ptolomy and Copernicus are largely discoursed of: And the Reasons, both Phylosophical and Physical, as well on the one side as the other, impartially and indefinitely propounded* (around 1633, Joseph Webbe, who had received a medical degree at the University of Padua in 1612, and conducted a language school in London for several years, prepared an earlier English translation of the *Dialogue*, which however remained unpublished, possibly because of the appearance of the 1635 Latin translation; it is available in manuscript at the British Library, Harley MS. 6320: *The Dialogues of Galileus Galilei one of the Academie of the Lincei, Superordinarie Mathematitian of the Vniversitie of Pisa, And Prime Philosopher and Mathematitian to the most renowned grand-Duke of Toscanie: Where In Foure Dayes meetings: Discourse is had vppon the two greatest Systems of the world; those of Ptolemæus, and Copernicus, Indeterminatly propounding Philosophicall and Naturall reasons aswell for the one part as the other*). There were no French or German translations during Viviani's lifetime. On May 16, 1634, Élie Diodati referred to Galileo that a friend of his (Pierre de Carcavy) had translated the *Dialogue* in a nondialogical form (*in discorso continuo*) and was going to publish it with some additional diagrams of his own (*OG* 16, no. 2942*, p. 96); the translation did not turn out as expected, though, and the project was aborted (Diodati to Galileo, 12 March 1635, ibid., no. 3090*, p. 231; this letter was copied by Viviani, who possibly still hoped a French translation was underway. Two years later, Diodati informed Galileo that de Carcavy was planning to publish the *Dialogue*, together with Galileo's other major works, and related works by Kepler and Scheiner: Diodati to Galileo, 11 June 1637, in *OG* 17, no. 3499, p. 109; see also Roberto Galilei to Galileo, 7 February 1635, in *OG* 16, no. 3072**, p. 206). The *Dialogue* was eventually translated into French by René Fréreux, in collaboration with François de Gandt: *Dialogue sur les deux grands systèmes du monde*, Paris: Éditions du Seuil, 1992. It was not until the end of the nineteenth century that a German translation, by Emil Strauss, appeared: *Dialog über die beiden hauptsächlichsten Weltsysteme, das ptolemäische und das kopernikanische*, Leipzig: B. G. Teubner, 1891.

134. Galileo, *Letter to Christina*: the original Italian text is printed facing its Latin translation by Robertus Robertinus Borussus (i.e., Élie Diodati).

135. This sentence is reminiscent of the full title of the first edition of Galileo's *Letter to Christina*: *Nov-Antiqua Sanctißimorum Patrum, & Probatorum Theologorum Doctrina, De Sacræ Scripturæ Testimoniis, in Conclusionibus mere Naturalibus, quæ sensatâ experientiâ, & necessariis demonstrationibus evinci possunt, temere non usurpandis.*

136. In fact, Galileo expressed profound gratitude both to Élie Diodati, by sending him Justus Suttermans's portrait (Galileo to Diodati, 23 September 1635, in *OG* 16, no. 3184*, p. 315), and to Matthias Bernegger (Galileo to Bernegger, 16 July 1634, ibid., no. 2966, pp. 111–112), also sending him new lenses (Galileo to Diodati, 15 August 1636, ibid., no. 3341, p. 474).

137. *S* has: "of Pythagoras."

138. In both *A* and *S* the sentence continues with these words: "and passed over in silence the new speculations he had not yet published; and he never refrained from praising her [i.e., Providence] generously and gratefully, always declaring her magnificence and greatness. According to such a grateful resolution, in 1636 he decided . . ."

139. New negotiations about Galileo's idea to use Jupiter's satellites for the computation of longitudes were started with representatives of the States General of Holland in August 1636, through the good offices of Galileo's friend Élie Diodati in Paris: see Galileo's letter to Huig de Groot (Hugo Grotius, 1583–1645) and the States General of the United Provinces of Holland, both on 15 August 1636 (in *OG* 16, no. 3337, pp. 463–468, and no. 3340, pp. 472–473). Already in 1627, from a letter of Alfonso Antonini (in *OG* 13, no. 1838, 25 October 1627, p. 377), Galileo knew about the prize established by the Holland Provinces for a solution to the problem, but at the time he was interested in the negotiations with the Spanish Court. The new negotiations dragged on for a while but were interrupted by the sudden death of astronomer Maarten van den Hove (Martinus Hortensius) in 1639, and were eventually suspended in 1640, also due to Galileo's health conditions: see Diodati's letter to Galileo on 15 June 1640 (in *OG* 18, no. 4021, pp. 203–204).

140. See the States General to Galileo, 25 April 1637 (in *OG* 17, no. 3468**, p. 66), as well as *OG* 19, pp. 539–540.

141. Laurens Reael (1583–1637) was Governor-General of the Dutch East Indies from 1616 to 1619. Maarten van den Hove (1605–1639) was a Dutch mathematician and astronomer, a pupil of Isaac Beeckman's at Leiden and Ghent, from 1615 to 1617; prior to his appointment in the commission negotiating with Galileo on the determination of longitudes, he lectured on navigation in Amsterdam (1636) and developed a method for measuring the diameters of planets based on the measured visual angle revealed by the telescope. Willem Janszoon Blaeu (1571–1638) was a renowned cartographer and atlas maker; between 1594 and 1596, he was a student of Tycho Brahe's at Uraniborg, Tycho's astronomical observatory on the island of Hven. Jacob van Gool (Golius, 1596–1667) was an Orientalist and mathematician based at the University of Leiden (his first name is mentioned only in *S*, while *A* does not mention him at all). Isaac

Beeckman (1588–1637), a pupil of Simon Stevin's, was a philosopher and scientist who developed a wide network of contacts with leading natural philosophers, including René Descartes, whom he introduced to Galileo's ideas.

142. See Diodati to Galileo, 7 July 1637, in *OG* 17, no. 3515, pp. 127–130; see also the States General to Galileo, 10 February 1638, ibid., no. 3675**, pp. 283–284, and *OG* 19, pp. 542–543.

143. *A* has: "six-year-long."

144. See Galileo to Fulgenzio Micanzio, 7 November 1637, in *OG* 17, no. 3595, pp. 214–215: "I made a truly amazing discovery in the face of the Moon, of which—although an infinite number of people observed it an infinite number of times—I do not think any change has ever been noticed, as we believe the Moon always shows us the same face. I do not think this is true, though. The Moon changes its appearance with all three possible variations; it shows to us the same changes shown by someone who, when offering his face directly to our eyes—in full swing, as we say—changes it in all possible ways, by turning it now to the right, now to the left, or else by lifting, or lowering it, or finally by tilting it toward the right or the left shoulder. We can observe all these changes in the face of the Moon, and the big and old spots that we observe on it make what I say evident and perceivable. Let me also add a second marvel: these three different changes have three different periods. One changes from day to day, and therefore has a daily period; the second changes from month to month, and has a monthly period; the third has a yearly period, after which the change is complete" ("Io ho scoperta una assai maravigliosa osservazione nella faccia della luna, nella quale, ben che da infiniti infinite volte sia stata riguardata, non trovo che sia stata osservata mutazione alcuna, ma che sempre l'istessa faccia nell'istessa veduta a gli occhi nostri si rappresenti; il che trovo io non esser vero, anzi che ella ci va mutando aspetto con tutte tre le possibili variazioni, facendo verso di noi quelle mutazioni che fa uno che esponendo a gli occhi nostri il suo volto in faccia, e come si dice in maestà, lo va mutando in tutte le maniere possibili, cioè volgendolo alquanto ora alla destra et ora alla sinistra, o vero alzandolo et abbassandolo, o finalmente inclinandolo ora verso la destra et ora verso la sinistra spalla. Tutte queste mutazioni si veggono fare nella faccia della luna, e le macchie grandi e antiche, che in quella si scorgono, ci fanno manifesto e sensato questo ch'io dico. Aggiugnesi di più una seconda maraviglia, et è che queste tre diverse mutazioni hanno tre diversi periodi: imperò che l'una si muta di giorno in giorno, e così viene ad haver il suo periodo diurno; la seconda si va mutando di mese in mese, et ha il suo periodo mestruo; la terza ha il suo periodo annuo, secondo il quale finisce la sua variazione"). See also Galileo to Alfonso Antonini, 20 February 1638, ibid., 3684, pp. 291–297.

145. On 6 June 1637, Galileo reports to Élie Diodati that he was "so much annoyingly affected by a fluxion in the right eye that not only I am unable to read or write, not a single syllable, but even those occupations that appeal to eyesight, neither more nor less than if I were completely blind" ("io mi trovo tanto molestamente aggravato dalla flussione nell'occhio destro, che non sola mente mi vien tolto il poter nè leggere nè scrivere una sillaba, ma il far ancora nessuno di quegli esercizi che ricercano l'uso della vista, né più né meno che se io fussi del tutto cieco": *OG* 17, no. 3495, p. 94; see also Vincenzo Renieri to Galileo, 9 July 1637, ibid., no. 3517, p. 135). As we know from another letter to Diodati, the situation worsened a few months later: "my lord—alas!—your dear friend and servant Galileo became irreparably blind a month ago" ("ahimè, Signor mio, il Galileo, vostro caro amico e servitore, è fatto irreparabilmente da un mese in qua del tutto cieco": 2 January 1638, ibid., no. 3635, p. 247).

146. Vincenzo Renieri (1606–1647), a member of the Olivetan Order, met Galileo while he was in Siena, in 1633, in the aftermath of the trial. Galileo asked him to attempt and improve his astronomical tables of the motion of Jupiter's satellites; this led Renieri to frequent visits in Arcetri, where he met and befriended Viviani. Before his death, Galileo decided to place all of the papers containing his observations and calculations in the hands of Renieri, who was to revise, complete, and publish them (see Galileo to Élie Diodati, 30 December 1639, in *OG* 18, no. 3953, pp. 132–133). According to Viviani, at the time of Renieri's premature death in Pisa, in 1647, the papers concerning longitude Galileo had entrusted to him were stolen; see also Riccioli, *Almagestvm Novvm* (1651), vol. I, Part I, p. 489. Galileo's and Renieri's manuscripts were eventually discovered by Eugenio Albèri in the Biblioteca Palatina (now part of the Biblioteca Nazionale Centrale), in Florence, and were (partly) published in *Opere 1842–1856*, vol. 5. They were eventually included in the appendix to vol. 3, Part 2, of the National Edition of Galileo's works: see *OG* 3.2, pp. 931–966 (Galileo's annotations) and 967–1054 (Renieri's computations); for a reconstruction and discussion of this material, see ibid., pp. 891–912.

147. By the time of Hortensius's passing (on August 7, 1639), three other members of the commission appointed by the States General of the United Provinces of Holland had died, in the span of less than one and a half years: Beeckman on May 19, 1637; Reael on October 21, 1637; and Blaeu on October 21, 1638.

148. A poet and composer, Constantijn Huygens (1596–1687) was secretary of Frederik Hendrik, the sovereign Prince of Orange, from 1625 to 1647; he also served in the same capacity for William II, who succeeded Frederik Hendrik and reigned until his own death, in 1650. Constantijn was the father of the prominent Dutch scientist Christiaan Huygens, who would later construct the pendulum clock (see Viviani 1659, below). Willem Boreel (1591–1668) was a Dutch jurist and diplomat, knighted by King

James I of England in 1618; as a lawyer, he worked for the Dutch East India Company, and was Pensionary of Amsterdam from 1627 to 1649. In 1655, Boreel intervened in the controversy over the invention of the telescope, giving credit to the Dutch spectacle-maker Zacharias Janssen over Hans Lipperhey. Boreel's high social status did not allow for objections, and the previously unknown Janssen was presented as the inventor of the telescope in Pierre Borel's *De vero telescopii inventore* (1655). The claim was to remain undisputed for many decades to come.

149. After "Finally," *A* has only: "when the same States resumed negotiations, once new commissioners had been elected . . .".

150. That is, Leopoldo de' Medici and his brother Giovan Carlo (1611–1663), as we know from a marginal note in *B*, which was later deleted, in which the sentence within brackets reads: "as Your Highness, too, claims he saw, together with the Most Serene Cardinal Giovan Carlo, your brother."

151. After several attempts, the problem was eventually solved by appealing not to the periods of Jupiter's satellites but by using a mechanical timepiece, to be carried on a ship, that would maintain the correct time at a reference location. It took several years to develop a clock that could maintain accurate time while being subjected to the rolling, pitching, and yawing of a moving ship, coupled with the pounding of wind and waves. A suitable timepiece, a marine chronometer, was first built by John Harrison (1693–1776), a Yorkshire carpenter (whose later and more advanced version was completed in 1759 and first used on a ship in 1761).

152. François de Noailles (1584–1645) was among Galileo's pupils in Padua, where he learned the use of the geometric and military compass (see *OG* 19, p. 154; see also Galileo to Diodati, 22 August 1637, in *OG* 17, no. 3547*, p. 174). He was appointed ambassador to the papal court in 1632 but reached Rome only in 1634. Together with Francesco Niccolini, the Tuscan ambassador, de Noailles unsuccessfully tried to mitigate Galileo's sentence to house arrest in Arcetri (see, for example, Castelli to Galileo, 9 December 1634, in *OG* 16, no. 3027, p. 171; and Roberto Galilei to Galileo, 3 February 1637, in *OG* 17, no. 3429, p. 26). On his way back to France, he met with Galileo in Poggibonsi (see Castelli to Galileo, 9 October 1636, *OG* 16, no. 3371, p. 500; and de Noailles to Galileo, 9 October 1636, ibid., no. 3372, pp. 500–501). There, on 16 October 1636, Galileo gave him the manuscript of the *Two New Sciences*, which was dedicated to François de Noailles and published in 1638 (*Two New Sciences*, pp. 5–6, and *OG* 8, pp. 43–44; see also de Noailles to Galileo, 1 January 1638, in *OG* 17, no. 3634**, p. 246; and 20 July 1638, ibid., no. 3763, pp. 357–358).

153. Filippo Salviati (1583–1614), a member of one of the weathiest and most influential Florentine families, was elected member of the Academy of the Lynxes in 1612. Galileo's *On Sunspots* was to a large extent written in Salviati's villa Le Selve, on the outskirts of Florence, and dedicated to him. Giovan Francesco Sagredo (1571–1620) was a Venetian mathematician and very close friend of Galileo's, with whom he discussed a number of scientific issues. In both the *Dialogue* and the *Two New Sciences*, Salviati is Galileo's spokeperson, arguing for the Copernican position, and Sagredo a learned and intelligent layman. Daniello Antonini (1588–1616) was a pupil of Galileo's in Padua and later pursued a military career. Paolo Aproino (1586–1638) studied under Galileo in Padua, where he graduated in 1608, and in 1612 he became canon of the cathedral of Treviso; he is one of the interlocutors (with Salviati and Sagredo) of the Sixth Day Galileo was planning to add to *Two New Sciences* ("On the Force of Percussion," pp. 281–306; *OG* 8, pp. 321–346). Paolo Sarpi (1552–1623) was a prominent Italian historian, prelate (he entered the Augustinian Servite order at the age of thirteen), canon lawyer and a statesman: see note 86. Also an experimental scientist and an advocate of the Copernican theory, he was a good friend and patron of Galileo's during his Paduan years.

154. Here Viviani is summarizing Galileo's dedicatory letter to Count de Noailles: see *Two New Sciences*, pp. 5–6 (*OG* 8, pp. 43–44).

155. Galileo was not pleased at the title Elsevier chose for the work. In his letters, Galileo refers to its parts as "the two works on motion and the resistance [of bodies]" ("le due opere del moto e delle resistenze": Galileo to Élie Diodati, 15 August 1636, in *OG* 16, no. 3341, pp. 473–474). When he received copies of the volume, in August 1638, he wrote to Diodati: "I was astonished and troubled by the liberty Elsevier took to change the title of my book, downgrading its noble title, which it deserved to have, to one which is way too ordinary, if not rough. And I am compelled to believe that someone who does not like me in Amsterdam advised him about it. And Your Most Eminent Lordship, as a true and sincere friend of mine, and master, should rightly see that the original title be restored" ("con maraviglia e travaglio son restato della libertà presasi il Sig.re Elzevirio di trasformare l'intitolazione del mio libro, riducendola di nobile, quale ella meritamente deve essere, a volgare troppo, per non dire plebea; et è forza, per mio credere, che qualche mio poco affetto in Amsterdam gl'abbia tenuto mano: e V. S. molto Ill.,ᵉ come mio vero e sincero amico e padrone, ben fa a procurare la reintegrazione di essa intitolazione": *OG* 17, no. 3780, p. 370; see also Galileo to Diodati, August 1638, ibid., no. 3781, p. 373).

156. Viviani was sixteen years old when he started visiting Galileo in Arcetri, on January 25, 1639: see *Gal.* 210, f. 244ᵇr.

157. In *A*, the beginning of the new paragraph reads: "About one year after this unexpected publication, Galileo granted that I be with him in his villa in Arcetri, under his guidance, to lead me—however blind his

body was, his mind was as sharp as ever—along the path of mathematics, to which I had earlier devoted myself, and was pleased that in my studies . . ." ("Circa un anno dopo questa inaspettata publicazione, compiacendosi il Sig.ʳ Galileo di havermi nella sua villa d'Arcetri appresso l'ottima sua disciplina per guidarmi, benché cieco, com'egli era, di corpo, di intelletto però lucidissimo, per il sentiero delle matematiche, alle quali poco avanti io m'era applicato, e contentandosi che nello studio ch'io facevo . . .").

158. After this, *S* has: "and for the novelty of the subject, which was physical in nature and therefore not entirely geometric."

159. "During the past several months, this youngster [i.e., Viviani], who is presently my guest and disciple, raised objections against the principle I assumed in my treatise on accelerated motion, which he had been studying with great diligence. They caused me to ponder on it so much, in order to convince him that such principle could be taken for granted and as true, that—to my and his great pleasure—I eventually found a conclusive proof of it [. . .]" ("L'oppositioni fattemi, son già molti mesi, da questo giovane, al presente mio ospite et discepolo, contro a quel principio da me supposto nel mio trattato del moto accelerato, ch'egli con molta applicatione andava allora studiando, mi necessitarono in tal maniera a pensarvi sopra, a fine di persuadergli tal principio per concedibile e vero, che mi sortì finalmente, con suo e mio gran diletto, d'incontrarne, s'io non erro, la dimostratione concludente [. . .]": Galileo to Castelli, 3 December 1639, in *OG* 18, no. 3945, p. 126). The principle Viviani is referring to is the postulate stated by Salviati in the section "On Naturally Accelerated Motion" in the Third Day of Galileo's *Two New Sciences*, p. 162: "I assume that the degrees of speed acquired by the same movable over different inclinations of planes are equal whenever the heights of those planes are equal" ("Accipio, gradus velocitatis eiusdem mobilis super diversas planorum inclinationes acquisitos tunc esse aequales, cum eorumdem planorum elevationes aequales sint": *OG* 8, p. 205). Following Viviani's remarks, Galileo wrote an additional section, dictated it to Viviani about October 1638, and revised it in November 1639: see *OG* 8, pp. 442–445; see also *On Mechanics*, pp. 171–172 (*OG* 2, pp. 180–181). Viviani put it into dialogue form and inserted it in the reprint of the *Two New Sciences* in the 1655–1656 edition of Galileo's collected works: see *Two New Sciences*, pp. 171–175. See also Viviani's own recollections in Viviani 1674, pp. 218–221.

160. *S* has: "geometric and mechanical."

161. After this, *A* has: "he asked me to write down the theorem, as his blindness prevented him from doing it properly whenever diagrams and letters were required; and sent copies to . . ." ("et imponendomi ch'io facesse il disteso del teorema, per la difficoltà che gli arrecava la sua cecità nell'esplicarsi dove bisognavano usar figure e caratteri, ne mandò più copie . . ."). Pappus of Alexandria (ca. 290–ca. 350), one of the last great mathematicians of antiquity. Galileo knew his work in the Latin translation by Federico Commandino: *Pappi Alexandrini Mathematicae Collectiones à Federico Commandino Vrbinate in latinvm conversæ, et commentariis illvstratae*, Pesaro: Girolamo Concordia, 1588. As noted above (note 45), Mersenne translated Galileo's *Le mecaniche*—then still unpublished—as *Les Méchaniques de Galilée*, Paris: Henri Guénon, 1634. See also Viviani 1674.

162. Definition 5: "Magnitudes are said to be in the same ratio, the first to the second and the third to the fourth, when, if any equimultiples whatever be taken of the first and third, and any equimultiples whatever of the second and fourth, the former equimultiples alike exceed, are likely equal to, or alike fall short of, the latter equimultiples respectively taken in corresponding order" ("Ἐν τῷ αὐτῷ λόγῳ μεγέθη λέγεται εἶναι πρῶτον πρὸς δεύτερον καὶ τρίτον πρὸς τέταρτον, ὅταν τὰ τοῦ πρώτου καὶ τρίτου ἰσάκις πολλαπλάσια τῶν τοῦ δευτέρου καὶ τετάρτου ἰσάκις πολλαπλασίων καθ᾽ ὁποιονοῦν πολλαπλασιασμὸν ἑκάτερον ἑκατέρου ἢ ἅμα ὑπερέχῃ ἢ ἅμα ἴσα ᾖ ἢ ἅμα ἐλλείπῃ ληφθέντα κατάλληλα": Euclid, *Elements*, Book V, vol. 2, p. 2.10–15). Definition 7: "When of the equimultiples, the multiple of the first magnitude exceeds the multiple of the second, but the multiple of the third does not exceed the multiple of the fourth, then the first is said to have a greater ratio to the second than the third has to the fourth" ("Ὅταν δὲ τῶν ἰσάκις πολλαπλασίων τὸ μὲν τοῦ πρώτου πολλαπλάσιον ὑπερέχῃ τοῦ τοῦ δευτέρου πολλαπλασίου, τὸ δὲ τοῦ τρίτου πολλαπλάσιον μὴ ὑπερέχῃ τοῦ τοῦ τετάρτου πολλαπλασίου, τότε τὸ πρῶτον πρὸς τὸ δεύτερον μείζονα λόγον ἔχειν λέγεται, ἤπερ τὸ τρίτον πρὸς τὸ τέταρτον": ibid., p. 4.1–5).

163. After this, *S* has: "when he was living with him." The fifth book of Euclid's *Elements* deals with the theory of proportions, and Viviani is here referring to Galileo's remarks and contributions to it, which he was planning to insert in a new edition of *Two New Sciences*. Together with Viviani and Torricelli, who assisted and worked with him in his last years, Galileo was drafting two additional days: the Fifth Day, on Euclid's definitions of sameness and compounding of ratios, and the Sixth Day, on the force of percussion. The former was first published in Viviani's *Qvinto libro degli Elementi d'Evclide* (1674), pp. 61–77, and was later included in *OG* 8, pp. 349–362 (English translation in Drake, *Galileo at Work* (1978), pp. 422–436). Galileo dictated the text to Torricelli between 10 October 1641 and early January 1642 (see Viviani 1674, pp. 220–223). However, from Torricelli's manuscript, in *Gal.* 75, ff. 10r-23v and 3r–9r (the title, on f. 10r, reads: "Trattato del Galileo Sopra la definizione delle Proport.ⁿⁱ d'Euclide. Giornata Quinta da aggiungersi nel lib. delle Nuove Scienzie," that is, "Galileo's treatise on the definition of Euclid's proportions: Fifth Day, to be added to the book on [Two] New Sciences"), it is impossible to ascertain Galileo's contributions from

Torricelli's own. Indeed, after Galileo's death, Torricelli kept working on proportions, and shortly before his sudden death (on October 25, 1647) he completed a short work, *De proportionibus liber*, which enjoyed a wide circulation in manuscript form, and was later employed by Viviani himself for his *Qvinto libro degli Elementi d'Evclide* (1674), and for a very popular edition of Euclid's elements (*Elementi piani, e solidi d'Evclide*, Florence: Cesare and Francesco Bindi, & Jacopo Carlieri, 1690, which includes a revised edition of Viviani's *Qvinto libro*): see Torricelli to Michelangelo Ricci, 20 July 1647, in *ODG* 1, no. 277*, pp. 386–387; and Viviani, in *Gal.* 77, f. 4r. Torricelli's *De proportionibus liber* was first published in Torricelli 1919–1944 (vol. I, Part I, pp. 295–327; a critical edition is now available in Giusti, *Euclides reformatus* (1993), pp. 301–340).

164. Leopoldo de' Medici to Galileo, 11 March 1639, in *OG* 18, no. 3981, p. 165. Fortunio Liceti (1577–1657) studied in Bologna and graduated in philosophy and medicine at the University of Genoa, in 1600. Later that year, he took up a lectureship at the University of Pisa, where he was awarded a chair in philosophy in 1605. In 1609, he was appointed a professor at the University of Padua; he taught at the University of Bologna from 1637 to 1645; and eventually returned to Pisa, to take up the first chair in theoretical medicine. While in Padua, he met and befriended Galileo, and extensively corresponded with him after the latter's departure. Liceti was a very prolific author, dealing with subjects ranging from genetics to stones, and animals. According to Bonaventura Cavalieri, however, "it takes him a week to write a book; so far, he told me, he must have published 37. Indeed, his works do not enjoy much circulation, or credit, among the scholars; rather, his writings—as a learned Father, public lecturer of metaphysics in Padua, told me—are referred to as jokes" ("[. . .] esso fa un libro in una settimana, e sin hora, per quanto mi disse, ne deve havere stampati da 37. Egli è ben vero che non hanno li suoi libri molto spaccio o credito appresso gl'intendenti; anzi le sue compositioni, come mi disse un valente Padre, lettore publico di metafisica in Padova, ivi sono chiamate barzellette": Cavalieri to Galileo, 5 June 1640, in *OG* 18, no. 4017, p. 201); indeed, at the end of his *Hieroglyphica* (1653), Liceti appended a list of 75 books. In his astronomical works, Liceti defended Aristotelian cosmology and geocentrism against the Copernican theory, advocated by Galileo. During the controversy that ensued after the appearance of a comet in 1618, which led to Grassi's and Guiducci's works, and eventually to Galileo's *The Assayer*, Liceti published a series of books upholding the Aristotelian view.

165. Between 1640 and 1642, Liceti had a friendly astronomical confrontation with Galileo. In the *Litheosphorvs, sive De lapide Bononiensi* (1640), Liceti examined the "Bologna stone," discovered by Vincenzo Casciarolo at the base of Mount Paderno, near Bologna, in 1603: when treated with heat, and exposed to sunlight, the stone glows for hours, sometimes days. Now we know that the stone is a type of barite with the property of becoming phosphorescent during the process of calcination; at the time of Galileo, however, it was believed that the phosphorescence was caused by the stone's absorption, and gradual release, of sunlight. Contrary to Galileo, who argued (in the *Sidereal Messenger*) that the illumination of the Moon is caused by the reflection of sunlight from the earth, Liceti argued that the Moon, analogously to the Bologna stone, released light absorbed by the Sun: see chapter 50 of the *Litheosphorvs*, titled "De Lunæ subobscura luce prope coniunctiones, & in deliquijs obseruata, digressio physico-mathematica" ("Physical and mathematical digression on the somewhat obscure light of the moon, when it is close to conjunction [with the earth and the Sun], and during an eclipse"), pp. 242–248. Liceti sent a copy of the book to Galileo, who wrote a letter to Prince Leopoldo de' Medici defending his views (see *OG* 8, pp. 489–545; see also pp. 549–556, and 483–486). This was Galileo's last scientific work. On July 14, 1640, Galileo sent a copy of the letter to Liceti; there followed a long and friendly exchange on the topic, and eventually Liceti published Galileo's letter, together with his own three-book, point-by-point response: *De lvnae svbobscvra lvce Prope Coniunctiones, & in Eclipsibus obseruata* (1642), Book II.

166. After this, *A* has: "until his very last days, and, I may say, the end of my own studies," *S* has: "until he breathed his last, which, I may say—due to all sorts of unexpected events, committments and works that occurred to me—was the last day of my most pleasant and carefree studies."

167. *A* has: "hardly allowed."

168. See the sketches for a Fifth Day (on Euclid's definitions of proportions) and a Sixth Day (on the force of percussion) in *OG* 8, pp. 349–362 (English translation in Drake, *Galileo at Work* (1978), pp. 422–436) and 321–346 (English translation in *Two New Sciences*, pp. 281–306), respectively.

169. Giovanni Alfonso Borelli (1608–1679), an eminent Italian mathematician, astronomer, and physiologist, wrote a lengthy anti-Aristotelian treatise, *De motv animalivm*, published posthumously in 1680–1681, in two volumes. Borelli was a pupil of Benedetto Castelli's and an acquaintance of Galileo's, whom he met in 1640.

170. Both Torricelli and Borelli worked on this topic: see Torricelli's *Lezioni accademiche* (1715), pp. 3–29 (lectures 2–4); and Borelli, *De vi percvssionis* (1667).

171. That is, January 8, 1642.

172. See Fulgenzio Micanzio to Galileo, 11 November 1634, in *OG* 16, no. 3011, p. 155: "The mind of Your Lordship is like the workshops of goldsmiths, where grids are in place, so that not even dust is wasted, because it has gold blended in it. I do not see this in anybody else" ("È l'ingegno di V. S. come le boteghe

degl'orefici; ove si fanno li cancelli, aciò che né anco la polvere si perda, perché ha mescolato oro. Io non trovo così in altri").

173. Before the beginning of the new paragraph, *A* has: "He died in his villa in Arcetri, and his body was brought to Florence and buried close to the sacristy of the church of Santa Croce, as requested by his Most Serene Master" ("Morì nella sua villa d'Arcetri, et il corpo suo fu trasportato in Firenze e depositato in luogo vicino alla sagrestia del tempio di S. Croce, di commissione del suo Ser.^{mo} Padrone"). See also next note.

174. Galileo made his first will on January 15, 1633, five days ahead of setting out for his journey to Rome, where he would have faced the inquisitors of the Holy Office (see *OG* 19, pp. 520–522). In his second (and last) will, dated 21 August 1638, Galileo expressed the wish to be buried in Florence, in the Santa Croce Basilica, where his father Vincenzo and other members of his family were buried (see ibid., p. 522). Accordingly, the day after Galileo's death, his body was quietly moved to Santa Croce and laid to rest—not with his father and ancestors, in the central nave of the basilica, though, but in the tiny vestry of a chapel (the Chapel of the Novitiate, otherwise known as the Medici Chapel), on the right hand side of the transept. Galileo died while under house arrest, having been found "vehemently suspected of heresy," and so no grand ceremony was held, nor official speeches or commemorations delivered. The Grand Duke wished to have a proper funeral monument erected in the basilica, but the plan lasted but a few days, and what was originally thought to be a temporary burial was to remain Galileo's tomb for nearly a century. Viviani was the undisputed champion of resurrecting the Grand Duke's plan: throughout his life, he tried his best to preserve the memory of his master's life, discoveries, and achievements, as well as to publish and circulate his works. The erection of a public funeral monument in Santa Croce was part of this project, for a proper monument would bear the double meaning of recalling the importance of Galileo's works, highlighting their superiority with respect to traditional scientific and philosophical ideas; and of claiming the liberty to build on them, extending them to other contexts. The image of Galileo heretic, as it sprang from the 1633 trial, was a clear obstacle to this—hence the key strategic, as well as psychological and political, relevance of a proper funeral monument for Galileo in Italy's most celebrated pantheon. As years went by, Viviani realized that his efforts were unlikely to succeed. In 1655–1656 he managed to publish (under the editorship of Carlo Manolessi, in Bologna) the first, two-volume edition of Galileo's collected works but failed to get approval to include the *Dialogue*. On December 7, 1689, aged 67, Viviani drew up his will, with which he bequeathed to his heirs the money and goods he thought would be required to erect a proper funeral monument for Galileo, opposite Michelangelo's, adorned with an inscription celebrating his discoveries and achievements (Viviani, *Testamento* (1735), pp. 3–4); as to his own body, he added, he wished it to be buried next to Galileo's remains. Toward the end of his life, as no public monument was to be erected, Viviani resolved to create a private one—one that, however, could be enjoyed, and benefited from, publicly. And so he decided to turn the façade of his own residence (in what is now via S. Antonino 13, in Florence, in proximity to the church of Santa Maria Novella) into a monument to Galileo's discoveries and achievements. The façade, designed by architect Giovanni Battista Nelli, presents, above the main entrance door, a bust of Galileo; two bas-reliefs celebrating Galileo telescopic discoveries and his achievements in physics and mechanics; and two enormous scagliola scrolls (hence the name by which the palace is referred to: "Palazzo dei Cartelloni"), one on each side of the façade, celebrating Galileo's achievements and calling Florence's citizens to honor him with a proper funeral monument, as was done for Michelangelo: see below, pp. 235–273. Viviani died in 1703 and never saw the erection of a proper funeral monument for Galileo in Santa Croce, nor the republication of his banned works. His lifelong efforts were not in vain, though, as they triggered a mechanism that led eventually to the fulfillment of his wish. In 1737, Galileo's body was exhumed and moved to the base of the new funeral monument (opposite Michelangelo's), which was completed on June 6, 1737.

175. Johannes Kepler was the first to compare Galileo to the Florentine explorer and navigator Amerigo Vespucci (1454–1512): see Kepler, *Dissertatio* (1610), p. 37, where Vespucci is referred to as the "Argonaut from Florence" ("Argonauta ille Florentinus"). Most interestingly, whereas here Viviani compares Galileo to Vespucci, in the long inscriptions on the façade of his house, as well as in their published version, he appeals to a stronger comparison with Michelangelo. Viviani argues that there was continuity—under the learned and enlightened rule of the Medici family, at the same time protecting and promoting the arts and the sciences—between the two. Indeed, he says, Michelangelo died on the same day in which Galileo was born, and so the loss of the restorer of painting, sculpture, and architecture was compensated for by Divine Providence's generous gift of the renovator of geometry and astronomy. In fact, Michelangelo died on February 18, 1564, whereas Galileo was born on February 15. Viviani knew it, but the alleged coincidence was a powerful rhetorical tool forcombining the exhortation to erect a proper funeral monument for Galileo, a celebration of the role played by the Medici family in the progress of the arts and the sciences, and the underscoring of Florence's continued cultural primacy.

176. Most likely, here Viviani refers to what happened in 1593 in Costozza, in the province of Vicenza (some 30 kilometers northwest of Padua), at the villa that belonged to Camillo Trento. In one of its rooms on the first floor, the opening of an air vent links the room with the nearby grottoes (the so-called Covoli

di Costozza, in Longare). These were used for centuries as cool storages for wine or foodstuffs, as the temperature range is a constant 10 °C–12 °C (50 °F–54 °F). The area was well known and attracted visitors from Vicenza, Padua, and Venice during hot summers. During his years in Padua, Galileo likely visited the area, possibly in the company of Count Marc'Antonio Bissaro, or else with other Venetian friends of his.

177. *A*, *B* and *S* have: "the letters and [*B*: of] the alphabet."

178. See the famous passage from *The Assayer*, pp. 183–184: "It seems to me that I discern in Sarsi a firm belief that in philosophizing it is essential to support oneself upon the opinion of some celebrated author, as if when our minds are not wedded to the reasoning of some other person they ought to remain completely barren and sterile. Possibly he thinks that philosophy is a book of fiction created by some man, like the *Iliad* or the *Orlando Furioso*—books in which the least important thing is whether what is written in them is true. Well, Sig. Sarsi, that is not the way matters stand. Philosophy is written in this all-encompassing book—I mean the universe—which stands continually open before our eyes, but it cannot be understood unless one first learns to understand the language and knows the characters in which it is written. It is written in mathematical language, and its characters are triangles, circles, and other geometric figures without which it is humanly impossible to understand a word of it; without these means, one is wandering about in a dark labyrinth" ("Parmi, oltre a ciò, di scorgere nel Sarsi ferma credenza, che nel filosofare sia necessario appoggiarsi all'opinioni di qualche celebre autore, sì che la mente nostra, quando non si maritasse col discorso d'un altro, ne dovesse in tutto rimanere sterile ed infeconda; e forse stima che la filosofia sia un libro e una fantasia d'un uomo, come l'Iliade e l'Orlando Furioso, libri ne' quali la meno importante cosa è che quello che vi è scritto sia vero. Sig. Sarsi, la cosa non istà così. La filosofia è scritta in questo grandissimo libro che continuamente ci sta aperto innanzi a gli occhi (io dico l'universo), ma non si può intendere se prima non s'impara a intender la lingua, e conoscer i caratteri, ne' quali è scritto. Egli è scritto in lingua matematica, e i caratteri son triangoli, cerchi, ed altre figure geometriche, senza i quali mezi è impossibile a intenderne umanamente parola; senza questi è un aggirarsi vanamente per un oscuro laberinto": *OG* 6, p. 232). See also *OG* 8, pp. 613–614, the *Dialogue*, pp. 11 and 103 (*OG* 7, pp. 35 and 128–129).

179. Ferdinando II de' Medici (1610–1670), Grand Duke of Tuscany from 1621. He was a patron and friend of Galileo's (as well as of Viviani's and Torricelli's), who dedicated the *Dialogue* to him.

180. Andrea Arrighetti (1592–1672) studied mathematics under Benedetto Castelli; he was elected a member of the Accademia della Crusca in 1613, and senator in 1649. A close friend and associate of Galileo's, he acted as the intermediary in the correspondence between Galileo and his daughter, Sister Maria Celeste. In a letter to Galileo, on 25 September 1633, he stated and proved a theorem on the resistance of solids that Galileo (then in Siena, where he resumed his work after the trial in Rome) very much appreciated, offering to add it alongside his own in the *Two New Sciences*: see *OG* 15, no. 2718, pp. 279–281; see also Galileo to Arrighetti, 27 September 1633, ibid., no. 2721, pp. 283–284. At the time of Galileo's death, it was Arrighetti who suggested the appointment of Torricelli as the new Chief Mathematician and Philosopher of the Grand Duke.

181. The only published work in his lifetime, Torricelli's *Opera Geometrica* (1644) included *De Sphæra Et Solidis Sphæralibus libri dvo*; *De Motv Gravivm naturaliter descendentium, Et Proiectorum libri dvo*; and *De Dimensione Parabolæ, Solidique Hyperbolici problemata dvo, cvm appendice De Dimensione spatij Cycloidalis, & Cochleæ*. The *Lezioni accademiche* was published posthumously, in 1715.

182. *S* has: "one day." In France, Blaise Pascal (1623–1662) worked and built on Torricelli's researches, clarifying the concepts of pressure and vacuum by generalizing his work.

183. In the margin, *B* has: "and for this reason he wrote in his mother tongue" ("e per ciò scrisse nella lingua materna"); the annotation was later crossed out.

184. Ludovico Ariosto (1474–1533) is the author of the romance epic *Orlando fvrioso* (1516), telling the adventures of Charlemagne, Orlando, and the Franks as they battle against the Saracens. Torquato Tasso (1544–1595) is the author of *Gerusalemme liberata* (1581), in which he depicts a highly imaginative version of the combats between Christians and Muslims during the siege of Jerusalem, at the end of the first crusade. Galileo preferred Ariosto's poem to the more fashionable Tasso's, and the *Orlando fvrioso* was arguably a major source of Galileo's prized literary style, at the same time elegant and biting, as displayed in *The Assayer* and the *Dialogue*. In one of the many side plots of Ariosto's poem, Astolfo, Orlando's cousin and fellow paladin, flies to the Moon in a chariot pulled by winged horses to look for Orlando's wits, and describes the Moon like a second Earth, with rivers, lakes, plains, valleys, mountains, and forests. See Galileo, "Considerazioni al Tasso," in *OG* 9, pp. 61–148; and "Postille all'Ariosto," ibid., pp. 151–194.

185. Jacopo Mazzoni (1548–1598) was Galileo's closest colleague at the University of Pisa. He considered himself primarily a philosopher, and in several works attempted to reconcile the views of Plato and Aristotle. A personal friend of Torquato Tasso's, he is most known for his work on literary criticism, particularly his defense of Dante's *Divine Comedy*: see Mazzoni, *Discorso in difesa della Comedia Del Diuino Poêta Dante* (1573), and *Della difesa della Comedia di Dante* (1587, 1688).

186. See Dante, *Rime*, XXXVIII 71–72: "Io non la vidi tante volte ancora / Ch'io non trovasse in lei nova bellezza" ("Never have I seen her so many times, / that I could not find new beauty in her").

187. See *Two New Sciences*, pp. 99–108 (*OG* 8, pp. 141–150).

188. Instead of "disciples," *A* has "pupils," and *S* has "pupils and followers." Francesco Nerli (1595–1670), Archbishop of Florence from 1652, was created a cardinal in 1669. Ascanio Piccolomini (1590–1671), Archbishop of Siena from 1629; a pupil of Bonaventura Cavalieri's, he hosted Galileo in Siena from July 9 to December 19, 1633, during which time Galileo recovered from the trial in Rome and resumed his work on what was to become the *Two New Sciences*. Giovanni Battista Rinuccini (1592–1653) was a noted legal scholar and became chamberlain to Pope Gregory XV, who made him the Archbishop of Fermo in 1625; he served as the papal nuncio to Ireland during the Confederate Wars (1645–1649). Giuliano de' Medici di Castellina (1574–1636) was ambassador of the Grand Duke of Tuscany to the court of Emperors Rudolf II and Matthias, from 1608 to 1618, during which he acted as an intermediary between Kepler and Galileo; he was created Archbishop of Pisa in 1620. Alessandro Marzi Medici (1557–1630), Archbishop of Florence from 1605. Giovanni Battista Ciampoli (1589–1643): Pope Gregory XV appointed him Secretary of Briefs in 1621, and in 1623 Pope Urban VIII promoted him to chamberlain; his influence was pivotal in ensuring that the *Dialogue* obtained the imprimatur; in 1632, when the book was published, Ciampoli was removed from his office in 1632 and exiled to The Marches. Filippo Pandolfini (1575–1655), senator from 1625: he translated into Latin Galileo's *On Floating Bodies*, *On Sunspots*, and *The Assayer*. Andrea Arrighetti (1592–1672) studied mathematics under Benedetto Castelli; he was elected a member of the Accademia della Crusca in 1613, and senator in 1649: see note 178. A historian and literary man, Tommaso Rinuccini (1596–1682) was ambassador of the Grand Duke of Tuscany in 1623, to congratulate the newly elected pope, and in 1625 was appointed secretary to Cardinal Francesco Barberini; he was a Knight of the Order of St. Stephen, from 1642. Pier Francesco Rinuccini (1592–1657), son of the poet Ottavio Rinuccini, was at the service of Grand Duke Ferdinand II, who nominated him his emissary in Milan in 1642, for thirteen years. Mario Guiducci (1585–1646) is the author of *Discorso delle comete* (1619), in response to Orazio Grassi's *De tribvs cometis* (1619): see note 116. Niccolò Arrighetti (1586–1639) was elected a member of the Accademia della Crusca in 1603; a pupil of Galileo's, Arrighetti succeeded him as Consul of the Florentine Academy in 1623. Braccio Manetti (1607–1652) was a member of both the Accademia della Crusca and the Florentine Academy, also serving in various capacities at the Grand Duke's court. A nobleman and poet, Niccolò Cini (d. 1638) was elected Canon of Florence in 1613. Piero de' Bardi (1564–1643) was a member of the Florentine Academy and the Accademia della Crusca, and contributed to bothe the first (1612) and second (1623) edition of the *Vocabolario degli Accademici della Crusca*, the earliest vocabulary of the Italian language. Filippo Salviati (1583–1614) was one of Galileo's closest friends, and the dedicatee of *On Sunspots*: see note 152. A pupil and friend of Galileo's, Jacopo Soldani (1574–1641) was Consul of the Florentine Academy in 1606, and was made senator in 1637; he is the author of "Capitolo contro i peripatetici" (1623), a satire of Aristotle's followers. Jacopo Giraldi (1576–1630) was a member of the Accademia della Crusca and of the Florentine Academy, of which he was elected consul in 1621. Michelangelo Buonarroti (1568–1646) was a Florentine poet and man of letters, known as "the Younger" to distinguish him from his famous granduncle. From 1586 to 1591, he studied mathematics at the University of Pisa, where he became friends with Galileo and Maffeo Barberini, the future Pope Urban VIII. He was elected to the Florentine Academy in 1585 and the Accademia della Crusca in 1589, and was one of the editors of the *Vocabolario degli Accademici della Crusca* (1612). Alessandro Sertini (1570–1632), a poet, was a member of the Florentine Academy, of which he was consul in 1602.

189. Albertino di Marcello Barisone (1587–1667) was one of the earliest pupils of Galileo's in Padua, where he studied law and philosophy. He was elected canon of the cathedral of Padua in 1610; after a few years in Germany (1623–1628), he returned to Padua, where he taught law (1628–1636) and, later, moral philosophy (1647–1653). From 1653 onward, he served as Bishop of Ceneda, in the province of Treviso.

190. *A* has: "of his *Compass* and other military instruments" ("del suo nuovo Compasso e di altri militari strumenti").

191. This alleged episode—confirmed by Niccolò Gherardini, whose biography of Galileo is independent from Viviani's (see pp. 154–155)—is often referred to as one of the many inaccuracies (possibly, the most notable one) of Viviani's biographical account of Galileo's life. In fact, Gustav II Adolf (1594–1632), the King of Sweden from 1611, was too young to attend Galileo's lectures in Padua (1592–1610). Galileo's pupil might well have been Prince Gustav (1568–1607), the son of Eric XIV, the King of Sweden from 1560 until his dethronement in 1568. After his father was deposed, Gustav traveled through Europe, and possibly visited Italy, stopping in Padua to attend Galileo's lectures (whose fame, as Viviani says, grew extraordinary, not only in Italy). Indeed, Gustav was very much interested in the natural sciences, especially chemistry, and was well versed in the Italian language. Also, as the son of a dethroned king, not only was he traveling incognito, under the name Monsieur Garse (the initial letters of *Gustavus Adolphus Rex Sueciæ*), but possibly aspired to regaining the throne, and the members of his retinue may have addressed him with the title of king. Accordingly, Galileo might well have been convinced he had King Gustav of Sweden among his pupils in Padua, and many years later referred the episode to Viviani.

192. Galileo tutored young Cosimo II de' Medici (1590–1621) between 1605 and 1608: see, for example, Girolamo Mercuriale to Galileo, 29 May 1601, in *OG* 10, no. 73, p. 84; Galileo to Cosimo, 29 December 1605, ibid., no. 131, p. 154; Cosimo's reply, on 9 January 1606, ibid., no. 132, p. 155; and Ferdinando Saracinelli to

Galileo, 12 January 1606, ibid., no. 133*, p. 156. In 1606, Galileo dedicated the *Compass* to Prince Cosimo, mentioning the time they spent together (see *Compass*, p. 39; *OG* 2, pp. 367–368); and in 1609, when Cosimo became the new Grand Duke of Tuscany, after his father Ferdinando I's death, Galileo once again referred to his great privilege and pleasure in tutoring the young Cosimo.

193. Francesco (1594–1614) and Lorenzo de' Medici (1599–1648), the fourth and seventh sons of Grand Duke Ferdinando I and Christina of Lorraine.

194. Benedetto Castelli (1578–1643) became full professor at the University of Pisa in 1613, and was appointed professor at the University La Sapienza, in Rome, by Pope Urban VIII, in 1625 (see also above, note 82). Niccolò Aggiunti (1600–1635) was appointed to the chair of mathematics in Pisa in 1626, as Castelli's successor. Dino Peri (1604–1640) was Aggiunti's successor, appointed in 1636. Vincenzo Renieri (1606–1647) was Peri's successor, appointed in 1640 (see also above, note 145). Bonaventura Francesco Cavalieri (1598–1647), a pupil of Castelli's in Pisa, was appointed to the chair of mathematics in Bologna, in 1629.

195. See, for example, Galileo to Belisario Vinta, 19 March 1610 and 7 May 1610 (in *OG* 10, no. 277, pp. 298–302; and no. 307, pp. 349–350); see also Alessandro Sertini to Galileo, 27 March 1610 (ibid., no., 282, p. 306) and Galileo to Kepler, 19 August 1610 (ibid., no. 379, pp. 421–422).

196. Odoardo I Farnese (1612–1646), fifth Duke of Parma and Piacenza. Maximilian I (1573–1651), called "The Great," a Wittelsbach ruler of Bavaria and prince-elector of the Holy Roman Empire; Galileo presented him with both a telescope and a microscope, and Maximilian I reciprocated with a gift: see Maximilian I to Galileo, 8 July 1610 (in *OG* 10, no. 354, p. 393). Vincenzo I Gonzaga (1562–1612), ruler of the Duchy of Mantua and the Duchy of Montferrat, from 1587; a friend of the poet Torquato Tasso's, he was a patron of the arts and sciences, and turned Mantua into a vibrant cultural center; in 1604 he presented Galileo with a medal and a neck chain: see *OG* 19, p. 155. Luigi I d'Este (1594–1664), grandson of Cosimo I de' Medici, the former Grand Duke of Tuscany. Leopold V (1586–1632), Archduke of Further Austria, and the younger brother of Emperor Ferdinand II; a patron and friend of Galileo's, he is the dedicatee of Mario Guiducci's *Discorso delle comete* (1619) and was presented by Galileo with two telescopes; see Mario Guiducci to Galileo, 18 October 1624 (in *OG* 13, no. 1672, p. 217), and Vincenzo Galilei's recollections, below, pp. 130–131. Charles of Austria (1590–1624), called "The Posthumous" (1590–1624), was a member of the House of Habsburg, and Prince-Bishop of Wrocław and Brixen; he was the youngest son of Charles II, Archduke of Austria. Galileo presented him with a telescope, and Charles of Austria reciprocated with a golden chain encrusted with jewels: see Vincenzo Galilei's recollections, below, p. 000. Instead of "Leopold and Karl," *S* has: "Ferdinand, Leopold, and Karl" ("Ferdinando, Leopoldo e Carlo").

197. For instance, Galileo received a golden chain from Cardinal Scipione Borghese (1576–1633): see Berlinghiero Gessi to Galileo, 30 June 1610 (in *OG* 10, no. 345*, p. 385); and gifts from Cardinal Francesco Barberini: see Galileo to Federico Cesi, 8 June 1624 (in *OG* 13, no. 1637, p. 182). In the last years of his life, when confined to Arcetri, Galileo received many gifts from Archbishop Ascanio Piccolomini, who had hosted him in Siena for six months, after the trial.

198. In appreciation of his showing the telescope to Doge, Leonardo Donà, and the Venetian Senate, Galileo's lectureship at the University of Padua was renewed for life, and his salary raised to three times the highest that had ever been paid to a lecturer of mathematics; see also Luca Valerio to Galileo, 29 May 1610 (in *OG* 10, no. 321*, pp. 362–363). The States General of the United Provinces of Holland offered Galileo a golden chain for his proposed solution to the problem of longitude (see the States General to Galileo, 25 April 1637, in *OG* 17, no. 3468**, p. 66; Laurens Reael to Galileo, 22 June 1637, ibid., no. 3506, pp. 118–119; and *OG* 19, pp. 396, 539–540 and 540–541). Galileo expressed gratitude but declined the offer: see his letter to Élie Diodati, August 1638 (in *OG* 17, no. 3780, p. 371).

199. Władysław IV Vasa (1595–1648), King of Poland and Grand Duke of Lithuania from 1632, a good friend and supporter of Galileo's: see, for example, Mario Guiducci to Galileo, 8 February 1625 (in *OG* 13, no. 1706*, p. 253). As to Gustav of Sweden, see note 191; he is not to be identified with Gustav II Adolf, King of Sweden but, rather, with Prince Gustav (1568–1607), the son of Eric XIV, King of Sweden.

200. It is unclear whether Viviani is here referring to Philip III (1578–1621), King of Spain and Portugal from 1598 (also called "Philip the Pious"), or to Philip IV (1605–1665), who succeeded his father at the time of the latter's death, and was a patron of the arts. The Holy Roman Emperors Rudolf II (1552–1612), Matthias (1557–1619) and Ferdinand II (1578–1637) were patrons of the arts and sciences.

201. In the margin, *B* has: "This may be seen from the universal approval with which his works, principles, and theories were accepted, as well as presented, taught, and embraced by the most eminent professors" ("Scorgesi dall'applauso universale con che sono state ricevute l'opere sue, i suoi dogmi e dottrine, e dalle più famose accademie e nelle più celebri catedre lette, insegnate et abbracciate").

202. *A* has: "who took care to meet him, and in case he was not in the city, they reached out to him at his villa, where he spent most of his time" ("che non procurasse conoscerlo, e non essendo in città, si trasferivano sino alla villa, dove ei dimorò più del tempo").

203. See Pliny the Younger, *Epistulae*, II 3, 8: "Have you ever read of a certain inhabitant of the city of Cadiz who was so struck with the illustrious name of Livy that he traveled from the ends of the earth

to see him, and as soon as he had seen him, departed?" ("Numquamne legisti, Gaditanum quendam Titi Livi nomine gloriaque commotum ad visendum eum ab ultimo terrarum orbe venisse, statimque ut viderat abisse?").

204. *S* has: "on September 8, 1638." See also *Vincenzo Galilei*, pp. 130–131.

205. From a number of letters of the time we know Galileo was unwell, and often in bed, for a few months: see, for example, Fulgenzio Micanzio to Galileo, 4 December 1638 (in *OG* 17, no. 3818, p. 409); from another letter Micanzio sent to Galileo, on 28 May 1638, we know that the Grand Duke often visited him in Arcetri (see *OG* 17, no. 3732**, p. 335). In the summer of 1638, John Milton (1608–1674) paid a visit to Galileo, as we know from the *Aeropagitica* (1644), p. 24: "There [i.e., in Florence] it was that I found and visited the famous Galileo grown old, a prisner to the Inquisition, for thinking in Astronomy otherwise then the Franciscan and Dominican licencers thought." Years before, in March 1618, Archduke Leopold V visited Galileo, who was ill in bed: see Galileo to Leopold V, 23 May 1618 (in *OG* 12, no. 1324, p. 389). In *The Assayer* (p. 176), Galileo describes the illness that prevented him from observing the comet that appeared between August 1618 and January 1619, and the visits he received from friends, with whom he discussed the phenomenon: "During the whole time that the comet was being seen, I was confined by illness to my bed, where I was often visited by friends. Discussions frequently took place there [...]" ("Per tutto il tempo che si vide la cometa, io mi ritrovai in letto indisposto, dove, sendo frequentemente visitato da amici, cadde più volte ragionamento delle comete [...]": *OG* 6, p. 225).

206. Both *A* and *S* have: "in political matters."

207. After comparing Galileo to Amerigo Vespucci (see note 173), in the preface to his projected biography Viviani compares him to Christopher Columbus (1451–1506), who discovered America and initiated the colonization of the new world. Later, Viviani would compare Galileo to Michelangelo (see pp. 266–267, below). The first to parallel Columbus and Galileo was Giovanni Battista Manso, in a letter to Galileo dated March 18, 1610, just a few days after the publication of the *Sidereal Messenger*: see *OG* 10, no. 275*, p. 296.

2

GIROLAMO GHILINI

GALILEO GALILEI

(CA. 1633)

Non è mai stato professore alcuno di Matematica, ch'habbia esposta così necessaria e nobil scienza con maggior chiarezza, e purità di quello, che fece il Galileo hoggidì[a] viuente in Fiorenza sua Patria, figliuolo di Vincenzo Galilei Gentilhuomo di belle lettere, assai famoso per diuerse opere, che ha scritto intorno alla Musica: Nessuno con più facile maniera, & efficacia risolse i difficili, & intricati dubbij in simile materia occorrenti; cosa che a lui riusciua facilissima per causa de' continui studij, e publiche letture di quella scienza, nella quale ha tenuto sempre impiegato il suo felicissimo, e sottilissimo ingegno, hauendola spiegata non solo in Pisa trè anni ad instanza di Ferdinando Primo Gran Duca di Toscana, a cui era benissimo noto il suo valore; ma anco in Padoua per lo spazio di 18. anni[b] continui con gran concorso di Scolari di varie nazioni, & in particolare di principalissimi Caualieri, e Signori grandi. Fratanto[c] essendosi egli trasferito a Vinezia sentì a leggere fra l'altre nuoue nella gazzetta di Fiandra, che vn maestro d'occhiali hauea presentato a Maurizio Prencipe d'Orange, vn'occhiale, che mostraua le cose lontane, come vicine, ond'egli la notte vegnente ne inuentò vno di propria industria; & hauendone fatto vn dono al Senato di quella Città, fu da quei Clarissimi con marauiglia grande riceuuto, e con grandissima generosità riconosciuta la sua ingegnosissima inuenzione; poiche per publico decreto li concessero con duplicato stipendio la Lettura, mentre viueua. Con l'vso di questo marauiglioso instrumento, comunemente chiamato Cannocchiale, che anco potrebbesi chiamare Segretario della Luna, e delle [69]Stelle, scoperse le nouità Celesti, & in particolare trouò i quattro pianeti giouiali, a quali diede il nome di Medicei; & ha similmente scoperti non solo i difetti della Luna, ma anco le vergogne del Sole, il quale non hà tanto potuto nascondersi sotto il velo della sua luminosa caligine, che il Galileo fatto nuoua Aquila con questo rinuouato e migliorato artifizio non habbia fissati in lui gli occhi, e scoperte le sue macchie. Finalmente il Gran Duca Cosimo II.[d] lo chiamò al suo seruigio in Fiorenza, oue con amplissimo[e] stipendio, senza obligo di leggere, attende a godere dopo la fatica della Lettura di molti anni, la felicissima quiete d'vna vita priuata, trouandosi di già nell'età di 73.[f] anni.

a hoggidì *H*] higgidì *D*
b 18. anni *H*] dieciott'anni *D*
c Fratanto *H*] Frattanto *D*
d II. *H*] secondo *D*
e amplissimo *H*] ampissimo *D*
f 73. *H*] settanta, e tre *D*

No professor of mathematics ever taught this noble and useful science with greater clarity and purity than Galileo, who is now living in Florence, his homeland. He is the son of Vincenzo Galilei, a learned nobleman, very well known for a number of works he wrote about music. Nobody ever solved more easily and effectively the difficult and complex problems presented by mathematics. Galileo did that most easily, thanks to his relentless studies and public lectures on that discipline with which he always engaged his most blessed and sharp mind. Indeed, he taught it not only in Pisa for three years, on behalf of the Grand Duke of Tuscany, Ferdinando I, who was very well acquainted with his merits, but also in Padua, for eighteen continuous years; and among his disciples were many students from various nations, and especially very important knights and great princes. Meanwhile, while he was in Venice, he learned from the Flanders gazette that a spectacle master had presented Maurice, Prince of Orange, with a spyglass that showed distant objects as if they were close to the eye. The following night, he built one with his own hard work,[1] and having presented it to the Senate of that city, those most eminent men received it with great amazement, and acknowledged his most ingenious invention with the greatest generosity: by public decree, they granted him a lifelong lectureship, with a doubled salary. By means of this amazing instrument, commonly referred to as telescope (and we might well name it the keeper of the Moon's and the [69]stars' secrets), Galileo made some celestial discoveries. In particular, he discovered the four planets around Jupiter, which he named after the Medici family. Similarly, he not only discovered the flaws of the Moon but also the blemishes of the Sun, which no longer managed to hide behind the veil of its own bright mist. Thanks to the device he updated and improved upon, Galileo—a new eagle—set his eyes on the Sun and discovered its spots. The Grand Duke Cosimo II finally called Galileo to his service in Florence, with a most generous salary and no teaching duties. There, after the burden of many years of teaching, he enjoys the most blissful rest of a retired life, being 73 years old already.[2]

Non passa Prencipe, ò gran Personaggio per quella sua Patria, che non si senta muouere da gran curiosità di vedere trà l'altre marauiglie di quella fioritissima Città il Galileo, cioè vn grandissimo ingegno trà i più famosi Letterati, & il più perfetto Matematico del presente secolo. Le opere, che ha publicate sono le seguenti, *Siderius Nuncius: Difesa contro le calunnie di Baldassar Capra: L'vso del compasso geometrico, e militare da lui ritrouato: Discorso delle cose, che stanno in sù l'acqua, ò che in quella si muouono: Istoria, e dimostrazione intorno alle macchie Solari, e loro accidenti compresi in trè lettere scritte a Marco Velseri: Il Saggiatore, nel quale con bilancia esquisita, e giusta si ponderano le cose contenute nella Libra Astronomica, e Filosofica di Lotario Sarsi: I Dialoghi diuisi in 4. giornate intorno a i due massimi Sistemi Tolemaico, e Copernicano.* Ha parimente scritte altre opere, le quali non sono ancora stampate, e trattano del moto, delle resistenze de' corpi ad essere spezzati, che sono dottrine nuoue, e del centro della grauezza de' solidi; Laonde porgendo questi dottissimi componimenti vtilità indicibile a studiosi ingegni, & a professori di quella scienza, vengono da essi con grandissime comendazioni esaltati, & a me porgono opportuna occasione di honorare i pregiati meriti suoi con darli principal luogo in questo mio Teatro.

No prince, or important person, ever passes through his homeland without feeling drawn by curiosity to see, among the other marvels of that most flourishing city, Galileo: a most eminent mind among the most famous learned men, and the best mathematician of the present century. The works he published are the following: *Sidereal Messenger*; *A Defense Against the Libels of Baldassarre Capra*; *The Use of the Geometric and Military Compass He Invented*; *Discourse on Bodies That Stay atop Water, or Move in It*; *History and Demonstration Concerning Sunspots and Their Properties, in Three Letters to Marc Welser*; *The Assayer, in Which with an Exquisite and Fair Balance Are Weighted the Things Contained in Lotario Sarsi's Astronomical and Philosophical Balance*; *Dialogues, in Four Days, Concerning the Two Chief [World] Systems, Ptolemaic and Copernican*. He also wrote other works, which have not yet been printed, that deal with motion, the resistance of bodies to being broken (which are new doctrines), as well as the center of gravity of solids. As these most learned works are of the highest value to scholars and teachers in that science, they are very highly praised and offer me the opportunity to honor Galileo's precious merits by giving him a central place in my *Theatre*.

NOTES

1. See *Racconto istorico*, pp. 20–21; and Siri, pp. 120–121.

2. In fact, when Ghilini was writing—in 1632 or early 1633, if my speculation about the publication date of the Milan edition of the *Teatro d'hvomini letterati* is correct (see the Editorial Note, p. LV)—Galileo was 68 or 69 years old. Ghilini does not say when Galileo was born (most likely, he did not know), and his estimate is wrong. Certainly, when Galileo was 73 (in 1637) he was not enjoying "the most blissful rest of a retired life," as he was under house arrest. Ghilini is talking about the years between 1610 and 1633, when Galileo was enjoying his time in Florence, with a very good salary and no teaching duties. All this changed in 1633, after the trial.

3

LEO ALLATIUS

GALILÆVS GALILÆVS GALILEO GALILEI

(1633)

[118]GALILÆVS GALILÆVS Florentinus, Lynceus, Philosophus, & Mathematicus edidit plura Hetrusco, Latinoque sermone.

1. *Sidereus Nuncius*, magna longèque admirabilia spectacula pandens, suspiciendaque proponens, maximè Philosophis, atque Astronomis, quæ perspicilli nuper à se reperti beneficio sunt obseruata in facie Lunæ, fixis innumeris, &c. apprimè verò in quatuor Planetis circà Iouis stellam disparibus interuallis, atque periodis celeritate mirabili circumuolutis, quos nemini huc vsque cognitos Auctor primus deprehendit, atque *Medicea Sidera* nuncupandos decreuit. Venetijs, 1610. in 4. & Francofurti apud Palthenios[1] in 8.

2. *Istoria, e Dimostrazioni intorno alle Macchie solari, e loro accidenti, composte in tre Lettere scritte al Sig. Marco Velsero.* Romæ apud Iacobum Mascardum 1613. in 4. Adduntur tres Epistolæ ad Marcum Velserum, *De Maculis solaribus.*

3. *Il Saggiatore, nel quale con bilancia esquisita, e giusta si ponderano le cose contenute nella Libra Astronomica, e Philosophica di Lotario Sarsi.* Romæ apud eundem 1623, in 4.

4. *L'Operationi del Compasso Geometrico, e Militare.* Patauij apud Petrum Marinellum 1606. in fol.

5. *Difesa contro alle calunnie, & imposture di Baldassarre Capra Milanese, vsategli nella consideratione Astronomica sopra la nuoua stella del 1604. come, & assai più nel publicare nuouamente come sua inuentione la fabrica, e gli vsi del Compasso* [119]*Geometrico, e Militare sotto il titolo di,* Vsus, & fabrica Circini cuiusdam proportionis, &c. Venetijs apud Baglionum 1607. in 4.

6. *Discorso intorno alle cose, che stanno in sù l'acqua, e che in quella si mouono.* Florentiæ apud Iunctam 1623. in 4.[2]

Audio penes eundem extare pleraque Opera Philosophica: inter ea non infima sunt,

Mechanica. Problemata Mechanica.
Elementa nouæ Scientiæ circà motus locales, Naturalem, & Violentum.

---

1 Palthenios *scripsi*] Phaltenios *R*2

2 *post* in 4. *scr.* 'Dialogo doue ne i congressi di quattro giornate si discorre sopra i due massimi Sistemi del mondo Tolemaico, e Copernicano tanto per l'vna, quanto per l'altra parte.' Florentiæ apud Io. Baptistam Landinum 1632. in 4. *R*1

[118]GALILEO GALILEI, Florentine, Lyncean, Philosopher, and Mathematician, published extensively in the vernacular and in Latin.

1. The *Sidereal Messenger,* unfolding great and very wonderful sights, and offering to the gaze of everyone, chiefly philosophers and astronomers, what was discovered with the help of a spyglass he newly invented about the surface of the Moon, countless fixed stars, etc., and especially four planets circling Jupiter at amazing speed, with unequal [space] intervals and revolution periods, and which were unknown to anybody until this day. The author was the first to detect them, and decided to name them *Medicean Stars.* Venice, 1610, quarto; Frankfurt: Palthenius, in octavo.[1]

2. *History and Mathematical Demonstrations Concerning Sunspots and Their Properties, in Three Letters to Marc Welser.* Rome: Giacomo Mascardi, 1613, in quarto; including three letters to Marc Welser, *On Sunspots.*

3. *The Assayer, Where with an Exquisite and Fair Balance the Things That Are Contained in Lotario Sarsi's Astronomical and Philosophical Balance Are Weighed.* Rome: the same, 1623, in quarto.

4. *Operations of the Geometric and Military Compass.* Padua: Pietro Marinelli, 1606, in folio.

5. *Defense against the Calumnies and Impostures of Baldassarre Capra, from Milan, Appealed to in the Astronomical Consideration on the New Star of 1604, as well as in Publishing Again, as if It Were His Own Invention, the Construction and Use of the* [119]*Geometric and Military Compass, under the Title* Use and Construction of a Certain Proportional Compass, etc. Venice: Baglioni, 1607, in quarto.

6. *Discourse on Bodies That Stay atop Water, and Move in It.* Florence: Giunti, 1623, in quarto.[2]

I am told that there are also several philosophical works by this same author. Among these, the most important are:

Mechanics. Mechanical Problems.
Elements of a New Science of Local Motions, Natural and Violent.

Ad librum Grassi, cui titulus fuit, Ratio ponderis, & Simbellæ, *responsio.*[3] Hunc aliorum sibi labores furto affertos, vt Tele-scopij inuentionem, & Sidera Medicea parum cautis diuenditasse, multi, & voce, & scripto palàm professi sunt. Illud tamen verum est, Tardinum Gallum inuentas à se de nouo maculas, Galilæi imitatione, *Sidera Borbonia,* inscripsisse. Tantique eum apud Gallos fieri, vt nonnulli illius tantùm videndi causa, iter in Italiam instituerint; & à fide digno Viro accepi, Diodatum quendam nobilem, literis, ac virtutibus clarum, Florentiam è Gallia eo solùm nomine contendisse; & cum per tredecim dies summa animi oblectatione, de varijs naturæ arcanis sermonem cum Galilæo habuisset, sat sibi esse propter eum Italiam vidisse ratus, cæteris alijs negotijs neglectis, in Galliam quàm maximis itineribus celerasse. De eo scribit Gualdus in vita Vincentij Pinelli, *Patauij in Academia res Mathematicas magna cum laude professum.* & Laurus in Orchestra, *Vibrauit Galilæus Galilæus, Medicearum stellarum inuestigator, & promus, tam non bellè quidem meritus de Solis, ac Lunæ orbibus, quos infuscat maculis, vtcumque lateat post Tabulam Apelles, & insulis modò prominentibus attollit, modò plagis depressioribus concamerat, quàm amenioris sapientiæ candidatis, aut trabeatis Principibus gratissimus habetur.* Iulius Cæsar Lagalla scripsit de Phænomenis in Orbe Lunæ noui telescopij vsu à Galilæo Galilæo nunc iterum suscitatis, Physicam Disputationem editam Venetijs apud Thomam Balionum 1612.[4]

APPENDIX

[68]CHRISTOPHORVS SCHEINER Societatis Iesu Sacerdos, Theologus, Sacræ linguæ, & scientiarum Mathematicarum Ingolstadij, & Fryburgi Brisgoiæ Professor, admirabile de Solis Maculis, & Faculis Phænomenon primus inuenit Ingolstadij anno Christi 1611. mense Martio, vt idemmet fatetur in volumine magno, quod Rosæ Vrsinæ, siuè Solis titulo præsigniuit ad Lectorem lib. 1. atque 3. passim, quas etiam primus sub Apellis nomine post tabulam latentis, Marci Velseri II. viri Augustani liberalitate in publicum edidit Augustæ Vindelicorum an. 1612. Non. Ian. perscriptis tribus ad eundem Velserum Epistolis, Itemque Idibus Septembris anno eodem missis ad eundem alijs Epistolis, cum obseruationum Iconismis, macularumque delineationibus. […] Elucubrauit præterea.

1. *Disquisitiones Mathematicas de controuersijs, & nouitatibus Astronomicis:* propugnatas à Ioanne Georgio Locher[5] discipulo, Præsidente Scheinero Magistro, Ingolstadij 1614. in quibus motum terræ Copernicanum eruditorum consensu prorsus elidit contrà omnes Nouatorum Paralogismos.

3 *post* responsio *scr.* Legi etiam ipsius, 'De motu terræ,' Discursum sermone Italico scriptum *R1*
4 Iulius—1612 *R1 post corr.*] Sed vbi clangunt Aquilæ, quid perstrepunt anseres? vnus illi Plato præ omnibus, Plato Philosophorum Deus, literatorum numen. Summus Princeps in adulatione perniciosa, raptum è luto, & humanis hisce Elogijs, vna secum ponit ad sidera, firmatque ne cadat, / 'Seù Scorpij cor, siue Canis facem / Miratur alter, vel Iouis asseclas, / Patrisuè Saturni, repertos / Docte tuo Galilæe vitro.' & inferiùs / 'Non semper extrà quod radiat iubar / Splendescit intrà: respicimus nigras / In Sole, quis credat? retectas / Arte tua, Galilæe, labes' *R1 ante corr.* Iulius Cæsar Lagalla scripsit de Phænomenis in Orbe Lunæ, noui Tele-scopij vsu, à Galilæo Galilæo nunc primùm suscitatis, Physicam Disputationem, editam Venetijs 1612 *R2*
5 Locher *scripsi*] Loher *R2*

Response to Grassi's book, titled *A Reckoning of Weights for the Balance and the Small Scale.*[3] Many openly declared, both in speech and in writing, that Galileo had sold to careless buyers the works by others, brought to him by stealth, such as the invention of the telescope and the discovery of the Medicean Stars. It is true, however, that the Frenchman Tarde named the spots he had first discovered *Sidera Borbonia*, in imitation of Galileo.[4] The latter achieved so much fame in France that some organized a trip to Italy only to see him. And I learned from an extremely reliable person that one nobleman, Diodati, well known for his erudition and virtues, traveled to Florence from France with the sole purpose of seeing Galileo. For thirteen days, Diodati had a most enjoyable exchange with him on various secrets of nature, and since Galileo was all he was interested in seeing in Italy, Diodati neglected all other commitments and rushed back to France in the quickest possible way.[5] Thus Gualdo writes about him in the *Life of Vincenzo Pinelli*: "he taught mathematics at the University of Padua, with great honor."[6] And Lauri, in his *Orchestra*: "Galileo Galilei shone: he investigated and spread the news of the Medicean Stars. However, he does not behave well with the orbs of the Sun and the Moon, which he tarnishes with spots (although Apelles hides himself behind the picture), and which he now elevates with bulges and now depresses with valleys. His merit consists rather in the fact that he is dearest to those who strive for beautiful knowledge, or to princes with power."[7] Giulio Cesare Lagalla wrote the *Physical Disputation on the Phenomena in the Orb of the Moon, Now Once More Produced by Galileo Galilei, Using a New Telescope*, published in Venice by Tommaso Baglioni, in 1612.[8]

Appendix

[68]CHRISTOPH SCHEINER, Jesuit theologian, professor of Hebrew and mathematics at Ingolstadt and Freiburg im Breisgau. He was the first to discover (in Ingolstadt, in March 1611) the remarkable phenomenon of sunspots and solar flares, as he himself says in the large volume titled *Rosa Vrsina, or the Sun*, in the "Preface to the Reader," and in Books I and III (*passim*).[9] Also, he published these discoveries for the first time under the name "Apelles hiding behind the canvas," printed in Augsburg, thanks to the generosity of the Augsburg duumvir Markus Welser, on January 5, 1612, in three letters to said Welser;[10] and, later, in a few additional letters, also addressed to Welser, and published on September 19 of the same year,[11] including pictures of the observations, and sketches of the spots. [...] Scheiner also worked on the following:

1. *Mathematical Disquisitions on Astronomical Controversies and Novelties*, defended by Johann Georg Locher, under the direction of Professor Scheiner (Ingolstadt 1614), in which, in accordance with scholars, he rules out the Copernican motion of the earth straight away, against all fallacious arguments of the renovators.[12]

2. *Solem Ellipticum.* Augustæ Vindelicorum 1615.

3. *Exegeses fundamentorum Gnomonicorum,* defensas a Io. Georgio Schonbergero discipulo, Præsidente P. Scheinero Professore Ingolstadij 1615.

4. *Refractiones Cœlestes,* siuè, *Solis Elliptici Phænomenon* illustratum. Ingolstadij 1617.

5. *Oculum,* hoc est, *fundamentum opticum.* Oeniponti 1619. in 4. liber est à multis commendatus, & nunc quoque expetitus.

6. *Rosam Vrsinam,* siuè, Solem ex admirando facularum, & macularum suarum Phænomeno varium, necnon circà centrum suum, & axem fixum ab occasu in ortum annua, circàque alium axem mobilem ab ortu in occasum conuersione quasi menstrua, super Polos proprios, lib. IV. mobilem ostensum. [69]Bracchiani apud Andream Phæum 1630. in folio.

In lib. I. Apellem suum ab iniurijs Galilæi Historica veritate auctor solidè tutatur, Galilæi circà hoc Phænomenon errores patefacit, seque Phænomeni solaris non inuasorem, sed primum repertorem, legitimum possessorem luculenter demonstrat.

In II. Difficultates omnes, quæ circà hanc apparitionem emergunt, amolitur, modos obseruandi, atque ordinandi maculas accuratè docet, multa præclara de oculo, de specierum visibilium radiationibus, de lentium tubi natura, tandemque ipsum tubum opticum, quòd hactenus desiderabatur, demonstrat.

In III. Obseruationes Macularum, & suas, & aliorum innumeras adducit, è quibus, & locum, & motum earum deducit.

In IV. Solis globum, circà eiusdemmet centrum gyrat annuè, & *menstruè,* gyrationis Theoriam proponit. Solem igneum, Cœlum liquidum, & corruptibile facit, ex ratione, experientia, Philosophis, Theologis, Sanctis Patribus penè omnibus, Scriptura sacra. Denique de lumine de terræ *Atmosphæra,* de Solis magnitudine visibili experimenta mirifica adfert. Opus laboriosum, subtilitate, doctrina, experientiaque varia, noua, eâque vera plenum, Auctore, & Societate Iesu dignum.

7. *Pantographicen,* seù, *artem delineandi,* res quaslibet per Parallelogrammum lineare, seù cauum Mechanicum mobile, libellis duobus explicatam, & Demonstrationibus Geometricis illustratam; quorum prior *Epipedographicen,* siuè planorum, posterior, *Stereographicen,* seù solidorum aspectabilium viuam imitationem, atque proiectionem edocet. Ars facillima, certissima totius graphices, atque proiectionis opticæ, seù perspectiuæ essentiam compendio complexa. Romæ apud[6] Grign. 1631. Idem Auctor in manibus nunc habet *Parhelia,* in quibus multa de Iridibus, Halonibus, Virgis, Chasmatis scitu digna tractat.

Maculas Solares aliquot reduces; earum genesim, & interitum nouum, eumque facillimum ipsas obseruandi, & cursus ordinandi modum depromit.

Scintillationes stellarum.
Oculum, seù fundamentum Opticum auctum.

---

6 apud] apnd *R2*

2. *The Elliptic Sun*, Augsburg, 1615.[13]

3. *Analyses of the Foundations of Gnomonics*, defended by Johann Georg Schönberger, under the direction of Father Scheiner, Professor at Ingolstadt, 1615.[14]

4. *Celestial Refractions*, that is, the explanation of the *Phenomenon of the Elliptic Sun*, Ingolstadt, 1617.[15]

5. *The Eye*, that is, *the Foundation of Optics*, Innsbruck, 1619, in quarto. This book is praised by many, and now also sought after.[16]

6. *Rosa Vrsina*, or that the Sun, from the remarkable phenomenon of its flares and spots, is variable and also shown to move about its center and fixed axis, from west to east, in a year, and about another movable axis, from east to west, in nearly monthly rotation about its own poles.[17]

   In Book I, the author, truthfully and completely defends his Apelles from Galileo's unfair attacks and exposes Galileo's errors about this phenomenon. Also, he brilliantly shows that he is not unjustly claiming priority over the discovery of this solar phenomenon but is its true discoverer, and rightfully has priority over it.

   In Book II, he removes all problems that arise about this phenomenon and explains how to observe and arrange the spots. He also shows many amazing things about the eye, the radiance of visible species, the nature of the lenses of the telescope, and finally about the telescope itself, which have remained so far undisclosed.

   In Book III, he presents countless observations of sunspots, both his own and others', from which he derives their position and motion.

   In Book IV, he advances a rotation theory that accounts for the yearly and monthly rotation of the globe of the Sun around its very center. He describes the Sun as a fire, and heavens as fluid and corruptible, in accordance with reason and experience, and with nearly all philosophers, theologians, the Holy Fathers, and the Holy Scriptures. Lastly, he offers astonishing experiments about light, the earth's atmosphere, and the visible size of the Sun. It is a complex work, full of insight and learning, and rich with various new and true experiences, worthy of its author and of the Society of Jesus.

7. *Pantograph*, that is, *the art of drawing whatever object by way of a linear parallelogram*, or a hollow mechanical movable tool, expounded in two booklets and illustrated by geometric diagrams. Of these, the first (*On drawing flat surfaces*) instructs how to draw a vivid copy and projection of planes; and the second (*On drawing solids*), of notable solids. It is the easiest and most reliable of the all the graphics techniques, and also includes a summary of the key elements of perspective. Rome: Lodovico Grignani, 1631.[18] Scheiner also authored these [unpublished] works: *Parhelia*, which treats many important issues about rainbows, vapors, lightning, and meteors.

Some Returning Sunspots, presenting their generation and recent destruction, and a very easy way to observe them and determine their course.

The Gleaming of Stars.
Oculus, or the Foundation of Optics, augmented edition.

Oculos, seù fundamentum Opticum secundum.
Opuscula Optica varia.

[70]Ante omnia autem *vnius Maculæ reducis triplicem cursum,* in lucem dabit vna cum cuiusdam alterius Maculæ cursu, ex quo Galilæus conatus est in suis recentibus Dialogis motum terræ annuum, & solis stationem deducere; vbi ostendet Scheiner ex illius Maculæ cursu, nihil eorum quæ vult Galilæus concludi: sed Galilæum verum Macularum motum ex Rosa Vrsina sibi hinc transmissa modo didicisse, atque hoc callide dissimulare, ideòque lectori imponere, Cœlo, Soli, Rosæ Vrsinæ, eiusque Auctori violentas manus inferre, & hæc quidem erit prælibatio quædam, quam mox sequetur ipsius operis,

Prodromus pro Stabilitate terræ contrà eundem Dialogistam, in quo compendiosè afferentur Galilæi errores Logici, errores Physici, errores Mathematici, errores Ethici, errores Theologici, atque sacri: adeòque ex omnibus his constabit detracta larua doctrinam hactenùs mentitam[7] imperitia.

Tum opus ipsum sequetur suo tempore, *pro motu Solis, pro statione Terræ,* elaboratum ex sacris, & profanis fontibus, ex sensu, & ratione stabilitum. Scheinerum clarissimè extulit Pineda in Ecclesiast. cap. I. vers. 4. illis verbis, *Vnum tandem pro multis vltimum,* refero, *sed in primis doctum nostri ordinis Christophorum Scheinerum, Sacræ linguæ, & Matheseos in Ingolstadiensi Academia Professorem, qui in Disquisitionibus Mathematicis, de Controuersijs, & nouitatibus Astronomicis disputatis Ingolstadij Anno 1614. penitùs disiecit, prostrauitque Copernicanum illud commentum, de motu terræ, pulcherrimis argumentis, tùm Physicis, tùm Mathematicis absurdam Copernicani Cœlorum Systematis falsitatem demonstrauit.* Sed quando iam Roma continuò Scheinerum, annoque superiori Galilæum perspexit oculis, accepitque auribus, concipiat etiam nunc animo absentem ordine tertium, virtute, studio, ac sapientia parem, quem honoris causa recolo, Petrum Gassendum Diniensis Ecclesiæ Theologum, virum primarium, qui nouo ingenij acumine, diserta orationis textura, & admirandorum monumentorum copia Europæ innotuit. Ea, quia Romana grauitate præcellunt, dignaque Romano aëre iudicantur ab omnibus in medium afferam: ne sapientia tantum, sed scriptis etiam editis, sicut reliqui præstantes viri in Aula Romana locum habeat. Vidi itaque ipsius.

[71]1. *Exercitationes paradoxas in Vniuersam Aristotelis Philosophiam,* Gratianopoli in 8.

2. *Epistolam Apologeticam pro P. Mersennio aduersùs Robertum Flud,* in qua ipsius Fludi Philosophia refellitur. Parisijs apud Sebastianum Cramoisy 1630. in 8.

3. *Parhelia, siuè, Soles quatuor, qui apparuerunt Romæ die xx. mensis Martij anni 1629.* ad Henricum Renerium. Parisijs apud Antonium Vitray 1630. in 4.

4. *Mercurium sub Sole visum, & Venerem inuisam.* Parisijs sumptibus Sebastiani Cramoisy 1632. in 4. Eiusdem,

Contrà Aristotelis Vniuersam Philosophiam, Volumina, & *Philosophiam Epicuream instauratam,* docti viri anxiè expectant.

---

7 mentitam *scripsi*] mentita R2

Oculus, or the Foundation of Optics, vol. 2.
Various Books on Optics.

[70]But before all other things, he will publish *The Threefold Course of One Returning Sunspot*, together with the course of another sunspot, from which Galileo tried to derive, in his recently published *Dialogue*, the earth's annual motion and the Sun's rest.[19] Here Scheiner will show that from the course of that sunspot we cannot derive any conclusions Galileo wants to draw. But Galileo learned the true movement of sunspots from the *Rosa Vrsina*, and shrewdly disguised it, and likewise imposed it to the reader, and did violence to the heavens, the Sun, the *Rosa Vrsina* and its author. This work will be a sort of prelude, which will soon be followed by

Preliminary Discourse for the Stability of the Earth against the Same Author of Dialogues,[20] in which will be summarized Galileo's logical errors, physical errors, mathematical errors, ethical errors, theological, and scriptural errors. Finally, from all these, once the mask is dropped, it will be clear that his doctrine is shown false because of his ignorance.[21]

Now, this work will follow, in due time, [a work] *in support of the Sun's movement, and of the Earth's rest*, drawn from both holy and profane sources, based on experience and reason. Pineda praised Scheiner most beautifully in his *In Ecclesiasten*, ch. 1, v. 4, with these words: "Finally", I will mention "one last—but first of all as to learning—member of our Order, Christoph Scheiner, Professor of Hebrew and Mathematics at the University of Ingolstadt. In the *Mathematical Disquisitions on Astronomical Controversies and Novelties* (Ingolstadt, 1614),[22] he thoroughly crushed and overthrew that Copernican fiction," about the earth's motion, "with very beautiful arguments, both physical and mathematical, and demonstrated the absurd falsity of the Copernican system of the heavens."[23] However, since Rome saw, and listened to, one after the other, Scheiner, and Galileo the previous year, let Rome imagine someone who has never been to Rome. Chronologically, he is the third, but equals [Scheiner and Galileo] in virtue, learning, and wisdom: Pierre Gassendi, theologian of the Church of Digne,[24] a prominent man, known in Europe for the extraordinary ingenuity of his mind, his elegant style, and the number of his admirable works; due to their Roman rigor, all regard them as worthy of Rome. I will mention them here, so that Gassendi may have his place at the Roman Court, like other excellent men, not only for his erudition but for his published works. I also saw these works of his:

[71]1. *Paradoxical Essays against the Whole of Aristotle's Philosophy*, Grenoble, in octavo.[25]
2. *Epistolary Essay, in Support of Marin Mersenne and against Robert Fludd*, in which Fludd's philosophy is shown false. Paris: Sébastien Cramoisy, 1630, in octavo.[26]
3. *The False Suns, That Is, the Four Suns, Which Were Observed in Rome on March 20, 1629*, to Henri Rener, Paris: Pierre Vitré, 1630, in quarto.[27]
4. *Mercury Seen in the Sun, and Venus Unseen*, Paris: Sébastien Cramoisy, 1632, in quarto.[28]

Scholars are looking forward to seeing his works *against the whole Aristotelian philosophy*, and on the *renovation of the Epicurean philosophy*.[29]

Notes

1. The 550 copies of the *Sidereal Messenger* printed in Venice were sold out in approximately a week. A second, most likely pirated, edition was printed in Frankfurt (*Siderevs Nvncivs Magna, longeqve admirabilia Spectacula pandens, suspiciendaque proponens vnicuique, præsertim vero philosophis, atque astronomis*, Frankfurt: Zacharias Palthenius, 1610). This included several typos and mistakes of various kinds; most notably, the original engravings of the lunar surface were replaced by much less accurate woodcuts.

2. In fact, *On Floating Bodies* was published in 1612. In the original version, here followed a reference to the *Dialogue*: "*Dialogue, in which the two chief systems of the world are debated over the course of four days, offering arguments in support of either views*, Florence: Giovanni Battista Landini, 1632, *in quarto.*"

3. Grassi, *Ratio pondervm libræ et simbellæ* (1626); Galileo annotated a copy of this work: see *OG* 6, pp. 377–500. Allatius must have known, from someone close to Galileo (possibly, Francesco Stelluti or Giovanni Ciampoli) that Galileo was planning to publish a new rejoinder to Grassi in book form. In agreement with Ciampoli and other *palatini* and *letterati*, Federico Cesi (the founder of the Academy of the Lynxes) tried to convince him to give up the project, however interesting the rejoinder was going to be, and devote his time to more important works: "Your Lordship is not competing in a carousel, and is not obliged to take the field, or enter the ring, so to speak, with anybody" ("V. S. è fuor di giostra, e che non è obbligato a discender in arena o entrar in steccato, come si dice, con alcuno": Cesi to Galileo, 9 September 1628, in *OG* 13, no. 1902, p. 448; but see the whole letter, pp. 448–449). Eventually, Galileo set the project aside. In the original version, here followed a reference to a clearly Copernican work of Galileo's: "I also read his *On the Motion of the Earth*, a discourse written in Italian." Allatius may be referring to Galileo's letters to Benedetto Castelli (December 21, 1613), Pietro Dini (February 16 and March 23, 1615), and Grand Duchess Christina of Lorraine (1615): see *OG* 5, pp. 281–288, 291–295, 297–305, and 309–348, respectively; or else to the *Discorso del flusso e reflusso del mare*, in the form of a letter to Cardinal Alessandro Orsini (January 8, 1616): ibid., pp. 376–395. It seems likely, however, that Allatius is here referring to one of the three papers Galileo wrote in 1615–1616 and which were collected by Antonio Favaro under the general title *Considerazioni circa l'opinione copernicana*: ibid., pp. 349–370 (see also Galileo to Piero Dini, 23 March 1615, ibid., p. 300). In particular, he might have had in mind Galileo's third paper (pp. 367–370), which offers a point-by-point rejoinder to Cardinal Bellarmino's remarks against the Copernican hypothesis, as presented in the letter to Paolo Antonio Foscarini of April 12, 1615 (in *OG* 12, no. 1110*, pp. 171–172), prompted by Foscarini's *Lettera . . . Sopra l'Opinione de' Pittagorici, e del Copernico. Della mobilitò della Terra e stabilità del Sole, E del nuouo Pittagorico Sistema del Mondo* (1615).

4. Jean Tarde (1561–1636) was a French historian and canon of the Sarlat cathedral, in the Dordogne department in Nouvelle-Aquitaine, in southwestern France. His curiosity for astronomy had been excited by Galileo's recent telescopic discoveries, and when he traveled to Rome in 1614, he met Galileo on November 12–13, and 15, in Florence (see *OG* 19, pp. 589–593). On his return to France in 1615 Tarde built a small observatory at La Roque-Gageac. Interested in the sunspots, which he thought were small planets, Tarde named them *Borbonia Sidera*, dedicated to Louis XIII Bourbon: see *Borbonia Sidera, id est Planetæ qvi Solis limina circvmvolitant motv proprio ac regulari* (1620).

5. This is the only source we have about Galileo's meeting with Élie Diodati, about which Allatius was likely told by the French scholar Gabriel Naudé (1600–1653). Diodati (1576–1671), a Swiss lawyer and jurist from a leading Calvinist family in Geneva, who had moved there from Lucca, in Italy, was always a loyal supporter of Galileo's and played a major role in the translation and circulation of his works abroad, after the 1633 trial. He was responsible for Matthias Bernegger's Latin translation of the *Dialogue*: *Systema cosmicvm . . . In quo Qvatvor Dialogis De Duobus Maximis Mundi Systematibus, Ptolemaico & Copernicano, Vtriusque rationibus Philosophicis ac Naturalibus indefinite propositis, disseritur*, Strasbourg: Bonaventura & Abraham Elzevier, at David Hautt, 1635: see Matthias Bernegger to Galileo, 10 October 1633, in *OG* 15, no. 2744, p. 299; and Pierre Gassendi to Galileo, 19 January 1634, in *OG* 16, no. 2851, pp. 20–21. The Latin translation of the *Dialogue* included an excerpt from Kepler's introduction to the *Astronomia nova* (1609) and Diodati's own (under the pseudonym Davides Lotæus: see *Systema cosmicvm* (1635), p. 459) Latin translation of Paolo Antonio Foscarini's *Lettera . . . Sopra l'Opinione de' Pittagorici, e del Copernico* (*Epistola . . . circa Pythagoricam, & Copernici opinionem de mobilitate Terræ, et stabilitate Solis*): see Bernegger to Wilhelm Schickardt, 9 June 1634, in *OG* 16, no. 2952*, p. 101. The *Systema cosmicvm* was reprinted a few times, notably in Lyon (1641) and Leiden (1700). Diodati also tried to have the *Dialogue* translated into French, but the project was abandoned: see Diodati to Galileo, 16 May 1634, in *OG* 16, no. 2942*, p. 96; and 12 March 1635, ibid., no. 3090*, p. 231. Most important, Diodati is responsible for the first edition of Galileo's *Letter to Christina*, published with facing Latin translation (by Robertus Robertinus Borussus, i.e., Élie Diodati himself) in 1636. Diodati's Preface (ff. A2r-A4r, a reprint of Diodati to Bernegger, 6 January 1635, in *OG* 16, no. 3058, pp. 194–196) is the earliest public defense of Galileo after the trial; see also Bernegger to Diodati, 28 December 1625, in *OG* 16, no. 3230*, pp. 366–367. Furthermore, the publication of Galileo's *Two New*

Sciences, in 1638, was made possible by Diodati's intercession with the publisher, Lodewijk Elzevir: see, for example, Galileo to Diodati, 27 October 1636, in *OG* 16, no. 3383*, p. 511; and Diodati to Galileo, 18 August 1637, in *OG* 17, no. 3546*, p. 173. Finally, Diodati mediated with the United Provinces of Holland for the adoption of Galileo's method for the determination of longitude, in 1635–1640.

6. Gualdo, *Vita Ioannis Vincentii Pinelli* (1607), p. 115.

7. Lauri, *Theatri Romani Orchestra* (1625), p. 39.

8. Giulio Cesare Lagalla (1576–1624) participated in the demonstrations of the telescope by Galileo when the latter went to Rome (in 1611) to submit his telescope and discoveries to Pope Paul V, eminent members of the Curia, and the Jesuits of the Roman College. Lagalla was professor of philosophy at the Roman College, and the two were introduced by Federico Cesi. Lagalla was not among those who doubted the ability of the instrument, but he did debate Galileo's three-dimensional representation of the Moon based on two-dimensional visual observations. Galileo also showed him the "Bologna stone" (see *Racconto istorico*, note 165). Lagalla wrote two books about these experiences, which he then published in a single volume as *De phænomenis in orbe Lvnæ* (1612). For some time, Galileo considered publishing a rejoinder to Lagalla but never actually began to write it. Later, upon the observation of a comet, in November 1613, Lagalla wrote the *Tractatus . . . de metheoro quod die nona novembris anni presentis 1613 in Urbe apparuit sopra collem Pincium* (which was never published; the manuscript is in *Gal.* 90, ff. 117r-129v). In it, Lagalla largely agreed with Galileo and was therefore accused of anti-Aristotelianism. After the death of Flaminio Papazzoni, Lagalla sought Galileo's and Cesi's support for replacing him at the University of Pisa. He did not get it, but the three of them remained on good terms. As explained in the Introduction (see p. xxxiii), this last sentence on Lagalla replaces the original ending of Allatius's entry on Galileo. It was inserted at the very last minute, in Allatius's handwriting, in the top margin of the previously printed sheet; later, it was typeset and printed in haste, as can be seen from the error of trivialization in the transcription of the title of Lagalla's 1612 book: the words *nunc iterum*, which Allatiusi correctly penned in the margin of the originally printed sheet, were replaced in the new version of the text by *nunc primum*; the compositor likely worked in a hurry and used a standard expression (which does not make sense in the title of Lagalla's work) instead of the actual words from Lagalla's title. Also, the name of the publisher (Tommaso Baglioni) was cut, possibly to save a line and fit the text in the space the compositor had before the beginning of the following entry. Further evidence of the haste with which the original text had to be revised is Allatius's different spelling of the publisher of Lagalla's work: *Balionum*, instead of *Baglionum*, as he had written before (when listing Galileo's *Defense* against Baldassarre Capra, at no. 5). In the present edition, I reinstated Allatius's holograph correction of the original text. The original ending of Allatius's entry reads: "But, where eagles resound, why do gooses honk? In his [Galileo's] eyes, Plato alone—Plato the god of philosophers, the divinity of men of letters—comes before all others. The Supreme Pontiff, in his *Adulatio perniciosa*, draws him from the mud of these human praises and takes him with himself, among the stars, and supports him, so that he does not fall: *Still others marvel at the heart of the Scorpion, / or the brightest star of the Dog, / or Jupiter's attendants, or his father Saturn's, which you discovered / through your telescope, learned Galileo*, and, farther down, *That which outwardly radiates brightness / does not always shine within: we gaze at black spots / in the Sun (who would believe it?) / laid bare by your art, Galileo*" (see the appendix to this book, pp. 304–305). The remark on Galileo's Platonism is striking. It was included in the original handwritten version (*Vat. lat.* 7075, f. 68v), and throughout the tormented publication history of the entry, and yet it seems out of place (it does not relate to either the comment on the quotes from Gualdo's and Lauri's works, which precedes it, or the quote from Urban VIII's *Adulatio perniciosa*—that is, *Perilous Flattery*—that follows it). Moreover, Allatius was a Byzantinist concerned with highlighting the links between the Eastern world and the Church of Rome, and Plato played no special role in his work or output. The oddity of the sentence, however, highlights its relevance: clearly, Allatius thought it important and wanted to include it, and gave it pride of place by putting it at the end of the entry, just before the words of the pope. Most likely, the remark came to Allatius directly from Galileo, or from someone close to him and well acquainted with his work. Indeed, Galileo's science was informed by his following Plato—and not Aristotle—on insisting that the principles of natural science be mathematical. This mathematical emphasis—which, of course, complemented the logical methodology of period Aristotelian philosophers (such as, for instance, Jacopo Zabarella, professor at Padua and the predecessor of Cesare Cremonini)—is quintessential to the Scientific Revolution and, at the time, formed the dividing line between the followers of Plato and those of Aristotle. See, for example, Francesco Buonamici (Galileo's professor of natural philosophy at the University of Pisa), *De motv* (1591), Book I, chapter X, pp. 54 H-55 A, and chapter IX, p. 56 H; and Jacopo Mazzoni, *In vniversam Platonis, et Aristotelis Philosophiam Præludia* (1597), pp. 187–197.

9. See, for instance, Scheiner, *Rosa Vrsina sive Sol* (1630), a 1vr-v1r.

10. Scheiner, *Tres epistolae de macvlis solaribvs* (1612).

11. Scheiner, *De macvlis solaribvs et stellis circa Iouem errantibus, accvratior disqvisitio* (1612).

12. Locher, *Disqvisitiones mathematicæ* (1614). Their 1612–1613 clash on sunspots notwithstanding, Scheiner and Galileo were still on good terms in 1615, and Scheiner sent him a copy of his pupil's book to

Galileo (see Scheiner to Galileo, 6 February 1615, in *OG* 12, no. 1077, p. 137). Things changed in 1619, when Galileo antagonized the Jesuits in the literary feud waged over the comets of the previous year. Galileo sarcastically refers to Locher's book in the *Dialogue*, where the book looms large in the discussions of the Second Day; see also *Dialogue*, pp. 91–92 and 318–319 (*OG* 7, pp. 117–118 and 346).

13. Scheiner, *Sol Ellipticus* (1615).

14. Schönberger, *Exegeses Fvndamentorvm Gnomonicorvm* (1615).

15. Scheiner, *Refractiones Coelestes, sive Solis Elliptici Phænomenon illvstratvm* (1617).

16. Scheiner, *Ocvlvs* (1619).

17. Scheiner, *Rosa Vrsina sive Sol* (1630).

18. Scheiner, *Pantographice* (1631).

19. See *Dialogue*, pp. 345–355, especially pp. 351–355 (*OG* 7, pp. 372–382, especially pp. 379–382). At the end of the Fourth Day, Galileo refers to this demonstration of the earth's motion as one of three "very convincing" ("assai concludenti") confirmations of the Copernican hypothesis, the others being the heliocentric explanation of retrograde motion, and the moving-Earth account of the tides: see ibid., p. 462 (*OG* 7, p. 487).

20. Of all the unpublished works listed by Allatius, this alone was eventually published (posthumously): Scheiner, *Prodromvs pro Sole mobili et terra stabili, contra Academicvm Florentinvm Galilævm a Galilæis* (1651). It is worth noting that whereas for most of Scheiner's works Allatius lists he describes their content by repeating (nearly *verbatim*) their subtitles, here Allatius refrains from explicitly mentioning Galileo.

21. Along the same line, in the entry on Melchior Inchofer, a prominent Austrian Jesuit who played an important role in the trial of Galileo, Allatius mentions Inchofer's forthcoming work, the treatise *Whether the Earth, according to faith, is at rest, to which a positive answer is shown to many* (*An sit de fide terram esse immobilem, vbi affirmatiua multis ostenditur*, Tractatus: Allatius, *Apes Vrbanae* (1633), p. 190). The book was indeed published later that very year: Inchofer, *Tractatvs syllepticvs* (1633).

22. Locher, *Disqvisitiones mathematicæ* (1614).

23. De Pineda, *In Ecclesiasten Commentariorum Liber unus* (1619), p. 131, col. 1.

24. The French philosopher, astronomer, and mathematician Pierre Gassendi (1592–1655). In 1612 the college of Digne called him to lecture on theology, and after he received his degree in theology at Avignon, he was elected Provost at the Cathedral Chapter of Digne, in the region of Provence. As explained in the full title of the book, Allatius's *Apes Vrbanae* includes entries about eminent people who visited Rome between 1630 and 1632 and published their works; Gassendi cannot be counted among them, so Allatius appends a short notice about him and his works at the end of the entry on Scheiner.

25. Gassendi. *Exercitationes paradoxicæ adversvs Aristoteleos* (1624).

26. Gassendi, *Epistolica exercitatio* (1630).

27. Gassendi. *Parhelia, sive Soles qvatvor* (1630).

28. Gassendi. *Mercvrivs in Sole visvs, et Venvs invisa* (1632).

29. It is difficult to single out a specific work about Aristotle's philosophy Allatius might have had in mind, as Gassendi critically engaged with Aristotle's philosophy, as well as early and late Aristotelianism, throughout his life. In his final years, Gassendi released a major portion of his Epicurean studies, publishing his Latin translation of Diogenes Laertius's Book X on Epicurus, along with ample commentary: *Animadversiones in decimvm librvm Diogenis Laertii* (1649). Two years earlier, he published a biography of Epicurus: *De vita et moribvs Epicvri* (1647).

GIAN VITTORIO ROSSI
(IANUS NICIUS ERITHRÆUS)

GALILÆVS GALILÆVS

GALILEO GALILEI

(1643)

Inter eos, qui bene atque præclare, virtute ingenii, maximarumque rerum scientia, nostra memoria, de Florentinæ civitatis nomine ac dignitate meruerunt, primum sine dubio locum ac numerum obtinet Galilæus, Florentiæ nobili ac vetere prosapia, non tamen legitimo toro, natus. Etenim quisnam est in toto orbe terrarum locus ita remotus, ita à nobis locorum intervallo disjunctus, quæ natio tam efferata, tam barbara, ubi aliquis sit bonis literis honor, in qua Galilæi nomen, omnium sermonibus ac literis, summa cum ejus patriæ, quæ talem virum genuit, honore ac laude non usurpetur? Sed quid miramur, tantam hominis virtutem, orbem omnium peragrasse terrarum, cum tubæ à se inventæ præsidio, tanquam curru invectus, atque in sublime elatus, per immensas cæli regiones evagaverit, clarissimum solis jubar adierit, atque umbris similes in eo maculas deprehenderit, intra Lunæ, aliena micantis luce, sphæram penetraverit, & latos in ea campos & colles & valles inspexerit, æthæreos omnium siderum orbes obierit, ac novas in eis stellas, quas Mediceas, ex Principum suorum cognomine, appellavit, invenerit? Atque utinam, cum res tantas, tantoque intervallo ab aspectus judicio remotas contempletur, imbecillitatem oculorum, qui in rebus propinquis atque perspi[280]cuis sæpenumero hallucinantur, in consilium adhibere voluisset; nunquam profecto, se vidisse ea quæ non vidisset, affirmasset, neque ea, quæ sacrarum literarum testimoniis, sanctorum Patrum consensu, ac fidei catholicæ veritate nituntur, libris editis convellere ac labefactare conatus esset; præterea non habuisset necesse, à Quæsitoribus fidei, Romam evocatus, palinodiam canere, ac cælum moveri, terram autem stare, contra id, quod docuerat multis audientibus, palam asserere: quamquam non id tum primum ab eo excogitatum & inventum sit; nam alii multo ante, fortasse ingenii acuendi declarandique causa, se in hac hæresi esse assimularunt. Verum, quamvis singulare hominis ingenium, ob ejus venustatem formæ amœnitatemque, simul, ut adspectum, adamatum sit etiam, tamen ubi suum illud admirabile oculare in medium attulit, cujus ope longissime oculorum acies intenderetur, ita ejus nomen celebrari omnium sermonibus cœpit, jam ut nihil sit eo illustrius, neque celebrius. Nam cum olim Patavio, ubi octingentorum aureorum stipendio mathematicas disciplinas juventuti tradebat, venisset Venetias, admonitus est à quodam patricii ordinis viro, in Germania inventum esse oculare, quod oculis admotum, res quantumvis remotissimas, ea qua essent magnitudine, aspicienti subjiceret, ille, qui fortasse jam diu hoc saxum volvebat, simul ac domum se recepit, fistulæ plumbeæ, ex organo detractæ, vitreos varii generis orbes ad certum intervallum accomodavit, unde eventum sibi ex sententia processisse cognovit: itaque alacris ad nobilem illum virum accurrit, à quo primum, tum ab aliis deinceps multis sui ordinis, summa eorum cum admiratione & Galilæi laude, factum est ejus ocularis periculum in turri campanaria S. Marci; qui omnes uno ore fuerunt illi auctores, ut Senatum adiret, deque suo illo tam admirabili invento ipse doceret; fore enim, ut patribus illis amplissimis rem gratissimam faceret; neque prædictionem fefellit eventus: nam adeo illo delectati sunt munere, ut ex S. C. veteri singulorum annorum stipendio ducentenos aureos nummos alios addiderint.

Among those who, in our recollections, deserve the name and the excellence of the city of Florence in the best possible way, for the strength of the mind and knowledge of the most important things, no doubt Galileo earns the first and foremost position. He was born in Florence of noble and old stock, even if not legitimately. Indeed, what place in the whole world where humanities are held in some consideration is so remote, so far away from us; what nation is so savage and barbarous in which the name of Galileo is not appealed to in every discourse or text, with honor and the highest praise of his homeland, which gave birth to so great a man? But why are we amazed that the eminence of such a person would travel the whole world? Thanks to the assistance of the tube he devised, he crossed the immense regions of heaven, approached the bright splendor of the Sun and captured its spots, similar to shadows, as if he were driven by a chariot and brought on high; he pierced through the sphere of the Moon, shining by reflected light, and contemplated its wide fields, its mountains and valleys; he met the heavenly paths of all the planets, and discovered among them new ones, which he called Medicean, after the household name of his princes. And while he was observing such enormous objects, very far removed from our eyes, if only he would have taken into account the weakness of these very eyes, which often get fooled by objects [280]close to them and very clear! Certainly he would not have declared to see what he never actually saw, nor would he have attempted, by publishing books, to overthrow and shake things that rest upon the testimony of the Holy Scriptures, the concord of the Holy Fathers, and the truth of the Catholic faith. Furthermore, summoned to Rome by the Inquisitors, he would not have been forced to recant, and publicly state that the heavens move and the earth is at rest, against what he had taught before a wide audience. Galileo was not the first to devise and conceive this, though. Others, much earlier, perhaps in order to sharpen and make their thoughts renowned, pretended to endorse this heresy.[1] However, although the extraordinary intelligence of that man was profoundly admired as soon as they saw him, due to the beauty and pleasantness of its appearance as soon as he presented his amazing spyglass, with whose help the intensity of eyesight was extraordinarily increased, Galileo's name began to be celebrated in all speeches—and nothing is nobler, or more famous. Indeed, when he moved to Venice from Padua, where he taught mathematics to young students for a salary of eight hundred golden coins, a certain nobleman told him that in Germany a spyglass had been discovered that, placed close to the eyes, offered to the beholders objects at their actual size, however very distant they actually were. Perhaps Galileo had already long been toying with that idea. He went back home immediately and set a few glass circles of various kinds at given distances in a leaden pipe which he had taken from an organ. As an outcome he found that the result was what he had anticipated. Hence, he eagerly went back to that nobleman. First, this nobleman, and later many other Venetian patricians, put the spyglass to practice on the bell tower of San Marco, with their greatest astonishment and praise for Galileo. They all advised him to introduce himself to the Senate and instruct it about his amazing invention. Indeed, they told him he would highly please those illustrious senators. And what they had predicted actually turned out to be true, for they were so happy with that gift that, upon deliberation, they increased his previous annual stipend by two hundred additional golden coins.

At reversus Patavium, totum, ad opus suum expolien ²⁸¹dum, se contulit. Iam in omnes Italiæ, Hispaniæ, Galliæ, atque adeo Europæ totius oras, Galilæi nomen emanarat; jam notum erat omnibus, ocularis à se inventi ope, novas in cælo stellas orbesque, omnibus ante sæculis obstructas ac reconditas, esse detectas, palamque prolatas; jam elegantissimus ille Nuncii Syderii titulo inscriptus ab eo liber exierat; cum hac fama compulsus Magnus Etruriæ Dux accersivit eum Pisas, ut mathematicarum artium Doctoris ac Magistri nomine, centenos singulis mensibus nummos argenteos magnos, quos laminas vocant, acciperet, alterum vero, quem vellet, muneris sui vicarium, & tanquam hypodidascalum sibi sufficeret: quam ipse tam luculentam conditionem, bonoᵃ Senatus Veneti cum venia, libens accepit; quæ tam insignis, liberalissimi in eum Principis benignitas, nunquam, quo ad vixit, clausa est; ejusque beneficio, nullius rei egens, patrio sermone, complures pertinentes ad Mathematicam libros composuit, in his, Dialogum de systemate mundi, in quo execrandam illam suam de terræ, circa cæli orbes, nullo agitatos motu conversione sententiam aperuit. Quatuor ferme & octuaginta complevit annos, quorum postremos, luminibus orbatus, in tenebris vixit. Obiit in villa agri Florentini; cujus fines ne excederet, Quæsitorum fidei sententia cautum fuerat, in pœnam, quod Sapientia ipsa, à qua à principio mundus constitutus est, sapientior esse affectasset.

a bono *K*] *vel delendum vel* animo bono *scribendum*

Then, once he got back to Padua, he fully devoted himself to improving the instrument. [281]The name Galileo had already reached all the regions of Italy, Spain, France, and even the whole of Europe. Everybody already knew that, aided by the spyglass he invented, Galileo discovered in the sky new stars and planets, which had hitherto remained hidden and concealed for centuries, and were now accessible to anybody; and he had already published that finest book, titled *Sidereal Messenger*. When the Grand Duke of Tuscany, driven by Galileo's fame, called him to Pisa so that he could receive a hundred large silver coins every month (which they call *scudi*), with the title of Doctor and Professor of Mathematics,[2] and appoint someone of his own choice as his assistant and substitute, Galileo gladly accepted such an excellent position, with the permission of the Venetian Senate. The benevolence of a renowned prince such as the Grand Duke, who was extremely generous with him, never ceased as long as Galileo lived. Thanks to the prince's support, Galileo was released from any need and wrote several books on mathematics in the language of his homeland. Among them was the *Dialogue on the [Two Chief] World Systems* in which he disclosed his abominable opinion on the revolution of the earth around the heavens that are at rest. Galileo lived nearly eighty-four years, [3] the last few of these in the dark, deprived of his eyesight. He died in a villa in the outskirts of Florence. He had been sentenced by the Inquisitors not to cross the boundaries of his villa as a punishment for demanding to be wiser than Wisdom itself, according to which the world was set up from the beginning.

NOTES

1. The Greek astronomer and mathematician Aristarchus of Samos (310–ca. 230 BC) advanced the first known model that placed the Sun at the center of the universe, with the earth revolving around it. He was influenced by the Pythagorean philosopher Philolaus of Croton (ca. 470–ca. 385 BC), who hypothesized the motion of the earth: Aristarchus identified the Sun with the "central fire" of the Pythagoreans and put the other planets in the correct order of distance around the Sun. Like Anaxagoras (ca. 510–ca. 428 BC), he thought that the stars were just other bodies like the Sun. He was also the first to deduce the rotation of the earth on its axis, influenced by some theories of Heraclides Ponticus (ca. 390–ca. 310 BC). We know of Aristarchus's theory from Archimedes, who referred to it in *The Sand Reckoner* (*Arenarius*, I 4–7). In the manuscript of *De revolutionibus*, Copernicus explicitly refers to this tradition: "Even if we admit that the motion of the Sun and Moon could be demonstrated by positing the immobility of the earth with respect to the other wandering bodies, this is less suitable. It may be for these and other similar reasons that Philolaus thought about the mobility of the earth, and Aristarchus of Samos was, according to some [authors], aloof the same opinion, without being disturbed by the argument alleged and criticized by Aristotle. But since these things are such that they cannot be understood without a keen mind and a long study, the vast majority of philosophers ignored them, and Plato does not conceal the fact that there were few people who, at his time, mastered the study of celestial motions" ("Et si fateamur Solis Lunæque cursum in immobilitate quoque terræ demonstrari posse: in cæteris vero errantibus minus congruit. Credibile est hisce similibusque causis philolaum mobilitatem terræ sensisse: quod etiam nonnulli Aristarchum samium ferunt in eadem fuisse sententia. non illa ratione moti: quam allegat reprobatque Aristoteles. Sed cum talia sint: quæ nisi acri ingenio et diligentia diuturna comprehendi non possent: latuisse tunc plerumque philosophos: et fuisse admodum paucos: qui eo tempore sydereorum motuum calluerint rationem, a platone non tacetur": excerpt from sections of the holograph manuscript of Copernicus's *De revolutionibvs*, which were not included in the printed edition; see the 2015 critical edition of Copernicus's work, edited by Lerner, Segonds and Verdet, vol. 2, p. 476.1–9.

2. In fact, the salary was 1,000 scudi. Actually, Galileo explicitly asked the Grand Duke's secretary, Belisario Vinta, to be appointed Mathematician and Philosopher: "[...] as to the title and reason for my service, I would wish, beside the title of Mathematician, that His Lordship add that of Philosopher, having spent more years studying philosophy than months studying pure mathematics" ("[...] quanto al titolo et preteso del mio servizio, io desidererei, oltre al nome di Matematico, che S. A. ci aggiugnesse quello di Filosofo, professando io di havere studiato più anni in filosofia, che mesi in matematica pura": Galileo to Vinta., 7 May 1610, in *OG* 10, no. 307, p. 353). The Grand Duke granted Galileo's request, and on 5 June 1610 Vinta sent Galileo the full description of his appointment: "These Highnesses ordered to offer Your Lordship the title of Chief Mathematician of the University of Pisa and Philosopher of the Most Serene Grand Duke, with no obligation to teach at the University of Pisa nor to live in the city of Pisa, and with a salary of 1,000 scudi, Florentine currency, so as to provide you with all the means to pursue your studies and finish your works" ("Hanno queste AA. deliberato di dar titolo a V. S. di Matematico primario dello Studio di Pisa et di Filosofo del Ser.ᵐᵒ Gran Duca, senz'obligo di leggere et di risedere né nello Studio né nella città di Pisa, et con lo stipendio di mille scudi l'anno, moneta fiorentina, et con esser per darle ogni commodità di seguitare i suoi studii et di finir le sue compositioni": ibid., no. 327, p. 369).

3. In fact, Galileo lived nearly 78 years: 15 February 1564–8 January 1642.

5

VITTORIO SIRI

EXCERPT FROM

DEL MERCVRIO OUERO HISTORIA DE' CORRENTI TEMPI

(1647)

[1720]Farò per auuentura il pregio dell'opera, quando ad imitatione de' più lodeuoli scrittori raccolga le memorie de gli huomini più illustri, che dentro i periodi di quel tempo, delle cui occorrenze intrapresi il racconto, chiusero i loro vltimi giorni. Sotto la penna caderà dunque in primo luogo Galileo Galilei Filosofo, e Mathematico il più celebre non solo nel secolo presente, ma che giustamente garir può di maggioranza con i più rinomati dell'età passate. Nacque in Pisa il giorno de' 19. di Febraro dell'anno 1564. Il Padre suo hebbe nome Vincenzo Galilei. Portato da' geniali suoi inchinamenti si diede tutto allo studio d'Euclide, e poscia de gli altri Mathematici di maggior grido. Due parti fràl altre marauigliosamente rilusero in lui, e che di rado insieme s'accoppiano;[a] Chiarezza, & Acutezza d'ingegno. I professori di questa scienza, gli antichi particolarmente, e trà moderni il Keplero, & il Vieta riescono tanto oscuri ne' loro scritti, che pare vogliano far pompa del lor sapere con non lasciarsi intendere: facendo passare le Mathematiche per Oracoli Sibillini. Ladoue il Galileo fù dotato d'vna espressione, facilissima, e tanto chiara, che valeua ad illuminare, & ammaestrare i più rozzi intelletti. L'acutezza del suo ingegno fecondissimo di speculationi appare principalmente dell'inuentione di tanti instromenti vtilissimi al viuere ciuile. Seppe addattare di maniera le Mathematiche alla Filosofia, che ben hà dato à diuedere la necessaria correlatione trà loro, e che per capire la profonda intelligenza de' sensi de' Filosofi, sia addibisogno di valersi delle notitie Mathematiche.

La prima carica, ch'egli sostenne fù quella di Lettore delle Mathematiche nello Studio di Pisa, e poscia in quello di Padoua; richiamato finalmente à Firenze dal Gran Duca Cosmo Secondo, con stipendio amplissimo, e con ti[1721]tolo di Filosofo, e Mathematico primario di quell'Altezza. Trovandosi in Venetia, riseppe, che in Olanda erano state ritrouate le Lunette, col cui beneficio gli oggetti visibili si rendeuano indistanti all'occhio, benche fossero in sito lontano. Senza vedere la forma di questo Instrumento, si mise à specularne la struttura, e come potesse essere formato, e finalmente gli sortì di rinuenire il Telescopio, vulgarmente chiamato il Canocchiale del Galileo, onde meritò testimonianze di stima, e d'aggrandimento dalla munificenza del Senato. Fù il primo, che drizzasse il Telescopio verso il Cielo, scoprendo la superficie Lunare non tersa, ma aspra piena di prominenze, e di cauità. Osservò vn nuouo moto di trepidatione, mostrandosi la Luna à noi hora più da vna parte, hora più da vn'altra. Trouò, che Venere imitaua gli aspetti delln Luna, apparendo tonda tal volta, dimezzata, e falcata. Manifestò la sensibilissima mutatione di grandezza ne' diametri apparenti di Venere, e di Marte; cosa di conseguenza molto rileuante, e cotanto necessaria nelle Theoriche de' due grandi Astronomi Copernico, e Ticone.

a s'accoppiano *scripsi*] s'accopiano C

[1720]I might now end in style, in imitation of the most praiseworthy authors, by gathering the stories of the most eminent men who lived their lives in the timespan whose events I recounted. In the first place, then, I will write about Galileo Galilei, who was not only the most eminent philosopher and mathematician of the present century, but also one who could rightly compete in greatness with the most renowned ones of the past. He was born in Pisa on February 19, 1564.[1] His father was Vincenzo Galilei. Driven by his brilliant talent, Galileo devoted himself completely to the study of Euclid, and later of the other most eminent mathematicians. Two qualities, among others, splendidly stood out in him: clarity and sharpness of mind, which are rarely present at once. The professors of this science—especially ancient ones, and Kepler and Viète[2] among the modern—turn out to be so obscure in their works that they seem to show off their knowledge by preventing others from understanding them, passing off mathematics as *Sibylline Oracles*.[3] By contrast, Galileo was gifted with such a very simple and clear expression that he managed to illuminate and teach the coarsest intellects. The sharpness of his mind, most prolific with speculations, is primarily manifested in the invention of many instruments very useful for civic life. He succeeded in adapting mathematics to philosophy so as to show clearly to us the necessary relationship between them, demonstrating that in order to deeply understand the philosophers' theories it is necessary to appeal to mathematical knowledge.

His first appointment was that of professor of mathematics at the University of Pisa, and later at that of Padua. He was eventually called back to Florence by the Grand Duke Cosimo II, with an extremely high salary and the [1721]title of Philosopher and Chief Mathematician of that Highness. While in Venice,[4] Galileo happened to learn that in Holland some spectacles had been invented, thanks to which visible objects appeared to be close to the eye, their distance from it notwithstanding. Without seeing the shape of this instrument, he began speculating about its structure and what parts it might have been made of, and eventually he happened to build the telescope, commonly known as Galileo's spyglass. For this, Galileo earned expressions of respect and an increase in the Senate's generosity. He was the first to point the telescope to the sky, where he discovered that the Moon's surface is not smooth but rough and full of bulges and hollows. He observed a new motion of libration by which the Moon shows itself to us sometimes more on one side and sometimes more on the other. Galileo found that Venus imitates the phases of the Moon, appearing at times circular, at times semicircular, and at times crescent. He showed the significant change in size of the apparent diameters of Venus and Mars, which had considerable implications, and was required by the theories of the two great astronomers Copernicus and Tycho.[5]

Hà fatto vergognare il Sole scoprendoli quelle macchie, che per tanti secoli haueua nella sua luminosa caligine sepellite; & queste macchie vide non già fisse, et eterne, come quelle della Luna, ma generabili, e corruttibili, aggirandosi intorno il Sole. Rinuenne, che intorno Gioue girauano altri quattro Pianetti non mai veduti dall'antichità, quali in honore della Serenissima Casa de' Medici Mecenate de gli huomini letterati, e tanto sua benefatrice, battezzò col nome di Stelle Medicee. Dalle frequentissime Ecclissi delle predette Stelline s'imaginò di ritrouare la longitudine della Geografia molto meglio, che con gli Ecclissi Lunari, onde ne compose le Tauole de' loro moti, lasciandole al P. D. Vincenzo Renieri Mathematico Pisano, il quale hauendole ricorrette, e perfettionate, si troua sù' procinti di darle alla Stampa.

S'accorse, che la stella di Saturno era tricorporea, composta, cioè, di tre corpi, vno sferico e principale nel mezzo; e di due altri minori laterali. Manifestò, che la via Lattea, e le Stelle nubilose altro non erano, ch'vna moltitudine di Stelline fisse tanto vicine frà di loro, e tanto minute, che la nuda vista non poteua distinguerle separatamente. Tutte queste osservationi furono fatte dal Galileo in pochi anni, non essendosi in tutto il corso di tanto tempo doppo la sua cecità scoperta altra nouità, se non in Gioue, che si mostra macchiato da alcune fasce, ò zone, che lo cingono: vedendosi hoggidì queste macchie molto bene con i Telescopij lauorati dal Torricelli in Firenze con sì isquisita perfettione, che si vede in quei vetri consumato lo sforzo dell'arte.

Tra l'Opere composte dal Galilei vna è il *Nuncio Sidereo*, in cui tratta dell'osseruationi da lui fatte in Cielo. *Le Galleggianti*, ò sia Di[1722]scorso delle cose, che stanno nell'acqua, & in essa si muouono, facendo vedere con questa Opera, che 'l nostro secolo non haueua ad inuidiare all'età passate il lor Archimede. Vn'altro Libro *delle macchie Solari*. Altro intitolato, *il Saggiatore* intorno il moto delle Comete. Vn compendio *Delle mecaniche*: Vn'altro *delle Fortificationi*. *Il Dialogo* sopra i due gran Sistemi del Mondo Tolemaico, e Copernicano.[b] Il suo *Compasso Geometrico, e Militare*, col quale si riducono alla pratica le più belle, e più necessarie operationi della Geometria. L'vltima fatica delle sue vscita alla luce è il libro *delle dimostrationi Mathematiche, attinenti à due nuoue scienze intorno alla Mecanica, & a' mouimenti locali.*

Visse gli vltimi otto anni della sua vita fuori di Firenze, parte in alcune ville conuicine alla medesima Città, e parte in Siena. Per le continue osseruationi del Cielo, e per molti patimenti dalle varie impressioni dell'aria notturna, debilitò di maniera la vista, che trè anni auanti la sua morte diuenne affatto cieco colui, che haueua insegnato di vedere à tutto il Mondo, e doppo hauer egli veduto più, che tutti gli huomini, e le nationi di tutti i secoli insieme. Sopportò la sua cecità con animo forte, e veramente filosofico, solleuandosi da cotal miseria con vna non interrotta speculatione, hauendo già preparata vna gran massa di materie, e principiato à dettare i suoi concetti, quando con vna malatia di trè mesi insensibilmente mancando, finì christianissimamente i suoi giorni alli 8. Gennaro 1642. in età di settantasette anni, mesi dieci, e giorni venti nella villa d'Arcetri.

b Copernicano *scripsi*] Copernicorno C

Galileo had the Sun embarrass himself by disclosing the spots it had been hiding behind its bright mist for so many centuries.[6] And, as he saw them, they are not fixed and everlasting, like those of the Moon; rather, they are generated and corrupted wandering around the Sun. Galileo found that four planets, never seen in antiquity, revolve around Jupiter; he baptized them with the name Medicean Stars, to honor the Most Serene House of the Medici, patron of learned men, from which he greatly benefited. And from the very frequent eclipses of these little stars he figured out a way to find longitude in geography, which he considered to be better than appealing to lunar eclipses. Hence, he compiled tables of their motions, which he bequeathed to Father Vincenzo Renieri, a Pisan mathematician, who revised and improved them and is about to have them published.

Galileo realized that the star Saturn is tricorporeal, that is, three-bodied: a main, spherical body at the center, and two smaller ones on the sides. He showed that the Milky Way and the nebulae are none other than a multitude of little fixed stars, so close to one another and so tiny that the naked eye is unable to tell them apart. Galileo made all these observations in the span of a few years. Since the time he went blind, no other discovery has been made, with the exception of Jupiter, which appears to be spotted by some strips, or belts, which wrap it.[7] Nowadays, we can observe these spots very clearly, thanks to the telescopes made by Torricelli in Florence, which are so perfectly polished that we see in those spectacles the culmination of this art.[8]

Among Galileo's works, the *Sidereal Messenger* deals with his observations of the heavens; *On Floating Bodies*, that is, a [1722]discourse on bodies that stay atop water, and move in it; by this work he shows that our century need not envy antiquity for Archimedes. Another book is *On Sunspots*. Another is titled *The Assayer*, on the motion of comets. Then there is a synopsis of *Mechanics*; a synopsis of *Fortifications*; the *Dialogue on the Two Chief World Systems, Ptolemaic and Copernican*; and his *Geometric and Military Compass*, by way of which the most beautiful and necessary geometric operations are turned into practice. His last work is a book on *Mathematical Demonstrations, Concerning Two New Sciences, Related to Mechanics and Local Motions*.

Galileo lived the last eight years of his life outside Florence, partly in some villas close to the city, and partly in Siena.[9] Owing to his long-lasting observations of the sky, and the many ills caused by [cold] air at night, his eyesight was so affected that three years before he died, he became completely blind—he, who taught the whole world how to see, and after seeing more than all the men and nations put together. Galileo endured blindness with a strong and truly philosophical spirit, overcoming that misery by way of relentless speculation; he had already gathered a large amount of material and begun to dictate his ideas. And then he died without pain, after a three-month long illness, ending his days in a most Christian manner, on January 8, 1642, aged seventy-seven years, ten months, and twenty days, in his villa in Arcetri.

Fù di statura piccola più tosto, ma d'aspetto venerabile, e di robusta complessione. Visse sempre giouiale, e faceto, e la sua conuersatione fù amabilissima. Hebbe gran gusto dell'Architettura, e di Pittura. Disegnava più, che ordinariamente; e suonaua con isquisitezza il Leuto. Si dilettò d'agricoltura particolarmente quando, soggiornando nelle ville non haueua altri trattenimenti, nè più grato essercitio di questo. Alla memoria di quest'Huomo l'Italia più d'ogn'altra Provincia è obligata, hauendo sparsa fra' suoi popoli la professione delle Mathematiche cotanto necessarie all'intelligenza dell'altre scienze, & al viuere civile, che per l'auanti erano oscure, & ignote.

Galileo was rather short, but his appearance commanded respect; he had a strong constitution. He was always jovial and witty, and very pleasant in conversation. He very much enjoyed architecture and painting. He drew better than most people, and played the lute exquisitely. He took delight in gardening, especially when he was living in a villa; he had no other recreation that pleased him more. To a greater extent than any other land, Italy is indebted to the memory of this man, as he spread among its people the study of mathematics, which is so necessary to understanding other sciences, as well as to civil life, and which had been hitherto obscure and ignored.

Notes

1. In fact, he was born on February 15, 1564.

2. The German astronomer Johannes Kepler (1571–1630), discoverer of the three laws of planetary motions that bear his name; he was Imperial Mathematician at the court of Rudolph II and his two successors, Matthias and Ferdinand II. François Viète (1540–1603), French mathematician who made fundamental contributions to algebra, introducing the use of letters as parameters in equations.

3. The *Sibylline Oracles* (sometimes called *pseudo-Sibylline Oracles*) is a corpus of oracular prophecies written in Greek hexameters—the meter used both in historical oracle centers, at Delphi and Asia Minor, and for the literary depiction of legendary prophecy—in which Jewish or Christian doctrines are allegedly confirmed by sibyls, prophetesses who uttered divine revelations in a frenzied state. Two collections of *Sibylline Oracles* survive from late antiquity, one dating to the end of the fifth century AD, the other to the period just after the Arab conquest of Egypt, in the seventh century. The material in these collections is extremely diverse; some is manifestly Christian, other passages are almost certainly Jewish, and yet other material is pagan. The subject matter ranges from Christian doctrine and predictions of woe for cities and peoples, to Roman history and sibylline biography.

4. A similar expression appears in the *Racconto istorico*, pp. 20–21, which was probably the source for Siri's biography of Galileo, as can be seen in a number of very similar expressions below.

5. Galileo first observed the full planetary phases of Venus in late September or early October 1610, and announced the discovery to Cristopher Clavius and Benedetto Castelli on December 30, 1610 (*OG* 10, nos. 446–447, pp. 499–500 and 503), and to Giuliano de' Medici on January 1, 1611 (*OG* 11, no. 452, pp. 11–12). He published the discovery only in 1613, in the first letter to Markus Welser on sunspots: see *On Sunspots*, pp. 92–94 (*OG* 5, pp. 98–100). The phases of Venus could not be accounted for by the Ptolemaic system, for should Venus and the Sun orbit the earth, it would be impossible for Venus to be fully lit from the perspective of the earth, as this would require it to be on the far side of the Sun. The Ptolemaic model predicted only crescent and new phases of Venus, whereas Galileo observed that it exhibits all phases of the Moon, which, by contrast, were predicted and accounted for by a heliocentric system (like Copernicus's), or a geo-heliocentric system (like Tycho's). Also, Galileo made use of observations of the distance of Mars from the earth to support his claim that the planets move around a fixed point that is not the earth: see Galileo's second letter to Piero Dini, 23 March 1615, in *OG* 5, pp. 298–299. Galileo further discussed the retrograde motion of Mars, attributing its rare retrogressions to its faster motion around the Sun, compared with the motions of Jupiter and other planets, which present more pronounced retrogressions. See the *Dialogue*, pp. 342–345 (*OG* 7, pp. 370–372).

6. For a similar remark, see Allatius's quotation from Lauri's *Orchestra*, at the end of Allatius's entry on Galileo (pp. 104–105).

7. Thanks to a number of Keplerian telescopes (that is, telescopes that use a convex lens as the eyepiece instead of Galileo's concave one; they allow for a much wider field of view and greater eye relief, although the image for the viewer is inverted) he himself constructed and progressively improved, Francesco Fontana (ca. 1585–1656), a Neapolitan lawyer and self-taught astronomer, was able to make very accurate observations. In particular, he suggested the rotation of Mars on its axis, and observed the belts of Jupiter: see Fontana, *Novae cælestivm terrestrivmqve rervm observationes* (1646), p. 106 and 107–125, respectively. He also included detailed images of the Moon, Venus, Saturn, and the Pleiades, greatly improving on Galileo's own.

8. In 1642–1647, while in Florence, Evangelista Torricelli (1608–1647), a pupil of Benedetto Castelli's, worked on lens making. He was the first to understand in scientific terms the wear process that produces spherical surfaces during grinding, and to realize the importance of the relative size of the tool plate with respect to that of the lens; he also recommended hot glues, instead of cold ones, to attach the lens to the holder. For the final polishing, he used an almost plane tool plate made of slate, slightly worn to concavity with some pumice. In so doing, Torricelli achieved the mastery of lens fabrication technology; for Torricelli's own account of his technique, see his letters to Raffaello Magiotti, 4 December 1643, in Torricelli, *Opere* (1919–1944), vol. 3, pp. 150–156 (including separate instructions about the use of the telescope and the requirements for the lenses); see also Torricelli to Magiotti, 6 February 1644, ibid., pp. 165–166. Whereas Galileo used to purchase lenses, choosing from them the best ones, a few years later—thanks to second-generation manufacturers of optical instruments such as Torricelli, Eustachio Divini (1610–1685), and the brothers Matteo and Giuseppe Campani (1620–1678 and 1635–1715)—the lens fabrication process had become a well-defined technology, in which the individual production steps were properly understood and brought to perfection.

9. After the trial, Galileo was granted permission to leave Rome and move to the residence of Ascanio Piccolomini, Archbishop of Siena and a good friend of Galileo's. Galileo arrived in Siena on July 9, 1633, and remained there for a few months, waiting for the plague to end. He was back in Arcetri by December 17, and remained there for the rest of his life.

6

VINCENZO GALILEI

ALCUNE NOTIZIE INTORNO ALLA VITA DEL GALILEO

NOTES ON GALILEO

(CA. 1654)

[594]Il Galileo nacque in Pisa l'anno 1563, a' 19 Febbraio.

Compose Vincenzio Galilei, padre del Galileo, e mandò in luce diverse opere, ma specialmente un Dialogo dottissimo Della musica antica e moderna.

Tra l'opere d'intavolatura di liuto composte dal padre del Galileo, è alla stampa il primo libro de' contrapunti a quattro voci; ma molte altre non sono andate in luce. Fu Vincenzio Galilei vuomo singolare in detto strumento, come anco in gioventù il Galileo, e non solo nel liuto, ma nello strumento di tasti ancora.

Il Galileo, impiegato nello studio della medicina, per qualche tempo si mostrò alieno dalle matematiche, benché il padre, ch'era in esse valoroso, ve lo esortasse; finalmente, per sodisfare al medesimo suo padre, vi applicò l'animo: ma non sì tosto cominciò a gustare la maniera del dimostrare e strada di pervenire alla cognizione del vero, che lasciando andare ogni altro studio, si diede tutto alle matematiche.

Il Galileo ebbe la lettura delle matematiche nello Studio di Pisa circa l'anno 1590, nel qual tempo cercandosi di Matematico degno di quella cattedra, il Sig.[r] Guido Ubaldo dal Monte, persona insigne in quella professione, propose al G. Duca il Galileo, affermando a S. A. che egli era tale che da Archimede sino a quel tempo niuno l'aveva pareggiato, non che avanzato.

L'anno 1592 fu eletto il Galileo lettore delle matematiche nello Studio di Padova: et avendo quivi l'anno 1609 inventato l'occhiale, presentò il detto strumento alla Ser.[ma] Republica in pieno Senato; dove essendo da quei Signori sommamente gradita sì nobile invenzione, ne conseguì il Galileo, oltre alle meritate lodi, uno stipendio a sua vita di fiorini 1000 l'anno, cioè molto maggior di quello che avesse mai avuto alcuno de' suoi antecessori in detto Studio.

[595]Continuando il Galileo la lettura nello Studio di Padova con onore e applauso grandissimo, nel tempo delle vacanze estive tornava in Firenze, dove benignamente ricevuto dal Ser.[mo] G. D. Ferdinando Primo, si degnò il Ser.[mo] D. Cosimo, Gran Principe di Toscana, di ascoltar dalla sua viva voce lezzioni di matematica e l'esplicazione dell'uso del compasso geometrico da lui inventato, con intera sodisfazione e gusto di S. A. L'anno poi 1610, essendosi già sparsa la fama del Galileo per tutta l'Europa, il medesimo Ser.[mo] D. Cosimo, già pervenuto Gran Duca, avendo caro d'avere appresso di sé un tant'vuomo e suo devotissimo vassallo, con sue benignissime lettere de' 10 Luglio di detto anno richiamò il Galileo a Firenze e al suo servizio, con titolo di Primario Matematico dello Studio di Pisa (benché esente dal carico di dover leggere) e suo Primo Filosofo, con stipendio amplissimo e conveniente alla somma generosità d'un tanto Principe.

Delle postille e risposte del Galileo a Antonio Rocco, impugnatore del suo Dialogo de i due Massimi Sistemi, solo una parte se ne trova appresso l'erede del medesimo Galileo, et anco quella di prima bozza e non ridotta al netto.

Fu il Galileo d'aspetto gioviale, massime in vecchiezza, di statura giusta e quadrata, di complessione robusta e forte, e tale che non ci voleva meno acciò ei potesse resistere alle fatiche veramente atlantiche da lui durate nelle continue osservazioni celesti; nondimeno fu travagliato, da circa 40 anni dell'età sua sino all'ultima sua vita, da dolori artetici o a quelli simili, i quali di quando in quando lo molestavano, or più or meno. Questi ebbero origine in lui da un soverchio fresco ch'ei patì una notte d'estate in una villa nel contado di Padova.

[594]Galileo was born in Pisa on January 19, 1563.

Galileo's father, Vincenzo Galilei, wrote and published several works, but particularly a most learned *Dialogue on Ancient and Modern Music*.

Among the works on lute tablature he wrote, the first book on the counterpoint of four voices is currently in print;[1] many others have never been published. Vincenzo Galilei was very talented at that instrument, as was the young Galileo; and he excelled not only at the lute but at the keyboard, too.

Galileo first studied medicine, and for some time showed no interest in mathematics, although his father, who was very good at it, encouraged him to consider the subject. He eventually studied mathematics, in order to comply with his father's wish; and as soon as he began to savor demonstrative procedures, and the way to know the truth, he gave up all other studies and gave mathematics his all.

Galileo became lecturer in mathematics at the University of Pisa around 1590. At that time, a mathematician worthy of that position was being sought, and Guidobaldo del Monte, an eminent figure in the profession, advised the Grand Duke to hire Galileo. Guidobaldo told the Most Serene Highness that nobody equaled Galileo, not to mention surpassed him, ever since the time of Archimedes.[2]

In 1592 Galileo was appointed lecturer in mathematics at the University of Padua. Here, in 1609, he invented the spyglass, which he presented to the Most Serene Republic, in front of the whole Senate. Those gentlemen appreciated Galileo's invention so much that, besides well-deserved praises, he was rewarded with a lifelong salary of 1,000 florins per year, much more than any of his predecessors had ever earned at that University.

[595]While lecturing at the University of Padua, for which he received the greatest honors and praises, during the holiday season Galileo went back to Florence, where the Most Serene Grand Duke Ferdinand I welcomed him warmly, and the Most Serene Cosimo, Great Prince of Tuscany, condescended to take private lectures in mathematics, as well as his explanations of the use of the geometric compass Galileo had devised, to the full satisfaction and pleasure of His Highness. In 1610, Galileo's reputation having spread all over Europe, the same Most Serene Duke Cosimo, who had by then become Grand Duke, decided to have near himself such a worthy man and most devout servant. Accordingly, with letters dated July 10, 1610, he called Galileo back to Florence, at his own service, with the title of Primary Mathematician of the University of Pisa (although released from the duty to teach) and his own Primary Philosopher, with a most generous salary, commensurate with the generosity of such a prince.[3]

Of Galileo's marginal annotations and rejoinders to Antonio Rocco, who challenged his *Dialogue on the Two Chief World Systems*, only part is to be found with Galileo's heir, and even that is a preliminary draft, yet to be polished.[4]

Galileo was jovial, especially in his later years; he was burly and of fair height, had a robust and strong physique. He had all that was necessary to withstand the efforts, truly worthy of Atlas, he had to endure during his tireless celestial observations. Nonetheless, from when he was about 40 years old until his last days, he was afflicted by arthritic pains or similar, which from time to time tormented him, with varying intensity. They originated from exposure to exceedingly crisp air, which he suffered during a summer night in a villa in the countryside near Padua.

Quanto fusse stimato il Galileo da grandissimi Principi e Signori, ne rendono certa testimonianza le lettere onorevoli che da essi riceveva, delle quali gran parte se ne conserva appresso il detto suo erede, e lo dimostrano apertamente i regali e gl'onori non ordinari da essi ricevuti. Nessun Principe, nessun personaggio di portata, passò mai per Firenze, che non volesse vedere e conoscere il Galileo. Il Ser.ᵐᵒ Arciduca Leopoldo, circa l'anno 1618, trovandosi il Galileo indisposto, lo visitò insino al letto. Il Ser.ᵐᵒ Arciduca Carlo, suo fratello, circa l'anno 1625 essendo in Firenze di passaggio per Spagna, lo regalò d'una bellissima collana gioiellata. Ma quanto fusse caro al suo natural Signore, al Ser.ᵐᵒ G. Duca Ferdinando IIº e a tutti li Ser.ᵐⁱ Principi suoi fratelli e di sua Casa, non si può facilmente esplicare. Veramente i favori e gli onori da questi ricevuti, in numero e qualità passarono ogni segno. Ma tacendosi delli altri, non è da passarsi sotto silenzio questo solo: che trovandosi il Galileo indisposto, l'anno 1638, il G. D. Ser.ᵐᵒ suddetto si degnò di visitarlo in persona al letto, trattenendosi per più di due ore a discorrer seco; esempio raro di affezzione di generosissimo e benignissimo Principe verso un gradito suo vassallo e servidore. Mostrò quel gran Principe, ⁵⁹⁶con tal atto, segno dell'infinita sua magnanimità, e di quanto onore sia degna e quanto si deva stimare una virtù straordinaria.

Simili onori di visita ebbe spesso il Galileo da' Ser.ᵐⁱ Principi fratelli del Gran Duca, e principalmente dal Ser.ᵐᵒ Leopoldo; ad instanza del quale scrisse il Galileo una lettera al Peripatetico Liceti, nella quale, contro all'opinione di detto filosofo, si dimostra come la luce secondaria nella luna procede dal reflesso del lume del sole dalla terra nella medesima luna. Si vede la detta lettera stampata nel libro stesso del suddetto filosofo, ch'ei fa in replica alla medesima lettera.

Molto si dilettò il Galileo di stare in villa, nella quale dimorò circa 30 anni, riconoscendo in gran parte la sanità e la lunghezza di sua vita dall'aria aperta e salubre della campagna, e così ritirandosi ancora dalli strepiti della città, per poter con più quiete attendere alle speculazioni e per esser di natura dedito alla solitudine, se ben tra gli amici fu di soavissima e gentilissima conversazione. La sua eloquenza et espressiva era mirabile; discorrendo sul serio era ricchissimo di sentenze e concetti gravi; ne i discorsi piacevoli l'arguzie et i sali non gli mancavano. Facilmente si muoveva all'ira, ma più facilmente si placava. Ebbe memoria esquisita, sì che oltre alle moltissime cose attenenti a' suoi studi aveva a mente gran quantità di poesie e specialmente gran parte dell'Orlando Furioso dell'Ariosto, che tra i poeti fu il suo favorito e l'autor suo esaltato da lui sopra tutti i poeti latini e toscani. Non era appresso di lui vizio più detestabile della bugia, forse perché mediante le scienze matematiche troppo ben conosceva la bellezza della verità. Si dilettava dell'agricoltura, la quale gli porgeva materia di filosofare e passatempo insieme; e spesse volte per suo diporto attendeva alla coltura delle piante e specialmente delle viti, potandole e legandole di propria mano con diligenza esquisita. Con tutto che fosse moderatissimo nel suo vitto ordinario, e specialmente nel bere, tuttavia si dilettava di vari vini, de' quali gliene venivano di diversi luoghi e specialmente dall'istessa cantina del G. Duca, così volendo la somma benignità di S. A.

The extent to which most eminent princes and lords kept Galileo in high esteem is clearly shown by the letters he received, the vast majority of which are kept by said heir, which showed great admiration; and the extraordinary gifts and honors he received from them openly prove this. No prince, no eminent person ever visited Florence without expressing the wish to meet or get to know Galileo. About 1618, the Most Serene Archduke Leopold visited Galileo as he was lying ill in bed.[5] About 1625, while he was in Florence on his way to Spain, the Most Serene Archduke Karl, Leopold's brother, presented him with a most beautiful chain adorned with jewels.[6] But it is not easy to say how much he was dear to his natural Lord, the Most Serene Grand Duke Ferdinando II, as well as to all the Most Serene Princes his brothers, and members of his lineage. The favors and honors Galileo received from them cannot be overstated; among many others, I wish to mention only the following one. As Galileo was lying ill, about 1638, the Most Serene Grand Duke visited him at his bedside, stopping to talk with him for more than two hours—a rare example of a most generous and benevolent prince's affection for a subject and servant he appreciated. That great prince showed [596]with this gesture—a sign of his limitless magnaminity— both how honorable extraordinary virtues are and how much they should be prized.

Galileo often received similar honors from the Most Serene Princes, brothers of the Grand Duke, and especially from the Most Serene Leopoldo. Upon his request, Galileo wrote a letter to the Aristotelian philosopher Liceti, which—contrary to the latter's opinion—shows how the secondary light of the Moon is produced by the earth's mirroring the light of the Sun on the Moon itself. This letter was included in the book by Liceti, with which he offers a rejoinder to that very letter.[7]

Galileo greatly enjoyed living in the country, in which he spent approximately thirty years, ascribing the health and length of his life mostly to the fresh and salubrious air of the country. And so he withdrew from the clamors of the city in order to attend to his speculations more quietly. Also, Galileo naturally enjoyed being by himself, although he very much relished most pleasant and agreeable conversations with friends. His eloquence and ability to express himself were amazing: on serious issues, he offered plenty of thoughtful opinions and ideas; in pleasant exchanges, he was quick-witted and cutting. He lost his temper easily but calmed down more easily. He had an excellent memory, so that besides the very many things related to his studies he knew a large number of poems by heart. In particular, he knew most of the *Orlando Furioso*: of all poets, Ariosto was his favorite, the one he praised above all other Latin and Tuscan poets. More than anything else, Galileo despised lying, possibly because, thanks to mathematics, he knew the beauty of truth all too well. He enjoyed agriculture, which offered him occasions to philosophize and to be at leisure at the same time. Frequently, for his own amusement, he looked after plants, and especially grew grapevines, which he personally pruned and tied, with exquisite assiduousness. However moderate with everyday food, and especially with drinking, Galileo very much enjoyed various wines, which came to him from different places, and especially from the very cellar of the Grand Duke, in accord with His Highness's great generosity.

Morì nella villa d'Arcetri, l'anno 1642, a dì otto di Gennaio a h. 4 di notte; né fu di piccolo pregiudizio la morte sua alli intelligenti della sua professione, poiché morendo egli si persero insieme moltissime proposizioni filosofiche e matematiche, che ben egli aveva digerite e resolute nella sua mente, ma non ancora deposte in carta.

Fu il corpo del Galileo depositato nella chiesa di S. Croce di Firenze, non già nell'antica sepoltura de' suoi antenati, che è in detta chiesa, ma in una stanza dietro alla sacrestia; non senza ferma speranza che sì come il suo generosissimo Signore l'onorò in vita, abbia a onorar parimente la sua memoria con qualche degno deposito.

Galileo died in his villa, in Arcetri, on January 8, 1642, at 4 a.m. And his death gravely affected those who intelligently practiced his profession, for with his passing very many philosophical and mathematical propositions, which he had devised and determined in his mind but not yet put on paper, were lost.

Galileo's body was deposited in the church of Santa Croce, in Florence—not in the ancient tomb of his ancestors, which is in that church, but in a room in the back of the sacristy. It is fervently hoped that just as his most generous Lord honored him while he was alive, so he would equally honor his memory with a proper tomb.[8]

Notes

1. Vincenzo Galilei, *Canto de contrapvnti a dve voci* (1584) and *Tenore de contrapvnti a dve voci* (1584).

2. See, for example, Guidobaldo del Monte to Galileo, 16 January 1588, in *OG* 10, no. 10, p. 25.

3. See Grand Duke Cosimo II to Galileo, 10 July 1610: "The eminence of your doctrine and of your treasured skills, accompanied by the excellence of your mathematics and philosophy, as well as by the most devoted affection, obedience, and service you have always showed us, caused us to wish to have you with us. And you, in turn, have always conveyed to us that by being back in your homeland, you would have enjoyed and welcomed, to the highest degree, to come and constantly be at our service, not only as Chief Mathematician at our University of Pisa, but in the capacity of our personal Chief Mathematician and Philosopher. Hence, we decided to have you with us, and we have chosen and appointed you Chief Mathematician of the aforementioned university, and as our personal Chief Mathematician and Philosopher. As such, we ordered and order whoever of our ministers is responsible for this, to provide you with a yearly salary of 1,000 scudi (Florentine currency), beginning with the date you arrive in Florence to be at our service. The corresponding installments will be paid every six months. You will not be required to live in Pisa, nor to teach there, unless on honorary occasions, whenever you might want to do so; or we might ask you explicitly, on special occasions, for our pleasure or that of foreign princes or lords who might come [to Florence]. Otherwise, you will customarily reside here in Florence and pursue the development of your studies and works, with the obligation, whenever we call you, to come to us wherever we are, even outside Florence. And may God Our Lord preserve and please you" ("L'eminenza della vostra dottrina et della valorosa vostra suffizienza, accompagnata da singular bontà nelle matematiche et nella filosofia, et l'ossequentissima affezzione, vassallaggio, et servitù che ci havete dimostra<to> sempre, ci hanno fatto desiderare di havervi appresso di noi; et voi a rincontro ci havete fatto sempre dire che, ripatriandovi, havereste ricevuto per sodisfazione et grazia grandissima di poter venire a servirci del continuo, non solo di Primario Matematico del nostro Studio di Pisa, ma di proprio Primario Matematico et Filosofo della nostra persona: onde, essendoci risoluti di havervi qua, vi habbiamo eletto et deputato per Primario Matematico del suddetto nostro Studio, et per proprio nostro Primario Matematico et Filosofo; et come a tale habbiamo comandato et comandiamo a chiunque s'appartiene de' nostri Ministri, che vi diano provisione et stipendio di mille scudi, moneta fiorentina, per ciascun anno, da cominciarvisi a pagare dal dì che arriverete qui in Firenze per servirci, sodisfacendovisi ogni semestre la rata, et senza obligo d'habitare in Pisa, né di leggervi, se non honorariamente, quando piacesse a voi, o ve lo comettessimo espressa et estraordinariamente noi, per nostro gusto o di Principi o Signori forastieri che venissino; risedendo voi per l'ordinario qui in Firenze, et proseguendo le perfezzioni de' vostri studii et delle vostre fatiche, con obligazion però di venir da noi dovunque saremo, anche fuor di Firenze, sempre che vi chiameremo. Et il Signore Iddio vi conservi et contenti": *OG* 10, no. 359*, pp. 400–401.

4. Antonio Rocco (1586–1652), a pupil of Cesare Cremonini's at the University of Padua, was an Italian priest and philosophy teacher. In 1633, he published the *Esercitationi filosofiche*, dedicated to Pope Urban VIII and aimed at providing a refutation of the main tenets of Galileo's *Dialogue*. After reading Rocco's book, Fulgenzio Micanzio, a friend and collaborator of Paolo Sarpi's, wrote to Galileo that "[t]he author is here [in Venice] regarded as a great Peripathetic *monoculus* [that is, extremely committed and loyal to Aristotle]. Actually, he seems to me that when dealing with gossip and words, he carries himself valiantly; when, by contrast, he is dealing with real matters, he runs away, either by avoiding them, or else by understanding in a way that allows him to gossip about them" ("L'autore qui è stimato un gran peripatetico *monoculus*, e mi pare in vero che mentre si sta in cianze et termini si porti da valente, ma quando si viene a cose, scappi con non le toccare o prenderle in senso che possi sopra ciarlare": Fulgenzio Micanzio to Galileo, 28 January 1634, in *OG* 16, no. 2861, p. 30). Later, through a common friend, Micanzio met with Rocco and described him as "[i]n fact, a nice man, polite, very well disposed. Except for the fact that he believes everything Aristotle said as true, more than the Gospel (a new Simplicius), he is definitely a nice man, with no malice, a true gentleman. I see him regretful of the criticisms of your book. He speaks about Your Lordship as a live oracle, except when dealing with Aristotle, about whom he does not question a single iota. Let this not deter Your Lordship from writing the commentary [on Rocco's book], for it may be possible to leave out the criticisms, but I see new and rare things being discussed, and I cannot be happier" ("Veramente è huomo di garbo, civile, pieno di buon affetto, e, levatole questo che crede tutto vero il detto da Aristotele più del Vangelo, un altro Simplicio, certo è huomo di garbo, senza malignità, in fatti un galanthuomo. Lo veggo pentito delle punture del suo libro; parla di V. S. come dell'oracolo vivo, eccetto che ove entra Aristotele *iota unum non praeteribit* [Matthew 15, 18]": Micanzio to Galileo, 19 August 1634, in *OG* 16, no. 2974, pp. 125–126). And again, a few days later: "Rocco seems to me a perfect gentleman. I cannot say the consideration with which he speaks about Your Lordship with everyone. If at all possible, he would retract all his criticisms—but, as far as Aristotle is concerned, don't hold on to me: he is Simplicius himself" ("Il Sig.^r Rocco mi riesce un compitissimo huomo. Non si può esprimere

con che honore a tutti parli di V. S. Se sapesse come, ritrattaria tutte le punture; ma ove entra Aristotele, *noli me tangere* [= John 20, 17]: *ipsissimus Simplicius*" (Micanzio to Galileo, 26 August 1634, ibid., no. 2981*, p. 127). Micanzio repeatedly encouraged Galileo to publish his critical annotations on Rocco's book, but they remained unpublished (they are now available in *OG* 7, pp. 712–750).

5. See Galileo to Leopold V, 23 May 1618, in *OG* 12, no. 1324, p. 389.

6. Charles of Austria, Bishop of Wroclaw (1590–1624). In 1624, he traveled to Madrid, at the request of his nephew King Philip IV of Spain, who planned to appoint him Viceroy of Portugal. Galileo presented him with a telescope (see Galileo to Giovanfrancesco Buonamici, 19 November 1629, in *OG* 14, no. 1967, p. 53), and Charles reciprocated with a chain adorned with jewels, as both Galileo's son and Viviani report.

7. Contrary to Galileo, who argued (in the *Sidereal Messenger*) that the illumination of the Moon is caused by the reflection of sunlight from the earth, Liceti argued that the Moon, analogously to the Bologna stone, released light absorbed by the Sun: see *Litheosphorvs, sive De lapide Bononiensi* (1640), ch. 50, "De Lunæ subobscura luce prope coniunctiones, & in deliquijs obseruata, digressio physico-mathematica" ("Physical and mathematical digression on the somewhat obscure light of the Moon when it is close to conjunction [with the earth and the Sun], and during an eclipse"), pp. 242–248. In what would have been his last scientific work, Galileo defended his own views in a letter to Prince Leopoldo de' Medici (see *OG* 8, pp. 489–545; see also pp. 549–556, and 483–486). On July 14, 1640, Galileo sent a copy of the letter to Liceti; there followed a long and friendly exchange on the topic, and eventually Liceti published Galileo's letter, together with his own point-by-point response, in Book II of the *De lvnae svbobscvra lvce Prope Coniunctiones, & in Eclipsibus obseruata* (1642).

8. In his (second and last) will, dated 21 August 1638, Galileo expressed the wish to be buried in the Santa Croce Basilica, where his father Vincenzo and other members of his family were buried (see *OG* 19, p. 523). Accordingly, the day after his death, the body was quietly moved to the basilica. No official ceremonies were held, or speeches given, and only a few relatives and friends attended the funeral. Since Galileo died under house arrest, under vehement suspicion of heresy, his body was not interred with those of his father and ancestors, in the central nave of the basilica, but was entombed behind a wall in the tiny vestry of the Chapel of the Novitiate (otherwise known as the Medici Chapel), on the right-hand side of the transept; the inscription on this tomb can be read at the end of Lorenzo Crasso's biographical sketch: see pp. 196–197. The Grand Duke's plan for a sumptuous funeral monument was stopped by the Church authorities, and what was meant to be a humble temporary burial remained Galileo's burial place for nearly a century. It was only in March 12, 1737 that, thanks to Viviani's previous efforts and the new political situation, Grand Duke Gian Gastone (the last Grand Duke of the Medici family) was able to have Galileo's body exhumed and moved to the base of the new funeral monument, opposite Michelangelo's. The monument was eventually completed on June 6, 1737.

7

NICCOLÒ GHERARDINI

VITA DI GALILEO GALILEI

LIFE OF GALILEO GALILEI

(1654)

WITH VINCENZO VIVIANI'S
ANNOTATIONS

[634]Non prima che nell'anno 1633 cominciai a pigliar prattica del S.ʳ Galileo Galilei: imperoché dimorando io in quel tempo nella città di Roma, dov'egli parimente si ritrovava per giustificarsi da certe accuse per causa et occasione de' Dialogi sopra 'l sistema Tolommeicano e Copernicano, da lui poco prima dati alle stampe; et havendo io qualche famigliarità con uno de' principali ministri del S. Offizzio, offersi l'opera mia in suo aiuto, quale veramente non potea consistere in altro che in avvisarlo di qualche particolare avvertimento per suo governo. A far ciò fui animito dal medesimo Prelato, come quello che non solamente per l'efficaci raccomandazzioni che gli venivano fatte da chi protegeva la causa e la persona del S.ʳ Galileo, ma per far contrappeso ancora in parte alla maligna intenzione d'un altro personaggio che sosteneva grand'auttorità in quel Tribunale, inclinava di sottrarlo dall'imminente e troppo severa mortificazzione. Mostrò di gradire allora il S.ʳ Galileo l'offerta e l'offizzio mio; ma, o perch'egli stimasse debole il soggetto, o perché sospettasse di qualch'artifizzio, o pure perché egli confidasse troppo nella sua innocenza, come egli diceva, si mostrò poco pieghevole a credere alcuni avvertimenti soggeritimi da quel Prelato, di cui non potevo nominare la persona per non rompere il sigillo. Da questa taciturnità procedé forse la durezza del Sig.ʳ Galileo in prestar orecchio agl'avvisi per altro salutari, onde ne sortì l'effetto che ad ogn'uno è noto. Fu però assai meno dì quello che nell'animo havea conceputo chi sapeva l'origine di sì fiera persecuzzione: in una parola, fu picciola la ferita fatta dalla saetta, se si considera la forza con la quale fu teso l'arco; effetto della singolar protezzione con la quale lo assistè il Ser.ᵐᵒ Gran Duca N. S., per mezzo del suo Ambasciatore allora residente in Roma.

Terminata che fu la causa del Sig.ʳ Galileo, e deliberata la partenza per venirsene a Siena, invitato dal S.ʳ Piccolomini, Arcivescovo di quella città,ᵃ ne' medesimi giorni comparve avviso della vacanza di S.ᵃ Margherita a Montici; dal che prese occasione il S.ʳ Galileo di persuadermi a lasciar la Curia e procurare la presentazione alla vacante Prioria da' miei compatroni, lodandomi assai la bella situazzione del luogo e l'amenità del paese. Non riuscì difficile la persuasiva per indurmi a questa resoluzzione, considerata la mia poca attitudine agli esercizzi della Corte et agli strepiti del foro, e motivato dalla natural inclinazzione al viver solitario; ma più d'ogn'altra cosa potendo in me il sapere la vicinanza della abitazzione mia a quella del S.ʳ Galileo, deliberai di lasciar gli studii legali, [635]eleggendo l'ozzio e la solitudine della villa, dove dimorai quasi del continuo per tutto quel tempo che sopravvisse il S.ʳ Galileo, con il quale per lo spazio di sette anni praticai con familiarità e domestichezza grande; per la qual cosa hebbi vantaggio di sapere, mediante gli spessi colloquii, alcune singolarità con accidenti occorsi nel tempo di sua vita.

---

a Non fu il Sig.ʳ Galileo invitato da Mons.ʳ Piccolomini, ma gli fu stabilita in Roma la casa di Monsignore in Siena fino a nuovo ordine.

[634]It was not earlier than 1633 that I began to spend time with Galileo Galilei.[1] At that time, I was living in Rome, where he was as well, to clear himself of certain allegations caused and occasioned by his *Dialogues on the Ptolemaic and Copernican Systems*, which he had recently published. Having some acquaintance with one of the prominent members of the Holy Office, I offered Galileo my help—which, truly, could not be other than to warn him about some dangers, so that he could prepare accordingly. I was encouraged to do so by the aforementioned prelate, who wanted to help Galileo avoid the overly harsh punishment hanging over him. The prelate intervened not only to respond to specific exhortations he had received from people who were protecting Galileo's cause and person but also to help counter the malevolent intentions of another prominent person who wielded great authority in the Tribunal. Galileo seemed to appreciate my offer and my role, but either because he deemed the argument too weak, or because he suspected some scheming, or else because he relied too much on his own innocence, as he said, he seemed unwilling to believe some of the warnings conveyed to me by the prelate, whose name I had promised not to disclose. My withholding the name of the prelate possibly prompted Galileo's resistance to heeding his otherwise sound advice, and the outcome followed that everybody knows. It was much less severe, however, than what the person who knew the source of the fierce persecution had feared. In a word, the arrow caused a small wound, if we take into account the strength with which the bow was stretched. This was due to the very special protection that the Most Serene Grand Duke, Our Lord, provided Galileo through his ambassador, who was residing in Rome at that time.

Once the trial was over, Galileo set off for Siena, on the invitation of Piccolomini, the archbishop of that town.[a] At the same time the news came that the priory of Santa Margherita a Montici had become vacant. Galileo took the opportunity to convince me to leave the Curia and ask my patrons to obtain the vacant priory for me, highly praising its pleasant location and the agreeableness of the countryside. It was not difficult to persuade me to do so, given my poor talent as a courtier, and my inclination toward a solitary life. What motivated me above anything else, however, was knowing that I would live close to Galileo's villa. So I decided to give up my legal studies, [635]opting for the leisure and solitude of a country house. There I lived almost continuously, for as long as Galileo survived; for seven years I was very close to him, and visited him frequently. Thanks to this, through our frequent conversations, I was able to learn about some eventful incidents of his life.

---

a Monsignor Piccolomini did not invite Galileo; rather, when the latter was still in Rome, he was ordered to the monsignor's residence in Siena, until further notice.

È ben vero che hauto risguardo all'intervallo del tempo scorso d'anni tredici[b] e più, et al bisogno ch'ho havuto d'applicarmi di nuovo agli studii legali, ho smarrite le spezzie di molte cose per la fiacchezza della mia memoria, senza che alcune poche ch'io sono per referire riusciranno forse manchevoli, tronche et imperfette: ma è tale la contentezza che ha l'animo mio in udire che s'habbino a scrivere la vita e l'azzioni d'un huomo quale per l'eccellenza della sua virtù sarà sempre famoso al mondo, che io, tralasciata ogn'altra occupazzione, ho procurato nella miglior maniera restaurarmi la memoria di tutto ciò che mi parrà a proposito per condurre a fine una cotanto nobile e desiderata impresa.

Naccque dunque il nostro Sr Galileo negl'anni della Salute . . . [1] della nobil famiglia de' Galilei, nella città di Fiorenza. Il padre suo fu chiamato Vincenzio, gentilhuomo stimato e di qualche nome in letteratura. Questi della sua legittima consorte acquistò più figliuoli maschi e femmine; per ciò trovandosi ristretto dentro i termini angustissimi di beni di fortuna, havea deliberato d'applicare il S.[r] Galileo, suo maggior figliuolo, all'esercizzio della lana;[c] ma perché il cognobbe inclinato allo studio e d'alto intendimento per la somma sua docilità, permise ch'andasse alla scuola di grammatica appresso d'un tal professore, huomo assai dozzinale che insegnava in una casa di propria abitazione posta in Via de' Bardi, dove in brevissimo tempo, sopita quella repugnanza ch'haver suole l'età fanciullesca ad un sì noioso impiego, apprese quelle buone regole che sono più importanti per saper la lingua latina, e con la lettura d'alcuni libri, de' quali era avidissimo, imparò di essa i più bei segreti, con maraviglia indicibile del maestro, il quale, più tosto confuso, referì al padre non esser egli più idoneo per insegnar di vantaggio al fanciullo: dal che prese animo il Sr Vincenzio d'introdurlo nella città di Pisa, acciò che s'applicasse allo studio della medicina, stimando per questo mezzo poter assai più presto e meglio conseguir guadagno per sollevamento della sua povertà. Obbedì prontamente il giovanetto, et in casa d'un parente, al quale ne fu raccomandata la custodia, si collocò.

b Sono anni 12, non 13.
c Non si sa per alcuno che il padre volesse applicarlo all'arte della lana.

---

1 . . . NY] 1654 T

Since thirteen years[b] or more have passed, and I had to go back to my legal studies, I have forgotten details of many things, due to my feeble memory; and some of the few events I shall write about might turn out to be wanting or incomplete. But such is my joy at hearing that the life and works of someone whose extraordinary virtue will always be renowned worldwide are to be recorded, that I neglected all other occupations and tried, to the best of my ability, to remember everything I deem appropriate for accomplishing so noble and deserving an enterprise.

So Galileo was born in . . . AD[2] to the noble family of the Galilei, in the city of Florence. His father was called Vincenzo, a highly esteemed gentleman, whose name is of some scholarly repute. He had a few children, sons and daughters, from his legitimate spouse.[3] Finding himself in some hardship, given the very limited amount of goods fortune provided him with, he decided to have his elder son, Galileo, work at the manufacturing of wool.[c] Once he realized that Galileo was exceptionally bright and willing to study, he graciously allowed him to study grammar. Galileo went to study with a rather mediocre teacher who gave lectures at his own house, in via de' Bardi. Here, after a very brief time, once he had overcome the usual youthful disgust for such boring work, he learned the basic rules of the Latin language; then, by reading a few books, which he avidly consumed, he learned the most beautiful secrets of that language, to the unspeakable astonishment of his teacher. Bewildered, he told Galileo's father he was no longer able to teach anything to the young boy. After this, Vincenzo decided to enroll Galileo in Pisa, so that he might take up the study of medicine. He hoped that would enable Galileo to earn money much more quickly and in larger amounts, and so help him out of his difficulties. The boy promptly obeyed his father, and moved to the house of a relative, who was trusted to look after him.

b It is 12 years, not 13.
c No one else has ever reported that Galileo's father wanted him to work at the manufacture of wool.

Nel primo ingresso ascoltò i primi principii dell'arte della medicina, e nell'istesso tempo quegli della filosofia, alla quale si sentiva più inclinato. Nell'anno [636] seguente, che venne il secondo, nell'aprirsi di nuovo lo Studio, ritornò a Pisa, con animo di continuare la medesima lezzione; ma portò il caso che si transferì alla medesima città il Ser.^{mo} Gran Duca Francesco con tutta la Corte, il quale di poco prima havea condotto allo stipendio un tal prete, il quale non mi si ricorda 'l nome (credo che si cognominasse de' Ricci e fosse di nazzione marchigiano), per insegnare ai SS.^{ri} paggi che servivano S. A. S. Era questo soggetto d'assai buona letteratura e di non mediocre intelligenza nelle mathematiche. Non so come, nell'amicizia di lui s'insinuò il Sr Galileo,^d e, per quanto mi disse egli, casualmente; et andato per parlargli alcuna volta, 'l trovò sempre in esercitio d'insegnare e dichiarare Euclide, sì che non potendo esser ascoltato, ascoltava egli le lezioni: dalle quali pigliava tanto gusto e nodrimento 'l suo intelletto, che, invaghitosene sempre più, trascurava d'andare allo Studio, dove era consueto d'udire la lettione di medicina, et in quella vece andava alle stanze dov'il Sig.^r Maestro Ricci leggeva mathematica, senza protesto alcuno di parlare e con meno confidenza di poter star presente apparentemente, già che la lezzione era solamente per i SS.^{ri} paggi, o altri ch'havessero servitio in Corte, onde gli conveniva star fuori della stanza, in luogo dove difficilmente udiva.

Perseverò egli d'ascoltar lezzione di mathematica, così clandestinamente et alla sfuggita, quasi due mesi, e nel medesimo tempo con premura grande cercò egli per Pisa un Euclide; e trovatolo, l'applicazzione sua era grandissima allo studio di quest'autore, internandosi negl'arcani più difficili e più profondi, onde n'attinse grandissima intelligenza, con riuscir maggior d'ogni difficultà. Ben è vero che non del tutto si fidava di sé medesimo, e procurava occasione d'abboccarsi col soprannominato professore, per conferire con esso seco alcuna delle proposizioni o demonstrazioni, et interrogarlo a dirgli sinceramente la verità intorno al buon indirizo.

Si compiacque il Sr Ricci d'udirlo; e dopo che l'hebbe udito ragionare, stette alquanto sopra di sé con stupore, e dimandò al giovinetto Galileo chi fosse stato di tal professione il suo maestro. Sorrise allora e sospese per qualche tempo la risposta, stando a spettare se di nuovo gliene domandava; ma perché non proseguiva il discorso, se prima non era satisfatto della curiosa dimanda, deliberò il Sr Galileo di scuoprire che altro maestro conosciuto non havea fuori di quello che l'interrogava. Accrebbe una tal risposta maraviglia maggiore all'interrogante, spezialmente perché non l'havea veduto presente alle sue lezzioni; et in questa maniera fu necessitato il Sr Galileo a far racconto del modo con che havea goduta l'occasione d'ascoltarle. Non si può facilmente esplicare qual contentezza [637] sentisse il Sr Ricci allora e con quale affetto si voltasse ad amare e stimar la persona del sconosciuto scolare, a segno che invitollo non solamente a comparire alla scoperta nel tempo ch'egli leggeva, ma s'offerse ancora che ad ogni suo piacere gl'haverebbe data commodità di parlargli con ogni domestichezza.

---

d Non s'introdusse nelle matematiche come si dice dal Sig.^r Gherardini, e per conseguenza quanto ne segue è falso.

At first, Galileo attended lectures on the basic principles of the art of medicine, and at the same time on philosophy, toward which he felt more inclined. The year [636]following, his second, he went back to Pisa as soon as the university courses began, in order to continue with those lectures. It turned out that the Most Serene Grand Duke Francesco[4] had moved to Pisa with the entire court. He had just hired a priest, whose name I do not remember (I believe his family name was de' Ricci, from the Marches), to teach the pages at the court of His Most Serene Highness.[5] This man was very well read and with no mediocre grasp of mathematics. I do not know how, but Galileo befriended him.[d] He told me it was by chance: Galileo went to talk with him a few times, and always found him teaching and explaining Euclid; and since Ricci could not listen to Galileo, Galileo listened to Ricci's lectures. Galileo enjoyed them so much, drawing in such nourishment for his mind, that he developed a growing passion for mathematics, thereby neglecting lectures in medicine at the university. Instead, he went to the room where Master Ricci lectured on mathematics.He had no right to speak, and was even less sure that he had a right to attend those lectures, as they were meant to be for the pages only, or for those who were in service at the court; so he remained in the room, in a place from which he could hardly hear what was being taught.

Galileo persisted in attending the mathematics lectures, secretly and briefly, for nearly two months; and at the same time he sought a copy of Euclid in Pisa. As soon as he found one, he devoted himself to its study as much as he could, examining in detail the most difficult and profound mysteries of the text. Galileo managed to understand them in depth, overcoming all difficulties. In fact, he did not trust himself completely and took care to find opportunities to talk with the aforementioned professor, so as to exchange views about some of the propositions or proofs, and to ask Ricci to tell him sincerely how best to proceed.

Ricci gladly listened to him, and once he had heard his arguments, Ricci was truly amazed. He asked young Galileo who was his teacher; Galileo smiled and left the answer hanging for a while, waiting to see whether Ricci would ask him again. As Ricci remained silent until Galileo answered his question, Galileo decided to reveal that he had not had any other teacher besides the person who was asking the question. Galileo's answer aroused an even greater amazement in Ricci, especially since he had never seen Galileo at his lectures. And so Galileo felt compelled to confess the way in which he had attended them. It is not easy to describe how pleased [637]Ricci felt, and with what affection he began to love and prize the unknown pupil. He not only encouraged Galileo to openly attend his lectures but also offered to talk with him informally, whenever he wished to do so.

d Galileo did not take up mathematics, as Gherardini reports. As a consequence, what follows is false.

Questo sì cortese invito, sì come operò ch'il Sr Galileo ben spesso si rappresentasse, così cagionò una diversione quasi totale dall'incominciato studio della medicina; della qual cosa prese occasione chi si fosse d'avvisarne al padre, il quale subito che seppe, ne prese cordoglio tale che lo strasportò precipitosamente a Pisa, per riprendere, anzi per ritirare, il figliuolo e ricondurlo con esso seco, con tanto più di prestezza, quanto si credea che la diversione dallo studio procedesse d'altre cause. In vedendo poi ch'il figliuolo non si lasciava tirare da passatempi vani, giuochi o simili, di che ne veniva certificato dalla testimonianza di chi abitava in casa, sospettò di quello che non era punto lontano dal vero; cioè che qualch'altro studio lo divertisse; che però stette per alcuni giorni osservando i libri che tenea in camera o sopra 'l tavolino d'essa: di che accortosi il giovanetto, per non disgustare il padre, tolse via Euclide ed ogn'altra apparenza di studio diverso da quello della medicina.

Si mitigò in parte il dispiacere dell'impazziente padre, ma non del tutto, perch'haverebbe desiderato che quanto prima ricevesse la laurea del dottorato; di che non volse il figliuolo assicurarlo, ma procurò di persuaderlo a deporre tanta impazzienza, et a consolarlo con dire che tra poco tempo haverebbe veduto il frutto e potuto sperare utile bastante per il di lui mantenimento, senza far altra dichiarazione.

Dopo questo discorso, incontinente si partì il Sr Vincenzio non interamente satisfatto. Partorì buonissimo effetto questo poco di viaggio che con tanta sollecitudine era stato fatto, poscia che con ogni sforzo maggiore si mise in animo il Sr Galileo d'applicarsi alle mathematiche, e stabilì d'eleger questa per la sua professione nel più perfetto grado.

Per conseguir dunque il fine d'una così diffcil impresa, s'applicò a studiare Archimede, con il consiglio del menzionato Ricci, dal quale ancora gli fu dato in dono. È cosa impossibile a raccontare quanto incremento ricevesse dal studio di questo grand'huomo; certo è che con la scorta di lui stabilì saldissimi fondamenti e non dubitò poscia di sollevarsi in alto, con impennar l'ali della speculazione, investigando non solamente le cose più nascoste operate dalla natura in questo mondo inferiore e sublunare, ma di sapere ancora tutto quello che si trova di maraviglioso nel superiore e celeste: potersi, diceva egli, passeggiar sicuro e senza inciampo sì per la terra come per il cielo, mentre non si fossero smarrite le pedate d'Archimede; e stimava ciò esser permesso a chiunque l'intendea, ma che in questo consistea ogni difficultà.

[638]S'accoppiarono in lui lo speculare e l'operare, la teorica e la pratica; impercioché provistosi d'alcuni instrumenti geometrici, ciò ch'intendeva con l'intelletto, non solamente dimostrava, ma con inusitato modo il rendea percettibile dal senso: nel che fare hebbe tanta facilità, che, per testimonianza di chiunque l'udiva discorrere o vedea operare, non era conosciuta differenza alcuna dall'uno all'altro. E chi vuol negare che ciò non derivasse dalla felicità del suo ingegno e da una naturale espressiva, congiunta con proprietà di termini e similitudini tanto calzanti, che si rendea impossibile il non rimanere persuaso o convinto? Per la qual cosa acquistossi egli una maravigliosa reputazione, e del di lui straordinario talento cominciò la fama a spargerne qualche romore, quale arrivò all'orecchie dell'Ecc.^mo S. D. Giovanni de' Medici, signor di gran qualità et esperienza di guerra, se si considera principalmente l'intelligenza che hebbe singolarissima delle fortificazioni e delle macchine d'ogni sorte.

This kind invitation led Galileo to attend his lectures very often and at the same time to turn away almost entirely from the study of medicine he had begun. Somebody informed Galileo's father about this, and as soon as he heard about it, he was so displeased that he rushed to Pisa to take Galileo back with him—the quicker, the better—as he thought that Galileo was turning away from medicine for reasons different [from the real ones]. However, when he realized that Galileo was not driven away by idle distractions, pastimes, or other similar solaces, as he was assured by those with whom he boarded, he suspected something that was by no means far from the truth, namely, that some other study had turned him away. Vincenzo spent a few days there, checking the books Galileo kept in his room or on his table. The boy became aware of this, and in order not to displease his father, removed Euclid and all evidence that he was studying something other than medicine.

Galileo partly alleviated the displeasure of his impatient father, but not entirely, because he would have liked his son to get his degree in medicine as quickly as possible. The son did not want to promise his father anything concerning this but endeavored to convince him not to be so impatient, and comforted him by saying that he would soon see some result, and that he hoped it would be useful enough for him to earn his living, without saying anything more.

Vincenzo left Pisa immediately after this exchange, not entirely satisfied. The short trip, so hurriedly taken, had a most positive outcome, however: Galileo decided to devote himself entirely to mathematics with even more determination, and chose this as his profession, at the highest level.

In order to achieve this difficult goal, he began to study Archimedes under the guidance of Ricci, who presented him with a copy of Archimedes's works. It is impossible to say how much progress Galileo gained from the study of this great man. Certainly, it was thanks to Archimedes that Galileo laid down such solid foundations from which he later would rise, spreading the wings of speculation and inquiring not only into the most secret operations of nature in this lower and sublunar world but also discovering all that is wonderful in the higher, celestial world. Galileo claimed it was possible to walk safely, without stumbling, on earth as well as in heaven, as long as we remain in Archimedes's footsteps. This, Galileo said, was possible for anyone who understood him, but this was not easy to do.

[638]Galileo combined speculation and work, theory and practice. Aided by a few geometric tools, he not only demonstrated what he understood with his intellect but in an unusual way made it perceptible to the senses. He did this so easily that there was no difference between the one and the other, as anyone who heard Galileo talk or saw him at work can testify. And who would deny that such ease derived from his brilliant mind and his natural ability to express himself, coupled with a proper use of terms and analogies? These were so fitting that it would have been impossible to remain skeptical or unconvinced. This allowed Galileo to earn such an astonishing reputation that fame of his extraordinary talent began to spread. It reached the ears of the Most Excellent Lord Giovanni de' Medici, a gentleman of great adroitness and experience at war, if we consider his most peculiar knowledge of fortifications and machines of all sorts.[6]

La relazione a favor del S.ᵣ Galileo appresso di S. E.ᵃ fu fatta da un tal S.ᵣ de' Marchesi dal Monte, di cui non mi sovviene il nome, ma soggetto di stima grande appresso tutti. Con l'appoggio di questi ottenne il S.ᵣ Galileo, in età assai giovenile, la cattedra di Mathematica nello Studio pisano, dove egli per lo spazio di due anni fu ascoltato con gran sua lode; ma per accidente occorso, non stimò bene di continuare in quella lettura. La resoluzzione² hebbe questa causa. In quei giorni havea proposto il S.ᵣ D. Giovanni ch'in Pisa si facesse una certa fabbrica, non so già se di fortificazzione o d'altro edifizio. Per l'effettuazzione del disegno si era concluso di metter in opra alcune macchine, quali, con il parere de' periti, erano giudicate molto a proposito: solo il S.ᵣ Galileo s'oppose, e con ragioni forse troppo vive procurò impedirne l'esecuzzione. Quello che seguisse, io non lo so; so bene che la contradizzione non fu grata al S.ᵣ D. Giovanni, il quale con parole di molto sdegno ne mostrò risentimento: di che si intimorì il S.ᵣ Galileo di maniera, che stimò bene non dopo molto tempo domandar licenza da quella condotta, con disgusto grande di quel S.ᵣ dal Monte, quale procurò di distorlo dal pensiero, offerendosi per ogni buono offizio appresso di chiunque fosse bisognato; ma nol potè ottenere, perché il S.ᵣ Galileo havea stabilito di voler tentare altra fortuna.

Nel ritorno che fece a Fiorenza, fu accompagnato con una sola lettera di raccomandazzione, scritta dal medesimo S.ᵣ Marchese dal Monte, nella quale venivano assai lodate le qualità del S.ᵣ Galileo appresso del Sig.ᵣ Filippo Salviati,ᵉ **639**gentilhuomo di chiarissima fama. Fra l'altre cose che venivano asserite in quella lettera v'era che nell'accoppiamento di speculativa e di prattica nelle mathematiche, da' tempi d'Archimede in qua, si stimava non essersi scoperto ingegno pari a quello per cui era fatta la raccomandazione. Questo offizio, passato con tanta lode del Sig.ᵣ Galileo dal Sig.ᵣ Marchese dal Monte, fu molto gradito dal S.ᵣ Salviati, ma molto più la persona del raccomandato, il quale con parole cortesissime fu constretto a rimanere in casa del medesimo Signore: e venne in acconcio l'invito, perché di già il S.ᵣ Galileo havea deliberato di non andare alla casa paterna per non cagionar disgusto ai suoi domestici, in tempo forse che in niun altro luogo sicuro havea l'assegnamento di coricarsi. In tutto quel tempo che dimorò in casa del S.ᵣ Salviati hebbe campo di guadagnare la di lui grazia, a segno che del continuo mangiava alla medesima tavola e con esso seco il conducea alle ville,³ conpiacendosi fuor di misura della gioconda conversazione e godendo d'incontrar quella fortuna tanto desiderabile dai ricchi e dai grandi, che è di nodrire l'ingegni con sottrargli del duro giogo della povertà, onde havea egli ordinato ch'al S.ᵣ Galileo fosse somministrato tutto ciò ch'il bisogno suo richiedea. Ma la premura maggiore del Sig.ᵣ Salviati era ch'il S.ᵣ Galileo trovasse recapito in qualche Studio o Università, acciò se l'aprisse la strada a far cognoscer la sua gran virtù.

<hr/>

e Non credo che il Sig.ᵣ Marchese dal Monte fusse a Pisa senz'altro, e non scrisse lettera di raccomandazione al Sig.ᵣ Filippo Salviati per il Sig.ᵣ Galileo, perché il Sig.ᵣ Filippo in quel tempo era di 9 anni in circa, ma più tosto scrissela al Sagredo.

<hr/>

2 La resoluzzione *Y*] E la resoluzzione (per quello si disse) *NT*
3 *post* ville, *scr.* e particolarmente a quella deliziosissima delle Selve, dove il Sig.ᵣ Galileo fece la maggior parte delle sue dottissime et ingegnosissime osservazioni *N* e particolarmente a quella deliziosissima delle Selve, dove il Sig.ᵣ Galileo fece la maggior parte delle sue grandi osservazioni *T*

Someone of the [family of the] Marquises dal Monte, whose name I cannot recall[7] but who was highly regarded by everybody, prepared a report in support of Galileo for His Excellency. With the marquis's support, Galileo, still young, was awarded the chair of mathematics at the University of Pisa, where he taught for two years, and his lectures were highly praised. Due to a mishap, however, Galileo decided against going on with his lectures. The appointment ended for the following reason. At that time, Lord Giovanni suggested that a new building be erected in Pisa, either a fortification or some other kind of building. To realize the project it was agreed to employ a few machines that, experts agreed, were deemed appropriate. Only Galileo opposed it, and with arguments that were possibly too severe managed to prevent its construction. What happened next, I do not know. I do know, however, that Lord Giovanni did not like Galileo's opposition, and expressed his resentment in very harsh words. Galileo was so alarmed that not long afterward he asked to be released from his duty, to the great dismay of dal Monte, who tried to change Galileo's mind. He also offered to intervene on Galileo's behalf with anyone who might be concerned; but all was in vain, as Galileo had already decided to try his luck elsewhere.

On his way back to Florence, Galileo carried with him only one letter of recommendation, from Marquis dal Monte, highly praising his qualities [639]to the most eminent Filippo Salviati.[e] Among other things in that letter, dal Monte wrote that nobody since the time of Archimedes had a comparable intellect, in its combination of speculation and practicality, to that of the bearer. Salviati was very pleased with this recommendation, in which Marquis dal Monte praised Galileo so highly. But he liked the recommended person even more, whom he invited to stay in his house with the kindest words. The invitation was most welcome, as Galileo had no other place to go to spend the night; he had decided against going to his father's house, so as not to cause dismay to his relatives. During the time Galileo spent at Salviati's house, he earned his host's approbation, as is testified by the fact that they ate at the same table, and Salviati took Galileo with him to his country houses,[8] enjoying his guest's cheerful conversation. Furthermore, he was pleased to have the opportunity—cherished by rich and powerful people—to support ingenious people by releasing them from the heavy yoke of poverty. He ordered that Galileo receive whatever he needed, but above all he tried his best to secure a position for Galileo in some college or university, where he might have a chance to make his great gifts more widely known.

e I do not think Marquis dal Monte was in Pisa at the time. Certainly he did not write a letter recommending Galileo to Filippo Salviati, as Salviati was about 9 years old at the time; most likely, he addressed one to Sagredo.

Per l'adempimento di concetto così nobile non fu strana la fortuna: avvegnaché in passando per Fiorenza l'Illmo Sr Gio. Francesco Sagredo, gentilhuomo veneziano, ripieno di rarissime qualità, nel ritorno da un'ambasceria, non so di Roma o di Spagna, fu convitato in un giorno dal Sr Salviati, il quale con l'occasione d'un tal colloquio commendò molto la persona e 'l valore del Sr Galileo a quel Signore; ^f e pregandolo ancora a proteggerlo per ottenere una lettura nello Studio di Padova, fu promessa ogni assistenza di favore per impetrarne l'effetto, in corrispondenza all'eccessiva cortesia e generosità con la quale era stato trattato dal Sr Salviati. Appena giunto in Venezia, introdusse sopra di ciò la prattica con quegli SS.^{ri} Senatori, deputati o protettori dello Studio, dai quali ne fu riportato favorevole il rescritto in conformità di quanto era stato ricercato. Quasi subito sopraggionse lettera d'avviso dal Sr Sen.^{re} Sagredo, per la qual veniva sollecitato il Sr Galileo a partire quanto prima et invitato ad andare in casa sua. Fu accettato l'invito con gran giubbilo del Sr Galileo, il quale con non meno prestezza s'incamminò a Venezia. Nella partenza fu provisto dal Sr Salviati ⁶⁴⁰di vestiti, biancheria e d'ogni altro più opportuno arnese. È ben vero che, come più volte udii dire da lui istesso, il suo baule, nel quale si contenea allora tutto il suo patrimonio, non eccedea libre dugento[4] di peso.

---

f Dubito che il principio di conoscere il Sig.^r Sagredo non fusse tale, e che non passasse per Firenze di ritorno da ambasceria.

---

4 libre dugento Y] cento libbre N libbre cento T

Good luck aided the achievement of such a noble goal. It so happened that the Most Illustrious Giovanni Francesco Sagredo, a Venetian gentleman of rarest qualities, passed through Florence on the way back from an embassy, whether from Rome or Spain I cannot tell, and one day Salviati invited him for a meal. During the visit Salviati greatly praised Galileo, both as a man and as an intellect.[f] Salviati asked Sagredo to help Galileo obtain a lectureship at the University of Padua, and Sagredo promised his full support in the effort to achieve this goal, in exchange for the extraordinary kindness and generosity with which Salviati had treated him. As soon as he got back to Venice, Sagredo raised the subject with those senators, deputies, or patrons of the university, who wrote a favorable report, as wished. Almost immediately afterward, a letter arrived from Senator Sagredo, urging Galileo to leave [for Venice] as soon as possible, and inviting him to stay at Sagredo's house. Galileo was very pleased to accept the invitation, and set out for Venice promptly. At his departure, Salviati provided him [640]with clothes, linen, and whatever else might be required. In fact, as he told me himself several times, the chest that contained all of Galileo's belongings at the time did not exceed 200[9] pounds in weight.

f I doubt that this was the way in which Sagredo first learned about Galileo, and that he was passing by Florence in returning from an embassy.

Arrivato in Venezia, fermossi il S.ʳ Galileo in casa del S.ʳ Senatore Sagredo per lo spazio di quasi due mesi, dove con molte carezze fu intrattenuto, sin tanto che venisse il tempo nel quale è solita usanza d'aprirsi lo Studio. In questo intermezzo visitò molti Senatori di quella gran Republica, e prima quei che sono i promotori del medesimo Studio.⁵ Venuto il tempo opportuno, si transferì in Padova, et ivi si provide d'una picciola casetta per la sua abitazione, non molto distante dal famosissimo tempio di S.ᵃ Giustina. La vicinanza di questo luogo fu di molta commodità al S.ʳ Galileo, conciossiacosaché quel P. Abbate che allora reggeva il monasterio era gentilhuomo veronese, di maniere assai cortesi e non poco intendente delle mathematiche, con la quale occasione s'introdusse il S. Galileo nella di lui amicizia; a contemplazione di che fu provisto di qualche necessario utensile e supellettile, come di letti, seggiole et altre cose simili, delle quali era non poco bisognoso, tanto più quanto, scuoprendosi di dilettevole e manierosa conversazione, molti degli scolari, etiamdio d'altra professione, ben spesso andavano in casa per rimaner quivi a desinare e cenare con esso seco. Quindi accadde sovente, non haver egli tovagliolini a bastanza per il numero de' commensali, in tempo ancora che non si potevano così all'improviso provedere, onde più d'una volta gli fu d'uopo far nuova giunta alla tavola et apparecchiarla con carte fogli all'improviso.ᵍ Ma più d'ogn'altra cosa accrescea la frequenza dei giovani scolari la singolare facilità ch'havea il S.ʳ Galileo nell'insegnare e dichiarare le cose più oscure nelle scienze, nell'esaminare i varii sentimenti o axiomi de' più rinomati filosofi, in una parola nell'indagare i principii di tutta la natural filosofia;⁶ le quali cose tutte tanto si rendeano più maravigliose, quanto che veniano maneggiate non solamente con metodo straordinario, ma con modo di speculare assai diverso dagl'antichi e moderni professori: laonde acquistandosi ogni giorno più di reputazzione, da tutte le parti concorrevano⁷ huomini, etiamdio provetti, con frettolosi passi, curiosi d'ascoltare, anzi di vedere, cose in quella professione del tutto nuove e pellegrine.

⁶⁴¹Non giudicò già il S.ʳ Galileo di satisfare al concetto che cognoscea formarsi di sé né all'universal espettazzione, se alla lettura o insegnamento di quelle scienze, non havesse accoppiato lo scrivere, per consegnare qualcheduna delle opere sue alle stampe: perciò diede alla luce quella che fu la prima e s'intitola *Il Compasso Geometrico*.

Alla comparsa di questa s'avanzò in tal credito, che vista e considerata da' più eccellenti professori di tutta l'Europa e spezialmente della Germania, fu molto commendata. Molti di quella nobilissima nazzione vennero da diverse provincie in Italia, fermandosi in Padova, non per altro fine che d'haver per maestro il S.ʳ Galileo.

---

g Non piace quell'apparecchiar con carte e fogli, e allettar gli scolari col mangiare, pigliandogli per la gola.

---

5 *post* Studio *scr.* nel qual intermezzo egli non tralasciò di visitare molti Senatori, e particolarmente i protettori del medesimo Studio, e fu da tutti molto ben veduto e accarezzato, essendo già per fama il suo nome in quella città assai ben noto N

6 *post* filosofia *scr.* spianando con la felicità del suo maraviglioso ingegno tutte le più scabrose difficoltà, da altri sino a quel tempo o male o non punto intese N

7 da tutte—concorrevano YT] concorrevano da tutte le parti d'Europa N

Once he got to Venice, Galileo stayed at Senator Sagredo's house for nearly two months, and he was treated most handsomely until the usual time of the university's reopening. In the meanwhile, he had paid visits to many senators of that great republic, starting with those who supported the university.[10] When the time came, he moved to Padua, where he arranged for a small house to live in, not far from the most renowned Basilica of Santa Giustina.[11] The neighborhood suited Galileo very well: the abbot in charge of that monastery was a gentleman from Verona, very courtly, who knew mathematics in some detail. On that basis Galileo took the opportunity to strike up an acquaintance with him, in consequence of which Galileo was given some necessary tools and furnishings (such as beds, chairs, and similar objects) that he needed desperately—especially since his pleasant and mannerly conversation drew many of his pupils, and others from different faculties, to visit him and to stay for lunch or supper. And so he frequently ran out of napkins, which were not enough for the number of guests at his table. At times it was impossible to obtain them rapidly, so that more than once he had to extend the table at the very last minute and set it with papers or sheets.[g] What made pupils flock to Galileo's lectures, however, was his unique ability to teach and explain the most obscure scientific subjects, to analyze the various doctrines and assumptions of the most eminent philosophers: in a word, to investigate the principles of all natural philosophy.[12] And this was all the more amazing because Galileo treated them not only by an unusual method but also with a kind of speculation that was utterly different from that of ancient and modern professors, whereupon his reputation grew every day, and persons came to him from all places.[13] Some of them were very good [at mathematics], and they rushed to him, curious to listen to, and indeed to see, completely new and singular things in that field of study.

[641]Galileo did not think he could live up to the idea he knew was spreading about him [among his audience], nor the general expectations, had he not coupled his lecturing and teaching of those sciences with writing, so as to publish some of his works. And for this reason he published his first work, titled *The Geometric Compass*.

As soon as it was published, it was very much admired, and the most eminent professors in the whole of Europe, and especially in Germany, saw and read it, and highly praised it. And many from various provinces of that most noble nation came to Italy, and stopped in Padua, with no other aim than having Galileo as their teacher.

---

g I do not like this setting the table with papers and sheets, and enticing pupils with food, winning them over by their gluttony.

Contro l'opera del Compasso Geometrico scrisse in latino un tal S.ʳ Baldassar Capra,[h] gentilhuomo milanese, d'assai buona fama nella professione di geometria, impugnandolo in moltissimi luoghi. Quando che al S.ʳ Galileo pervenne notizia di questa impugnazione, e dopo che l'hebbe veduta, ne prese grandissimo gusto, e disse che molto restava obbligato al S.ʳ Capra impugnatore, poscia che gli porgeva occasione non solamente di difendersi con l'apologia, ma gl'apriva assai spazzioso il campo da potersi slargare in molte cose, tutte in corroborazione di quanto havea scritto, per confusione dell'avversario e per addottrinamento degl'altri. E nel vero riuscì tanto gagliarda e vigorosa la risposta, che né il S.ʳ Capra né altri di poi hebbero ardire d'opporsi.

In questo proposito soleva egli dire che grandissimo piacere sentiva quando alle sue opere incontrava contradittori, poscia che da questi gli veniva somministrato argumento e materia di speculare e di scriver in miglior forma; anzi che molte cose a bello studio havea date fuora, al suo giudizio imperfette, non per altro se non perché più facilmente trovassero opposizione:[i] il che stimerà esser verissimo chiunque leggerà l'opere di lui, perché troverà esser nelle repliche più ammirabile.

Ma poco nulla haverebbe stimato il S. Galileo d'haver guadagnato, se allo scrivere e stampare diverse opere, nelle quali scorgeasi profondissima speculazione, non havesse aggiunto qualche peregrin ritrovamento, mediante il quale non solamente la sovranità dell'ingegno suo venisse manifestata, ma l'humana condizione in un certo modo privilegiata. E perché non v'era cosa al mondo dalla quale potesse ricever maggior ingrandimento quanto dalla cognizione delle cose naturali, spezialmente delle celesti, e che questa non si potea ottenere senza l'aiuto de' sentimenti esterni, cioè senza l'aumento di quello del vedere, [642]si propose in animo di voler rinovar al mondo il disusato, anzi disperato, modo di far un instrumento per il quale venisse tanto avvantaggiata la facultà o potenza visiva, che ella non meno che da vicino potesse le maraviglie tanto più prodigiose quanto lontane dagl'occhi de' viventi contemplare.[j] Nella qual impresa, benché stimata per altro d'impossibile riuscita, hebbe tanta felicità nel saperla indirizzare, mercé del suo ingegno veramente divino, che condotto a perfezzione l'instrumento detto *telescopio* e volgarmente l'*occhiale*, osò con la vista trapassare in un attimo, poco o nulla curando l'ampiezza e l'immensità degli spazzi, questo mondo elementare, et osservare i viaggi de' globi celesti, i moti delle stelle ed affrontare l'istesso sole e la luna, prencipi dei pianeti, inaccessibili per altro e sicuri d'ogni ingiuria, con la quale l'humana curiosità pretendessi oltraggiargli.

Chi vuol ridire o, per dir meglio, chi può esprimere il suono con il quale la fama in un subito riempì tutta l'Europa, anzi isvegliò il mondo tutto e gl'abitatori d'esso ad inarcar le ciglia? Onde maraviglia non è se principalissimi cavalieri e prencipi venivano non solamente dalla Germania e dalla Francia, ma dalla Pollonia, Svezia, Ungheria e dalla Transilvania, etiamdio quegli che non professavano scienze, non per altro che per vedere e conoscere di vista il tanto rinominato S.ʳ Galileo, tra' quali si numerano molti che sortirono poscia gran nome nell'arte militare, che sarebbe troppa lunga serie e superflua il nominargli tutti.

h Il Capra non scrisse contro al Compasso, ma l'usurpò; e però è falso tutto ciò che ne segue in questo proposito.

i Non è vero che abbia dato fuori molte cose imperfette a posta per trovare opposizioni.

j L'invenzione dell'occhiale non fu come dice il Sig.ʳ Canonico, che egli medesimo pensasse di ritrovarla, ma fu nel modo che l'ho raccontato io.

One Baldassarre Capra, a gentleman from Milan who had a very good reputation as a geometrician, wrote a book in Latin against *The Geometric Compass*, taking issue with it on several points.[h] As soon as Galileo learned about these criticisms, and once he saw the book, he was greatly amused by it. He said he was very much in debt to the challenger, Capra, since not only did Capra offer Galileo a chance to defend himself but also opened a way to develop the text in different directions, always in support of what he had originally written, both to the opponent's confusion and for the instruction of others. And, indeed, Galileo's rejoinder was so bold and vigorous that neither Capra nor others dared to counter.

Galileo used to say that he greatly enjoyed criticism of his works, as it offered opportunity and material to elaborate and write better. In fact, he published many works that he deemed imperfect expressly in order to receive criticism more easily.[i] Anyone who reads Galileo's works will realize this is true, for he will see how Galileo's rejoinders are even more admirable [than his original formulations].

But Galileo would have thought he had achieved little had he not complemented the writing and publishing of his several works, in which we can find the most profound speculations, with some singular discovery in which not only his towering mind could be manifested but also the human condition, so to speak, could be given pride of place. There is nothing in the world from which we can derive more progress than from knowledge of natural things, and especially celestial ones. Since this cannot be achieved without the aid of external sense experience, that is, without improvement in vision, [642]Galileo undertook the task of renovating the old-fashioned and fruitless technique of constructing an instrument by which eyesight could be so improved and enhanced that it could contemplate marvels that are the more marvelous the farther they are from us as if they were close by.[j] Although it was deemed impossible, Galileo succeeded so well in this enterprise, thanks to his truly divine mind, that once he perfected the instrument called *telescope*, or *spyglass* (as it is commonly referred to), he dared to go beyond this elemental world in an instant, with little or no concern for the size and immensity of [the heavenly] spaces. And he observed the paths of the celestial globes and the motion of the stars, and confronted the very Sun and the Moon, princes of the planets, unreachable by anything and safe from any injury with which human curiosity might presume to violate them.

Who could repeat, or, better, express the clamor with which the news immediately filled all Europe, and indeed awakened the whole world and the curiosity of its inhabitants? It is no surprise, then, that most eminent noblemen and princes, even those who were not devoted to sciences, came not only from Germany and France but also from Poland, Sweden, Hungary, and Transylvania for no other reason than to see and meet with Galileo, whose name was so renowned. Among them were many who later made a name for themselves in the military art, whom it would be too long and unnecessary to list here.

h Capra did not write against the *Compass* but plagiarized it; hence, all that follows on this subject is false.

i It is not true that he published many imperfect works on purpose so as to inspire criticism.

j The invention of the spyglass did not take place as the Canon tells us, namely, that Galileo himself devised it. Rather, things happened as I said.

Non voglio già tacere che per questo medesimo fine venne a por domicilio in Padova, con nobilissime camerate, incognito però,[8] il Ser.^{mo} Gostavo Re di Svezia, quello, dico, che riuscì tanto formidabile in guerra, che la sola memoria di lui rende spavento all'universo. Questo gran personaggio, intrattenendosi per alcuni mesi in Padova, si trovò quasi sempre presente alle lezioni del Sr Galileo, il quale a richiesta di quei gran Signori fu persuaso a leggere in lingua toscana; conciossiacosaché essendo tutti questi assai ben instrutti nella latina favella et in qualche parte dirozzati nell'italiana, desideravano nell'istesso tempo ch'imparavano le scienze et acquistarne d'essa la perfezzione: e da indi in poi di rado e quasi mai fu udito il S. Galileo con altra lingua fuori che con la natia, etiamdio nella publica cattedra, favellare;[k] il che diede materia ad alcuni suoi poco amorevoli di tacciarlo come di poco pratico nell'idioma latino: ma la ve[643]rità è che ciò facea per compiacere alla voglia degli scolari, la maggior parte oltramontani, e per metter in reputazione il parlar toscano, con adattare acconciamente i termini d'esso alle conclusioni di filosofia e mathematica, senza dimandargli in presto o mendicargli da altra lingua che non fosse la propria, contro l'oppenione dei più, quali per addietro ciò stimavano inconvenevole, anzi impossibile. Et invero chi non l'havesse udito, non haverebbe creduto tanta proprietà di parole o di vocaboli, congiunta con eccessiva chiarezza nel'esprimere i suoi concetti; di che fanno testimonianza l'opere sue, nelle quali per il modo di scrivere si rende del tutto impareggiabile.

Avvenne, non dopo molto tempo che dal Sr Galileo fu data forma all'occhiale e che in moltissime parti d'Italia e fuori si cominciò a metter in uso, apparve nel cielo un nuovo cometa; di quello parlo, che si vide nell'anno mille secento quattro, non so già se nel segno di Cassiopea o del Cigno.[l] Questi, come per lo più accader suole, commosse tutto il mondo, e diede occasione agl'astronomi più celebri di qualsivoglia nazione di far osservazioni e discorsi; ma niuno ardiva manifestargli, se prima non avvisava il Sr Galileo per intenderne il suo sentimento: onde comparivano lettere da ogni parte, d'huomini insigni nella professione, come se da lui, quasi da oracolo, dovesse uscire la decisione delle controversie che nasceano tra di loro. In questa occasione rispose a tutti con lettere, per le quali significò ciò che con la scorta del suo telescopio havea in diversi tempi osservato.[m]

Non so veramente se sopra di questo particolar cometa, o nuova stella, ci sia discorso dato alle stampe: so bene che dalle repliche fatte a queste lettere, delle quali io ne vidi e lessi moltissime di quei grand'huomini ch'allora viveano, veniva ringraziato il Sr Galileo, confessando di restare del tutto appagati, e che quando dovessero sopra di ciò scrivere per sodisfazzione degl'altri, non si sarebbero niente partiti dal di lui parere, ma che volentieri se n'astenevano per non usurparsi la lode che alla virtù sua si conveniva; et appresso lo pregavano di prestezza nel dare alla luce qualche scrittura.

---

k Non è vero che leggesse in toscano in publico, ma al più qualche lezzione straordinaria.

l Non fu nuova cometa in Cassiopea o nel Cigno nel 1604, ma la nuova stella nello Scorpione; e quando apparve questa stella, non aveva ancora trovato l'occhiale, perché lo trovò nel 1610, però non potè osservarla.

m Non è per conseguenza vero ciò che si dice in questa faccia.

---

8 incognito però Y] om. NT

I do not want to conceal, however, that for this very purpose the Most Serene Gustav, King of Sweden, came to live in Padua (albeit undercover) with his noblest retinue[14]—he who succeeded so formidably in war that the mere memory of him terrifies the universe. This great man, stopping in Padua for a few months, almost regularly attended Galileo's lectures, which he and other nobles persuaded Galileo to deliver in the Tuscan language. They were all well versed in the Latin language and to some extent in Italian, but they wanted to learn Italian perfectly at the same time they acquired science. This is why, from then on, Galileo rarely, if ever, was heard speak in a language other than his mother tongue, even when giving public lectures.[k] This offered some of his detractors the opportunity to portray him as imperfect in Latin, but the [643]truth is that he did so at the request of his pupils, most of whom came from abroad, and to revive the Tuscan language by adjusting its vocabulary to the language of philosophy and mathematics—and to do so without borrowing or begging from languages other than his own. This went against current opinion, which held it to be inappropriate, and indeed impossible. In fact, those who did not hear him would not believe that appropriate terms and vocabulary could be found and used with exceptional clarity in the explanation of concepts: his own works, with their inimitable writing style, show that it can be done.

Not long after Galileo made the spyglass and many began to use it, both in Italy and abroad, a new comet appeared in the sky. I am referring to the one that was observed in the constellation Cassiopeia, or Cygnus, in 1604.[l] As usually happens, the comet upset the entire world, causing the most eminent astronomers of the various nations to produce observations and discourses—but nobody dared to go public before first notifying Galileo and asking his opinion. Accordingly, letters came from all over, from eminent people in the profession, as if Galileo—like an oracle—would decide the controversies among them. He replied to all of them in letters telling them what he had observed through the telescope at different times.[m]

Actually, I do not know whether there is any published discourse on this specific comet, or new star. What I do know, from the replies to Galileo's letters—very many of which I saw and read, from these great men (who were then still living)—is that they thanked Galileo. They said they were completely satisfied by his reply, and that if ever they wrote anything on the subject for others, they would conform to Galileo's opinion. But they were happy not to publish anything, so as not to take for themselves the praise Galileo deserved; and they asked him to hasten to publish something himself.

k It is not true that he lectured in Tuscan in public; at most, he did so in some extraordinary lectures.

l There was no new comet in the constellation Cassiopeia, or Cygnus, in 1604; rather, there was a new star in Scorpius. And when this new star appeared, Galileo had not yet invented the telescope. He invented it in 1610 and therefore could not observe the 1604 new star with it.

m As a consequence, it is not true what is said on this page [last two paragraphs of pp. 154–155].

A questi tali rescrisse indietro il S.ʳ Galileo (per quanto mi disse in proposito di questo discorso), che l'osservazioni fatte intorno alla nuova stella o cometa era scarso argomento e picciola occasione di dar fuori scrittura particulare, ma che sperava cumulare questa con altre osservazioni più prodigiose; e volse, cred'io, [644]alludere a quelle fatte da lui medesimo intorno alla luna, alle macchie solari et alle stelle intorno a Giove, non prima vedute né conosciute.[9]

Divulgatasi dunque la fama d'un ingegno così eminente e d'un soggetto per cotanta virtù ragguardevole, quale era il S.ʳ Galileo, il quale non solamente recava onore all'Italia, ma gloria e splendore alla sua patria, si compiacque il Ser.ᵐᵒ Gran Duca Cosimo di felice memoria d'invitarlo al ritorno e di richiamarlo, con provisione eccedente quella che havea in Padova e con il rimetter alla sua libertà il leggere e non leggere nello Studio pubblico.

Per gradir un così cortese invito stimò il S.ʳ Galileo di mostrar prontezza all'obbedire, e non senza gran disgusto e contrasto ottenne licenza dai SS.ʳⁱ Veneziani, quali gl'offersero notabilissimo augumento di provisione.[10]

In tutto quel tempo che dimorò in Padova, che fu per lo spazzio di anni diciotto, non si vide mai stare in ozio il S.ʳ Galileo: poscia che, oltr'allo studio che gli conveniva fare per la cattedra et oltre alle fatiche in scrivere sopra diverse cose, assai più di quello che se ne vedino stampate, delle quali fu liberalissimo in donare ai suoi amici e scolari, fu adoperata l'industria di lui in soprantendere a molti edifizii o fortificazioni che si fecero in diversi tempi nell'augustissimo dominio e stato della Ser.ᵐᵃ Republica di Venezia; onde egli ne riportò grosse recognizioni, oltre all'annuo stipendio, al quale niun altro professore in quella cattedra era mai arrivato d'ottenere: che se fosse stato (come dicea egli) inclinato al tener conto del denaro, haverebbe potuto cumulare non poca ricchezza; ma sì come fu sempre lontano da una certa affettazione di filosofo o di letterato, così si vide in ogni tempo dedito ai passatempi d'ogni sorte, spezialmente a quegli di ritrovarsi ai conviti con amici,[11] e difficilmente si accomodò di ridursi, se non negl'ultimi anni della sua vecchiezza, a mangiar solo.

Nella conversazione era giocondissimo, nel discorso grato, nell'espressiva singolare, arguto ne' motti, nelle burle faceto. Ben spesso havea in bocca i capitoli di Francesco Berni, del quale i versi e sentenze in molti propositi adattava al suo proposito, niente meno che se fossero stati suoi proprii, con somma piacevolezza. In lui era ammirabile la facilità con la quale sapeva accomodarsi all'inclinazione degl'amici, e dopo brevissimo tempo o discorso formava concetto dell'altrui capacità.

9 *post* conosciute *scr.* e tutto ciò per l'uso del suo telescopio o cannocchiale, di nuovo da lui ritrovato *N in marg.* Qui si può fare special mentione et ordinata di tutto quello c'ha lasciato il Sig. Galileo <su> [su *supplevi*] questa materia *add. Y*

10 *post* provisione *scr.* per persuaderlo a restare, al quale effetto usarono ancora tutti quei mezzi per loro possibili *N*

11 spezialmente—amici *YT*] e particolarmente a quelli di trovarsi con gli amici in conversazione a cena e desinari *N*

To these Galileo replied (as far as he informed me about it) that the observations made of the new star, or comet, made too little an argument and were an insufficiently interesting event to justify a specific publication. He hoped, however, to complement them with other, more wonderful, observations; he wished, I believe, [644]to hint at observations he himself had made of the Moon, sunspots, and the stars circling Jupiter, which nobody had yet seen or known about.[15]

The fame of so distinguished a man, and of such an extraordinary talent as Galileo was, spread, bringing not only honor to Italy but also glory and renown to his homeland. And so the Most Serene Grand Duke Cosimo, of blessed memory, decided to call him back. He invited him to return to Tuscany, with a salary that exceeded the one he had in Padua, and left it up to him to decide whether to lecture or not at the public university.

Galileo appreciated the kind summons and wanted to obey it promptly. And not without dismay and dispute did the Venetians, who offered him a most substantial increase of salary,[16] allow him to withdraw from his duties.

Throughout the time he lived in Padua, that is, for eighteen years, Galileo was never idle. Besides the work he did for his teaching, and his labor in writing on various issues (many more than those we now see in print), which he most generously donated to friends and pupils, he was asked to oversee the construction of several buildings and fortifications, erected at different times in the most august kingdom and state of the Most Serene Venetian Republic. He was highly rewarded for this assignment, adding to his annual salary, which thus amounted to more than any other professor in that chair had ever received. Had he had a propensity (so he used to say) for saving money, he could have made a significant amount of it—but as he never assumed the pretentions of a philosopher or a literary man, he was always ready for amusements of all sorts, especially parties with friends,[17] and only with difficulty could he adjust to dining alone, except in the last years of his old age.

He was most pleasant in conversation, a good talker, unrivaled in his ability to express himself effectively, witty in quips, funny in his jokes. He often quoted from Francesco Berni, whose verses and maxims on various subjects he most happily adapted to his purpose as if they were his own. His ability to accommodate himself to the wishes of his friends was admirable, as was his ability to assess other people's abilities in a short time or after they had spoken a very few words.

Con pochi, o con niuno, fuori de' suoi intrinseci, favellava di materie filosofiche o mathematiche; anzi per liberarsi alcuna volta da certe dimande che da [645]molti con curiosità poco opportuna gli venivano fatte, divertiva il discorso et applicavalo subito ad altro, tanto graziosamente, che se bene parea lontano, lo facea cadere a proposito per la satisfazione di chi l'interrogava, con far racconto di qualche paraboletta, caso seguito o frottola, delle quali cose era abbondantissimo.

Fu il Sr Galileo di pochissima presunzione, anzi di modesto sentimento di sé medesimo. Non s'udì mai iattanza propria in disprezzo degl'altri; solamente dicea in quest'ultimi anni, quando che ogni giorno più andava deteriorando nella vista, potersi nella sua disgrazia consolare, giaché de' figliuoli d'Adamo niun altro huomo havea veduto più di lui. È lontano parimente da ogni verità che degl'antichi filosofi, e nominatamente d'Aristotele, parlasse con poca stima e disprezzo, come alcuni che professano d'esser suoi seguaci scioccamente sparlano: dicea egli solamente ch'il modo di filosofare di quel grand'huomo non l'appagava, e che in esso si trovavano fallacie et errori. Lo lodava in alcune opere particolari, come ne' libri della Hypermenia e sopra tutte l'altre quegli della Rettorica e dell'Etica, dicendo che in quest'arte havea scritto mirabilmente. Esaltava sopra le stelle Platone, per la sua eloquenza veramente d'oro e per il metodo di scrivere e comporre in dialoghi; ma sopra ogn'altro lodava Pitagora per il modo di filosofare, ma nell'ingegno Archimede dicea haver superato tutti, e chiamavalo il suo maestro.

In tutte le scienze o arti fu pratichissimo, sì come degli scrittori o professori d'esse. Dilettossi straordinariamente della musica, pittura e poesia. Fu sempre partialissimo di Lodovico Ariosto, di cui l'opere tutte sapeva a mente e da lui era chiamato divino. Il poema d'Orlando Furioso e le Satire erano le sue delizie: in ogni discorso recitava qualcheduna dell'ottave, e vestivasi in un certo modo di quei concetti per esprimere, in diversi ma spessi proposti, i proprii. Non potea tollerare che si dicesse, Torquato Tasso entrare in paragone: dicea egli sentire l'istessa differenza tra l'uno e l'altro, che al gusto o palato suo gli recava il mangiar citrivuoli, dopo ch'havessi gustato saporiti poponi; e per escludere affatto questa comparazione, si cimentò di fare alcune note o postille alla margine assai spazziosa d'un suo Furioso, in quei luoghi appunto nei quali s'era ingegnato il Tasso immitarlo. Questa sua fatica haverebbe desiderato che fosse stata letta e vista; perciò deplorava assai la disgrazia d'haverla smarrita senza speranza di ritrovarla. Fu familiarissimo d'un libro intitolato 'l *Ruzzante*, scritto in lingua rustica padovana, pigliandosi gran piacere di quei rozzi racconti con accidenti ridicoli.

Abitò quasi del continuo in alcune ville suburbane,[12] per trovar maggior quiete et occasione di speculare. Non si vedé però mai star sequestrato dal commerzio degl'altri, anzi che la casa di sua abitazione era mai sempre frequentata da nobi[646]lissime persone, la maggior parte forastieri d'ogni nazione, i quali, viaggiando per l'Italia, apposta venivano per vederlo e cognoscerlo, credendosi in un certo modo di non dover tornar alla propria patria con reputazione se havessero tralasciata l'occasione di visitarlo.[13]

12 suburbane—trovar *Y*] suburbane, e più che altrove in quella d'Arcetri, luogo detto al Piano de' Giullari, a fine di trovar *T*

13 *post* visitarlo *scr*. Il Serenissimo Grand Duca Ferdinando Secondo, oltre alle continovate missioni che gli faceva di tempo in tempo per intendere lo stato di sua salute, non sdegnò di trasferirsi in persona più volte per visitarlo, essendo sovranissima la stima che faceva questo gran Principe del suo alto ingegno, godendo sommamente de' suoi discorsi, i quali non sempre erano di filosofa ed intorno ai maravigliosi discoprimenti da lui fatti nel cielo, ma bene spesso d'altre scienze, dilettandosi talvolta di mescolarci alcune piacevolezze, il tutto però in maniera che sempre ugualmente faceva apparire, con stupore universale, il suo grand'intelletto *T*

He talked philosophy or mathematics with few people, if anyone, other than those closest to him. In fact, in order to avoid some questions that [645]many, with inappropriate curiosity, asked him, he changed the subject and turned it into something else—so graciously, though, that however different the new subject was, he made the change look like a natural way to satisfy those who were questioning him; and he did so by appealing to some metaphor, accident, or tale, of which he had very many in store.

Galileo was far from full of himself—in fact, he had no high opinion of himself. He never boasted of himself to the depreciation of others. He only used to say, in his last years, when his eyesight was failing more and more as days went by, that he had found this solace in his misfortune: of all of Adam's children, nobody had seen farther than he. Equally far from the truth is the charge that he despised or thought little of ancient philosophers, and notably Aristotle, as some who claim to have been his disciples foolishly state. He merely said he could not rest content with that great man's way of philosophizing and that in it he found fallacies and errors. Galileo praised him for some of his works, such as the books of the *Hypermenia*,[18] and above all those on Rhetoric and Ethics, saying that in this art Aristotle had written most admirably. He praised Plato above the stars, for his truly golden eloquence and for his talent to write and structure dialogues; but above all others he praised Pythagoras, for his way of philosophizing, although as far as genius is concerned, he claimed Archimedes had exceeded everybody else, and called him his master.

Galileo was an expert in all sciences and arts, just like those who taught or wrote about them. He thoroughly enjoyed music, painting, and poetry. He was always very partial to Ludovico Ariosto, whose works he knew by heart, and whom he called divine. The poem *Orlando Furioso* or the *Satires* were Galileo's delights:[19] whenever he spoke, he recited some of the verses, and clothed himself in them, so to speak, to express his own mind on different subjects. He could not abide hearing Torquato Tasso[20] put on a par with Ariosto: to him, he said, they were as different as tasting cucumbers after flavorful melons is to the palate—and in order to counter any such comparison, he made annotations or comments in the very wide margin of a copy of his of the *Furioso*, next to passages Tasso had striven to imitate. Galileo would have loved this work of his to be seen and read, and he deeply deplored losing it, with no hope of recovering it. He also knew a book titled *Ruzzante* very well. It is written in a rustic Paduan language, taking great pleasure in rough stories and hilarious incidents.[21]

He lived in some country houses almost continuously,[22] so as to enjoy more rest and opportunity to speculate.[23] However, he never kept himself from exchanges with others: in fact, the house where he lived was always visited by [646]very noble people, most of whom were foreigners from all nations, who, while traveling in Italy came purposely to see and meet him. They were convinced that had they missed the opportunity to pay a visit to Galileo, once back home they would have lost their reputation.[24]

Hebbe pochissima quantità di libri, e lo studio suo dependea dalla continua osservazione, con dedurre da tutte le cose che vedea, udiva o toccava, argomento di filosofare; e diceva egli ch'il libro nel quale si dovea studiare era quello della natura, che sta aperto per tutti.

Gustò fuor di modo dell'agricultura, asserendo che pochi erano quegli che sapeano metter in pratica i suoi precetti. Nel tempo del potare e rilegar le viti, si trattenea molte ore continue in un suo orticello, e tutte quelle pergolette ed anguillari voleva accomodare di sua mano, con tanta simetria e proporzione ch'era cosa degna d'esser veduta. E perché s'adoperava in questo esercizio in quei giorni ne' quali il sole ha molta attività nel smuovere, s'attribuisce a questo disordine, come a causa, la cecità del già vecchio S.ʳ Galileo: e fu negl'ultimi anni assai travagliosa, poscia che era congionta con dolori di tal sorte, che gl'havevano tolto affatto il sonno. Se ne lamentava egli cruccioso, ma non s'asteneva però di dire qualche arguzia, secondo che ne veniva il proposito. Ma non potendo resistere né al disagio né al peso degl'anni, gli convenne, dopo alcuni giorni di lenta febbre, lasciar la vita nell'età sua . . . ,[14] con pianto e cordoglio degl'amici e conoscenti. Huomo, se si risguarda la perspicacità dell'ingegno, l'eccellenza di quello ch'ha lasciato scritto, e le dote singulari concessele dalla natura, a niun altro degl'antichi inferiore; veramente degno d'esser annoverato tra i più famosi, e senza dubbio in questo nostro seculo, già più di mezzo transcorso, senza pari.

Fu il S. Galileo d'aspetto grave, di statura più tosto alta, membruto e ben quadrato di corpo, d'occhi vivaci, di carnagione bianca e di pelo che pendea nel rossiccio.

Questo è quanto ho potuto raccogliere della vita et azioni del S. Galileo, somministratomi da ciò ch'udii dire da lui medesimo in diverse occasioni et colloquii, lasciando ch'altri aggiunga, levi o correggha, conforme sarà giudicato più opportuno o necessario.

Niccolò Gherardini.

14 . . . Y] di 77 anni T

Galileo had very few books; his study consisted of relentless observation, deducing philosophical arguments from everything he saw, heard, or touched. He used to say that the book we have to study is the book of nature, which lies open in front of us.

He greatly enjoyed agriculture, claiming that there were few people who could put its principles into practice. He pruned and tied the grapes, and spent many hours in his little vegetable garden, without interruption.He wanted to set all the shoots and lines with his own hands, with such symmetry and proportion that it was worth seeing. And since he practiced this activity on days when the sun was particularly powerful, old Galileo's blindness is attributed to it. This blindness was so painful in his later years that it deprived him of sleep. Worried, he complained about these pains, but did not refrain from uttering some apt jokes as they came to his mind. However, he could not resist the inconvenience and burden of his years, and he left this life after a few days of a persistent fever, at the age of . . . ,[25] to the sorrow and grief of his friends and those who knew him. If we consider the perspicacity of his mind, the excellence of his writing, and the unique gifts nature granted him, he was inferior to none of the ancients, and in our century, half of which has already passed, he was undoubtedly without equal.

Galileo had a serious look, was rather tall, with strong limbs and well built; he had bright eyes, a light complexion, and hair tending to reddish.

This is what I have been able to assemble about the life and works of Galileo as I gathered it from what I heard from him personally, on several occasions and in different conversations. I leave it to others to add, leave out, or correct, according to what is regarded as appropriate or necessary.

Niccolò Gherardini

1. Gherardini belonged to a noble Florentine family. His mother, Faustina Popoleschi, was a distant relative of Pope Urban VIII: her mother, Ginevra, was the daughter of Carlo Barberini, the pope's grandfather. Gherardini studied law in Pisa and then moved to Rome, where he probably was drawn by his connections to the Barberini family. This would explain his meeting with Galileo in 1633, at the time of the trial.

2. *T* has "1564."

3. This is a reference to the allegation—spread by Gian Vittorio Rossi: see pp. 114–115—that Galileo was the illegitimate son of Vincenzo Galilei and Giulia Ammannati. See also Viviani's and Crasso's remarks, pp. 268–269 and 194–195, respectively.

4. Francesco I de' Medici (1541–1587), Grand Duke of Tuscany from 1574.

5. Ostilio Ricci (1540–1603), from Fermo, in the Southern Marches. He was not a priest. A friend of Galileo's father, he introduced Galileo to the study of mathematics and geometry, and especially to the works of Euclid and Archimedes, who were to prove very influential in Galileo's later work. Beginning with 1580, Ricci was the Court Mathematician to the Grand Duke Francesco I.

6. Giovanni I de' Medici (1567–1621), natural son of the Grand Duke of Tuscany, Cosimo I. Gherardini explicitly refers to him in the clash over the construction of a hydraulic machine in the port of Leghorn, which eventually led to Galileo's resignation from chair of mathematics at the University of Pisa. Viviani, by contrast, refrains from mentioning his name (the *Racconto istorico* is a letter addressed to Prince Leopoldo, a member of the Medici family): see *Racconto istorico*, pp. 14–15.

7. Guidobaldo del Monte (1545–1607), one of the most eminent Italian mathematicians, was one of the earliest supporters of Galileo's.

8. Here both *N* and *T* add: "and particularly that of the most pleasant Villa delle Selve, where Galileo performed most of his very learned and clever [*T*: his great] observations."

9. *N* and *T* have: "one hundred."

10. *N* has: "In the meanwhile, he did not refrain from paying visits to many senators, and especially to those who supported the university; they all were very well disposed toward him, and cherished him, as the fame of his name was already well known."

11. When Galileo moved to Padua, in 1592, he was hosted by Gian Vincenzo Pinelli (1535–1601), in via del Santo, very close to the Basilica of Saint Anthony of Padua. Pinelli had one of the largest private libraries of the time and hosted scholars from all over Europe who came to work in the library or visited the university. Soon after Pinelli's death, Galileo moved to Borgo Vignali (which would be renamed after him later on). Both these two residences are close to the Benedictine Abbey of Santa Giustina, which dates to the tenth century; the abbey is attached to the Basilica of Santa Giustina, which was built in the sixth century.

12. *N* adds: "by explaining, thanks to the sharpness of his amazing mind, all the greatest difficulties that other scholars had heretofore misunderstood, or failed to understand."

13. *N* has: "from all over Europe."

14. Whereas Viviani and Gherardini differ on several issues, as evidenced by Viviani's critical annotations, both report this episode, whose source was probably Galileo himself. See Viviani, *Racconto istorico*, footnote 191. Both *N* and *T* lack the words in parentheses, "albeit undercover."

15. *M* adds: "and all this thanks to the telescope, or spyglass, he devised in the first place." In the margin, *Y* has: "Here, specific and ordered mention may be made to whatever Galileo wrote [on] this subject."

16. *N* adds: "to persuade him to stay, to which aim they appealed to all means they could possibly resort to."

17. *N* has: "and especially conversing with friends at lunch or dinner." *M* abruptly ends here.

18. *Hypermenia* has no meaning in Greek, and no similar title is attested in the ancient lists of Aristotle's works. Most likely, Gherardini wished to refer to Aristotle's Περὶ Ἑρμηνείας (*Peri Hermeneias*, or *De Interpretatione*; in the Middle Ages, the work was usually referred to as *Peryermeneias, Peri Hermenias*). The error might be the copyist's, as it would be surprising if Gherardini made it. The fact that the author speaks of "the books of Περὶ Ἑρμηνείας," however, shows that he did not know what he was talking about, as the *De interpretatione* is very short (16a–23b Bekker) and is not divided into books.

19. Ludovico Ariosto (1474–1533) was one of the major Italian poets of all time, and one of the most celebrated and influential authors of his own time. The *Orlando fvrioso* was first published in 1516 (in 40 cantos); a second, revised edition appeared in 1521 and was followed by several reprints; the definitive edition was published in 1536 (in 46 cantos). Within ten years, it enjoyed some 36 reprints and was translated into several languages by the end of the sixteenth century. The poem tells the feats and love affairs of Orlando, a Frankish military leader under Charlemagne, against the background of the war between the Christian paladins and the Saracen army that has invaded Europe and is attempting to overthrow the Christian empire; it mixes realism and fantasy, humor and tragedy, and is staged across the entire world (plus a trip to the Moon). The *Satire* was first published in 1534, after Ariosto's death, and is modeled after

Horace's *Sermones*; it includes seven satires, written between 1517 and 1524, and dealing with moral issues: the dignity and independence of the writer, criticism of ecclesiastical corruption, the need to refrain from ambition, marriage, selfishness, and education.

20. Torquato Tasso (1544–1595) is best known for his epic poem *Gerusalemme liberata* (1581, in 20 cantos); the work had circulated since 1575, and an unauthorized version in 14 cantos was published in Venice, in 1580, under the title *Il Goffredo* (after Godfrey of Bouillon), which was likely the title Tasso himself had chosen for his work. The poem presents a highly imaginative version of the First Crusade, in which the Christian knights, led by Godfrey of Bouillon, battle the Muslims in order to take Jerusalem.

21. Angelo Beolco (1502–1542), nicknamed "Il Ruzzante," was an Italian actor and playwright known by his rustic comedies in the Pavan language of Padua (related to the Venetian language), featuring a peasant called "Ruzzante", and painting a vivid picture of Paduan country life in the sixteenth century. Galileo spent eighteen years in Padua, and knew the local language quite well; most likely, he contributed to the writing of Cecco di Ronchitti's (that is, Girolamo Spinelli, a Paduan Benedictine monk) *Dialogo in perpvosito De La Stella Nvova*, published in 1605 and written in the Pavan language, on the 1604 nova.

22. *T* adds: "and especially in that of Arcetri, in the area called Piano de' Giullari."

23. Besides his villa in Arcetri (Il Gioiello), in which he lived from 1631, Galileo spent several years in the countryside. While in Padua, between 1592 and 1611, Galileo used to return to Florence for the summer, living either in one of the summer residences of the Grand Duke (such as the Medici villas in Pratolino, or La Ferdinanda in Artimino) or with his sister Virginia. From the time he took up his position as the Chief Mathematician and Philosopher to the Grand Duke, he spent extensive periods on the outskirts of Florence: at the Villa L'Ombrellino, in Bellosguardo; at the Medici Villas La Petraia and in Marignolle; at Villa Michelangelo, in Settignano, hosted by Michelangelo Buonarroti the Younger; and especially at Filippo Salviati's Villa Le Selve, in Lastra a Signa.

24. *T* adds: "Every now and then, the Most Serene Grand Duke Ferdinando II repeatedly sent people to visit him in order to gather information about his health conditions, and the Grand Duke did not disdain to visit him personally several times. This great Prince held Galileo's mind in high esteem and greatly enjoyed conversing with him. They did not talk only about natural philosophy and Galileo's amazing new discoveries in the sky but also about other sciences. Sometimes, Galileo enjoyed adding some jokes—which, anyway, always showed the greatness of his mind, to everyone's amazement."

25. *T* has: "aged 77."

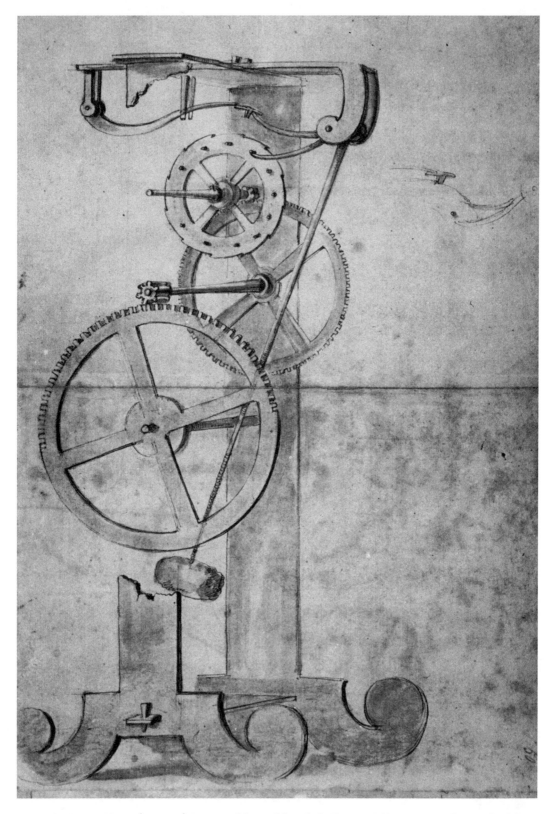

FIGURE 1. Vincenzo Viviani (1622–1703). Drawing of the pendulum clock. Florence: Biblioteca Nazionale Centrale, *Gal.* 85, f. 52*r* (by kind permission).

8

VINCENZO VIVIANI

LETTERA AL PRINCIPE LEOPOLDO DE' MEDICI INTORNO ALL'APPLICAZIONE DEL PENDOLO ALL'OROLOGIO

LETTER TO PRINCE LEOPOLDO DE' MEDICI ON THE APPLICATION OF PENDULUM TO CLOCKS

(1659)

Il Sig.^r Principe Leopoldo di Toscana, mio Signore.

Ser.^{mo} Principe,

Mi comanda l'A. V. S., sempre intenta a nobilissime e giovevoli speculazioni, ch'io debba ordinatamente mettere in carta quelle notizie che si ànno circa all'invenzione et usi del maraviglioso misurator del tempo col pendolo di Galileo Galilei d'eterna e gloriosa fama, e principalmente circa all'applicazione del medesimo pendolo alli usati orivuoli. Obbedisco, non già con quella evidente ed ornata narrativa, e quale si richiederebbe avendo a comparire avanti al purgatissimo giudizio dell'A. V., ma ben sì con quella sincerità che è mia propria, cavando il tutto da quel sommario racconto che, d'ordin pure di V. A., io scrissi, già sono 5 anni, intorno a vari accidenti ed azzioni della vita di sì grand'Vuomo, e da quanto io so aver sentito dalla di lui viva voce.

Sì come adunque è notissimo, per le tradizioni pervenuteci, che a niuno degli antichi o moderni filosofi è stato permesso dal sommo, incomprensibil Motore l'investigare pur una minima parte della natura del moto e de' suoi ammirandi accidenti, fuori che al nostro gran Galileo, il quale con la sublimità del suo ingegno seppe 'l primo sottoporlo alle strettissime leggi della divina geometria, così non si revoca in dubbio, il medesimo Galileo essere stato il primo a regolare con semplicissimo e, per così dire, naturale artifizio la misura del tempo dall'istesso moto misurato. E per ridurre il tutto distintamente a memoria, l'origine ed il progresso di questa utilissima invenzione fu tale.

Trovavasi il Galileo, in età di venti anni in circa, intorno all'anno 1583 nella città di Pisa, dove per consiglio del padre s'era applicato alli studi della filosofia e della medicina; et essendo un giorno nel Duomo di quella città, come curioso ed accortissimo che egli era, caddegli in mente d'osservare dal moto d'una lampana, che era stata allontanata dal perpendicolo, se per avventura i tempi delle andate e tornate di quella, tanto per gli archi grandi che per i mediocri e per i minimi, fossero uguali, parendogli che il tempo per la maggior lunghezza dell'arco grande potesse forse restar contraccambiato dalla maggior velocità con che per esso vedeva muovere la lampana, come per linea nelle parti superiori più declive. Sovvennegli dunque, mentre questa andava quietamente movendosi, di far di quelle andate e tornate un esamine, come suol dirsi, alla grossa per mezzo delle battute del proprio polso e con l'aiuto ancora del tempo della mu⁶⁴⁹sica, nella quale egli già con gran profitto erasi esercitato; e per allora da questi tali riscontri parvegli non aver falsamente creduto dell'ugualità di quei tempi. Ma non contento di ciò, tornato a casa pensò, per meglio accertarsene, di così fare.

Legò due palle di piombo con fili d'egualissime lunghezze, e da gli estremi di questi le fermò pendenti in modo, che potessero liberamente dondolare per l'aria (che per ciò chiamò poi tali strumenti dondoli o pendoli); e discostandole dal perpendicolo per differenti numeri di gradi, come, per esempio, l'una per 30, l'altra per 10, lasciolle poi in libertà in un istesso momento di tempo: e con l'aiuto d'un compagno osservò che quando l'una per gl'archi grandi faceva un tal numero di vibrazioni, l'altra per gl'archi piccoli ne faceva appunto altrettante.

To the Most Serene Prince
Leopoldo of Tuscany, my Lord

Most Serene Prince,

Your Most Serene Highness, who is always absorbed in the noblest and most beneficial reflections, asked me to put neatly on paper what we know about the invention and use of the amazing way of measuring time using the pendulum of Galileo Galilei—whose fame is eternal and glorious—and, specifically, the application of said pendulum to commonly used clocks. I obey, but not in that obvious and ornate style expected from works to be submitted to Your Highness's most sharp judgment. Rather, with my usual frankness, I will draw everything from the sketchy account I wrote, five years ago, at Your Highness's request, about the various events and deeds in the life of this great man, and from what I heard from his own voice.

It is very well known, from what has come down to us through tradition, that the highest, baffling Mover allowed no ancient or modern philosopher's inquiry into the least part of the nature of motion and its amazing features, but He granted this possibility to our great Galileo. Thanks to his towering mind, he was the first to subjugate motion to the strictest laws of divine geometry. Likewise, it cannot be doubted that he was the first to regulate, with a very simple and, so to speak, natural device, the measurement of time, which is measured by motion itself. In order to record everything clearly, here are the origin and development of this very useful invention.

In 1583, when he was about twenty years of age, Galileo was in Pisa, where he was studying philosophy and medicine, following his father's advice. One day, he was in the town's cathedral, and being curious and watchful, he happened to focus on the motion of a swinging lamp.[1] He wanted to see whether the times of its back-and-forth swings—either along large, medium, or small arcs—were equal. It seemed to him that the time required to travel a longer distance along an arc might be compensated by the greater velocity with which he saw the lamp moving along it, as if along a line whose upper parts were steeper. As the lamp was quietly swinging, it occurred to him to measure the back-and-forth swings—a rough measure, as it is commonly said—by referring to the beats of his own pulse and to the aid of time in music, **649**in which he was very well trained. In that period, judging from this evidence, it seemed to him that he was not wrong in believing those times were equal. Being dissatisfied with this, once he got back home he decided to make sure about his discovery, in the following way.

He tied two lead bobs with precisely equal strings, letting them hang so that they could freely swing in the air (that is why he called them swings, or pendulums). He pulled them sideways by a different number of degrees—say, for instance, one by 30 degrees, and the other by 10—and then let them swing freely, at the very same moment in time. Aided by an assistant, he observed that if one swung a certain number of times along the longer arcs, the other swung an equal number of times along the shorter ones.

In oltre formò due simili pendoli, ma tra loro di assai differenti lunghezze; ed osservò che notando del piccolo un numero di vibrazioni, come, per esempio, 300, per i suoi archi maggiori, nel medesimo tempo il grande ne faceva sempre un tal istesso numero, come è a dire 40, tanto per i suoi archi maggiori che per i picco-lissimi: e replicato questo più volte, e trovato per tutti gl'archi et in tutti i numeri sempre rispondere l'osservazioni, ne inferì ugualissima esser la durazione tra l'andate e le tornate d'un medesimo pendolo, grandissime o piccolissime che elle fossero, o non iscorgersi almeno tra loro sensibile differenza, e da attribuirsi all'impedimento dell'aria, che fa più contrasto al grave mobile più veloce che al meno.[a]

S'accorse ancora, che né le differenti gravità assolute, né le varie gravità in ispecie delle palle, facevano tra di lor manifeste alterazioni, ma tutte, purché appese a fili d'uguali lunghezze da i punti delle sospensioni a i lor centri, conservavano una assai costante ugualità de' lor passaggi per tutti gl'archi; se però non si fusse eletta materia leggierissima, come è il sughero, il di cui moto dal mezzo dell'aria (che al moto di tutti i gravi sempre contrasta, e con maggior proporzione a quello de' più leggieri) vien più facilmente impedito, e più presto ridotto alla quiete.

Assicuratosi dunque il Galileo di così mirabile effetto, sovvennegli per allora d'ap-plicarlo ad uso della medicina per la misura dell'accelerazioni de' polsi, come pur tuttavia communemente si pratica.

Indi a poch'anni applicatosi agli studi geometrici, ed agli astronomici appresso, vedde l'importante necessità ch'essi avevano d'uno scrupuloso misuratore del tempo per conseguire esattissime l'osservazioni; che perciò fin d'allora introdusse il valersi del pendolo nella misura de' tempi e moti celesti, de' diametri apparenti delle fisse e de' pianeti, nella durazione de gli eclissi ed in mill'altre simili [650] operazioni, princi-palmente ottenendo da tale strumento, più e più accorciato di filo,[b] una minutissima divisione e suddivisione del tempo, ancora oltre a i minuti secondi, a suo piacimento.

Guidato poi dalla geometria e dalla sua nuova scienza del moto, trovò le lunghezze de' pendoli esser fra loro in proporzione duplicata di quella de' tempi d'ugual nu-mero di vibrazioni. Ma perché il Galileo nel communicare le sue speculazioni, come abbondantissimo che egli ne era, ne fu insieme liberalissimo, quindi è che questi usi e le nuovamente da esso avvertite proprietà del suo pendolo a poco a poco divul-gandosi, trovaron talvolta o chi con troppa confidenza se le adottò per propri parti, o chi nella publicazione di qualche scritto, artifiziosamente tacendo il nome del lor vero padre, se ne valse in tal guisa, che almeno da quei che ne ignoran l'origine po-trebbero facilmente credersi invenzioni di essi, se a ciò non avesse abbondevolmente provveduto la sincerità de i ben affetti, tra i quali è il Sig.[r] Cristiano Ugenio olandese, che nel proemio dell'Orivuolo, da esso publicato nel 1658, fa di queste invenzioni gratissima testimonianza a favore del medesimo Galileo.

a e da attribuirsi—meno GZ] *om.* P
b più e più—filo GZ] *om.* P

Furthermore, he also constructed two similar pendulums, whose lengths, however, were very different. And he observed that if the shorter one swung, say, 300 times, along its longer arcs, in the same interval the longer pendulum swung the same number of times, say, 40, along either the longer or the shorter arcs. He repeated the experiment several times and found that the results were always in agreement, whatever the length of the arcs and the number of swings. Therefore, he concluded that the back-and-forth swings, whether very long or very short, always took the very same time, for the same pendulum; no significant differences could be noticed, besides that due to the resistance of air, which opposes the fast object more than the slow one.

Galileo also noticed that neither the different absolute weights nor the various weights of the individual bobs affected the process significantly: all of them, if hanging from strings of equal length (from the hanging point to the center of the bob), preserved a very constant swinging period along the arcs—unless very light matter, such as cork, was chosen, as air (which opposes the motion of all bodies, and the more so the lighter the bodies are) obstructs its motion more easily and the sooner causes the bob to come to rest.

Once Galileo ascertained this amazing result, it occurred to him that he could apply it to medicine, for the measurement of heartbeats, as is still the practice.

A few years later, while studying geometry and later astronomy, he saw the extent to which a careful measurement of time was crucial to obtaining very accurate observations. Accordingly, from then on Galileo began to make use of the pendulum to measure times and motions of celestial bodies, the apparent diameters of the fixed stars and the planets, the duration of eclipses, and in a thousand other similar [650]operations. He did so especially by cutting the string of the instrument shorter and shorter, so as to achieve a very accurate division and subdivision of time, well beyond seconds, as he pleased.

Guided by geometry and by his new science of motion, Galileo discovered that the length of pendulum strings is in duplicate proportion to times, given an equal amount of swings. However, since Galileo was as generous in communicating his thoughts, as he was prolific in producing them, these new uses of the pendulum and its newly discovered properties gradually circulated. Some people appropriated them as if they were their own discoveries; still others, artfully omitting the name of the actual discoverer, made use of them in some publications, so that those who did not know the source might well believe they were the discoverers—had it not been for the generous sincerity of men of goodwill, among whom was Christiaan Huygens, a Dutchman, who gave a most welcome testimonial on Galileo's behalf about these inventions, in the preface of his *Horologium*, published in 1658.[2]

Non terminò già qui l'applicazione delli usi di questa semplice macchina, poiché doppo avere il Galileo scoperto per mezzo del telescopio, nell'anno 1610, i quattro pianeti intorno al corpo di Giove, da lui denominati Medicei, subito dall'osservazioni de' variati loro accidenti di occultazioni, di apparizioni, d'eclissi e d'altre simili apparenze di brevissima durazione, caddegli in mente di potere valersene per universal benefizio de gli vuomini ad uso della nautica e della geografia, sciogliendo per ciò quel famoso e difficil problema che indarno aveva esercitato i primi astronomi e matematici de i passati e del presente secolo, che è di potere in ogni ora della notte, o almeno più frequentemente che con gl'eclissi lunari, in ogni luogo di mare e di terra, graduare le longitudini. Per ciò ottenere diedesi allora ad una assidua osservazione de' periodi e de' moti di tali Stelle Medicee; ed in meno di 15 mesi dal primo discoprimento ne conseguì tanto esatta cognizione, che arrivò a predire le future costituzioni di ciaschedun satellite, comparate fra loro et col corpo stesso di Giove, publicandone un saggio per i due mesi avvenire di Marzo et Aprile dell'anno 1613, come si vede in fine della Storia delle Macchie Solari. Ma conoscendo che in servizio della longitudine richiedevasi molto maggior perfezione per potere calcolare le tavole ed effemeridi, e che ciò non era possibil avere che doppo gran numero d'osservazioni e tra loro assai distanti di tempo, non prima che dell'anno 1615 si risolvé di proporre [651]questo suo ammirabil pensiero a qualche gran Principe d'Europa, che fosse potente in mare principalmente; e conferendo ciò col Ser.^{mo} Gran Duca Cosimo II, suo Signore, volle questi per sé medesimo muoverne allora trattato con la Maestà Cattolica di Filippo Terzo, Re di Spagna. Fra le invenzioni del Galileo concorrenti all'effettuazione di così grande impresa (oltre all'offerirsi dal medesimo di somministrare ottimi telescopi già fatti; il modo di fabbricarli, atti all'osservazione di Giove e suoi satelliti, e di poter facilmente usarli in nave, benché fluttuante; le tavole et effemeridi per la predizione delle future costituzioni di quei Pianeti), eravi ancora quella dell'orivuolo esattissimo, consistente in sustanza nelle ugualissime vibrazioni del suo pendolo. Questo trattato, da vari accidenti interrotto, fu poi in diversi tempi riassunto, ma in fine, del 1629, non so per qual fatalità, abbandonato.

This was not the end of the applications of this simple device, though. After the discovery, in 1610, with the aid of the telescope, of the four planets revolving around Jupiter, which he named after the Medici family, it immediately dawned on Galileo, from the observation of the varying features of their occultation and apparitions, of their eclipses and other extremely short-lived appearances, that he could make use of them for the universal benefit of mankind, in navigation and geography. Hence, he solved that famous and difficult problem with which the most eminent astronomers and mathematicians of the past, and of the present century, had struggled in vain, namely, that of computing longitudes at any time of the night, or at least more frequently than by appealing to lunar eclipses, in any location at sea as well as on earth. To that end, Galileo devoted himself assiduously to observing the times and motions of those Medicean Stars, and less than fifteen months after their first discovery, he achieved such a detailed knowledge of them that he succeeded in predicting the future positions of each satellite with respect to one another and to the body of Jupiter and publishing a sample for the two forthcoming months of March and April 1613, as we can see at the end of the *History of Sunspots*.[3] He knew that in order to compute longitudes he would have to be much more precise, though, so as to compute the tables and ephemerides; and this was not possible, unless a large number of observations, made at very distant moments in time, were made. It was not before 1615, therefore, that he resolved to propose [651]this amazing idea of his to some European grand prince, somebody whose power was especially exercised at sea. Galileo talked about it with the Most Serene Grand Duke Cosimo II, his Lord, who became interested in drawing up a treaty with His Catholic Majesty Philip III, King of Spain. Among Galileo's inventions that contributed to the achievement of his great undertaking (besides very good telescopes, ready-made by Galileo, and suitable for observation of Jupiter and its satellites, together with instructions for constructing and using them easily on ships, even when unsteady; and tables and ephemerides to foresee the future positions of those planets) was that most exact clock, that is, basically, the most precise swinging of the pendulum. Due to several mishaps, the treaty was suspended, and later resumed at different times, but was eventually given up in 1629. I do not know why.

Stimando per tanto il Galileo che il maggior ostacolo e la massima dell'eccezzioni che forse avesse incontrato la sua proposta, fosse stata il far credere di averla esibita per quel premio di facultadi e di onori che da tutti i re di Spagna e da altri potentati veniva promesso a chi di tale invenzione fosse stato l'autore; volendo pur far conoscere che egli già mai da stimolo così vile era mosso, ma bensì dalla sicurezza del suo trovato, e con l'unica brama d'arricchire il mondo di cognizione cotanto necessaria e profittevole all'umano commercio, e sé medesimo ornare della gloria per ciò dovutagli, stabilì finalmente di farne libera e generosa offerta a i Potentissimi Stati Generali delle Provincie Confederate: onde nel 1636, mediante l'opera incessantissima del Sig.^{r} Elia Diodati, celebre iureconsulto di Parigi e Avvocato del Parlamento, amico suo carissimo e confidentissimo, e col patrocinio del Sig.^{r} Ugon Grozio, allora ambasciador residente in Parigi per la corona di Svezia, venne all'attual proposta del suo trovato alli Sig.^{ri} Stati d'Olanda, diffusamente spiegando con più e diverse scritture e lettere colà inviate (tanto a i Sig.^{ri} Stati suddetti, quanto al Sig.^{r} Lorenzo Realio, presidente eletto da i medesimi all'esamine di questa proposizione, ed agl'altri Sig.^{ri} Commessari a ciò deputati, che furono i SS.^{ri} Martino Ortensio, Guglielmo Blaeu, Iacopo^c Golio ed Isaac Becchmanno) ogni suo particolar segreto e modo attenente all'uso della propria invenzione, sì quanto alla difficultà oppostagli del ridurre praticabile il telescopio nell'agitazione della nave, quanto circa al valersi del suo pendolo per misuratore del tempo; suggerendo al Sig.^{r} Lorenzo Realio con lettera de' 5 Giugno 1637 un pensiero sovvenutogli intorno al toglier il tedio del numerar le vibrazioni del pendolo, adombrandogli brevemente la fabbrica d'uno orivuolo o macchinetta, la quale, mossa nel passaggio dal medesimo pendolo (che servir doveva in luogo di quel che vien detto il tempo dell'orivuolo), mostrasse il numero delle vibrazioni, dell'ore e delle mi^{652}nute loro particelle decorse; come tutto può vedere A. V. S. dal seguente capitolo, qui di parola in parola trascritto dalla suddetta lettera del Galileo al Sig.^{r} Realio:

c Iacopo GZ] om. P

SECTION 8

Galileo thought that the greatest obstacle and the most important objection his proposal might encounter would be that people would think he had submitted it for the rich reward and honors promised by all the Spanish kings and other princes to whoever succeeded with the invention. He wished, however, to make known that he had not been driven by such a vulgar incentive but, rather, by confidence in his invention. Galileo merely wanted to enrich the world with necessary knowledge from which human commerce could have profited, as well as to revel in the glory he deserved for it. He eventually decided to offer it freely and generously to the Most Powerful States General of the Confederate Provinces of Holland. Accordingly, in 1636, by the relentless good offices of Élie Diodati, eminent lawyer in Paris and Attorney of the Parliament, as well as dearest and closest friend of Galileo's, and with the support of Hugo Grotius, ambassador of the Swedish Crown in Paris, Galileo submitted a proposal for his invention to the delegates of the States of Holland. He explained at length, in several papers and letters he sent them (either to the aforementioned States, and to Laurens Reael, appointed by the States General as the president of the commission charged with the examination of Galileo's proposal, as well as to the other commissioners who had been appointed, that is, Maarten van den Hove, Willem Blaeu, Jacob van Gool, and Isaac Beeckman), every specific secret and technique for the use of his invention, both in order to overcome the difficulty in using the telescope on a pitching ship, and in making use of the pendulum to measure time. In a letter dated June 5, 1637, Galileo suggested to Laurens Reael an idea he had for doing away with the chore of counting a pendulum's swings, via a clock, or small mechanism, which when struck by the passage of the pendulum (which supplied what is usually referred to as the *tempo* of the clock) showed the number of swings, hours, and [652]minutes that elapsed. Your Most Serene Highness may see all from the following section, which I copy word for word from the aforementioned letter Galileo sent to Reael:

Da questo verissimo e stabil principio traggo io la struttura del mio misuratore del tempo, servendomi non di un peso pendente da un filo, ma di un pendolo di materia solida e grave, qual sarebbe ottone o rame; il qual pendolo fo in forma d'un settore di cerchio di 12 o 15 gradi, il cui semidiametro sia 2 o 3 palmi; e quanto maggior sarà, con minor tedio se gli potrà assistere. Questo tal settore fo più grosso nel semidiametro di mezzo, andandolo assottigliando verso i lati estremi, dove fo che termini in una linea assai tagliente, per evitare quanto si possa l'impedimento dell'aria, che essa sola lo va ritardando. Questo è perforato nel centro, per il quale passa un ferretto in forma di quelli sopra i quali si volgono le stadere; il qual ferretto, terminando nella parte di sotto in un angolo, e posando sopra due sostegni di bronzo, acciò meno si consumino per lo continuo muoversi del settore, rimosso esso settore per molti gradi dallo stato perpendicolare (quando sia ben bilicato), prima che si fermi, anderà reciprocando di qua e di là numero grandissimo di vibrazioni; le quali per potere andare continuando secondo il bisogno, converrà che chi gl'assiste gli dia a tempo un impulso gagliardo, riducendolo alle vibrazioni ample: e fatta, per una volta tanto, con pazienza la numerazione delle vibrazioni che si fanno in un giorno naturale, misurato con la revoluzione d'una stella fissa, si haverà il numero delle vibrazioni d'un'ora, d'un minuto e di altra minor parte.

Potrassi ancora, fatta questa prima sperienza col pendolo di qualsivoglia lunghezza, crescerlo o diminuirlo, sì che ciascheduna vibrazione importi il tempo d'un minuto secondo; imperoché le lunghezze di tali pendoli mantengono fra di loro duplicata proporzione di quella de' tempi, come per esempio: Posto che un pendolo di lunghezza di 4 palmi faccia in un dato tempo mille vibrazioni, quando noi volessimo la lunghezza d'un altro pendolo che nell'istesso tempo facesse doppio numero di vibrazioni, bisogna che la lunghezza di [653] questo pendolo sia la quarta parte della lunghezza dell'altro; et in somma, come si può vedere colla sperienza, la moltitudine delle vibrazioni de' pendoli da lunghezze diseguali è sudduplicata di esse lunghezze.

I draw the structure of my time measurer from this truest and solid principle: I make use not of a weight hanging from a string but of a pendulum made of solid and weighty matter, such as brass or copper. I make this pendulum in the shape of a circular sector of 12 or 15 degrees, whose semidiameter is two or three spans; the larger the size, the less the trouble of tracking it. I make the sector thicker at the middle semidiameter, sharpening it at the extremities, which I make very sharp, in order to avoid the resistance of air as much as possible, because this is the only thing that slows it down. The pendulum is pierced at the center [of the circumference of which the pendulum is a sector], and through the hole I insert a metal piece, like those around which scales revolve. The lower part of this piece of metal is in the shape of an angle and lies upon two supports made of bronze so that they will not wear out, due to the continuous movement of the sector. Once the sector is pulled sideways by many degrees (if it is well balanced), it will swing back and forth very many times before its rests. In order to have it swing as needed, its attendant will have to push it smartly, making it swing widely. When the number of swings is patiently counted once, throughout a calendar day, measured by the revolution of a fixed star, we will be able to deduce the number of swings in one hour, in one minute, or in any of its smaller parts. Once this first experience, with a pendulum of any length, is made, we might lengthen or shorten it, so that each swing takes a second; the lengths of these pendulums are in duplicate proportion to the times. For instance: let us assume that a pendulum four spans long swings a thousand times in a given period of time; if we wish to find the length of another pendulum that swings twice as many times in the same period, the length of [653] this pendulum would have to be one-fourth of the length of the other. To sum up: as can be gathered from experience, the number of swings of pendulums with different lengths is in subduplicate proportion to these lengths.

Per evitar poi il tedio di chi dovesse perpetuamente assistere al numerare le vibrazioni, ci è un assai comodo provedimento, in cotal modo: cioè facendo che dal mezzo della circonferenza del settore sporga in fuori un piccolissimo e sottilissimo stiletto, il quale nel passare percuota in una setola fissa in una delle sue estremità, la qual setola posi sopra' denti d'una ruota leggierissima quanto una carta, la quale sia posta in piano orizontale vicina al pendolo, et avendo intorno intorno denti a guisa di quelli d'una sega, cioè con uno de' lati posto a squadra sopra il piano della ruota e l'altro inclinato obliquamente, presti questo offizio, che nell'urtare la setoletta nel lato perpendicolare del dente, lo muova, ma nel ritorno poi la medesima setola sopra il lato obliquo del dente non lo muova altrimente, ma lo vadia strisciando e ricadendo a piè del dente susseguente: e così nel passaggio del pendolo si muoverà la ruota per lo spazio d'uno de' suoi denti, ma nel ritorno del pendolo essa ruota non si muoverà punto; onde il suo moto ne riuscirà circolare, sempre per l'istesso verso, et havendo contrassegnati con numeri i denti, si vedrà ad arbitrio nostro la moltitudine de i denti passati, et in conseguenza il numero delle vibrazioni e delle particelle del tempo decorse. Si può ancora intorno al centro di questa prima ruota adattarne un'altra di piccolo numero di denti, la quale tocchi un'altra maggiore ruota dentata, dal moto della quale potremo apprendere il numero delle intere revoluzioni della prima ruota, compartendo la moltitudine de i denti in modo che, per esempio, quando la seconda ruota haverà dato una conversione, la prima ne abbi date 20, 30 o 40 o quante più ne piacesse. Ma il significar questo alle SS. LL., che ànno vuomini esquisitissimi et ingegniosissimi in fabbricare orivuoli et altre macchine ammirande, è cosa superflua, perché essi medesimi sopra questo fondamento nuovo, di sapere che il pendolo, muovasi per grandi o per brevi spazii, fa le sue reciprocazioni egualissime, troveranno conseguenze più sottili di quel che io possa immaginarmi. E siccome la fallacia delli orologii consiste principalmente nel non si esser potuto sin qui fabbricare quello che [654]noi chiamiamo il tempo dell'orivolo, tanto aggiustatamente che faccia le sue vibrazioni eguali; così in questo mio pendolo semplicissimo, e non soggetto ad alterazione alcuna, si contiene il modo di mantener sempre egualissime le misure del tempo. Ora intenda V. S. Illma, insieme col Sig.^r Ortensio, quale e quanto sia grande il benefizio di questo strumento nelle osservazioni astronomiche, per le quali non è necessario fare andare perpetuamente l'orivuolo, ma basta, per l'ore^d da numerarsi a meridie ovvero ab occasu, sapere le minuzie del tempo sino a qualche ecclisse, congiunzione o altro aspetto ne' moti celesti.

d l'ore PZ] le cose G

To avoid the trouble of having to constantly watch the pendulum and count its swings, there is a very convenient expedient. It is as follows. Let's make sure that a very little thin pin sticks out halfway through the circumference of the sector, and strikes a bristle fixed at one of its own ends; that the bristle dangles over the teeth of a wheel, as light as a piece of paper, which is set horizontally, next to the pendulum; and that the teeth around the wheel resemble those of a chainsaw (that is, with one of the edges perpendicular to the plane of the wheel, and the other tilted obliquely). It works like this: by striking the bristle on the side perpendicular to the tooth, it moves it; but on the way back the same bristle, on the obliquely tilted side of the tooth, does not move it, but slides along it, falling at the bottom of the next tooth; in so doing, each time the pendulum swings forth, the wheel will move by the distance of one of its teeth; when it swings back, by contrast, the wheel will not move at all. Accordingly, the motion of the wheel will be circular, always in the same direction; by marking the teeth with numbers, it will be possible to see how many teeth passed by, as we wish, and therefore count the number of swings and fractions of time elapsed. Further, we may adapt another wheel to the center of the first wheel, with a smaller number of teeth, which touches a bigger-toothed wheel, from whose movements we might draw the number of whole revolutions of the first wheel (by dividing the number of teeth, so that, for instance, when the second wheel completes a turn, the first will have completed 20, 30, or 40, or as many as we wish). I need not tell these things to Your Lordships, as you have available very competent and clever men to construct clocks and other amazing mechanisms: on the basis of this new basic principle, namely, knowing that the time of the swings of a pendulum is always the same, whether it travels larger or shorter distances, they will draw subtler consequences than I have been able to do. And since the main flaw of clocks is that we have not yet been able to set what [654]we call the tempo of the clock so precisely that all its swings are equal, this very simple pendulum of mine, not subject to any alteration, is the way to keep measurements of time always equal. Your Most Serene Highness will surely understand, together with Hortensius, what and to what extent astronomical observations might benefit from this instrument. It is not necessary to set the clock in motion forever, but suffice it to know, for the number of hours elapsed since noon or sunset, the fractions of time until some eclipse, conjunction, or other aspect of celestial motions.[4]

E conseguentemente in appresso fu da esso comunicato alli altri SS.^{ri} Commessarii ed agl'altri SS.^{ri} Olandesi^e che successivamente s'adoprarono con i SS.^{ri} Stati a favor del Galileo, fra' quali fu un tal Sig.^r Borelio, Consigliere e Pensionario della città d'Ansterdam, et un Sig.^r Constantino Ugenio di Zulichen, allora primo Consigliere e Segretario del Sig.^r Principe d'Oranges e padre del sopranominato Sig.^r Cristiano.

Vedendo il Galileo che il dover trattare questa sua proposizione per lettere, in tanta distanza di luoghi, richiedeva gran lunghezza di tempo, nel rimuovere quelle difficoltà che per altro con la presenza in pochi giorni egl'averebbe sperato di superare, e che dopo averle spianate gli conveniva tornar da capo a informar nuovi Deputati (come gli era succeduto, dopo 5 anni continui di negoziati, per la morte di tutti e quattro i SS.^{ri} Commessari destinati all'esamine della sua proposta); da che l'età sua cadente di 75 anni e la sua cecità non gli permetteva il trasferirsi in Ansterdam, come in altro stato volentierissimo averebbe fatto; desiderando pure per publico benefizio che, se non in vita sua, almeno in vita di quelli che già ne erano consapevoli, si venisse quanto prima alla sperienza del suo trovato, che egli reputava esser l'unico mezzo in natura per conseguire la cercata graduazione delle longitudini; stabilì d'inviar colà amico suo fidatissimo et intelligentissimo delle cose astronomiche, il quale s'era dimostrato assai pronto di trasferirvisi, ed al quale il medesimo Galileo aveva già, doppo ⁶⁵⁵la perdita della vista, ceduto tutte le proprie fatiche, osservazioni e calculi, attenenti a i Pianeti Medicei, e conferito la teorica per fabbricare le lor tavole et effemeridi. Questi fu il Padre Don Vincenzio Renieri, monaco Olivetano, stato insigne Matematico nello Studio di Pisa, il quale s'era con tanto gusto applicato a continuare le dette osservazioni e talmente impadronitosene, che, come è benissimo noto all'A. V., prediceva per molti mesi avvenire ogni particolar accidente intorno a i detti Pianeti; e nel 1647 fece vedere all'A. V. et al Ser.^{mo} Principe Cardinal Gio. Carlo le tavole et effemeridi formate per molti anni, quali stava in punto di publicare, quando piacque a Dio, che tutto a miglior fine dispone, indi a pochi mesi togliercelo quasi repentinamente di vita. Non so già per qual disgrazia attraversandosi il caso a così profittevole cognizione, mentre egli se ne stava moribondo, fu da taluno ignorante o pur maligno spirito, ch'ebbe l'adito nelle sue stanze, spogliato lo studio de' suoi scritti, tra' quali era la suddetta opera perfezionata e la serie ordinata^f di tutte l'osservazioni e calculi del Galileo dal 1610 al 1637, con gl'altri successivamente notati dal detto Padre Renieri fin al 1648; e così in un momento si fece perdita di ciò che nelle vigilie di 38 anni, con tante e tante fatiche, a pro del mondo s'era finalmente conseguito.

e E conseguentemente—Olandesi GZ] Queste stesse notizie ed altre molte s'avranno in breve nella pubblicazione che intende fare l'A. V. di tutte le scritture che intorno al negozio delle longitudini ultimamente ella ottenne dalla liberalità del Sig.^r Elia Diodati, il quale di tutte, come di prezioso tesoro, avea tenuto particolarissima cura, come quegli che solo poté farne raccolta, essendo che tanto le lettere del Galileo che quelle de' SS.^{ri} Stati e de' lor SS.^{ri} Commessari, che scambievolmente passarono dal 1635 fino al 1640, erano di comun consenso inviate al sudetto Sig.^r Elia per il lor recapito, avendo questi facoltà d'aprire 'l tutto e prendersene copia, per restar pienamente informato di tale affare. Da questa medesima pubblicazione, oltre all'autentica storia di questo fatto, chiaramente vedrassi come 'l concetto di cavar dal pendolo un orivuolo fu prima del nostro Galileo, e come appresso fu da esso comunicato alli sopranominati SS.^{ri} Commessari, e conseguentemente agl'altri SS.^{ri} Olandesi P

f la serie ordinata di GZ] om. P

After this, Reael immediately reported to the other commissioners and Dutch gentlemen,[5] who later agreed to support Galileo with the States. Among them were Boreel, councilor and stipendiary from Amsterdam, and Constantijn Huygens, from Zuilichem, who was then first councilor and secretary of the Prince of Orange, and father of the aforementioned Christiaan.

Galileo realized that dealing with these topics by letter, at such great distance, would require a huge amount of time: he had to overcome difficulties that, had he been there, he would have hoped to resolve in a few days. And once he had smoothed everything out, he would have to restart from the beginning with new deputies (as it happened, all four commissioners charged with examining his proposal died, after five years of continuous negotiations). His advanced age of 75 years and his blindness prevented him from traveling to Amsterdam, as he would have been happy to do in other circumstances. He also wished, for the public benefit, that his invention be tested, if not while he was still alive, at least while those who knew about it were, as he thought his invention was the only available means, in nature, to obtain the much-sought-after computation of longitudes. For all these reasons, Galileo decided to send a friend of his, whom he trusted completely, and who knew astronomy very well. He had showed himself more than willing to move to Holland, and Galileo, [655]after the loss of eyesight, had entrusted to him all his works, observations, and computations about the Medicean Planets, also providing him with the theories required to compile their tables and ephemerides. He was father Vincenzo Renieri, an Olivetan monk, formerly an eminent mathematician at the University of Pisa, who had devoted himself entirely to continuing the aforementioned observations; he had learned how to do it so well, that, as Your Highness very well knows, he could predict any specific event concerning these planets for several months in the future. In 1647, Renieri showed Your Highness and the Most Serene Prince, Cardinal Giovanni Carlo, the tables and ephemerides he had compiled for many years to come, which he was about to publish. However, God, who decides everything for the best, wanted to take Renieri from us almost unexpectedly, a few months later. I cannot say for what unfortunate reasons fate got in the way again, thus preventing such a useful and profitable advance in knowledge. While Renieri lay on his deathbed, someone who did not know what he was doing, or else was led by evil intentions, entered his rooms and stripped his studio of all his papers, among which were the aforementioned complete work and the ordered series of all of Galileo's observations and computations, from 1610 to 1637, together with all other data recorded by Father Renieri up to 1648. And so, what the world had gradually achieved, through thirty-eight years of watchful nights, with so many efforts and pains, was suddenly lost.[6]

Ma tralasciando le digressioni, intendeva il Galileo d'inviare alli SS.^ri Stati d'Olanda questo Padre Renieri, e forse ancora in sua compagnia il Sig.^r Vincenzio, proprio figliolo, giovane di grand'ingegno et all'invenzioni mecchaniche inclinatissimo, i quali insieme fossero provveduti et istrutti a pieno di tutte le cognizioni necessarie all'effettuazione di sì grand'opera. Mentre dunque il Padre Rinieri attendeva alla composizione delle tavole, si pose il Galileo a speculare intorno al suo misurator del tempo; et un giorno del 1641, quando io dimorava appresso di lui nella villa d'Arcetri, sovviemmi che gli cadde in concetto che si saria potuto adattare il pendolo a gl'orivuoli da contrapesi e da molla, con valersene in vece del solito tempo, sperando che il moto egualissimo e naturale d'esso pendolo avesse a correger tutti i difetti dell'arte in essi orivuoli. Ma perché l'essere privo di vista gli toglieva il poter far disegni e modelli a fine d'incontrare quell'artifizio che più proporzionato fosse all'effetto concepito, venendo un giorno di Firenze in Arcetri il detto Sig.^r Vincenzio suo figliolo, gli conferì il Galileo il suo pensiero, e di poi più volte vi fecero sopra vari discorsi; e finalmente stabilirono il modo che dimostra il qui aggiunto disegno, ^657 e di metterlo in tanto in opera per venire in cognizione dal fatto di quelle difficoltà che il più delle volte nelle macchine con la semplice speculativa non si sogliono prevedere. Ma perché il Sig.^r Vincenzio intendeva di fabbricar lo strumento di propria mano, acciò questo per mezzo de gl'artefici non si devulgasse prima che fosse presentato al Ser.^mo Gran Duca suo Signore et appresso alli SS.^ri Stati per uso della longitudine, andò differendo tanto l'esecuzione, che indi a pochi mesi il Galileo, autore di tutte queste ammirabili invenzioni, cadde ammalato, et agl'otto di Gennaio del 1641, *ab Incarnatione*, mancò di vita; per lo che si raffreddarono talmente i fervori nel Sig.^r Vincenzio, che non prima del mese d'Aprile del 1649 intraprese la fabbrica del presente orivuolo, sul concetto somministratoli già, me presente,^g dal Galileo suo padre.

g   me presente GZ] *om.* P

Leaving digressions aside, however, Galileo meant to send Father Renieri to the States of Holland, and possibly have his own son Vincenzo, a brilliant young man with a talent for mechanical inventions, accompany him, for both would have been instructed and fully provided with all the knowledge required for the fulfilment of so great an enterprise. While Father Renieri was compiling his tables, then, Galileo began to ponder his time measurer. One day in 1641, while I was living with him at his villa in Arcetri, I remember that it occurred to him that the pendulum could be adapted to clocks driven by weight or spring, by using it instead of the customary way of measuring time, in the hope that the very uniform and natural motion of the pendulum would correct all the defects intrinsic to clock-making. But because his being deprived of sight prevented his making drawings and models that would be required for the determination of the device that would be best adapted to the desired effect, Galileo told his idea to his son Vincenzo, who came up from Florence one day. Several discussions followed on the subject, and they eventually decided on the model shown in the accompanying drawing. [657]They agreed to proceed at once with its production, in order to discover the difficulties that, as a rule, are not easily foreseen in the theoretical design of mechanisms but appear in the constructed samples. However, since Vincenzo wanted to construct the instrument by himself, lest the makers should divulge it before it had been presented to the Most Serene Grand Duke, his Lord, and to the States General for the computation of longitude, he deferred the construction so long that Galileo, the author of these amazing inventions, fell ill a few months later, and died on January 8, 1641, *ab Incarnatione*.[7] As a consequence, Vincenzo's enthusiasm cooled so much that it was not until April 1649 that he undertook to make the present clock, in accordance with the ideas Galileo had imparted to him, in my presence.

Procurò dunque d'avere un giovane, che vive ancora, chiamato Domenico Balestri, magnano in quel tempo al Pozzo dal Ponte Vecchio, il quale aveva qualche pratica nel lavorare orivuoli grandi da muro, e da esso fecesi fabbricare il telaio di ferro, le ruote con i lor fusti e rocchetti, senza intagliare; ed il restante lavorò di propria mano, facendo nella ruota più alta, detta delle tacche, n.° 12 denti, con altrettanti pironi scompartiti in mezzo tra dente e dente, e col rocchetto nel fusto di n.° 6, et altra ruota, che muove la sopradetta, di n.° 90. Fermò poi da una parte del bracciuolo, che fa croce al telaio, la chiave o scatto, che posa su detta ruota superiore, e dall'altra impernò il pendolo, che era formato d'un filo di ferro, nel quale stava infilato una palla di piombo, che vi poteva scorrere a vite, a fine d'allungarlo o scorciarlo secondo il bisogno d'aggiustarlo con il contrapeso. Ciò fatto, volle il Sig.ʳ Vincenzio che io (come quegli che era consapevole di quest'invenzione e che l'avevo ancora stimolato ad effettuarla) vedessi così per prova e più d'una volta, come pur vedde ancora il suddetto artefice, la congiunta operazione del contrapeso e del pendolo: il quale stando fermo tratteneva il descender di quello,[h] ma sollevato in fuori e lasciato poi in libertà, nel passare oltre al perpendicolo, con la più lunga delle due code annesse all'impernatura del dondolo alzava la chiave che posa e incastra nella ruota delle tacche, la qual tirata dal contrapeso, voltandosi con le parti superiori verso il dondolo, con uno de' suoi pironi calcava per disopra l'altra codetta più corta, e le dava nel principio del suo ritorno uno impulso tale, che serviva d'una certa accompagnatura al pendolo, che lo faceva sollevare fin all'altezza donde s'era partito; il qual ricadendo naturalmente e trapassando il perpendicolo, tornava a sollevar la chiave, e subito la ruota delle tacche, in vigor del contrapeso, ripigliava il suo moto, seguendo a volgersi e spignere col pirone susseguente il detto pendolo: e così in un certo modo si andava perpetuando l'andata e tornata del pendolo, fino a che il peso poteva calare a basso.

[658]Esaminammo insieme l'operazione, intorno alla quale varie difficoltà ci sovvennero, che tutte il Sig.ʳ Vincenzio si prometteva di superare: anzi stimava di potere in diversa forma e con altre invenzioni adattare il pendolo all'orivuolo; ma da che l'aveva ridotto a quel grado, voleva pur finirlo su l'istesso concetto che n'addita il disegno, con aggiunta delle mostre per le ore e minuti ancora; perciò si pose ad intagliar l'altra ruota dentata. Ma in questa insolita fatica sopraggiunto da febbre acutissima, gli convenne lasciarla imperfetta al segno che qui si vede; e nel giorno XXII del suo male, alli 16 di Maggio del 1649, tutti gl'orivuoli più giusti, insieme con questo esattissimo misurator del tempo, per lui si guastarono e si fermarono per sempre, trapassando egli (come creder mi giova) a misurar, godendo nell'Essenza Divina, i momenti incomprensibili dell'eternità.

---

h il descender di quello Z] 'l moto del contrappeso P 'l moto del contrapeso G

Vincenzo then engaged a young man who is still alive, named Domenico Balestri. At that time he was a locksmith at the well of Ponte Vecchio, who had some experience in the construction of large wall clocks. Vincenzo caused Balestri to make the iron frame, the wheels, with their arbors and pinions, without cutting their teeth; the rest of the work he executed with his own hands. Vincenzo cut 12 teeth in the higher wheel (the notched one, so called), with an equal number of pins between the teeth; 6 teeth with the pinion in the arbor; and another wheel, which moves the aforementioned one, with 90 teeth. Then, he fixed the detent, or stop, that lies upon the higher wheel, at one end of the bracket which crosses the frame; at the other end he hinged the pendulum, made of an iron wire inserted in a lead bob. This could be moved along the wire by twisting, in order to lengthen or shorten the pendulum as needed, to balance it with a counterweight. After this, Vincenzo more than once demonstrated to me—in view of my acquaintance with this invention, and of my having urged him to construct it—and wanted me to see and confirm, as did the aforementioned maker, the combined working of the counterweight and the pendulum. When the latter was at rest, it prevented the fall of the former[8] when it was pulled sideways and let move freely as it swung across the perpendicular, with the longer of the two arms mounted on the pivot of the pendulum, it raised the detent, which acted upon and engaged with the notched wheel; pulled by the counterweight, the latter revolved with its upper parts toward the pendulum, so that one of its pins pressed the shorter arm above, giving it at the beginning of its return an impulse in such a manner as to convey a definite motion to the pendulum, causing it to rise to the height from which it had begun. When it swung back naturally and crossed the perpendicular, the pendulum again raised the detent, and the notched wheel, by the force of the counterweight, immediately resumed its motion, continuing to revolve and impel the pendulum by the next pin. And so the back-and-forth swinging of the pendulum was maintained, until the weight had run down to the bottom.

[658]We examined the functioning of the mechanism together and noticed several difficulties, all of which Vincenzo intended to overcome. In fact, he thought he could adapt the pendulum to the clock in various forms and with other inventions, but inasmuch as he had brought it to this condition, he wanted to finish it according to the design shown in the drawing, with the addition of the hands to show hours and minutes; thereupon he set himself to cut the teeth of the other toothed wheel. However, in the course of this labor, to which he was unaccustomed, he was affected by very acute fever, and had to leave it unfinished, at the point that is seen here. And on the twenty-second day of his illness, on May 16, 1649,[9] all the most accurate clocks, together with this most precise time measurer, broke down for him, and stopped forever, as he passed away (as I like to believe) to measure, enjoying the Divine Essence, the moments of eternity that surpass all understanding.

Questo, Ser.^{mo} Signore, è il progresso, o, per così dire, questa appunto è stata la vita, del misuratore del tempo, degno parto del gran Galileo. Come ha sentito, egli nacque nell'antichissimo e famoso tempio di Pisa intorno all'anno 1583, con tutto che il fondamento della sua concezzione fosse eterno, mentre eterno è l'effetto dell'egualissime vibrazioni e reciprocazioni del pendolo, benché non prima osservato che dal perspicacissimo nostro Linceo; principio invero semplicissimo, e dal quale chiaramente s'apprende la verità di quel gran detto del medesimo Galileo, che *la natura opera molto col poco*, e che *tutte le sue operazioni sono in pari grado maravigliose.* Questo parto nella sua infanzia fu di vaga scorta alla Medicina. Nutrito poi dalla robustissima Geometria, e per la vigilante educazione di quella cresciuto, s'applicò in servizio dell'altissima Astronomia, e non men atto e pronto si dimostrò all'arte Nautica ed alla Geografia. Si preparò a maggior uso intorno all'anno 1641, quando nella idea del suo genitore Galileo si vestì d'altra forma; e finalmente 8 anni doppo, quando per mano del Sig.^r Vincenzio Galilei stava per ricevere l'ultima perfezione, nell'età sua più matura, restò per allora infelicemente abbandonato.

Quanto al rimanente, non tralascerò di ricordare all'A. V. come sono intorno a 4 anni che il Ser.^{mo} G. Duca, perspicacissimo promotore sempre di cose utilissime e nuove, si dimostrò curioso di qualche modo per havere senza tedio e con sicurezza il numero delle vibrazioni del pendolo, ma però del pendolo libero e naturale, che non havesse (come nell'orivolo del Galileo) connessione o dependenza da altro estraneo motore; che allora io feci vedere a S. A., col sopra riferito capitolo di lettera del medesimo Galileo, che questi l'aveva stimato fattibile, e descrittone un modo di propria invenzione, con inviarlo in Olanda; che Filippo Treffler augustano, ingegnosissimo e perfettissimo artefice, degno in vero di tanto Principe, da questa apertura animato, fabbricò quella galante macchinetta, la quale, sottoposta all'imo punto del verticale del pendolo, per via d'una alietta di ⁶⁵⁹essa, che nell'andata, ma non già nel ritorno, della palla veniva mossa da un acutissimo stile fissato nella parte inferiore d'essa palla, dimostrava, per mezzo di leggierissime ruote, il numero preciso delle vibrazioni e delle minutie del tempo, secondo che più si aggradiva; che per conservare il

This, Most Serene Lord, is the development—or, this was indeed the life, so to speak, of the time measurer, an idea worthy of the great Galileo. As you heard, it was conceived in the most ancient and famous cathedral of Pisa, about 1583. The foundation of its design was eternal, though, as eternal is the effect of the most equal swinging back-and-forth of the pendulum, although it was our most insightful Lynx[10] who first observed it. Its principle is, indeed, very simple, and from it we learn the truth of that great saying of Galileo himself: *nature accomplishes much with little*, and *all her accomplishments are equally amazing*. At the beginning, this idea was very valuable in medicine; nourished later by the most robust geometry,[11] it grew with geometry's help, and became useful in sublime astronomy. Also, the idea was no less suitable and fitting for the art of navigation and geography. It was preparing for a better use around 1641, when, in the mind of its parent Galileo, it took another form; and eventually, eight years later, when it was about to be perfected once and for all by the hands of Vincenzo Galilei, at its maturest age, it was sadly abandoned.

As to the aftermath, I will not refrain from mentioning to Your Highness that it has been about four years since the Most Serene Grand Duke, who has always been a most perceptive supporter of new and very useful things, became curious to learn some way to count the swinging of a pendulum, without trouble and with certainty. He was interested in the free and natural pendulum, though, without any connection with or dependence on (as in the case of Galileo's clock) an external mover. I then told His Highness, by showing him the passage from the letter by Galileo I quoted above, that Galileo had thought it possible, and sent to Holland a way of doing so that he had devised himself. And that Philipp Treffler, of Augsburg, a most ingenious and excellent maker, truly worthy of so great a prince, starting from this insight, constructed that lovely little mechanism.[12] Placed just beneath the lowest point of the pendulum, it presented a light appendage [659]that could be engaged by a very thin pin attached to the bottom of the bob during the forward part of the pendulum's swing, but not on the return.

moto di questo pendolo per un medesimo verticale si proposero e messero in opera varie invenzioni; che, per comandamento pure del medesimo Serenissimo, si specularono et inventarono diverse macchine, le quali, alquanto prima che il pendolo si riducesse verso la quiete e cessasse di sollevare l'alietta del detto numeratore, riconducevano il pendolo a quell'altezza di gradi dalla quale era stato lasciato da principio, e così perpetuavasi in un certo modo il suo moto, e conseguentemente la numerazione delle sue vibrazioni; che in questo medesimo tempo fu presentato a S. A. dall'ingegner Francesco Generini un modello di ferro, nel qual però era unito al pendolo il contrapeso, in modo simile a quello che 14 anni avanti s'era immaginato il Galileo, ma sì bene con diversa e molto ingegnosa applicazione; che Filippo soprannominato adattò l'invenzione a un orivuolo da camera per S. A., il qual mostrava l'ore ed i minuti, e che di poi n'ha fabbricati per LL. AA. de gl'esattissimi, i quali dimostrano il tempo assai più minutamente diviso, e nel corso di molti giorni non variano tra di loro di un sol minuto; che, d'ordine di S. A. medesima, l'istesso Filippo, togliendo dall'una e dall'altra invenzione, ha ridotto a questa foggia l'orivuolo pubblico della Piazza del Palazzo dove abitano LL. AA.; e finalmente che i mesi a dietro[i] fu inviato di Parigi all'A. V. la già nominata scrittura, in dichiarazione del disegno d'un simile orivuolo, del sopradetto Sig.^{r} Ugenio. Ma ne i particolari de' fatti fin qui narrati non istarò a diffondermi con maggior tedio di V. A., già che tutto ha per sé stessa veduto e a tutto si è trovata presente; onde profondamente inchinandomele, bacio all'A. V. la veste.

Di Casa, li 20 Agosto 1659.   Umiliss.^{mo} e Devotiss.^{mo} Oblig.^{mo} Servo
Di V. A. Ser.^{ma} Vincenzio Viviani.

i i mesi a dietro *PZ*] circa a quattro mesi fa *G*

The motion of this appendage turned very light wheels in the mechanisms, which recorded the number of swings and fractions of time as precisely as one wished. In order to preserve the motion of the pendulum through the same vertical, it was proposed to put in operation various devices. By order of the Most Serene Highness, various mechanisms were devised and invented, which, some time before the pendulum came to rest and ceased to lift the small wing of the counter, lifted the pendulum back to the height (in degrees) from which it dropped in the first place, thereby perpetuating its motion, so to speak, and hence the counting of its swings. At this same time, engineer Francesco Generini[13] presented to His Highness a model in iron, in which, however, a counterweight was applied to the pendulum, in a manner similar to that which Galileo had devised 14 years earlier, although with a different and very clever application. The said Philipp [Treffler] adapted the invention to a chamber clock for His Highness, which indicated hours and minutes, and from then onward he had other, most accurate types constructed for Their Highnesses, which show time with even more accurate divisions and do not vary from one another by even a single minute over the course of many days. Furthermore, upon the order of His Highness, Philipp [Treffler], borrowing from one or another invention, produced in this shape the public clock of the square of the palace in which Their Highnesses live. Finally, in the past months[14] the writing referred to above was sent to Your Highness from Paris, describing a design of a similar clock, by the aforementioned Huygens. But I will not go into the details of the facts told here, to the boredom of Your Highness, because he has seen everything for himself and was always present. Therefore, with a deep bow, I kiss Your Highness's robe.

From home, 20 August 1659.
The Most Humble, Devout, and Obliged Servant
of Your Most Serene Highness

Vincenzo Viviani

1. The lamp—traditionally referred to as 'Galileo's lamp'—allegedly still sways in the Pisa cathedral. The identification is controversial, though: Viviani dates the observation to 1583, while Galileo was still a student at the University of Pisa; Magalotti agrees with Viviani on this: see *Saggi di natvrali esperienze* (1667), p. 100; Nelli, by contrast, dates the discovery to 1588: *Vita e Commercio Letterario di Galileo Galilei* (1793), vol. 2, p. 691. The lamp, however, dates to 1587. See also *Racconto istorico*, pp. 8–9. Galileo's first account of his discovery of the isochronism of pendulums, according to which the oscillation period of pendulums of equal length is constant, regardless of the amplitude of the oscillation, is in a letter to Guidobaldo del Monte, on November 29, 1611 (in *OG* 10, no. 88, pp. 97–100).

2. Huygens completed a prototype of his first pendulum clock by the end of 1656 (see Huygens to Ismaël Boulliau, 26 December 1657, in Huygens, *Œuvres complètes*, vol. 2: *Correspondance 1657–1659*, no. 443, pp. 109–110; and first described it in a letter to Jean Chapelain, 28 March 1658 (ibid., no. 477, pp. 161–162). He then hired a clockmaker of The Hague, Salomon Coster, to construct others. He patented the device on June 16, 1657: see Huygens, *Horologivm* (1658), p. 3 (possibly, this is the Latin translation of a work published in Dutch in 1657, and now lost). Here Viviani refers to p. 4 of this work: "Without doubt, accustomed to the faults in water clocks and automata of all sorts used for observations, at last, from the new teaching of that most wise man, Galileo Galilei, the astronomers initiated this method: they impelled manually a weight suspended by a light chain, and by counting the individual vibrations they derived a matching number of time units, all equal to one another" ("Nimirum fallentibus clepsydris automatisque quibuslibet, quæ inter observandum adhibere consueverant, tandem, docente primum Viro sagacissimo Galileo Galilei, hunc modum inierunt, ut è catenula tenui pondus appensum manu impellerent, cujus vibrationibus singulis dinumeratis, totidem colligerentur æqualia temporis momenta"). See also Huygens, *Horologivm oscilla- torivm* (1673), p. 13: "To be sure, if those who wish to refer the origins of this matter to Galileo say that he tried but failed to complete this invention, it seems that they detract praise from him rather than from me, since indeed in that case I would have carried out this same investigation more successfully than he did. On the other hand, if they maintain that this invention was actually made by Galileo, or by his son, as one scholar has recently claimed, and that these clocks were actually produced, then I do not know how they could hope to be believed, since it is hardly likely that such a useful discovery would remain unnoticed for eight full years prior to my publishing it" ("Qui vero Galileo primas hic deferre conantur, si tentasse eum, non vero perfecisse inventum dicant, illius magis quam meæ laudi detrahere videntur, quippe qui rem eandem, meliore quam ille eventu, investigaverim. Cum autem vel ab ipso Galileo, vel à filio ejus, quod nuper voluit vir quidam eruditus, ad exitum productum fuisse contendunt, horologiaque ejusmodi re ipsâ exhibita, nescio quomodo sibi creditum iri sperent, cum vix verisimile sit adeo utile inventum ignoratum manere potuisse annis totis octo, donec à me in lucem ederetur"). The scholar (*vir quidam eruditus*) Huygens is referring to is Lorenzo Magalotti (1637–1712), secretary of the Accademia del Cimento from 1660 to 1667. In his *Saggi di natvrali esperienze*, including reports of experiments performed by members of the Accademia between 1662 and 1667, Magalotti wrote on p. 100: "[. . .] they decided to apply the motion of the pendulum to the clock, an idea first devised by Galileo and realized by his son Vincenzo Galilei in 1649. According [to the model they devised], the pendulum is forced by a spring or by its own weight to fall always from the same height. Whereby, with mutual advantage, not only do the times of the oscillations perfectly match one another, but also the flaws of the other devices in the clock somewhat correct one another" ("[. . .] fu stimato bene applicare il Pendolo all'oriuolo su l'andar di quello che prima d'ogni altro immaginò il Galileo, e che dell'anno 1649 messe in pratica Vincenzio Galilei suo figliuolo. Così è necessitato il Pendolo dalla forza della molla o del peso a cader sempre dalla medesima altezza; onde con iscambievol benefizio non solamente vengono a perfettamente uguagliarsi i tempi delle vibrazioni, ma eziandio a correggersi in certo modo i difetti degli altri 'ngegni di esso oriuolo"); the summary printed in the margin is even clearer: "Galileo was the first to apply the pendulum to the clock; Vincenzo Galilei was the first to put his idea into practice" ("Galileo il primo, che pensasse di adattare il Pendolo all'oriuolo. Vincenzio Galilei il primo che mettesse in pratica tal pensiero"). Strictly speaking, Magalotti's passage is not inaccurate, but it is phrased so that Galileo and his son Vincenzo are ascribed theories that, in fact, were devised by Huygens: see Huygens, *Horologivm* (1658), pp. 11–15. Indeed, in a letter to Leopoldo de' Medici on 22 May 1673, Huygens rightly remarks: "I do know that there are some who refuse to ascribe to me the invention of the pendulum clock; among them, the author of the Experiments of the Florence Academy [that is, the *Saggi di natvrali esperienze*], who attributes the discovery to Galileo and to his son, and disguises my endeavors, so that, quite patently, it seems I am charged with plagiarism" ("Scio enim non deesse, qui ipsum quod trado Horologij oscillatorij inventum nobis adscribi nolint, inter quos Exper- imentorum Academiae Florentiae scriptor ita ad Galileum filiumque ejus illud refert, nostrosque conatus dissimulat ut non obscure plagij crimen mihi objecisse videatur": *Œuvres complètes*, vol. 7: *Correspondance 1670–1675*, no. 1940, pp. 279–280; see also Leopoldo's reply, ibid., no. 1941, pp. 281–286, and the the editor's

annotations, especially nos 3–4). See also the 1659 exchange between Ismaël Boulliau and Leopoldo de' Medici, in Huygens, *Œuvres complètes*, vol. 3: *Correspondance 1660–1661*, pp. 459–480; as well as Nicolaas Heinsius the Elder to Carlo Dati, April 1660, ibid., no. 749ª, p. 498.

3. See *OG* 5, pp. 241–245, and the postscript, pp. 247–249 (English translation of the latter in *On Sunspots*, pp. 297–299).

4. Galileo to Laurens Reael, [5] June 1637, in *OG* 17, no. 3496, pp. 101–103 (lines 179–252).

5. Instead of this sentence, *P* has: "This very information and many more will be made available soon in the publication that Your Highness is going to issue, gathering all documents related to the negotiation on longitudes, as You recently obtained them through the kindness of Elia Diodati. He very carefully preserved them all, as a precious treasure, and he was the only one who could gather them all, since both the letters from Galileo and those from the States [of Holland] and the commissioners, which they exchanged between 1635 and 1640, were all delivered—by everyone's agreement—to Diodati. He was granted permission to open the letters and make a copy for himself, so that he might keep up to date on that matter. From this very publication not only will the true story of the events be clear but also that the idea of drawing a clock from a pendulum was first devised by Galileo; he then told it to the commissioners, and then to the other Dutch gentlemen."

6. In fact, they were found in the nineteenth century by the editor of the Florentine edition of Galileo's collected works, Eugenio Albèri, and included in *Opere 1842–1856*, vol. 5. They are also available in the appendix to vol. 3, Part II, of the National Edition of Galileo's works: see *OG* 3.2, pp. 931–966 (Galileo's annotations) and 967–1054 (Renieri's computations).

7. That is, 1642, according to our calendar.

8. Instead, both *G* and *P* have "the motion of the counterweight."

9. Vincenzo Galilei (1606–1649).

10. Lynxes were said to be the animals with the sharpest eyesight.

11. That is, solid geometry.

12. Johann Philipp Treffler (1625–1698), from Augsburg, moved to Florence in the early 1650s and remained there as the Grand Duke's clockmaker until 1664.

13. A pupil of renowned sculptor Pietro Tacca's, Francesco Generini (1593–1663) worked on mechanics and hydraulics for the Grand Duke of Tuscany. He was also interested in the construction of optical instruments, and about 1630 he attached the telescope to a goniometer, also providing it with a finderscope.

14. *G* has: "about four months ago."

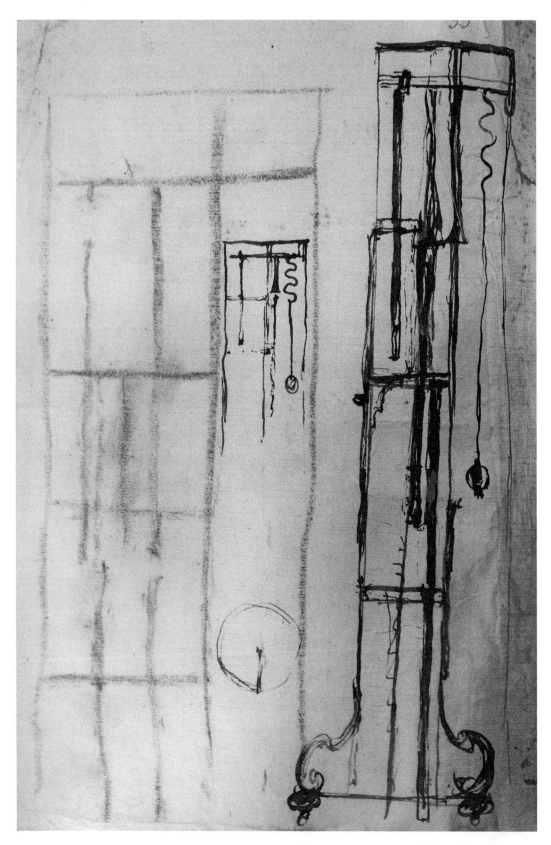

FIGURES 1A–E. Vincenzo Viviani (1622–1703). Sketches of the pendulum clock. Florence: Biblioteca Nazionale Centrale, *Gal.* 85, ff. 53*r*, 54*v*, 55*r*, 56*v*, 57*r* (by kind permission).

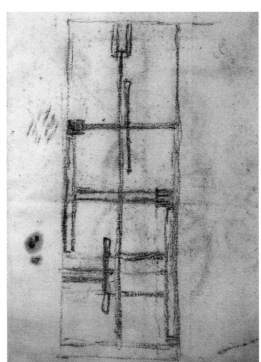

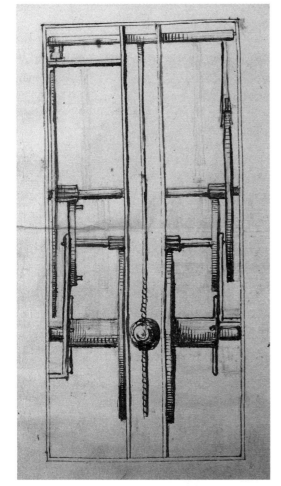

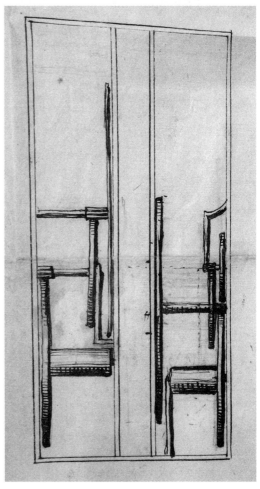

FIGURE 2. Anonymous. Portrait of Galileo; etching. Inscription: "GALILEO GALILEI." In Lorenzo Crasso, *Elogii d'hvomini letterati*, Venice: Sebastiano Combi & Giovanni La Noù, 1666, [Vol. I], p. 243. San Marino, CA: The Huntington Library (by kind permission). See *OGA* 1, D31, pp. 88–89.

9

LORENZO CRASSO

GALILEO GALILEI

(1666)

Chi sarà colui, che il glorioso Nome di Galileo Galilei non onori, quando egli dopo di hauere stanche l'ale della Fama per tutto l'ambito della Terra, portò le sue Glorie fin soura i Cieli, indagando Aquila de gli'Ingegni con voli inusitati, e peregrini delle Stelle, della Luna, e del Sole gli aspetti, i lumi, le densità, le macchie, osseruando le Pianete giouiali, la grandezza delle sfere, e le nuoue Stelle, che Medicee dal Cognome de' Principi della sua Patria le chiamò? Nacque Galileo in Firenze Figliuolo naturale, come scriue Giano Niccio, di Vincenzo [244]Galilei Nobile Fiorentino nelle amene lettere versato. La sua adolescenza non curò paterno sprone per incaminarsi alle Virtù, che tutto intento allo studio parea, che nel sapere precorrer gli anni volesse. Il Padre godendo dell'attiuità, e inclinazione del Giouane, l'accalorò à gli studi. Hauendo inteso la Filosofia, e la Matematica da più Maestri, e non sodisfacendosi appieno d'alcuni assiomi, e proposizioni, che forse la longhezza, e corruttela de' tempi l'haueuano stabilite per irrefragabili, cominciò à dubitare d'alcune cose delle già dette amendue professioni, fermandosi maggiormente nella sua opinione, da molte cognizioni, che trasse dalla lettura di Celio Calcagnino, e di Francesco Patrizio, Letterati famosi del passato Secolo, e fuor de gli ordinarij Ingegni cupidi indagatori di nouità. Ma temendo Galileo d'esser per temerario ripreso, se intempestiuamente vscisse in campo con nuoue dottrine, diè freno al desiderio procurando di cumulare più validi argomenti alle sue opinioni; e in tanto, per dar saggio al Mondo del saper suo, cominciò à far correre qualche fatica per osseruar, come da' Letterati venisse riceuuta: e vedendo, che già il suo Nome si propalaua per le bocche de gl'Intendenti, con franchigia di cuore seguì la traccia de' suoi pensieri, mandando alla luce parte de suoi scritti. Con tali maniere innalzatosi, venne ricercato dalle prime Cattedre dell'Italia. Passò à Venezia chiamato da quella Republica à legger la Matematica nella Città di Padoua, dove conferitosi con grosso stipendio la spiegò per lo spazio d'anni dieci otto con vasto numero d'Vditori, venendo molti Oltramontani solamente per vdire il Galileo, e vedere le nouelle sperienze da lui insegnate. Solito da Padoua trasferirsi in Venezia, auuenne vn giorno, che vdì ne' fogli de' rapporti, che à Maurizio Principe d'Orange era stato donato da vn Fiamengo vn'occhiale, con cui mal grado della distanza, all'occhio facea vicini gli Oggetti; onde acceso del [245]l'inopinata curiosità, non trouando quiete nell'animo, non penetrando l'ordine di sì nobile strumento, senza dar giammai riposo alla mente, tanto fè, che la vegnente notte con acume d'ingegno ne formò vn'altro, il quale hauendo presentato nel giorno al Senato, n'hebbe oltre gli applausi, l'accrescimento dello stipendio, e la lettura perpetua, benche da Giouan Battista della Porta prima d'amendue se ne fosse fatto vn'abbozzo.

Who will not honor the glorious name of Galileo Galilei, who, after exhausting fame throughout the globe, saw his glory rise beyond the skies? Eagle among brilliant minds, he inquired with uncommon and extraordinary flights the appearances, lights, densities, and spots of the stars, the Moon, and the Sun; he observed the Medicean Planets, the size of the spheres, the new stars, which he named Medicean, after the family name of the princes of his homeland.[1] Galileo was born in Florence, the natural son—as Giano Nicio writes[2]—of Vincenzo [244]Galilei, Florentine nobleman, well versed in the humanities. In his youth, Galileo did not need his father's spur to pursue the virtues: he seemed to devote himself completely to studies, as if he wished to anticipate years in knowledge. His father, glad to see the youngster's work and aptitude, encouraged him to study. Galileo learned philosophy and mathematics from several teachers and did not feel completely satisfied with some axioms and propositions which the corroding passage of time had established as irrefutable. He began to doubt some of the things stated in both said disciplines, ever more convincing himself of the validity of his own opinion, on the basis of a number of notions he drew from reading Celio Calcagnini and Franciscus Patricius (two famous learned men of the previous century, as well as extraordinary inquirers of novelties).[3] He was afraid of being reprimanded as a reckless spirit, though, had he untimely advanced new doctrines. Accordingly, he bridled his craving and turned to gather stronger arguments in support of his opinions. Meanwhile, in order to have the world assay his skills, he began to circulate a few works so as to assess how learned men received them. And seeing that his name spread from mouth to mouth among professional readers, no longer afraid, he followed the course of his thoughts and published some of his writings. Having become famous in this manner, Galileo was sought after by the foremost [university] chairs in Italy. He moved to the Republic of Venice, which appointed him to teach mathematics in the city of Padua. Here, provided with a large salary, he taught mathematics for eighteen years, to large audiences. Many came from all over Europe only to listen to Galileo and see the new experiments he taught. While in Padua he used to visit Venice. There, one day, he happened to hear reports that Maurice, Prince of Orange, had been given a spyglass as a gift from a Flemish man; with it, objects, however distant, appeared close to the eye. Hence, burning [245]with unexpected curiosity, his spirit was restless. Unable to grasp the structure of such a noble instrument, he could not put his mind to rest and worked so hard that the following night, with great perspicacity, he made another one. On the next day, Galileo presented it to the Senate, thereby receiving, besides approval, an increase in the salary and life-long tenure—although Giovan Battista Della Porta had sketched that instrument before either of them.[4]

Era la Casa del Galileo vn'Officina d'inuenzioni, vna continua Accademia di Virtuosi insigni, ne con titolo d'Huomo grande chiamaua colui, che non sapeua fuor dell'vso comune trouar nouità nelle Scienze. Ma da tanta specolatiua, e sottigliezza d'ingegno indotto ad asserire in materia della Terra, e del Cielo della fermezza, e del moto contra lo stabilimento della Romana Chiesa, passò non pochi trauagli, che vennero sopiti da' riceuuti consigli, e dall'aiuto de' Grandi. Chiamato dal Gran Duca alla Lettura della Matematica nello studio di Pisa, incontrò il gusto del suo Principe, quale sperimentò liberalissimo nel premiare la sua Virtù. Ne fù minor premio de gli altri il vedersi riuerito da' Letterati, e da' Principi, e che molti Forastieri venissero da lontane parti per ammirarlo. Hauendo compiuto l'anno settantesimo ottauo dell'Età sua in una Villa del tenimento di Firenze, l'implacabile Morte atterrò quella Vita, che nacque per trionfar della Morte nel 1642.

[246]GALILEO GALILEI FLORENTINO,
Philosopho, & Geometræ verè Lynceo,
Naturæ Oedipo,
Mirabilium semper inuentorum Machinatori;
Qui inconcessa adhuc mortalibus Gloria
Cœlorum Prouincias auxit,
Et Vniuerso dedit incrementum.
Non enim vitreos sphærarum orbes, fragilesque stellas conflauit,
Sed æterna Mundi corpora Mediceæ beneficentiæ dedicauit,
Cuius inextincta Gloriæ cupiditas,
Vt oculos Nationum, Seculorumque omnium videre doceret,
Proprios impendit oculos;
Cum iam nil amplius haberet Natura quod ipse videret.
Cuius inuenta
Vix intra rerum limites comprehensa
Firmamentum ipsum non solum continet,
Sed etiam recipit.
Qui relictis tot scientiarum monimentis,
Plura secum tulit, quàm reliquit.
Graui enim, sed nondum effœta[a] senectute
Nouis contemplationibus maiorem Gloriam affectans
Inexplebilem Sapientia Animam, immaturo nobis obitu,
Exhalauit.
ANNO M. DC. XLII. ÆTATIS SUÆ LXXVIII.

a effœta *scripsi*] effœcta *E*

Galileo's house was a workshop for inventions, a permanent academy of eminent virtuosos—a place where men were not deemed of great value unless they had been able to rise above common practice and find some novelties in the sciences. However, induced by his thoughtful and sharp mind to make statements about the rest and motion of the earth and the heavens against what the Roman Church had established, Galileo went through many troubles, which were assuaged by the advice and help he received from powerful people. Appointed by the Grand Duke to the professorship of mathematics at the University of Pisa, Galileo enjoyed the favor of this prince, who proved extremely generous in rewarding his talents. Nor was it a lesser reward to see his name revered by learned men and princes, and by many foreigners who traveled from faraway lands to pay their respects. In 1642, at the age of seventy-eight years, in a villa on the outskirts of Florence, merciless Death crushed that life, which was born to triumph over it.

[246]TO GALILEO GALILEI, FLORENTINE,
Philosopher and Geometer, truly Lyncean,[5]
nature's Oedipus.[6]
To the creator of ever new wonderful inventions,
Who, blessed with a glory never before granted to any mortals,
increased the heavenly provinces,
and added to the size of the universe.
He did not assemble the crystalline orbs of the spheres, nor the frail stars,
but dedicated eternal bodies of the universe to the Medicean generosity.
Due to his boundless desire for glory,
he sacrificed his own eyes
so that the eyes of all nations and generations might learn to see,
because Nature did not have anything else he could see.
His discoveries
are hardly confined within the limits of the universe;
the firmament itself not only contains them,
but it welcomes them.
And although he bequeathed us so many scientific monuments,
he took with him more than he left.
For, due to the burdensome yet always productive old age,
wishing to achieve more glory with new researches,
in 1642, aged 78 years,
he returned his soul, always eager for knowledge,
in what was for us a premature death.[7]

Infra tutti i Viuenti,
    Ergansi in aria a volo,
    Guizzin per l'onde algenti,
    O stampin l'orme in sul terreno suolo:
    L'Huom sol discorre; e solo
    Sà, vale, intende, e vede
    Sì, che a gli Angeli appena altero ei cede.

Emulo di Natura,
    Ciò ch'ella mai produce
    Fingerlo anch'ei procura
    E lo fà sì, che merauiglia adduce;
    L'Intelletto hà per Duce,
    Ed alto a la sua destra
    Somministra valor l'arte maestra.

S'a fera, od a se forma
    Imago somigliante.
    Trouare ei sa la norma
    Onde gli occhi, e le man muoua, e le piante.
    Fè per l'aria volante
    Vna Colomba Archita
    Di legno, e pur senso hauer parue, e vita.

E quel saggio, che grande
    Dal suo valor s'appella,
    Diè con arti ammirande
    La voce a finto labro, e la fauella.
    Forma destra nouella
    Simulato augelletto,
    E fa, che tragga alta armonia dal petto.

Di puro vetro e terso
    Altri a compor s'accinse
    Un picciolo Vniuerso;
    E poiche di più Cieli intorno il cinse,
    Le stelle entro vi finse;
    E lor diè moto, e giro
    Pari a quel, ch'i sublimi Astri sortiro:

TO GALILEO GALILEI FOR HIS TELESCOPE

By Francesco Stelluti[8]

Lyncean Academician

Among all living beings–
    either those that wing their way up through air
    or dart among chilly waters,
    or else tread the earth and leave their footprints–
    only Man converses; and he alone
    knows, has authority, discerns, and sees,
    so that he is just one step below the angels.

He emulates Nature,
    and whatever she produces
    he cares to reproduce,
    and does it so well that he causes wonder.
    Man has a mind to serve as a guide,
    and the master art gives high value
    to his right hand.

If he would portray some animal,
    or himself, in a true image,
    he knows how to do it so
    that it may move its eyes, its hand, or its feet.
    Archytas made a wooden dove
    fly through air,
    and still it seemed to have senses, and life.[9]

And that sage, whom everybody calls
    "the Great" for his virtue,[10]
    gave by wondrous art
    voice to lifeless lips, and speech.
    And a dextrous narrative creates
    an artificial little bird,
    and makes it sing harmoniously.

Another man set up
    to make a small model of the universe
    with pure and clear glass;
    and after he encircled it with many heavens,
    the stars he laid in it;
    and set them in motion, and made them whirl,
    with the same rotation the sublime planets were blessed with.[11]

La sù gli aperti, ed ampi
    Spatij del Ciel, sonori
    Strali di foco, e lampi
    [248]Scorron talor con tema alta de' cori.
    Forma volanti ardori
    L'Huom anco, e con rimbombo
    Sà fulminar da cauo ferro il piombo.

Ai confini d'Alcide
    Sicuro altri le spalle
    Riuolge, e senza guide
    Sù cauo legno per l'ignoto calle
    De la lubrica valle
    De l'Ocean profondo
    Vassene, e aggiunge un nuouo Mondo al Mondo.

E più di questi audace
    Oltre l'human costume
    Con la sua man sagace
    Ali Dedal si fà di lieui piume;
    E cotanto presume,
    Che a volo s'erge, e quale
    Veloce augel per l'aria vola, e sale.

E tù s'io ben riguardo
    Vigoroso, ed altero
    Ti festi in guisa il guardo,
    Che trapaßa il mirar d'human pensiero,
    Onde Talpa il Ceruiero
    Appo te, Galileo,
    Fora, & Argo senz'occhi, orbo Linceo.

Ne sol dela tua fronte[b]
    I Fortunati rai
    Quelle virtù sì conte,
    Han, ch'a lor tù co' tuoi cristalli dai:
    Ma quel Bel lume, c'hai
    Dentro la mente accolto,
    Quell'anco vince ogni veder di molto.

b fronte *scripsi*] fonte E

Up there, to our great awe,
    loud thunders and lightnings
    cross at times the open and wide
    [248]heavenly spaces.
    Man, too, creates flying fires,
    and resoundingly strikes
    with bullets from guns.

Another man self-confidently turns
    his back to the Pillars of Hercules,
    and guideless sails
    on a ship along the unknown path
    of the wavering valley
    of the deep ocean,
    and adds a new world to the world.[12]

Bolder than these,
    going beyond what men do,
    with his clever hand
    Daedalus creates wings out of light feathers,
    and he dares so much,
    that he soars, and like a swift bird
    flies in air, and raises upward.[13]

And you, if I am not mistaken,
    likewise turned your eyesight
    vigorous and proud,
    so that it surpasses what human thought may see.
    Beside you, Galileo,
    lynxes would be moles,
    Argus would be eyeless,[14] and Lynceus blind.[15]

And not only do the happy rays of sight
    from your forehead
    have the well-known power
    you give them by way of your spyglass;
    but the bright light
    you have in your mind—that, too,
    by far surpasses anybody else's visual ability.

Onde ciò, ch'altrui cela
    Natura entro del seno,
    Aperto si riuela
    A l'vno, e l'altro tuo sguardo sereno.
    Altri si crede appieno
    Col Saggio di Stagira
    Mirarlo ancor, ma vn'ombra sol ne mira.

Quello c'hor tu n'insegni,
    Non da le carte antiche,
    Non da i moderni ingegni
    L'hauesti nò, non da le Stelle amiche;
    Le tue lunghe fatiche,
    Le proue tue, gli studi
    [249]Fur, che tante destaro in te Virtudi.

Qualunque i sensi adopra,
    Se da te non l'apprende,
    Come l'odor si scopra,
    E man tocca, occhio mira, orecchio intende,
    Aperto ei nol comprende,
    E non ben sà la lingua
    Altrui ridir com'i sapor distingua.

Ne sà ben come il gelo,
    Come il Caldo altri senta,
    Come produca il Cielo
    Ciò che più di stupor sù n'appresenta;
    Tu v'hai sì l'alma intenta,
    Ch'o vicino, o remoto
    Oggetto alcun non miri à te mal noto.

Quei, che cercò là presso
    A la Calcidia riua
    Perche l'onda sì speßo
    Colà d'Euripo a variar veniua,
    Se la cagion n'vdiua
    Da te, cui non s'asconde,
    Sommerso non si fora entro quell'onde.

Accordingly, what Nature, within herself,
　　to anybody else conceals,
　　she openly discloses to either one
　　or the other of your serene gazes.
　　Others still believe, too,
　　they know it all, following the sage Stagirite,[16]
　　but merely get a pale sight of it.

What you teach now
　　you did not draw from ancient works,
　　nor got from modern minds,
　　nor received from good luck.
　　Your long labors,
　　your efforts, your studies:
　　249these are what aroused your powers.

Anybody who appeals to the senses
　　would not clearly understand
　　how smell is sensed,
　　hands touch, eyes see, and ears hear,
　　unless he learns all this from you,
　　nor would he be able to explain
　　how somebody else's tongue discerns different flavors.

Nor does he know how others
　　feel cold, or heat;
　　how heavens might produce
　　the wondrous things we see in the sky.
　　You so much devoted your soul to it
　　that no object you might observe,
　　either near or far, is unknown to you.

He who, by the Chalcidice shore,[17]
　　tried to understand
　　why the tides of the Euripus[18]
　　vary so often
　　would not drown in those waves
　　had he heard the reason from you,
　　from whom it is not hidden.

E quei di te pria nati
  Dotti Hipparchi, ed Atlanti
  S'intenti rimirati
  Teco hauesser quei seggi alti stellanti,
  Non detto haurian quei tanti
  Lumi, ch'in Cielo han loco,
  Passar di mille il numero di poco.

Né dato hauriano il dorso
  Adeguato, e polito,
  Ne dal ver lunge il corso,
  Ne il numero di sette stabilito,
  Ne concesso quel sito,
  Che non hanno, o confine
  A quei, ch'erran la sù con aureo crine.

Altri erranti aggiungesti
  A quegli tù sù, doue
  Per quei vani celesti
  Va[c] con rai si benigni errando Gioue.
  E vedi in forme nuoue
  Chi sù men pronto suole
  Mouersi in giro, e chi precorre il Sole.

E di quei rai sù fissi
  Tanti gir ne mirasti
  250Per quegli immensi Abissi,
  Ch'occhio non v'è che a numerargli basti.
  In quei confin sì vasti
  Tanti il Ciel ne contiene,
  Che pon del mar quasi adeguar l'arene.

Cedanti pure il vanto
  Quei noui Tifi arditi,
  Che glorioso han tanto,
  Perche scoprir mari nouelli, e liti:
  Poiche tù non additi
  Terre qua giù nouelle,
  Ma nel sublime Ciel lucenti Stelle.

c Va *scripsi*] Van *E*

And those who lived before you,
    the learned Hipparchuses,[19] and Atlases,[20]
    had they carefully observed with you
    those high starry seats,
    they would not have said
    that those lights in the sky
    number a little bit more than a thousand.

Nor would they have regarded their surface
    as even and polished,
    nor stated their path to be far from the true one,
    nor set their number to seven,
    nor would they have granted those bodies
    that travel up there with golden hair,
    a place, or boundary, they do not have.

More planets you added
    to those up there,
    in those celestial spaces,
    where Jupiter is erring with benign rays.
    And in new forms you see
    the slowest planet
    and the one that outstrips the Sun.

And of those fixed lights
    you observed so many whirl,
    250through those immense abysses,
    that to number them the eye hardly suffices.
    The sky contains such a great number of them,
    in its wide boundaries,
    that they nearly equate the ocean's sand in number.

And those daring modern Tiphyses,[21] too,
    so much celebrated for their discovery
    of new seas and lands,
    must yield their glory to you,
    for you do not simply show
    some new lands here,
    but bright new stars in the sublime skies.

Nouelli solo a Noi
   Quei discopriro Imperi,
   Non già noui a gli Eoi,
   Che là per gli ondeggianti lor sentieri
   Giunti v'eran primieri;
   Ma scopri tù più scaltro
   Orbi a ciascun nouelli, e pria d'ogni altro.

Molto a te l'Huom per tali
   Trouati obbietti deue;
   Ei co' tuoi vetri frali
   Sen và fin preßo al Ciel spedito, e lieue.
   Molto il Ciel, che riceue
   Da te beltà più chiare,
   Più nel sen luci, e in maggior forma appare.

E s'à spiar la via
   Non giuan gli occhi tui
   Del' alto Ciel, qual pria
   Ei fora ancor: tù sei, che i globi sui
   Celati prima a nui
   Orni con auree chiome,
   E lor dai moto, e loco, e vanto, e nome.

Onde se da la vista
   De le tue Luci accorte
   Tante il Ciel pompe acquista,
   Ei non permetterà, ch'vnqua t'apporte
   Il fosco Oblio la morte;
   Ma fin che gira intorno,
   Splenderai tù d'illustre gloria adorno.

They discovered empires
   new only to us,
   and not to eastern people,
   who had arrived there first,
   following the wavering paths.
   You, far more sage, discover orbs
   unknown to anybody, and before anybody else.

For these newly discovered objects,
   men are much in debt to you:
   they peruse the sky quickly and easily,
   thanks to your frail glasses.
   And the sky greatly owes to you, too,
   for it shines with brighter beauties,
   contains more stars, and appears to be larger.

Had your eyes not soared up high,
   spying out wide heaven's ways,
   it would appear
   just as it did before: its globes,
   previously hidden from us,
   you dress with golden crowns,
   and give them motion, place, glory, and name.

Therefore, if, from the view
   of your attentive eyes,
   heaven gains much new glory,
   it will not allow death
   to bring you its grim oblivion.
   As long as heavens whirl,
   adorned with bright glory you shall shine.

À GALILEO GALILEI
DEL CAVALIER GIO: BATTISTA MARINO.

Osò già d'Argo intrepido Nocchiero,
  Romper il mar con baldanzoso abete,
  E con l'oro appagò l'auara sete,
  Ch'il trasse l'onde à violar primiero.

Varcò poscia il Ligustico Guerriero
  Del forte Alcide le prescritte mete,
  E scouerse per vie strane, e secrete
  Nouo Ciel, noua terra, e nouo impero.

Ma Tù maggior del primo, e del secondo,
  I campi innaccessibili, e remoti
  Gisti à spiar delo stellato mondo.

Et internato in que' recessi ignoti,
  Trouar sapesti entro il suo sen profondo,
  Noui orbi, noui lumi, e noui moti.

OPERE STAMPATE.

Nuncius Siderius.
L'Vso del Compasso geometrico, e militare.
Difesa contra Baldassar Capra.
Discorso intorno le cose sù l'acque.
Dimostrazione delle macchie Solari.
Il Saggiatore, Ponderazione sopra Lotario Sarsi.
Dialoghi ne' Sistemi di Tolomeo, e di Copernico.

[251]TO GALILEO GALILEI
By Cavalier Giovan Battista Marino[22]

The brave pilot of Argo[23]
>dared to cut through the sea with a bold ship,
>and with gold he quenched his covetous thirst,
>which led him first to break the waves.

Later, the Ligurian warrior[24]
>crossed the pillars strong Hercules had set,
>and discovered, by unusual and unknown ways,
>a new sky, a new land, and a new empire.

Greater than the former and the latter,
>you went to sneak on the inaccessible
>and remote fields of the starry world.

And once you entered into those unknown recesses,
>in that inward bosom you were able to find
>new orbs, new lights, and new motions.

PUBLISHED WORKS

Sidereal Messenger
Use of the Geometric and Military Compass
Defense against Baldassarre Capra
Discourse on Bodies that Stay atop Water
Demonstrations Concerning Sunspots
The Assayer, Reflection on Lotario Sarsi
Dialogues on the Ptolemaic and Copernican Systems

1. Quite surprisingly, Crasso seems to believe that "the Medicean Planets" differ from "the new stars, which he named Medicean." In fact, the *Medicea sidera* ("Medicean Stars") were Jupiter's moons, first observed in early January 160, and announced in the *Sidereal Messenger*, p. 38 (*OG* 3.1, p. 60).

2. See Giovan Vittorio Rossi (Grecized as Janus Nicius Erythraeus), Erythraeus, pp. 114–115.

3. Celio Calcagnini (1479–1541) was an Italian humanist and scientist from Ferrara. One of the most celebrated scholars of the Renaissance, he is mentioned twice in Ariosto's *Orlando furioso* (XLII 90, 5–8; and XLVI 14, 8). In his *Qvod caelvm stet, terra moveatvr. Vel de perenni motv terræ* (which seems to have been written between 1518 and 1524, but was published only in the posthumous edition of Calcagnini's collected works: *Opera aliqvot* (1544), pp. 388–395), he proposed to explain the apparent daily motion of the stars by attributing to the earth a rotation from west to east, in one sidereal day. He declared that the earth, originally in equilibrium in the center of the universe, received a first impulse that imparted to it a rotary motion; and, as nothing opposed it, it was indefinitely preserved by virtue of the principle set forth by Buridan and accepted by Albert of Saxony and Nicholas of Cusa. According to Calcagnini, the daily motion of the earth was accompanied by two oscillations, one explaining the precession of the equinoxes, and one setting the waters of the sea in motion, thereby determining the ebb and flow of tides. See Carlo Conto to Galileo, 7 July 1612, in *OG* 11, no. 723, p. 354; and Paolo Antonio Foscarini to Galileo, 1615–1616, in *OG* 12, no. 1159*, p. 216. Francesco Patrizi (Franciscus Patricius, 1529–1597) was a philosopher and scientist from the Republic of Venice, known as a defender of Platonism and an opponent of Aristotelianism. He was appointed the first ever professor of Platonic philosophy at the University of Ferrara, in 1577, and later held the same chair in Rome (1592). His *Nova de vniversis philosophia* (1591) was one of the most impressive and influential philosophical work produced by Italian Renaissance nature philosophers (also including Bernardino Telesio, Giordano Bruno, and Tommaso Campanella). In Patrizi's cosmology, the universe is geocentric, but the earth is at the center of an infinite expanse of light-filled space beyond the material realm; furthermore, Patrizi accepts the earth's diurnal rotation. See Castelli, *Risposta alle opposizioni Del S. Lodovico delle Colombe, e del S. Vincenzio di Grazia* (1615), p. 674; and Federico Cesi to Galileo, 12 January 1615, in *OG* 12, no. 1071, p. 130.

4. Giovan Battista Della Porta (1535–1615) was a Neapolitan polymath and playwright. Efforts to understand how a combination of two lenses, or of a lens and a mirror, could make distant objects seem large dated to the second half of the sixteenth century. The practical effect was rather well known, and there were numerous allusions to it in print. An explicit description of the property of convex and concave lenses, and of their magnification power, can be found in Book XVII of the second edition of Della Porta's *Magiae natvralis* (1589). It took no special talent or inventiveness to come up with the idea that combining two different lenses (for example, at the ends of a tube) would allow faraway objects to be seen: the real problem, however, was that this combination of lenses failed to provide sharp images, at least until Hans Lippershey (1570–1619), who succeeded in transforming a widely known idea into an instrument that could perform well. In a sense, the telescope (or a potential telescope) existed before Lippershey, but all it produced were magnified but blurred images. This is why Della Porta, in the summer of 1609 and for the rest of his life, would call Galileo's spyglass a hoax, and claim the invention of the instrument for himself, alleging that all the pertinent optical details are to be found in earlier works of his: "I saw the secret of the spyglass, and it is a hoax: it is drawn from Book IX of my *De refractione*" ("Del secreto dell'occhiale l'ho visto, et è una coglionarla, et è presa dal mio libro 9 *De refractione*": Della Porta to Federico Cesi, 28 August 1609, in *OG* 10, no. 230, p. 252; a very rough sketch of the spyglass is also provided); see also Della Porta to Cesi, 1610, ibid., no. 450**, p. 508. In another letter, written in 1613, Della Porta refers both to Book XVII, chapter 10, of the second edition of his *Magiae natvralis* (1589), and to Book VIII of the *De refractione* (1593): however, neither book VIII or IX of the *De refractione*, nor Book XVII of the *Magiae natvralis*, presents anything that can be interpreted as a reference to a telescope (see *OG* 11, no. 962*, pp. 611–612). See also Kepler, *Dissertatio Cum Nvncio Sidereo* (1610), pp. 291–294, 296.

5. The lynx was supposed to be the animal with the sharpest eyesight, and for this reason it was chosen as the symbol of the Lyncean Academy, founded by Prince Federico Cesi in Rome, in 1603. Galileo became a member of the Academy in 1611. He is described as "truly Lyncean" because, through the telescope, he saw more and farther than anybody else had previously seen.

6. Just like Oedipus, who solved the Sphinx's riddle, defeating it and winning the throne of the dead king of Thebes (see Sophocles, *Oedipus Tyrannus*, 390–398), so Galileo solved the riddles of nature, succeeding where many others before him had failed. Indeed, Oedipus's words may well have been spoken by Galileo: "it was I that came [to solve the riddle], Oedipus who knew nothing, and put a stop to her; I achieved the result by his own judgment, not by what I learned from birds" ("ἀλλ᾽ ἐγὼ μολών, / ὁ μηδὲν εἰδὼς Οἰδίπους, ἔπαυσά νιν, / γνώμῃ κυρήσας οὐδ᾽ ἀπ᾽ οἰωνῶν μαθών": lines 396–398).

7. This is what was inscribed, and can still be read, on the east wall of the vestry of the Chapel of the Novitiate, in the Santa Croce Basilica, in Florence, where Galileo was buried at the time of his death. His

body was to remain there until March 12, 1737, when it would be moved to its present location, in the left nave of the basilica, part of a new funeral monument opposite Michelangelo's.

8. Francesco Stelluti (1577–1653) was an Italian polymath, cofounder (with Federico Cesi, Anastasio de Filiis, and Johannes van Heeck) of the Academy of the Lynxes, in 1603. In 1625, Stelluti made the first microscopic observations to be published, probably with an instrument that Galileo had sent to Federico Cesi (see Galileo to Cesi, 23 September 1624, in *OG* 13, no. 1665, pp. 208–209). In 1630, Stelluti published his translation, with commentary, of the satires of Persius (dedicated to Cardinal Francesco Barberini). As the coat of arms of the Barberini family included three bees, a reference to Arezzo, in which the Barberini family supposedly originated, was pretext enough to insert a section, titled "Descrizzione dell'ape" ("Description of the bee"), illustrated by woodcuts based on Cesi's *Apiarium*), and the coat of arms of the Barberini family. Persius's allusion to a grain weevil is illustrated by a microscopic representation (magnified 10 times), the tip of the snout with its mandibles (magnified 20 times), and a view of the whole (life-size): see Stelluti, *Persio tradotto in verso sciolto e dichiarato*, pp. 51–54. Stelluti also published another poem in praise of Galileo: see *On Sunspots*, pp. 379–380 (*OG* 5, p. 92).

9. See Aulus Gellius, *Noctes Atticae*, X 12, 9–10: "[. . .] that which the Pythagorean Archytas is reported to have envisioned and made ought to seem neither less remarkable nor equally groundless, for many celebrated Greeks and in particular the philosopher Favorinus, a very erudite man concerning ancient traditions, have written most positively that a model of a dove, which was made by Archytas out of wood and specially constructed in accordance with the discipline of mechanics, flew. Evidently it was poised just so by counterweights and was set in motion by a puff of air concealed inside. In the case of somethingso hard to believe, I certainly want to quote the words of Favorinus himself: 'Archytas of Tarentum, who was an expert in the field of mechanics, made a wooden dove which flew; whenever it alighted, it rose no more'" ("[. . .] id, quod Archytam Pythagoricum commentum esse atque fecisse traditur, neque minus admirabile neque tamen vanum aeque videri debet. Nam et plerique nobilium Graecorum et Favorinus philosophus, memoriarum veterum exsequentissimus, affirmatissime scripserunt simulacrum columbae e ligno ab Archyta ratione quadam disciplinaque mechanica factum volasse; ita erat scilicet libramentis suspensum et aura spiritus inclusa atque occulta concitum. Libet hercle super re tam abhorrenti a fide ipsius Favorini verba ponere: Ἀρχύτας Ταραντῖνος τὰ ἄλλα καὶ μηχανικὸς ὢν ἐποίησεν περιστερὰν ξυλίνην πετομένην· ὁπότε καθίσειεν, οὐκέτι ἀνίστατο": ed. Huffman, *Miscellaneous Testimonia*, no. A10a, pp. 570–571).

10. See, for example, Cicero, *Cato Maior de senectute*, XII, 39.

11. Aristotle (384–322 BC) proposed that the heavens are composed of concentric, crystalline spheres in which celestial objects are nestled and which rotate at different uniform velocities. Aristotle's universe is finite, spherical, and geocentric: outside there can be no body, nor any location or vacuum; within it, there can be no vacuum. Celestial spheres are made of ether and carry the heavenly bodies: they are set in motion by divine intelligences, and the primary cause of their motion is a prime mover, set beyond the outermost sphere of the fixed stars. The apparent motion of the fixed stars is represented by the rotation of one sphere about its diameter, while those of the Sun, Moon, and the five known planets (Mercury, Venus, Mars, Jupiter, and Saturn) are each represented by a different nest of concentric spheres. In between the spheres carrying the heavenly bodies Aristotle intercalated reagent spheres designed to insulate the movement of each celestial body from the complex of motions propelling the body next to it. See Aristotle, *Metaphysica*, XII 8, 1073 a 14–1074 b 14 (the crystalline spheres); ibid., XII 7, 1072 a 19–1073 a 13 (the unmoved mover); *Physica*, VIII 9, 265 a 13–266 a 9 (circular motion is primary and eternal); and *De caelo*, I 2–3, 268 b 11–270 b 31 (properties of bodies moving with circular motion). Within a century, astronomers turned from this theory to one involving epicycles, superseding Aristotle's physical structure of concentric nonoverlapping spheres. His basic model of a geocentric universe, however, retained its authority and can be seen again in the introductory chapters of Ptolemy's *Almagest*. In the second century AD, the Alexandrian astronomer, mathematician, and geographer Claudius Ptolemy (ca. AD 100–170) developed a system that standardized geocentrism: in the Ptolemaic system, planets are moved by a system of two (or more) spheres, one being the deferent and the other the epicycle. The deferent is a circle whose center point (the eccentric) is removed from the earth; the epicycle is a circle whose center lies on the deferent. A given planet moves around the epicycle while, at the same time, the center of the epicycle moves along the deferent. Ptolemy's system could account for retrograde motions, which was problematic for the Aristotelian model.

12. Christopher Columbus (ca. 1451–1506), who discovered the North American continent on October 12, 1492.

13. According to the myth, Daedalus fled to Crete after he was sentenced for the murder of his sister's son. In Crete, his fame as sculptor and architect earned him the friendship of Minos, who asked him to design the labyrinth at Cnossus, in which the Minotaur (a monstrous creature with the head of a bull and the body of a man) was kept. Once the labyrinth was completed, Daedalus was imprisoned to prevent the knowledge of the labyrinth from spreading to the public. Because Minos had seized all ships on the coast of Crete, Daedalus constructed wings for himself and his son Icarus: he tied feathers together and secured them at their midpoints with string and at their bases with wax, and gave the whole a gentle curvature.

When they both were ready to take flight, Daedalus warned Icarus not to fly too high, because the heat of the Sun would melt the wax, nor too low, because the sea foam would soak the feathers. While flying over the sea Icarus forgot his father's instructions and began to soar upward, toward the Sun. The heat softened the wax that held the feathers together, and they came off. Icarus quickly fell into the sea and drowned, whereas Daedalus flew safely over the Aegean and reached Sicily. See Ovid, *Metamorphoses*, VIII 183–235.

14. In Greek mythology, Argus Panoptes ("all-seeing") was a hundred-eyed giant of Argolis, in the Peloponnese. When Zeus was consorting with the Argive nymph Io, his jealous wife Hera suddenly appeared on the scene; Zeus quickly transformed Io into a white heifer, but the goddess was not deceived. She demanded the animal as a gift and set his servant Argus to watch her. Zeus sent Hermes to rescue his lover: Hermes lulled the giant with his music and slew him with his sword. See Apollodorus, *Bibliotheca*, II 1, 2–3; and Ovid, *Metamorphoses*, I 588–721; and Valerius Flaccus, *Argonautica*, IV 366–422.

15. Lynceus was one of the Argonauts and participated in the hunt for the Calydonian Boar; he was said to have excellent sight, even able to see through trees, walls, and underground. See Apollonius Rhodius, *Argonautica*, I 151–155.

16. That is, Aristotle, who was born in the city of Stagira, in central Macedonia, near the eastern coast of the Chalcidice peninsula.

17. Chalcidice (Chalkidiki, or Halkidiki) is a peninsula in northern Greece, part of the region of central Macedonia.

18. The city of Chalcis, the chief town and principal port of the island of Euboea, on the Euripus Strait (in the Middle Ages, Chalcis itself was known by the name Euripus). The strait is a narrow channel separating the island from Boeotia, in mainland Greece; it is subject to strong tidal currents that reverse direction approximately four times a day. Whereas tidal flows are usually quite weak in the eastern Mediterranean, the Euripus Strait is a remarkable exception, and lesser vessels are often incapable of sailing upstream; near flow reversal, sailing is even more precarious because of vortex formation. See Diodorus Siculus, *Bibliotheca Historica*, XIII 47, 3–5.

19. Hipparchus of Nicea (ca. 190–ca. 120 BC) was a Greek astronomer and geographer, most famous for his discovery of the precession of the equinoxes. According to Pliny (*Naturalis Historia*, II 95), Hipparchus noticed a new star and, because it moved, began to wonder whether other fixed stars move; he therefore decided to number and name the fixed stars for posterity, inventing instruments to mark the positions and sizes of each. The original catalog has not survived, but apparent excerpts found in Late Greek and Latin sources give the total number of stars in each constellation. These suggest that Hipparchus obtained the positions of at least 850 stars; it is disputed which coordinate system (or systems) he used.

20. The son of the Titan Iapetus and the Oceanid Asia, or Clymene, Atlas was a leader of the Titans in their war against Zeus; after their defeat, he was condemned to carry the heavens upon his shoulders. According to others he was instead (or later) appointed guardian of the pillars that held the earth and sky asunder. Atlas was also the god who instructed mankind in the art of astronomy, a tool used by sailors in navigation and farmers in measuring the seasons. These roles were often combined, and Atlas became the god who turns the heavens on their axis, causing the stars to revolve. See Hesiod, *Theogonia*, 746–750; Ovid, *Metamorphoses*, VI 174–175; and Diodorus Siculus, *Bibliotheca Historica*, IV 27, 4–5.

21. Tiphys was the helmsman of the ship *Argo*. He was known for his skill and knowledge of winds, currents, and the course of the stars: see Apollonius Rhodius, *Argonautica*, I 105–110; and Valerius Flaccus, *Argonautica*, I 418–419, 481–483 and 689–692.

22. Giovan Battista Marino (1569–1625), Italian poet best known for his epic *L'Adone* (1623). Just like Astolfo, in Ariosto's *Orlando Fvrioso* (1516), in Canto X of Marino's poem (titled "Le Maraviglie," "The Marvels") Adonis visits the Moon, where, after a discussion of the possible causes for the spots we observe on the lunar surface, he praises Galileo and his telescope (X, 42–47).

23. See footnote 21.

24. Christopher Columbus (1451–1506), who was born in Genoa, in the region of Liguria.

10

VINCENZO VIVIANI

RAGGVAGLIO DELL'VLTIME OPERE DEL GALILEO

REPORT ON GALILEO'S LATER WORKS

(1674)

[86]Rimarrebbe ora da sapersi, quali, e fino a che segno delle cose promesse in questa Nota restassero scritte alla morte del Galileo, e quel che poi ne sia stato. Per informazione di che, e di altro ancora, che grato riuscirà, procederò nella narratiua con l'ordine di que' paticolari che mi son noti.

Ma prima sappiasi che dopo tal giorno de' 23. Gennaio 1638. sopravvisse il Galileo intorno a 4. anni, dentro al qual tempo patì [87]d'una continua flussione di occhi molestissima; cadde più volte gravemente ammalato, e fu spesso travagliato da dolori artritici; oltre ad altre indisposizioni solite accompagnar la decrepità: ond'è ch'ei non poté applicar di proposito a dettar, e distendere questo residuo delle sue speculazioni; massimamenteché, dovendosi egli servire degli occhi altrui, non quegli di ciascheduno eran atti a supplire alla di lui impotenza, ma si richiedevano quei di Persona, la quale, non solamente gli fosse amorevole, ma in istato libero a segno di poter conviver con lui dov'ei dimorava, ed ancora (quanto ogn'altra cosa) erudita, e ben instrutta nelle Matematiche, e nelle Filosofiche Discipline, affinché, appena ch'egli avesse spiegato il concetto suo, l'Amico poi nel distenderlo fosse abile a dargli forma convenevole, e perfezione.

NON ostante però la mancanza di tal Soggetto, e l'assidue afflizioni d'animo, e di corpo, quella in ispecie del trovarsi privo della vista, continuando egli, quasi per un altr'anno, a valersi della penna del suddetto R. P. Ambrogetti impiegata nel tradurre in latino l'Opere sue, dettò a questo la Relazione di quell'ultimo suo scoprimento celeste della titubazione della faccia Lunare, indirizzandola al già Signor Conte Alfonso Antonini di Udine Commissario Generale della Cavalleria della Serenissima Signoria di Venezia, con ispiegarla in una lettera de 20. Febbraio 1638. l'original della quale alcuni anni sono mi fu consegnata dal nostro sapientissimo Socrate il Sig. Priore Orazio Rucellai, d'immortal gloria degno, in nome dell'Eminenza Reverendissima del Sig. Cardinale Delfino, il quale (consapevole della mia applicazione in andar facendo raccolta di ciò che vada attorno del mio Maestro) per incomparabil sua Vmanità, e per sommamente onorarmi si degnò d'impiegar in ciò i suoi favori coll'impetrarmela dal Sig. Conte Danielle Antonini, degnissimo nipote del sopraddetto Sig. Alfonso, insieme con altra lettera del Sig. Paolo Aproino de 27. d'Ottobre 1612. di Treviso al già Sig. Conte Danielle, dove si parla del Galileo nel far menzione d'uno strumento da multiplicar l'udito immaginato, e fabbricato dal medesimo Sig. Aproino; le quai lettere originali io conservo appresso di me in memoria di tanto onore.

SBRIGATOSI il detto R. Ambrogetti dalle traduzioni di tre dell'Opere del Galileo, cioè del Saggiatore, delle Macchie Solari, e delle Galleggianti (le quali anco per proprio esercizio aveva con somma chiarezza tradotte in latino il Sig. Senator Filippo Pandolfini Amico intrinseco del Galileo, e nelle Matematiche versatissimo) se ne tornò in Firenze intorno alla fine dell'anno 1638.

[86][...] It remains now to make known which of the works promised in this note had, in fact, been written at the time of Galileo's death, to what extent they were completed, and what happened to them. In order to provide useful information about all this, and much more, I will now go on with my report, following the order of the details known me.

Before that, however, let it be known that after January 23, 1638, Galileo lived for about four years, during which he suffered [87]from a persistent and most annoying discharge from the eyes; he repeatedly fell gravely ill, and was often afflicted with arthritic pains, besides other ailments that usually accompany old age. As a consequence, he could not set himself to dictate and write up these last speculations of his, especially since he had to rely on someone else's eyes. Not just anyone's eyes could compensate for his deficiency, though: he needed the eyes of someone who not only cared for him but was available to live with him in his house. Furthermore, and most important, he needed someone learned and well trained in mathematics and philosophical disciplines, so that whenever Galileo explained some idea to him, he would be able to write it down in the most appropriate and perfect way.

Galileo suffered from the lack of such a person, as well as from his persistent afflictions, both psychological and physical, especially being left without eyesight. All this notwithstanding, he continued for almost another year to avail himself of the pen of that Reverend Father Ambrogetti whom he appealed to, to translate his works into Latin.[1] Galileo dictated to him the report on his latest celestial discovery, on the libration of the face of the moon, addressing it to the late Count Alfonso Antonini, from Udine, Commissioner General of the Cavalry of the Most Serene Republic of Venice, with a letter dated February 20, 1638.[2] A few years ago, I had the original version of this letter from our most learned Socrates, Prior Orazio Rucellai, worthy of immortal glory, on behalf of the Most Reverend Eminence, Cardinal Delfino.[3] The latter knew I had undertaken the task of gathering all materials related to my Master, and by his incomparable generosity did me the very great honor of using his influence to have this letter given me by Count Daniele Antonini, most virtuous nephew of the aforementioned Alfonso, together with another letter from Paolo Aproino, from Treviso, dated October 27, 1612, to the late Count Daniele. In it, Galileo is referred to in connection with an instrument to enhance hearing, devised and constructed by Aproino himself.[4] I keep the original letters with me, in memory of so great an honour.

Once he had completed the translations of three works by Galileo, namely, the *Assayer*, *On Sunspots*, and *On Floating Bodies* (which, as an exercise, Senator Filippo Pandolfini,[5] a close friend of Galileo's, very learned in mathematics, had also translated into Latin), Reverend Ambrogetti went back to Florence about the end of 1638.

[88]ET essendoché pochi mesi prima, in età mia di circa anni 16. io fossi assiduamente esortato, e quasi dissi infestato dal mio Maestro di Logica (il P. Lettor Sebastiano da Pietrasanta, gravissimo Teologo, e Confessore al presente di quest'ALTEZZA REVERENDISSIMA) à studiar anche la Geometria, asserendomi che da questa una continua, e perfettissima Logica si praticava, mi lasciai in fine persuadere a pigliarne qualche lezione dal P. Clemente di S. Carlo, sacerdote delle Scuole Pie per dottrina, e per bontà amabilissimo, che in quel tempo era qui solo a insegnarla, ed era stato Discepolo del P. Francesco di S. Giuseppe della stessa Religione, il quale attualmente instruiva allora nelle Matematiche la medesima ALTEZZA, e ne fu poi Lettor pubblico a Pisa, e Autore di quell'ingegnoso Trattato della Direzzione de' Fiumi, che si vede fuori sotto nome di D. Famiano Michelini.

GVSTATA appena ch'io ebbi l'evidenza delle prove Geometriche, ben mi accorsi quanto vere fossero le Massime di que' due miei Maestri (a quali io conservo tuttavia gratissima obbligazione) del primo cioè, che nella sola Geometria sia riposto ogni vero scibile, per mezzi dimostrativi, dall'Vmano intelletto: e dell'altro, che qualunque mediocre ingegno può molto felicemente intender l'Opere, e le proprietà dimostrate da' Geometri senza aiuto d'alcun Maestro, come che questi non possa a gli Scolari giovar in altro, che in mostrar loro a principio la regola del leggerle, e l'ordine, e 'l modo dello studiarle. Et in vero, fondandosi le Dimostrazioni Matematiche sopra alcuni pochi principi, la scienza de quali nasce con noi medesimi, e camminando quelle con discorso ordinato d'una Logica rigorosa, e per mezzo di necessarie conclusioni dependenti l'una dall'altra, è forzato ancora il Maestro, s'e' non vuol confonder l'intelletto dello Studente, di spiegarle in quel modo appunto, in che le spiega l'Autore stesso, essendoché ogni poco di più sia superfluo, e difettivo ogni meno. Altro dunque non fa il Maestro che risparmiare a Discepoli l'affaticarsi gli occhi nella lettura, e la mente, e la testa nel dover applicar interrottamente, ora al discorso, & ora a' caratteri, & a' segni delle figure, dal qual risparmio però non così spesso adiviene, ch'altri ne ritragga profitto vero, conciosiecosaché l'unico mezzo di ben apprendere, e di possedere le Dimostrazioni geometriche sia quello del proprio studio, e non dell'altrui; per esser, al creder mio, fra queste due maniere d'erudirsi molto maggior divario di quel ch'e' sia fra l'andar da se stesso con particolar curiosità, ed attenzione vedendo, e osservando 'l Mondo, e lo starsene semplicemente alle carte di Geografi, quantunque esatti, & a relazione di scrittori più che fedeli.

[...]

[88]A few months earlier, when I was about 16 years old, my teacher of logic (professor Father Sebastiano da Pietrasanta, a very important theologian, currently confessor of this Most Reverend Highness[6]) strongly encouraged me—almost annoying me, I might say—to study geometry. He claimed that by doing geometry I would continuously practice logic, at the highest level, and so I was persuaded to take some lessons from Father Clemente di San Carlo, a priest of the Pious Schools.[7] A most lovely man, both for his doctrine and for his kindness, he was the only one who taught geometry in Florence, at that time; he had been a pupil of Father Francesco di San Giuseppe's and belonged to the same religious order. Father Francesco was then teaching mathematics to the aforementioned Highness and later became public lecturer of mathematics in Pisa. He wrote the ingenious *Trattato della direzione de' fiumi* (Treatise on the management of rivers), published under the name Famiano Michelini.[8]

As soon as I savored the clarity of geometric proofs, I realized how true were the maxims of those two teachers of mine (to whom I still feel in very great debt). The former said: in geometry lies all that may be demonstrably known by the human mind; the latter said: any common mind can happily understand the works and the properties proved by geometricians, without the help of any teacher, as the teacher cannot help his pupils otherwise than by showing them, from the very beginning, the rule for reading them, and the order and method of studying them. In fact, mathematical proofs are grounded on just a few principles, the knowledge of which we have from birth, and proceed by following a structured argument, governed by rigorous logic, by way of necessary deductions, each dependent on the preceding one. Hence to avoid confusing the mind of his pupils the teacher cannot help but explain them in the way in which the author himself explains them. Whatever he adds to this, however little it is, is superfluous; whatever he omits, cannot be done without. The teacher does nothing but save his pupils the effort of straining their eyes in reading, and of focusing incessantly their mind and head one moment on the argument, another on the symbols, and the next on the diagrams. It is uncommon, however, that from this economy the pupil might really profit, for the only way to learn geometric proofs correctly, and to master them, is personal study, not the study under someone else's guidance. Indeed, I believe that the difference between these two kinds of study is much greater than that between traveling around alone, looking and observing the world with particular curiosity and attention, and resting content with the maps of geographers, however precise they might be, and with the stories told by writers, however faithful.

[...]

[99]Invaghitomi io dunque, per divina grazia, di scienza così sublime assai più godevo di farvi studio per me medesimo: ma appena ebbi scorsi i primi Elementi, che impazziente di vederne l'applicazione, passai alla scienza de' moti naturali nuovamente promossa dal Galileo, e che allora appunto era uscita in luce: ed arrivato a quel principal supposto, che le velocità de' mobili naturalmente discendenti per piani d'una medesima elevazione sieno uguali tra loro, dubitai, non già della verità dell'assunto, ma dell'evidenza di poterlo supporre come noto: Onde, per che mediante il sopraddetto Padre Clemente, mi s'era già aperto l'adito di trasferirmi spesso in Arcetri a godere de' suavissimi, e saggi ammaestramenti di quel buon Vecchio, il quale mi porgev'ardire di ricorrere a lui per la soluzione di quelle difficultà che (sì per fiacchezza del mio ingegno, sì per la novità di quell'argomento di Natura Fisico, e perciò non interamente sottoposto all'infrefragabili evidenze Geometriche) io fossi andato incontrando, lo richiesi un giorno di qualche più chiara confermazione di esso principio, con che porsi a lui occasione che in una delle seguenti notti, solite passarsele con modesta vigilia, egli ne ritrovasse la dimostrazione Geometrica Meccanica, deducendola dalla dottrina da lui stesso dimostrata già contro ad'una Proposizione di Pappo Alessandrino, la quale egli aveva confutato in quell'antico suo Trattato di Meccaniche dato fuori per la prima volta dal Padre Marin Mersennio Celebre Matematico Franzese.

Permettendo dipoi la BONTÀ SVPREMA, che dopo quattro mesi di studio di Geometria verso il principio del 1639. il Galileo mi volesse appresso di se come suo Ospite, e Disceplo, per guidarmi così cieco ch'egli era, co' suoi amorevoli insegnamenti per quel sentiero, che egli ogni giorno più mi dav'animo di proseguire, volle che quivi io facessi il disteso della dimostrazione di quel Teorema per supplire alla di lui cecità, che gli toglieva il così bene spiegarsi, dove occorrevano far figure, & appor caratteri; e di tal disteso mandò egli copia subito al P. Abate Don Benedetto Castelli Monaco Cassinese, e nobil Bresciano, uno de' suoi più antichi, e devoti Discepoli, & insigne per l'egregia sua Opera della misura dell'acque correnti; Trattato Elementare da esso nuovamente promosso. Di questo Teorema stesso mandò poi copia il Galileo a diversi altri Amici per l'Italia, e fuori, & è quel medesimo, che io con altre cose non più stampate somministrai all'ultima impressione di tutte l'opere di lui fatta in Bologna nel 1656. come qui[100]vi si vede a facce 132. del Terzo Dialogo. Questa medesima proprietà la confermò dipoi in varj modi il degnissimo Successore del Galileo, Evangelista Torricelli, nel suo Trattato de' Moti, quando non aveva avuto notizia ancora di quella di esso Galileo, con valersi però, in alcuno di que' modi, di certe altre proprietà dimostrate già da questo in quel suo antico Trattato di Meccaniche, poco avanti quì nominato. La medesima passione volle ancora con sottilissimo progresso autenticare quel sublime ingegno di Cristiano Vgenio nell'opera sua due anni sono pubblicata, e con stupor de' Matematici applaudita, trattante del moto de' Pendoli; e l'istessa pure si prese ultimamente a confermare, & a stabilire l'ingegnosissimo Sig. Alessandro Marchetti Filosofo Ordinario nella celebre Accademia Pisana.

[99]By divine grace, I developed a true passion for this sublime science, and great enjoyment in studying it by myself. However, once I went through the first books of the *Elements*, and was eager to see their content applied, I switched to the science of natural motions, newly advanced and just published by Galileo. I arrived at the basic assumption according to which the velocities of bodies naturally descending along planes with equal elevation are equal. I was in doubt not about the truth of the assumption but about the possibility of taking it for granted. Now, thanks to the Father Clemente mentioned earlier, I often had had the chance to visit Arcetri and enjoy the most pleasant and wise teachings of that good old man; he granted me the opportunity to ask his help in solving the difficulties that (either due to the weakness of my mind or to the novelty of the matter, which was physical in nature and therefore not entirely subject to unquestionable geometric clarity) I might be encountering, and one day I asked him for some clearer confirmation of the assumption in question. In so doing, I offered him the opportunity, during one of the following nights, which he usually spent tediously awake, to find its geometric and mechanical proof, by drawing it from the theory he himself had proved against a proposition by Pappus of Alexandria, which he had refuted in that old treatise *On Mechanics*, first published by the eminent French mathematician Father Marin Mersenne.[9]

Later, the Supreme Goodness decreed that after four months studying geometry, at the beginning of 1639, Galileo wanted me with him, as a guest and disciple, to lead me—blind as he was—by way of his caring teachings, along that path he increasingly encouraged me to follow as days went by. Once there, he wanted me to write down the proof of the aforementioned theorem, to compensate for his blindness, which prevented him from explaining himself clearly by drawing diagrams with letters. He immediately sent a copy of the proof to Father Benedetto Castelli, a Benedictine monk and gentleman from Brescia, one of his oldest and most devout disciples, well known for his excellent *Elementary Treatise on the Mensuration of Running Waters*, newly published by him.[10] Next, Galileo sent copies of that theorem to several other friends of his, both in Italy and abroad, and it is this very theorem, with other things Galileo left unpublished, that I contributed to the latest printing of Galileo's collected works, published in Bologna in 1656 (as [100]can be seen there, on p. 132 of the third dialogue).[11] This same property was later confirmed in various ways by the most deserving successor of Galileo, Evangelista Torricelli, in his treatise *On Motions*,[12] when he had not yet heard of the property discovered by Galileo. In some of the procedures he followed, however, Torricelli made use of some results that Galileo had previously demonstrated in his old treatise *On Mechanics*, which I mentioned above.[13] The very same property was confirmed, by an ingenious procedure, by the sublime intellect of Christiaan Huygens, in the work he published two years ago, on the motion of pendulums,[14] which was very well received by mathematicians. Lately, the same property was confirmed and demonstrated by the most ingenious Alessandro Marchetti, full professor of philosophy at the famous University of Pisa.[15]

Per una simile occasione di dubitare intorno alla quinta, ed alla settima difinizione del Quinto d'Euclide mi aveva per avanti conferito il Galileo le dimostrazioni di quelle difinizioni del Quinto Libro senza però applicarle a figure, che, fermatomi poi in Arcetri, egli mi dettò in Dialogo assai prima della venuta quivi del Torricelli quando ancor il Galileo non aveva risoluto di porla nella quinta Giornata, ma pensava tuttavia d'aggiugnerla alla quarta a facce 153. dell'impressione di Leida, dopo la prima Proposizione de' Moti equabili nel caso del ristamparsi con l'altre opere sue quell'ultima delle due nuove Scienze. Questa tal dettatura diede poi qualche facilità al medesimo Galileo, ed al Torricelli per fare quel più ampio disteso in Dialogo, che si è veduto: e la medesima, come inutile, rimase a me, & ancora la conservo. Mi restò in oltre quella breve lettera indirizzata dal Galileo al Sig. Conte Piero de' Bardi in soluzione del Problema, onde avvenga che d'estate l'acqua del fiume, a chi v'entra, appaia prima fredda, e poi calda assai più di quella stessa aria temperata, che prima, trovandosi bagnato, fredda appariva. Dettommi dipoi quella lunghissima lettera in data di 25. Marzo 1641. scritta allora al Serenissimo Signor PRINCIPE LEOPOLDO di Toscana, oggi l'Alt. Reverendissima del Sig. CARDINAL DE' MEDICI, il quale col solito stimolo d'erudirsi l'aveva richiesto del suo parere intorno al Cap. 50. del Litosforo del Famoso Filosofo Fortunio Liceti, dove questi confutava l'opinione del Galileo riferita nel suo Trattato delle Macchie Solari, & altrove ancora: ma questa lettera fu poco dopo stampata dal Liceti stesso in una sua replica, e di nuovo nell'impressione Bolognese dell'Opere del Galileo insieme con la sopraddetta lettera al Sig. Conte Bardi, e con l'altra ancora al Sig. Conte Alfonso Antonini.

[101]Nel susseguente mese d'Aprile 1641. giunse di Roma in Firenze, per passare a Venezia al suo Capitolo Generale il predetto Padre Abate Castelli, che si trasferì di subito dal Galileo, dove io pur mi trovava, & avendo egli appresso di se il Manoscritto di quel Trattato del Moto, composto allora da Evangelista Torricelli, il quale 10. anni indietro era stato suo Scolare nelle Matematiche, fece sentire in ristretto al Galileo il contenuto, e la diversa maniera che in varj luoghi aveva praticato quegli per ampliare la di lui maravigliosa Scienza del Moto naturalmente accelerato, e del violento. Si rallegrò questi che in vita sua avesse già preso così grand'agumento, e favore quella dottrina da se nuovamente promossa, e di qui, e dalle relazioni dategli da quel Padre dell'altre singulari qualità del Torricelli, fece egli di questo concetto altissimo, né s'ingannò. Con tale occasione considerando il Padre Abate Castelli, che per la compassionevole cecità, e per l'età oramai cadente del Galileo si correva pericolo di perder quel residuo delle di lui speculazioni non pubblicate, che egli sapeva non essere ancora poste in carta, prese animo di proporgli il Torricelli per Aiuto a farne il disteso, & il Galileo ben volentieri accettò uomo così degno, e per Aiuto, e per Compagno, e restò col Padre Abate, che al suo arrivo in Roma l'avrebbe potuto incamminar liberamente a questa volta. Si trattenne questi in Venezia assai più del credutosi, e perciò non prima che il dì 10. d'Ottobre 1641. seguì in Arcetri la nobil Copula di questi due gran Lumi del Sistema Filosofico, e Matematico.

In a similar circumstance, when I happened to doubt the fifth and seventh definitions of the fifth book of Euclid, Galileo clarified those definitions of the fifth book for me, though without the use of diagrams. When I moved to Arcetri, Galileo dictated these clarifications to me in the form of a dialogue, well before Torricelli went to Arcetri, when Galileo had not yet decided to insert it in the Fifth Day but, rather, considered inserting it in the Fourth Day (on p. 153 of the Leiden edition), after the first proposition on uniform motions, for a possible reprint of *Two New Sciences*, the last of his published works, together with his other works. This dictation offered Galileo and Torricelli the opportunity to put in dialogue form the full version, as we have seen.[16] I have with me the dictated version, now obsolete, and still preserve it. Also, I have the brief letter that Galileo sent to Count Piero de' Bardi, offering a solution for the following problem: how is it that, in summer, a person who enters a river at first feels its water cold and then much warmer than the temperate air, which he felt cold when he was wet.[17] Afterward, he dictated to me this very long letter, dated March 25, 1641, to the Most Serene Prince Leopoldo of Tuscany, now the Most Reverend Cardinal de' Medici. Following his usual bent for learning, the prince had asked Galileo's opinion about chapter 50 of *Litheosphorus*, by the famous philosopher Fortunio Liceti, [18] in which the author refuted Galileo's opinion as expressed in his treatise *On Sunspots*, and elsewhere.[19] This letter, however, was later printed by Liceti himself in a rejoinder of his, and again in the Bologna edition of Galileo's collected works, together with the aforementioned letter to Count Bardi, and another one to Count Alfonso Antonini.[20]

[101]In the following month of April 1641, Father Abbot Castelli came to Florence from Rome, on his way to his General Chapter in Venice.[21] Castelli came immediately to stay with Galileo, with whom I was staying, too; he had with him the manuscript of that treatise *On Motion*, written by Evangelista Torricelli, who had been his pupil in mathematics ten years earlier. Castelli privately told Galileo about the content of the manuscript, also telling him the different procedures Torricelli had followed, in various passages, to develop Galileo's amazing science of naturally accelerated and violent motions. Galileo was pleased to know that the novel theory he had held was so well received and improved upon while he was still alive; on this basis, as well as on the basis of Castelli's other reports of Torricelli's exceptional qualities, Galileo had a very high opinion of him; and he was not mistaken. On that occasion, Father Abbot Castelli thought that, due to Galileo's pitiful blindness and his advanced old age, there was the risk of losing the part of his speculations that he had not yet published and that, Castelli knew, he had not even put on paper. Hence, he thought it wise to suggest that Torricelli help Galileo write them down, and Galileo gladly accepted such a worthy man as his assistant and friend; and he agreed with Father Castelli, once he returned to Rome, to tell Torricelli he could move in with Galileo. Castelli remained in Venice much longer than he had anticipated, and so it was not until October 10, 1641, that there took place, in Arcetri, the noble meeting of these two towering figures in philosophy and mathematics.

Immantinente cominciò il Galileo a comunicar al Torricelli ciò che allora ei meditava di spiegar in Dialogo in altre Giornate: ma, iniqua sorte invidiando a gli uomini acquisti, e cognizioni maggiori nelle Scienze (appena scorsi tre mesi, dopo la congiunzione di questi Pianeti al Mondo Letterato così benefici) interposesi fra di loro, eclissandoci per sempre il maggiore concedutoci da DIO Sommo Sole per discoprir ne' Cieli, e nella Natura maraviglie non più vedute, e verità ammirande state occulte a tutta l'Antichità.

Dentro sì breue tempo, e del quale la malattia stessa del Galileo portò via la parte maggiore, altro non poté fare il Torricelli, che la bozza del disteso della Quinta Giornata quì avanti riferita (la quale egli, seguìta la morte del Galileo, si ritenne per ridurla al segno che s'è veduta) e non so quali cose a parte intorno alla forza della percossa.

Erede del Galileo fu il dottor Vincenzio suo figliuolo, uomo di non volgar letteratura, d'ingegno perspicace, e inventivo di strumenti [102]Meccanici, & in particolare musicali, e fra gli altri d'un Liuto con tal'arte fabbricato, che sonandolo egli per eccellenza, cavava ad arbitrio suo dalle corde le voci continuate, e gagliarde, come se uscissero dalle canne d'un'Organo: & in vero con suavissima armonia, come più volte io l'udj nel trovarmi in sua casa; imperciocché quell'amica corrispondenza, che seco io aveva contratto vivente il Padre, la medesima continuò fra 'l Figliuolo, e me fin ch'ei visse. Nelle mani di questo, (il quale col Torricelli, e con me aveva assistito alla malattia, ed alla morte del Galileo suo Padre seguita a gli 8. di Gennaio 1642.) veddi oltre alle bozze Originali dell'Opere già stampate, quelle ancora di varie lettere, e discorsi, scritti dal Galileo in diversi tempi in occasioni di ragguagliare, o di rispondere, o di dir pareri sopra quesiti fattigli, o simili, che di tutte si contentò, ch'io ne avessi copia, dettandomene molte da se stesso, quando, e bene spesso mi ritrovavo da lui: se bene ò veduto poi che della maggior parte di queste vanno attorno altre copie pur manoscritte.

Tra le dettatemi, tre ve n'erano, ch'io sapeva di certo non esser ancora fuori in istampa, ma non sapeva già il Sig. Vincenzio, né meno io, se ne fossero copie altrove, credendosi allora più tosto che nò.

La prima contiene il disteso di sei Operazioni Astronomiche, di quelle, mi cred'io, mentovate in quest'ultima nota dal Galileo, dall'introduzion delle quali manifestamente apparisce, che tali Operazioni sarebbero state molte più in numero. So bene che queste poche, lette da me, si meritaron l'applauso d'uno degli Eminenti Letterati della famosissima Adunanza Reale di LVIGI IL GRANDE mio Sig. Clementissimo, che fu il Sig. Gio: Domenico Cassini, celebre Astronomo, quand'un'Estate molti anni sono egli fu quì di passaggio.

Galileo began immediately to tell Torricelli what he was planning to expound, in dialogue form, in the course of other Days. But evil fate, envying men for great achievements and knowledge in the sciences, came between them: and only three months after the conjunction of these planets, so beneficial to the world of letters, it eclipsed forever the greatest and highest sun God ever granted to us so as to bring about the discovery in the heavens and in nature marvels never to be seen again, and amazing truths that had been concealed to the whole of antiquity.

Within such a brief period of time, most of which was consumed by Galileo's illness, Torricelli could not do more than draft the Fifth Day (which, as we have seen,[22] he decided to reshape after Galileo's death), and I do not know what other unrelated things on the force of the percussion.

Galileo's heir was Dr. Vincenzo, his son, a learned man, with an insightful mind, who invented mechanical [102]and especially musical instruments. Among the latter, a lute constructed so well that when he played it excellently, he was able to obtain from the strings, when he wished, strong continuous sounds, as if they came from organ pipes. He truly played with the most pleasing harmony: I listened to him many times while I was staying at his house. Consequently, that friendly intimacy I had developed with his father, when he was still alive, I continued to have with Galileo's son as long as he lived. In Vincenzo's hands (who, together with Torricelli and me, had attended his father Galileo's illness and death, on January 8, 1642) I saw, besides the original drafts of the works that had been already printed, those of several letters and discourses written by Galileo at different times, when he was asked to inform, or reply, or else offer his opinion about questions asked him, and so on. Vincenzo gladly agreed to provide me with copies of them, and he dictated many of them to me when I happened to be with him (which was often)—although I later saw other copies of most of these same original drafts circulating in manuscript form.

Among those he dictated to me there were three which I knew had not been published; but Vincenzo did not know, nor did I, whether there were other copies elsewhere, although we believed there were not.

The first letter contains the text of six *Astronomical Operations*,[23] those, I believe, referred to in this last annotation by Galileo;[24] from the introduction, it is clear that such operations would have been in much greater number. I know certainly that the few I read earned the praise of one of the eminent scholars of the Royal Society of Louis the Great, my Most Benevolent Lord, Giovanni Domenico Cassini, celebrated astronomer, when he stopped here during summer many years ago.[25]

La seconda consiste in numero 12. Problemi, o Questioni spezzate del medesimo Galileo, parte delle quali si vedono risolute in alcuna dell'Opere sue sin quì stampate, e l'altre son forse di quelle della nota sopra riferita. Questi Problemi erano di mano del Sig. Vincenzio, che dissemi avergli distesi lui medesimo su le soluzioni spiegategli dal proprio Padre, già cieco, in alcuni giorni, ne' quali avanti al mio stanziare in Arcetri egli andava colà a visitarlo: E tanto le sopradette Operazioni Astronomiche, quanto questi Problemi, insieme con quei più (che impossibil'è indovinar quali, e quanti) parmi che dovevan comprendersi nella continuazione della quinta Giornata scritta dal Torricelli, dopo qualche esplicazione, & aggiunta ad alcuna delle cose dette nelle precedenti quattro Giornate; e nella medesima Quinta si avevan ad esaminare, e risolvere que' Problemi diversi, [103]e particolarmente d'Aristotile, & in specie del Trattato del muoversi degli Animali.

La terza scrittura dettatemi, è un'altro principio di nuovo congresso, intitolato *Vltimo*, forse così detto dal Galileo avanti che gli venisse concetto di ridurre anco le postille a' suoi Oppositori in forma di Dialogo. In questo Congresso il Galileo introduce (al solito) per Interlocutori il Salviati, ed il Sagredo, escludendo Simplicio, e ponendo per terzo quel soprannominato Sig. Paolo Aproino stato già suo Vditore delle Matematiche in Padova. Tal principio è disteso in Dialogo in sei fogli in circa, dove si spiegano alcune sperienze fatte dal Galileo fin ne' tempi, ch'egli era colà Lettore, allora che andava investigando la misura della forza della percossa (che in ultimo egli considerò come infinita) e questa materia, dopo spiegate le sperienze, voleva il Galileo trattar matematicamente in tutto 'l restante di tal Congresso, come terza Scienza, dopo le due già promosse da lui medesimo, e con questa finir di pubblicare il rimanente delle sue più elaborate fatiche, quale sarebbe stata questa, intorno alla quale egli medesimo disse aver consumato molte migliaia d'ore speculando, e filosofando, & avervi in fine conseguito cognizioni lontane da' nostri primi concetti, e però nuove, e per la loro novità ammirande.

Finalmente per quanto si cava dalla suddetta nota del Galileo de' 23. Gennaio 1638. di ciò che gli rimaneva da scrivere, e pubblicare, dovevansi comprender in un'altro Dialogo, che sarebbe stato il settimo (oltre a' primi quattro de' due massimi sistemi) tutte quelle note, osservazioni, e repliche da lui chiamate postille, fatte intorno a' luoghi più importanti di coloro che gli avevano scritto contro.

Immensa dunque è stata la perdita delle preziose speculazioni rimaste entro sì ricca miniera d'un tanto Filosofo, e Matematico; ma siccome quella della percossa è stata poi egregiamente trattata dal celebratissimo Sig. Gio: Alfonso Borelli, così è da aspettarsi che segua dell'altra da esso promessaci de' Moti degli Animali, in quella guisa ancora che dall'acutissimo Sig. Lorenzo Bellini insigne Anatomista nel famoso Studio Pisano si attende di veder matematicamente trattata la materia fin'ora oscurissima della respirazione, che egli stesso ci fa sperar di godere in breve, per la quale ben si vedrà (come nella sua Miologia lo dimostrò pure il Dottissimo, e Candidissimo Sig. Niccolò Stenoni) quanto vaglia, e quanto sia necessaria al Filosofo, all'Anatomista, & al Medico la nobile, ma negletta Geometria.

The second letter consists of 12 problems, or distinct questions, by Galileo himself; part of them we see solved in some of the works published so far, and the others are perhaps those in the note mentioned above. These problems were in Vincenzo's handwriting; he told me he wrote them himself, on the basis of solutions expounded to him by his father. Galileo was already blind, and Vincenzo wrote them in the course of a few days during a visit to his father in Arcetri, before I moved there.[26] If I am not mistaken, both the aforementioned *Astronomical Operations* and these *Problems*, together with those additional ones (it is impossible to guess which ones they were, and how many), should have been included in the continuation of the Fifth Day, written by Torricelli, following some explanations and additions to what is said in the previous four Days.[27] In that same Fifth Day, a number of different problems were to be examined and solved, [103]and especially those discussed by Aristotle, specifically those in his treatise *On the Motion of Animals*.[28]

The third paper dictated to me is the introductory part of still another exchange, titled *The Last [Day]*; perhaps Galileo chose this title before he had the idea of turning his annotations on his opponents into a dialogue. In this exchange, Galileo introduces (as usual) Salviati and Sagredo as interlocutors, but excludes Simplicius, presenting Paolo Aproino, a former pupil of his in mathematics at Padua, as the third interlocutor. This introductory part is a dialogue on approximately six sheets of paper. In them Galileo explains a few experiments he made while he was a lecturer in Padua, when he was investigating the measure of the strength of percussion (which he eventually deemed infinite). After describing the experiments, Galileo wanted to treat this subject mathematically throughout the remainder of the exchange, as a third science, after those two he had fostered. In so doing, he wanted to publish what was left of his most refined works, such as this, about which he himself said he spent several thousand hours reflecting and philosophizing, eventually attaining knowledge far removed from our first ideas, and hence new, and amazing for their novelty.[29]

Lastly, as far as it is possible to gather from the aforementioned annotation by Galileo (dated January 23, 1638) concerning what remained to be written and published,[30] all those notes, observations, and rejoinders he called annotations, about the most important passages of the books written by those who had criticized him, were to be part of yet another dialogue, which would have formed the Seventh Day (besides the first four, on the chief world systems).[31]

The loss of the precious speculations that remained within such a rich philosopher's and mathematician's mine was immense. But since those on percussion were masterfully treated by the most celebrated Giovanni Alfonso Borelli, it is to be expected that the other one, which he promised on the motions of animals, will soon follow.[32] Similarly, we also await from the very sharp Lorenzo Bellini, eminent anatomist of the celebrated University of Pisa, the mathematical treatment of the otherwise very unclear subject matter of respiration.[33] Bellini himself gives us hope that we may see it soon,[34] and from it we will see clearly (as the most learned and straightforward Nicolas Steno, in his work on myology, showed) how important, and necessary to philosophers, anatomists, and physicians is noble—but neglected—geometry.[35]

[104]Ma tornando alla copia, ch'io mi ritrovo della Scrittura intitolata *Vltimo Congresso*, questa, in alcuni luoghi dov'io aveva qualche difficultà, mi fu in aiuto in riscontrarla col proprio suo Originale, il Molto Reverendo Sig. Cosimo figliuolo del suddetto Sig. Vincenzio, e degno Nipote del Galileo, prima che egli partisse di Firenze per passare al servizio suavissimo dell'Eminenza Reverendissima del Signor Cardinal Barbarigo mio Benignissimo, e Riveritissimo Sig., & in quell'occasione dissemi averne egli medesimo già dato fuori altre copie. Col di lui aiuto riscontrai ancora la mia copia col suo Originale delle Operazioni Astronomiche, e nel margine di quello sovviemmi ch'io feci di mia mano una certa nota. Ebbi finalmente di mano del medesimo Sig. Cosimo copia d'un frammento di parere, o risposta del Galileo a quesito Meccanico, e mi permesse il copiare certe postille da' Libri d'alcuni de' Contradittori alle di lui prime Opere. So in oltre che esso Sig. Cosimo aveva un'esamina, & alcuni calculi fatti in proposito di que' del Chiaramonti in materia della Stella nuova, siccome altre simili postille, e risposte a varj degli Oppositori più moderni, delle quali cose mi son poi meco stesso più volte doluto di non m'esser fatto dar copia, per esser il Sig. Cosimo, già son due anni, passato a miglior vita in Napoli, dove egli era Superiore di quella Congregazione della Missione, e per diligenze fatte allora da me colà, & a Roma, d'ordine ancora del Sig. Carlo, fratello (per la Dio grazia) vivente del medesimo Sig. Cosimo, si ricevè per risposta, che un'anno avanti, prima di tornare a stanziare a Napoli, egli aveva stracciato, e abbruciato in Roma gran quantità di Scritture, tra le quali non si sà se vi erano i sopraccennati Originali, & i libri postillati, &c. giacché non erano tra quelle Scritture che furono riceuute quattr'anni sono da me per mano del detto M. Reuerendo Sig. Cosimo l'ultima volta ch'egli se ne tornò di qui a Roma per passar a Napoli, com'apparisce dall'Inventario, che fatto da esso, e da me sottoscritto, rimase allora nelle mani del soprannominato Sig. Carlo suo fratello ultimo de' tre felici Nipoti del Galileo.

Le scritture del sopraddetto Inventario consistono, (fuori d'alcuni discorsi, e lettere di Altri) o in bozze dell'Opere stampate del Galileo, o in discorsi, e lettere del medesimo, che di già si vedono fuori sparse; e solo tra le cose del Galileo, ch'io non so che' ne vada copia attorno, due ve ne sono.

La prima, un manoscritto del Galileo in più quinternetti in ottavo intitolato fuori sulla coperta *De Motu antiquiora*, il quale si riconosce esser de' primi giovenili studj di lui, e per i quali nondime[105]no si vede, che fin da quel tempo non sapev'egli accomodare 'l libero 'ntelletto suo all'obbligato filosofare della comune delle Scuole. Quello però di più singulare, che è sparso in tal manoscritto, tutto, come si vede, l'incastrò poi egli stesso opportunamente a' suo' luoghi nell'opere, che egli stampò.

[104]But going back to the copy I have with me, of the work titled *The Last Exchange*, in deciphering passages, I found somewhat difficult to understand, I was helped by the most reverend Cosimo, son of Vincenzo, the worthy grandson of Galileo,[36] who checked it against the original copy in his possession, before he left Florence to work most pleasantly for the Most Reverend Eminence, Cardinal Barbarigo,[37] most kind and dear to me. On that occasion, Cosimo told me he had already circulated other copies. With his help I checked again my own copy of the *Astronomical Operations* against the original copy in his possession, and I remember that in one of its margins I had penned an annotation. Eventually, I also had a copy, in the same Cosimo's handwriting, of a part of Galileo's opinion, or reply, to a question on mechanics, and he allowed me to copy a number of annotations from books criticizing Galileo's early works. Furthermore, I know that Cosimo had an examination and some computations [of Galileo] concerning what Chiaramonti wrote on the new star,[38] as well as other similar annotations and rejoinders to a number of more modern critics. Of all this, time and again I have regretted not asking Cosimo for copies, for he passed away two years ago in Naples, where he was superior of the Congregation of the Mission.[39] I made careful searches there, as well as in Rome, on behalf of Carlo, Cosimo's brother,[40] who (by God's grace) is still living; and I was told that one year earlier, in Rome, before moving back to Naples, Cosimo had torn up and burned a large quantity of papers, among which might have been the original versions, the annotated books, etc. I conjecture that because these were not among those I received four years ago from the Most Reverend Cosimo, the last time he left Florence for Rome on his way to Naples, as is evident from the inventory he drew up and I signed, and which remained in the hands of Carlo, his brother, the last of Galileo's blessed grandsons.

Besides some discourses and letters by other people, the works mentioned in this inventory are either drafts of Galileo's printed works, or else his discourses or letters, which have circulated; only two of these items, both by Galileo, have not been in circulation yet (as far as I know).

The first of them is a manuscript by Galileo, in a few small quires of twenty pages, *in octavo*, bearing the title *De Motu antiquiora* on the front cover; it can be seen that this belongs among Galileo's early works.[41] However early they were, from them [105]we gather that Galileo could not accommodate his free mind to the constrained philosophizing commonly taught at universities. As can be seen, however, Galileo himself inserted the most interesting parts of this manuscript in works he published, at those points he thought more appropriate.

L'altra è un libretto in foglio di mano del Padre Don Benedetto Castelli intitolato *Errori del Signor Giorgio Coresio, raccolti dalla sua Operetta del galleggiar della figura,* ma con qualche postilla, e rimessa in margine di mano del Galileo; dal che, siccome dal vedere che le bozze delle Risposte, e Considerazioni di esso Padre Castelli contro al Grazia, & alle Colombe sono, per la maggior parte, di mano del medesimo Galileo, io prendo occasione di credere, che, e quell'Opere, e queste fossero dettate, se non in tutto, almeno in qualche parte da esso Galileo al detto Padre, e poi da lui fatte pubblicare, & a lui attribuitele, forse per non dar onor di soverchio col proprio nome a' suoi così deboli Oppositori. Non sò già per qual cagione questa risposta al Coresio non uscisse allora in luce coll'altre due, giacché, e per esser coll'approvazione de' Superiori, non restav'altro che metterla sotto 'l Torcolo: ma forse di ciò ne dà, benché oscuramente, qualche cenno il medesimo Padre Abate Castelli nella Dedicatoria di quelle sue Considerazioni stampate.

Restami ora a dir quant'io sò intorno all'uso delle catenuzze promesso dal Galileo nel fine della quarta Giornata, riferendolo quale egli me l'accennò quando, presente lui, io stava studiando la sua scienza de' Proietti. Parmi dunque che egl'intendesse[a] di valersi di simili catene sottilissime pendenti dall'estremità loro sopra un piano, per cavar dalle diverse tensioni di esse la regola, e la pratica di tirar coll'artiglieria ad un dato scopo. Ma di questo a sufficienza, e ingegnosamente scrisse il nostro Torricelli nel fine del suo Trattato de' Proietti, onde tal perdita rimane risarcita.

Che poi la sacca naturale di simili catenuzze s'adatti sempre alla curvatura di linee Paraboliche, lo deduceva egli, se mal non mi sovviene, da un simile discorso.

Dovendo i gravi scender naturalmente colla proporzione del momento, che essi hanno da' luoghi dove e' son'appesi, & avendo i momenti de' gravi uguali attacati a' punti d'una libra sostenuta nell'estremità, la medesima proporzion de' Rettangoli delle parti di essa libra, come il Galileo stesso dimostrò nel Trattato delle resistenze, e questa proporzione essendo la medesima che quella tra le linee rette, che da' punti di tal libra, come base d'una Parabola, si tirano pa[106]rallele al diametro di tal Parabola (per la dottrina de' Conici) tutti gli anelli della catenuzza, che son come tanti pesi uguali pendenti da' punti di quella linea retta, che congiugne l'estremità dove essa catena è attaccata, e che serve di base alla Parabola, dovendo in fine scendere quant'è loro permesso da' lor momenti, e quivi fermarsi, fermar si douranno in que' punti, dove le scese loro son proporzionali a' proprj momenti da' luoghi di dove pendono essi anelli nell'ultimo stante del moto; che poi son que' punti, che s'adattano ad'una curva Parabolica lunga quanto la catena, & il di cui diametro, che si parte dal mezzo di detta base, sia perpendicolare all'Orizonte.

a egl'intendesse *scripsi*] egll'intendesse Q

The second item is a booklet in folio by Father Benedetto Castelli, titled *Errors of Giorgio Coresio, Drawn from his Little Work on Floating Bodies*, with some annotations and rejoinders Galileo penned in its margins.[42] From this fact, and since the drafts of Castelli's rejoinders to and remarks on Di Grazia and Delle Colombe are largely in Galileo's own handwriting, I am tempted to believe that both the *Errors* and the rejoinders to Di Grazia and Delle Colombe were dictated—if not in full, at least to some extent—by Galileo to Father Castelli, and that Galileo had Castelli publish them under his name, possibly in order not to honor too much, with Galileo's name, the weak works of his critics.[43] I do not know why this rejoinder to Coresio was not published then, together with the other two; having received the imprimatur, all that was left to do was to print it. Possibly, Father Abbot Castelli himself offers some reasons for that, however obliquely, in the dedicatory letter of the printed version of his remarks.

Now I only have to say what I know about the use of thin chains promised by Galileo at the end of the Fourth Day:[44] I will report what he briefly told me, when I was studying the science of projectiles in his presence. I believe he meant to make use of such very thin chains, hanging from their two ends over a plane, in order to draw from their different tensions the rule and practice of aiming a cannon at a target. On this, however, our Torricelli wrote at length and insightfully at the end of his treatise *On Projectiles*, so that this loss is compensated for.[45]

If I remember correctly, Galileo deduced that the natural bending of such thin chains resembled the curvature of parabolic lines. He did so by the following argument.

Since heavy bodies naturally fall in proportion to the momentum they have at the point from which they hang, and since the momentums of equal bodies attached to the points of a balance fixed at its end are in the same proportion as the rectangles of the parts of the balance, as Galileo himself showed in his treatise on resistances, and since this proportion is the same as the one between the straight lines that from the points of that balance—as if it were the base of a parabola—are drawn [106]parallel to the diameter of the parabola (according to the theory of conics), all the links of the thin chain, which are just like several equal weights hanging from the points of the straight line that links the ends at which the chain is attached, and which serves as the basis of the parabola, have to descend as much as they are allowed to by their momentums, and then stop; those links also have to stop where their descending lines are proportional to their own momentums at the points where those links hang at the last moment of their motion. These, in other words, are the points that resemble a parabolic curve of the same length as the chain, and whose diameter, which begins from the middle point of that base, is perpendicular to the horizon.[46]

Sappiasi finalmente, che del riferito, e scritto fin quì resta appieno informata l'Altezza Reverendissima del Sig. PRINCIPE CARDINAL DE' MEDICI, alla cui straordinaria affezzione alle scienze, da essa ad alto grado non men possedute, che protette, dovete, o Lettori, aver tutto l'obbligo delle ricevute notizie, assicurandovi per la mia parte del continuato mio buon volere di far pubblico tutto ciò, che del Gran Galileo mio riverito Maestro per ora si stà privato, e sparso in diverse mani, e da me raccolto, di quello cioè, non solo, ch'io ricevei dal di lui Figliuolo, e dal predetto Nipote, ma di quell'ancora, che mercé alla protezione, e favore della prefata Altezza Reverendis., & alla cortesia d'Amici, e Padroni di quì, e fuori, dopo una particolare attenzione, e diligente ricerca m'è riuscito d'andar di quà, e di là rispigolando. Et affinché segua ciò in forma la più copiosa che possibil sia, supplico tutti quegli, a' quali perverrà notizia di questi miei grati sentimenti, a voler essermi liberali in farmi pervenir i Trattati, o' discorsi, o le lettere ch'essi trovansi del Galileo non ancora pubblicate, o in proccurarmele da altre parti, perché, oltre al non tacere il nome di chi a così nobil opera avrà contribuito, vi sarà in ricompensa, non dico il mio gradimento, che nulla vale, ma quello di tutta la Repubblica Letterata.

Finally, I want anybody to know that the Most Reverend Highness, Cardinal de' Medici,[47] is fully informed about everything I have reported and written about here. To his extraordinary care for the sciences, which he knows at the same high level to which he fosters them, you, readers, owe all the gratitude for the information you have received. I ensure you that, as far as I am concerned, I will pursue further my intention to make public whatever is still unpublished by the great Galileo, my revered master—scattered, as it is, in various hands, and which I have gathered. I am referring, that is, not only to what I received from his son and grandson but also to what I have managed to glean here and there, by paying special attention and through a diligent research, thanks to the protection and support of the Most Reverend Highness, and to the courtesy of friends and patrons, both here and abroad. And in order to gather all material as exhaustively as possible, I beg all those who receive news of these grateful intentions of mine to be generous with me and let me have unpublished treatises, discourses, or letters they may happen to have by Galileo, or else to let me have them from others. For not only will I explicitly mention the names of all those who contribute to such a noble enterprise, but they will be awarded not so much with my gratitude, which is worthless, but with that of all the Republic of Men of Letters.

Notes

1. A Florentine priest, Marco Ambrogetti lived with Galileo in Arcetri from June 1, 1637, to January 25, 1639. Galileo dictated letters to him, of which a large number still remain. As Viviani reports a few lines below, Ambrogetti translated into Latin *The Assayer, On Sunspots*, and *On Floating Bodies*. After 1639 he continued to assist Galileo as amanuensis.

2. Galileo to Alfonso Antonini (1584–1657), 20 February 1638, in *OG* 17, no. 3684, pp. 291–297.

3. Orazio Ricasoli Rucellai (1604–1674), a man of letters and philosopher, author of several philosophical dialogues, most of which remained unpublished. Cardinal Giovanni Dolfin, or Delfino (1617–1699), a playwright and former senator of the Venetian Republic, was created a cardinal in 1667.

4. Both Daniele (or Daniello) Antonini (1588–1616) and Paolo Aproino (1586–1638) were pupils of Galileo's in Padua. Daniele, in fact, was Alfonso Antonini's brother, not his nephew. The archive of the Aproino family was destroyed at the end of the eighteenth century, and so was Andrea Aproino's letter to Daniele Antonini, referred to by Viviani; only seven letters from Aproino to Galileo survive. Both Aproino and Antonini are mentioned at the beginning of what was going to be the Sixth Day ("On the Force of Percussion") of the *Two New Sciences*: see pp. 281–282 (*OG* 8, pp. 321–322). On Aproino's hearing instrument, see Aproino to Galileo, 1 June 1613, in *OG* 11, no. 885**, pp. 518–519, and 27 July 1613, ibid., no. 905, pp. 540–544.

5. A wise and learned man, Filippo Pandolfini (1575–1655) was elected a member of the Accademia della Crusca in 1601 and of the Academy of the Lynxes in 1614, and was consul of the Florentine Academy in 1639.

6. Cardinal Leopoldo de' Medici (1617–1675).

7. Clemente Settimi (b. 1612), who took vows as a Scolopian in 1632. He was repeatedly denounced to the Holy Office as a follower of Galileo's, but was always acquitted. He assisted and took good care of Galileo in the last years of his life. We still have a number of letters in Settimi's handwriting, which Galileo wrote to Bonaventura Cavalieri and Evangelista Torricelli.

8. Famiano Michelini (ca. 1600–1666) took vows as a Scolopian, after which he was known as Father Francesco di San Giuseppe. He succeeded Vincenzo Renieri as professor of mathematics at the University of Pisa.

9. See *Racconto istorico*, pp. 14–15 and 44–45.

10. Castelli, *Della misvra dell'acqve correnti* (1628).

11. See *Opere 1655–1656*, vol. 2, pp. 132–134.

12. See Torricelli, *De motv gravivm Naturaliter descendentium, Et Proiectorum libri dvo*, in *Opera geometrica* (1644), pp. 104–106.

13. See *On Mechanics*, pp. 171–175 (*OG* 2, pp. 180–183).

14. See Huygens, *Horologivm oscillatorivm* (1673), pp. 31–32.

15. See Marchetti, *Fvndamenta vniversæ scientiæ De motu vniformiter accelerato* (1674), pp. 9–15. Marchetti first announced the proof in a letter to Antonio Magliabechi, 28 December 1673: see Tondini, ed. (1782), vol. 1, no. 115, pp. 120–121.

16. The Fifth Day of *Two New Sciences*, on Euclid's definitions of sameness and compounding of ratios, was first published in Viviani, *Qvinto libro degli Elementi d'Evclide* (1674), pp. 61–77; it is reprinted in *OG* 8, pp. 349–362 (English translation in Drake, *Galileo at Work*, pp. 422–436).

17. *OG* 8, pp. 595–597 ("Intorno la cagione del rappresentarsi al senso fredda o calda la medesima acqua a chi vi entra asciutto o bagnato").

18. Contrary to Galileo, who argued that the illumination of the Moon is caused by the reflection of sunlight from the earth, in chapter 50 of the *Litheosphorvs, sive De lapide Bononiensi* (1640), pp. 242–248, Liceti argued that the Moon, analogously to the Bologna stone, released light absorbed by the Sun. Liceti sent a copy of the book to Galileo, who wrote a letter to Prince Leopoldo de' Medici, defending his views (see *OG* 8, pp. 489–545; see also pp. 549–556, and 483–486). On July 14, 1640, Galileo sent a copy of the letter to Liceti; there followed a long and friendly exchange on the topic, and eventually Liceti published Galileo's letter, together with his own three-book, sentence-by-sentence response: *De lvnae svbobscvra lvce Prope Coniunctiones, & in Eclipsibus obseruata* (1642), Book II.

19. See *On Sunspots*, pp. 283–286 (*OG* 5, pp. 221–225), and the *Sidereal Messenger*, pp. 53–57 (*OG* 3.1, pp. 72–75). Galileo's argument was anticipated by Michael Mästlin, in the *Disputatio de eclipsibvs solis et lvnæ* (1596), nos. XXI-XXIII, pp. 7–8, and supported by Johannes Kepler, in the *Dissertatio Cum Nvncio Sidereo* (1610), pp. 32–33 (*KGW* 4, p. 301); see also Kepler's *Paralipomena ad Vitellionem* (1604), pp. 263–268 (*KGW* 2, pp. 221–225).

20. See *Opere 1655–1656*, vol. 2, pp. 71–93 (Galileo to Leopoldo de' Medici, 31 March 1640, in *OG* 8, pp. 489–542), 124–126 (Galileo to Piero de' Bardi, undated, in *OG* 8, pp. 595–597; the same subject is treated in *OG* 8, p. 599), and 54–59 (Galileo to Alfonso Antonini, 20 February 1638 [1637, according to the Florentine calendar], in *OG* 17, no. 3684, pp. 291–297).

21. Antonio Castelli (1578–1643) entered the Benedictine Order in 1595, and adopted the name Benedetto. The General Chapter is the general assembly of the Order, typically composed of representatives from all the monasteries.

22. See Viviani, *Qvinto libro degli Elementi d'Evclide* (1674), p. 60.

23. Galileo, *Operazioni astronomiche*, in *OG* 8, pp. 453–466. See also Galileo to Fulgenzio Micanzio, 5 November 1637, in *OG* 17, no. 3593, p. 212; and Galileo to Élie Diodati, 23 January 1638, ibid., no. 3653, p. 262.

24. See Viviani, *Qvinto libro degli Elementi d'Evclide* (1674), p. 86.

25. Giovanni Domenico Cassini (1625–1712) was an Italian (naturalized French) mathematician, astronomer, and engineer. In the early 1660s he was named correspondent of the Accademia del Cimento and had frequent exchanges with its members, among whom was Viviani. Jean-Baptiste Colbert invited him to France, to join the Académie des sciences. Cassini arrived in Paris in 1669, where he supervised the construction of the observatory and coordinated the observations of the mathematicians, astronomers, and geographers of the Académie. He observed surface markings on Mars, determined the rotation periods of Mars and Jupiter, and discovered four satellites of Saturn: Iapethus (1671), Rhea (1672), and Tethys and Dione (1672), which he called—following Galileo's example—*Sidera Lodoicea*, in honor of King Louis XIV of France (1638–1715), also known as Louis the Great (*Louis le Grand*) or the Sun King (*le Roi Soleil*).

26. Galileo, *Problemi*, in *OG* 8, pp. 598–607.

27. Indeed, the *Astronomical Operations* opens with the words: "The exchanges of the past few days [. . .]" ("I ragionamenti che ne i giorni passati sono occorsi [. . .]": *OG* 8, p. 453).

28. See Galileo's annotations in *OG* 8, pp. 610, 612, and 615–617. See also Galileo to Belisario Vinta, 7 May 1610, in *OG* 10, no. 307, p. 352; and Galileo to Élie Diodati, 23 January 1638, in *OG* 17, no. 3653, p. 262.

29. "On the Force of Percussion," in *Two New Sciences*, pp. 281–306 ("Della forza della percossa", in *OG* 8, pp. 321–346). The dialogical part ends on p. 300 (*OG* 8, p. 339).

30. See Galileo to Élie Diodati, 23 January 1638, in *OG* 17, no. 3653, p. 262; Viviani, *Qvinto libro degli Elementi d'Evclide* (1674), p. 85.

31. Various sets of annotations pertaining to the *Two New Sciences* are available in *OG* 8, pp. 365–448.

32. See Giovanni Alfonso Borelli (1608–1679), *De vi percvssionis liber* (1667), and *De motv animalivm* (1680–1681).

33. Lorenzo Bellini (1643–1704) graduated in philosophy and medicine at the University of Pisa, in 1663, and taught there until 1703. He was a disciple and collaborator of Borelli's, and a pupil of Francesco Redi's and Alessandro Marchetti's. The fifth of his treatises on anatomy, published posthumously, concerns respiration: see *Discorsi di anatomia* (1741), pp. 109–128. See also Bellini's letter to Antonio Vallisneri, 14 January 1701, also published posthumously (*Giornale de' letterati d'Italia*, IV, 1710).

34. See Bellini, *Gratiarvm actio* (1670), pp. 9–17.

35. Both Nicolas Steno's (1638–1686) *De musculis & glandulis observationum specimen* (1664) and *Elementorvm myologiæ specimen, sev Musculi descriptio Geometrica* (1667), and Bellini's *Gratiarvm actio* (1670), share a geometric approach to anatomy.

36. Cosimo Galilei (1636–1672), third-born of Vincenzo, Galileo's son (1606–1649), and Sestilia Bocchineri (d. 1669).

37. Gregorio Giovanni Gaspare Barbarigo (1625–1697) studied mathematics, history, and philosophy at the University of Padua and obtained a doctorate in *utroque iure* (that is, in both canon and civil law) in 1655; later that year, he was ordained a priest by the Cardinal Patriarch of Venice, Gianfrancesco Morosini. He served as the Bishop of Bergamo (1657) and later of Padua (1664), where he was counselor of the university, and in this capacity he sought to appoint Viviani to the chair of mathematics. Barbarigo was elevated to the cardinalate in 1660, and one of his episcopal acts was to consecrate as a bishop Nicolas Steno (Niels Stensen), on September 19, 1677. He was canonized by Pope John XXIII, in 1960.

38. Scipione Chiaramonti (1565–1652), a professor of natural philosophy at the University of Pisa, was a staunch advocate of Peripatetic astronomy, and a steadfast opponent of Galileo's. After he published his first work, *Discorso della cometa pogonare dell'anno MDCXVIII* (1618), Chiaramonti strenuously defended the incorruptibility of heavens, arguing that comets are sublunary phenomena. After the condemnation of the Copernican theory in 1616, Chiaramonti challenged Tycho's views in the *Antitycho* (1621), especially criticizing Orazio Grassi's views. Whereas Galileo briefly mentioned Chiaramonti's *Antitycho* in *The Assayer*, p. 83 (*OG* 6, p. 231), positively referring to his arguments against Tycho, Kepler mercilessly called attention to Chiaramonti's mistakes in the *Hyperaspistes* (1625), and in a letter to Peter Crüger, 9 September 1624, wrote that in the *Antitycho* "boldness struggles with vanity" ("certat audacia cum futilitate": *KGW* 18, no. 993, p. 211). Later on, after a careful consideration of Chiaramonti's arguments, Galileo poked fun at them in the *Dialogue*: see, for example, pp. 280–281 (*OG* 7, p. 304). Chiaramonti replied to Kepler's criticism in the *Apologia . . . pro antitychone svo adversvs Hyperaspistem Ioannis Kepleri* (1626), but Kepler refrained from crossing swords with him again. In a letter to Paul Guldin, 24 February 1628, he wrote: "I do not think I should respond to Chiaramonti if not in a manner appropriate to the position he holds. I did not

know, for sure, that he is a senator, an ambassador, a legislator; I thought he was a young man, who had just gotten out of school" ("Claramontio non puto respondendum, nisi fortè moraliter. Nescivi nimirum, Senatorem, legatum, legislatorem esse, putavi juvenculum esse hominem, qui de schola prodeat": *KGW* 18, no. 1072, p. 331). Chiaramonti reiterated his "ridiculous and impossible" thesis (Benedetto Castelli to Galileo, 5 August 1628, in *OG* 13, no. 1898, p. 444) in the *De tribvs novis stellis qvæ Annis 1572. 1600. 1604. Comparuere libri tres* (1628), aiming at a refutation of the Copernican theory, and criticizing Brahe and all those astronomers who regarded the three novae of 1572, 1600, and 1604 as stars in the heavens and not as sublunary phenomena. Galileo rejected the arguments in his letter to Cesare Marsili, 7 April 1629, in *OG* 14, no. 1943**, p. 31. The publication of the *Dialogue* in 1632, and of its Latin translation in 1635, was a big blow to Chiaramonti's reputation. In the *Dialogue*, not only is the Aristotelian worldview met with irrefutable evidence to the contrary, "[b]ut you, Simplicio," asks Salviati, "what have you thought of to reply to the opposition of these importunate spots which have come to disturb the heavens and, worse still, the Peripatetic philosophy?" ("Ma voi, Sig. Simplicio, che cosa vi sete immaginato di rispondere all'opposizione di queste macchie importune, venute a intorbidare il cielo, e più la peripatetica filosofia?": *Dialogue*, p. 53; *OG* 7, p. 77). But Chiaramonti's own arguments are ridiculed: "Truly," says Salviati, "it was with too scant a store of ammunition that this author rose up against the assailers of the sky's inalterability, and it is with chains too fragile that he has attempted to pull the new star down from Cassiopeia in the highest heavens to these base and elemental regions. Now, since the great difference between the arguments of those astronomers and of this opponent of theirs seems to me to have been very clearly demonstrated, we may as well leave this point and return to our main subject" ("Veramente che con troppo scarsa provisione d'arme s'è levato quest'autore contro a gl'impugnatori della inalterabilità del cielo, e con troppo fragili catene ha tentato di ritirar dalle regioni altissime la stella nuova di Cassiopea in queste basse ed elementari. E perché mi pare che assai chiaramente si sia dimostrata la differenza grande che è tra i motivi di quelli astronomi e di questo loro oppugnatore, sarà bene che, lasciata questa parte, torniamo alla nostra principal materia": *Dialogue*, p. 318; *OG* 7, p. 346). Chiaramonti retorted immediately, in the *Difesa . . . al svo Antiticone, e Libro delle tre nuoue Stelle dall'oppositioni dell'avtore de' Due massimi Sistemi Tolemaico, e Copernicano* (1633). The dedication of this work to Cardinal Francesco Barberini (who was the dedicatee of Chiaramonti's 1628 *De tribvs novis stellis*, too), and the difficult situation in which Galileo found himself in the aftermath of the publication of the *Dialogue*, prevented him from responding. Easily dismissed by Galileo's pupils and friends—see, for example, Antonio Nardi to Galileo, 20 July 1633, in *OG* 15, no. 2590, p. 185: "ridiculous and disdainful argument" ("Materia ridicola e di sdegno"); Mario Guiducci to Galileo, 20 August 1633, ibid., no. 2649, p. 231: "blad pedantry" ("Insipida pedanteria"); and Niccolò Aggiunti to Galileo, 1 February 1634, in *OG* 16, no. 2863, p. 31: "filthy little thing" ("porcheriola"); and 23 February 1634, ibid., no. 2891, p. 50: "substantial boondoggle" ("palpabil castroneria")—Chiaramonti's argument could not be dealt with in public by Galileo: "[. . .] it is worth considering that some, seizing the very big opportunity of being able to benefit from adulation in order to increase their own interests, allowed themselves to write things that, in different circumstances, would be easily deemed excessive, if not thoughtless" ("[. . .] quello che è degno di considerazione, alcuni, vedendosi un larghissimo campo di poter senza pericolo prevalersi dell'adulazione per augumento de' proprii interessi, si son lasciati tirare a scriver cose, che fuori delle presenti occasioni sarebbero facilmente reputate assai esorbitanti, se non temerarie": Galileo to Élie Diodati, 25 July 1634, in *OG* 16, no. 2970, p. 118; see also Galileo to Fortunio Liceti, January 1641, in *OG* 18, no. 4106, pp. 294–295). Chiaramonti left Pisa in 1636, and the opportunity of a chair in Padua having vanished, he spent the last years of his life in Cesena, his hometown.

39. The Congregation of the Mission (*Congregatio Missionis*) is a Roman Catholic society of apostolic life of priests and brothers who take vows, founded by Vincent de Paul (1581–1660) in Paris, in 1625. Its members are popularly known as Vincentians, Paules, or Lazarists (from the enclos Saint-Lazare, which housed the Congregation in 1632).

40. Carlo Galilei (1632–1675), second-born of Vincenzo Galilei and Sestilia Bocchineri.

41. Galileo's *On Motion*, written largely between 1589 and 1592, before Galileo moved from Pisa to Padua. The original title was *De motu antiquiora* (namely, "the older writings on motion"), but Favaro decided to adopt the more general title *De motu*, because the folio bearing *De motu antiquiora* was missing: see *OG* 1, pp. 245–246.

42. Giorgio Coresio was lecturer of Greek language at the University of Pisa, from 1609 to 1615. In 1612 he published the *Operetta intorno al galleggiare de corpi solidi*. Galileo wished to offer a rebuttal and asked Benedetto Castelli to list the errors in Coresio's work. Castelli's *Errori dei più manifesti commessi da Messer Giorgio Coresio, lettore di lingua greca in Pisa, nella sua Operetta del galleggiare della figura* passed the revision of the Florentine canon Francesco Nori on September 9, 1613, but was never published. It is available in *OG* 4, pp. 247–286, with Galileo's annotations. See Castelli to Galileo, 28 October 1612, in *OG* 11, no. 787*, p. 419.

43. Lodovico delle Colombe (1565–1616) and Vincenzo Di Grazia held that ice was condensed water and floated because of its flat shape, whereas Galileo (whose views were informed by those of Archimedes) maintained that ice was rarified water and floated because it was less dense than the water supporting it. All

bodies denser than water sink, he claimed, while all lighter float, regardless of their shape. Galileo formally set out his views in *On Floating Bodies* (1612), to which delle Colombe swiftly replied with the *Discorso apologetico d'intorno al Discorso di Galileo Galilei, Circa le cose, che stanno sù l'Acqua, ò che in quella si muouono* (1612). The following year, Di Grazia published the *Considerazioni sopra 'l Discorso di Galileo Galilei Intorno alle cose che stanno su l'acqua, e che in quella si muouono* (1613). Galileo did not wish to respond in person and asked Benedetto Castelli to do it. Although the *Risposta alle opposizioni del S. Lodovico delle Colombe, e del S. Vincenzio di Grazia* (1615) was published anonymously, Castelli signed the dedication to Enea Piccolomini d'Aragona (*OG* 4, p. 453; see also Galileo to Paolo Gualdo, 16 August 1614, in *OG* 12, no. 1038, ppp. 95–95). As we gather from the manuscript version, Galileo contributed to the text and actually wrote a large part of it (beginning with p. 599, line 24: see *OG* 4, pp. 13–16). Castelli and Galileo's detailed rejoinder to delle Colombe and Di Grazia goes well beyond an accurate refutation of the latter's thesis: indeed, Castelli and Galileo denounce their opponents' ignorance of mathematics, which prevented them from properly understanding the Archimedean arguments advanced by Galileo and Castelli. In so doing, Galileo turned the dispute on floating bodies into a clear statement of the relevance of mathematical analysis in the study of physical phenomena, thus marking the distance between his own approach and that of the Peripatetic tradition. Speaking in the third person—as, officially, Castelli is the author of the book—Galileo offers a true manifesto of the new science: "if [Galileo]'s views are different from the received ones, this is because he realized, after long observations, that the received views are ill-founded and unable to account for the problems that rise about the causes of nature's effects; and that he refused always to subordinate the liberty of arguments to the authority of the naked words of this or that author, who had the senses and a brain just like many others sons of nature. Accordingly, he covered his wings with the feathers of mathematics (without which it is impossible to rise from the ground, not even by an arm's length) and attempted to discover at least some particles of the infinite recesses of natural science, which he deems so difficult and vast that he grants that many individuals perfectly commanded either one or another ability, and one ability more than others; but he believes that all men who lived so far, or will live in the future, never knew, or will ever know, more than a tiny part of natural philosophy" ("se egli ha delle opinioni diverse dalle comuni, ciò è nato dall'aver, per lunghe osservazioni, conosciute queste mal fondate e inabili a sciòr le difficoltà che nascono circa le cause degli effetti di natura, e dal non voler mantener sempre sottoposta la libertà del discorso all'autorità delle nude parole di quest'o di quell'autore, uomo di sensi e di cervello simile a molt'altri figliuoli della natura; e però doppo l'aversi impennate l'ali con le penne delle matematiche, senza le quali è impossibile sollevarsi un sol braccio da terra, ha tentato di scoprir almeno qualche particella de gl'infiniti abissi della scienza naturale, la quale egli stima tanto difficile ed immensa, che, concedendo lui molti uomini particolari aver saputo perfettamente chi una e chi un'altra e chi più d'una dell'altre facoltadi, crede che tutti gli uomini insieme, stati al mondo sin ora e che saranno per l'avvenire, non abbino saputo né forse sien per sapere una piccola parte della filosofia naturale": *OG* 4, p. 653). Other relevant texts are available in *OG* 4, pp. 443–447.

44. See *Two New Sciences*, pp. 256–260 (*OG* 8, pp. 310–313).

45. See Torricelli, *De motv gravivm Naturaliter descendentivm, Et Proiectorvm libri dvo*, in *Opera geometrica* (1644), pp. 226–243.

46. *Two New Sciences*, pp. 256–257: "[…] the cord thus hung, whether much or little stretched, bends in a line that is very close to parabolic. The similarity is so great that if you draw a parabolic line in a vertical plane surface but upside down—that is, with the vertex down and the base parallel to the horizontal—and then hang a little chain from the extremities of the base of the parabola thus drawn, you will see by slackening the little chain now more and now less, that it curves and adapts itself to the parabola; and the agreement will be the closer the less curved and the more extended the parabola drawn shall be. In parabolas described with an elevation of less than 45°, the chain will extend almost exactly along the parabola" ("[…] la corda così tesa, e poco o molto tirata, si piega in linee, le quali assai si avvicinano alle paraboliche: e la similitudine è tanta, che se voi segnerete in una superficie piana ed eretta all'orizonte una linea parabolica, e tenendola inversa, cioè col vertice in giù e con la base parallela all'orizonte, facendo pendere una catenella sostenuta nelle estremità della base della segnata parabola, vedrete, allentando più o meno la detta catenuzza, incurvarsi e adattarsi alla medesima parabola, e tale adattamento tanto più esser preciso, quanto la segnata parabola sarà men curva, cioè più distesa; sì che nelle parabole descritte con elevazioni sotto a i gr. 45, la catenella camina quasi ad unguem sopra la parabola": *OG* 8, p. 310).

47. Leopoldo de' Medici.

FIGURE 3. Nicolas III de Larmassin (ca. 1640–1725). Portrait of Galileo; woodcut. Inscription: "GALILÆVS GALILÆI | LYNCEVS PHILOSOP[HVS] | ET MATHEMA[TICVS] | N. de Larmessin sculp[sit]." // "Galileo Galilei | Lyncean Philosopher | and Mathematician | Engraved by N. de Larmassin." In Isaac Bullart, *Académie des Sciences et des Arts*, Paris: Jacques Ignace Bullart, 1682, Vol. II, Book II, p. 131. San Marino, CA: The Huntington Library (by kind permission). See *OGA* 1, D33, pp. 91–93.

11

ISAAC BULLART

GALILÉE GALILEI

GALILEO GALILEI

(1682)

Il n'y a point de Nation dans le monde, où les Lettres sont en quelque recommandation, qui ne connoisse le nom de GALILÉE, & qui ne luy accorde la gloire qu'il a meritée par les belles inventions de son esprit. Il peut avec raison tenir un rang fort considerable entre les Sçavans hommes qui ont rendu celebre la Ville de Florence, dans laquelle il prit naissance, d'un sang noble; mais d'une couche illegitime; puisque par le moyen du Tube, dont il est l'Inventeur, il a vagué parmy les regions immenses de l'air, a regardé sans ébloüissement la lueur éclatante du Soleil, a penetré la Sphere de la Lune, [132]y a découvert des taches, & des ombres, a trouvé dans le firmament des Etoilles nouvelles, & inconnuës, auxquelles il a donné le nom de Medicis, du surnom de ses Princes.

Cét Homme ingenieux enseignoit les Mathematiques en la ville de Padoüe, où il se faisoit aimer de tout le monde par la beauté de son esprit, autant que par son humeur charmante. Estant allé à Venise, il apprit d'un Patricien de cette Ville que l'on avoit inventé en Allemagne une lunette, qui appliquée aux yeux portoit la veuë jusqu'aux objets les plus éloignez, & les faisoit paroistre dans leur naturelle grandeur. Luy qui meditoit peutestre desia ce dessein, ne voulant pas que l'Italie fust redevable de cette invention à un étranger, retourna à son logis, resolu de n'en point partir qu'il n'eust construit cette petite, & merveilleuse machine. Il prit à cét effet un tuyau de plomb, tiré de quelques orgues; il appliqua des verres à diverses distances, & consulta si heureusement les regles de l'Optique, qu'il tomba enfin dans la connoissance parfaite de ce qu'il cherchoit: il courut aussi-tost tout joyeux en faire part à ce Gentilhomme; qui accompagné de plusieurs monta avec luy sur la tour de Saint Marc, & ayant fait l'épreuve de cette Lunette au grand étonnement des assistans; luy conseilla de la presenter au Senat; l'asseurant que ce beau secret y seroit tres-bien receu. GALILÉE ne fut point trompé dans ses esperances: car ces graves Senateurs prirent tant de plaisir à cette agreable nouveauté, qu'ils adjoûterent par Decret public deux cens escus annuels à ses gages de Professeur en l'Université de Padoüe.

There are no countries in the world in which humanities are held in some consideration that ignore the name of Galileo and do not render him the glory he earned thanks to the beautiful discoveries of his mind. He may rightly hold a very high position among the learned men who made the city of Florence famous; there he was born, from a noble family, but as an illegitimate child.[1] Thanks to the telescope, which he invented, he crossed the immense regions of air, he looked undazzled at the bright light of the Sun, and pierced through the sphere of the Moon, [132]and there he discovered some spots and shadows; he found new and unknown stars in the firmament, which he named after the Medici, the name of his Princes.

This ingenious man taught mathematics in Padua, where everybody liked him for the nobility of his mind, as well as charming spirit. Having gone to Venice, he learned from a nobleman of that city that in Germany someone had invented a lens that, applied to the eyes, allowed one to see very distant objects, making them appear at their full size. Galileo, who had perhaps already thought of this idea, did not want Italy to owe this discovery to a foreigner; so he went back home and resolved not to leave until he had constructed that little and amazing instrument. In order to do that, he took a leaden pipe, which he had taken from some organ, and applied [to it] a few glasses at different distances. He consulted the laws of optics, so that he ended up with a perfect knowledge of what he was looking for. He happily hastened back to that nobleman to inform him about it; accompanied by many others, the latter climbed up the bell tower of San Marco with Galileo and tested the spyglass to the great amazement of those who were attending. He advised Galileo to present the instrument to the Senate, assuring that it would welcome such a beautiful secret with great pleasure. Galileo's hopes were not disappointed, for those serious senators so happily welcomed such a pleasant novelty that, by public decree, they added 200 scudi to Galileo's yearly wage at the University of Padua.

Ce Sçavant Homme retourna puis aprés instruire la Jeunesse qui estoit sous sa discipline; & achevant de perfectionner ce qu'il avoit inventé à la haste dans Venise, il dressa ces Instrumens avec tant d'industrie & de proportion, qu'il perça par la conduite de ces petits tuyaux cette vaste étenduë qu'il y a de la Terre aux Cieux, & remarqua dans ces voutes crystallines des astres que la foiblesse de nostre veuë nous avoit cachez jusqu'alors, & qu'il a fait connoistre aux hommes dans son Livre intitulé, *Nuntius Sydereus*. Cét ouvrage excellent porta sa reputation dans toute l'Europe. Le grand Duc de Toscane luy offrit une Chaire de Docteur és Mathematiques dans Pise; laquelle GALILÉE accepta avec d'autant plus de joye, que cette condition honorable l'engageoit au service de sa Patrie. Se trouvant dans l'opulence par les grandes liberalitez de son Prince, il mit en lumiere plusieurs Traittez de Mathematiques en Italien: entre autres celuy de la nouvelle Composition du monde; dans lequel il découvre son opinion touchant le mouvement du globe terrestre, & la stabilité des Cieux. Il véquit heureusement jusqu'à l'âge de quatre-vingts quatre ans, auquel il perdit l'usage de la veuë par un excez de veilles & d'estudes; après l'avoir étenduë si loin par la subtilité de son esprit, & par les secrets de sa Science: sa mort arriva l'an mille six cens quarante-deux, dans une metairie voisine de Florence, où il avoit esté relegué par Sentence des Inquisiteurs de la Foy, en punition de ce qu'il affectoit d'égaler la Sagesse increée, qui a creé le Ciel & la terre.

[133]SONNET.

Pour un Esprit si pur la Terre estoit impure;
Cét Homme tout celeste est monté dans les Cieux;
Il y void clairement ces flambeaux radieux,
Dont nous n'avons icy qu'une lumiere obscure.

De ces voutes d'azur la noble Architecture
Ravit également son esprit & ses yeux;
Et l'élevant plus haut que sa propre nature,
Luy fait connoistre enfin la nature des Dieux.

Il me semble desia qu'au travers de ce verre,
Dont son art approchoit le Ciel, l'Onde, & la Terre,
Je le vois éclater au front du Firmament:

Et si l'on se transforme en la chose qu'on aime;
Comme il fut amoureux des Astres seulement,
Que le grand GALILÉE est un Astre luy-mesme.

COLLETET.

After that, the learned man went back to teach the young people who were study-ing under him, and continued to improve what he had hastened to invent in Venice. He made these instruments with such industry and proportion that through these little tubes he overcame the large stretch between the earth and the heavens, and in these crystalline vaults he observed stars that, owing to the weakness of our sights, had remained hitherto hidden from us. He made them known by way of his book, titled *Sidereal Messenger*; this excellent work spread his fame throughout Europe. The Grand Duke of Tuscany offered him the chair of Professor of Mathematics in Pisa,[2] which Galileo was happy to accept, and the more so because such an honorable position would put him at the service of his homeland. He became rich, thanks to the great generosity of his prince, and published several treatises of mathematics in Italian, among them that on the new system of the world, in which he discloses his views about the motion of the terrestrial globe, and the stability of the heavens. He happily lived to the age of 84 years,[3] when he lost his eyesight due to an exceedingly high number of vigils and studies. After having extended his sight so far by the inge-nuity of his mind and the secrets of his science, he passed away in the year 1642, in a villa on the outskirts of Florence, where he had been confined by a sentence of the Inquisitors of Faith, as a punishment for his attempt to equal the uncreated Wisdom, which created the heaven and Earth.

[133]SONNET

Earth was impure for such a pure spirit;
this all-heavenly man raised up to the skies;
and there he clearly saw these bright flames,
of which we have, here, but a feeble light.

The noble architecture of those celestial vaults
enchanted his mind and his eyes,
and by rising higher than his [human] nature,
it caused him to know the nature of gods.

It seems to me that through that glass,
whose art approached the heaven, sea, and Earth,
I see him bursting at the threshold of the Firmament;

and if we turn into what we love,
since he loved dearly the stars alone,
the great Galileo is now himself a star.

COLLETET.[4]

1. See Rossi, pp. 114–115.

2. In fact, the position of the Grand Duke's Chief Mathematician and Philosopher allowed Galileo to spend all his time doing research, with no obligation to teach.

3. In fact, at the time of his death Galileo was 78 years old.

4. Guillaume Colletet (1598–1659) was a French poet, essayist, and translator. He belonged to the *Illustres Bergers*, a literary group of poets and scholars established in 1625, and was a founding member of the Académie française. He spent all his life in Paris, achieved a great reputation among his contemporaries, and enjoyed the patronage of several important people, including Cardinal Richelieu. This sonnet for Galileo was not included in Colletet's *Epigrammes* (1653) and was thought to be unpublished (the manuscript is in Rome, at the Biblioteca Angelica, Cod. 1984, f. 112r); among many others, the collection includes an epigram for Tommaso Campanella (p. 215), one in praise of Urban VIII's poetry (p. 199), and one lamenting his passing (p. 200).

ILLUSTRATIONS

FIGURE 4. Giuliano Traballesi (1727–1812), draftsman; Francesco Allegrini (1624–1684), engraver. Portrait of Vincenzo Viviani with his family's crest; burin and etching. Inscription: "VINCENZIO DI IACOPO VIVIANI FRANCHI PATRIZIO FIOREN.NO MATTEMATICO DE SEREN.MI G[RAN] D[UCHI] DI TOSCANA, GEOMETRA, E FILOSOFO CELEBERRIMO. | nato il dì V Aprile MDCXXII. morto il dì XXII Settembre MDCCIII. | Al Nobil Giovane il Sig:re Sinibaldo Nelli Patrizio Fiorentino. | Preso da un Quadro in Tela di Giusto Subtermans appresso il Sig:re Gio[vanni] Battista Nelli Erede del sud[detto] Viviani | Giuliano Traballesi del[ineavit] Fran[cesco] Allegrini inci[dit] 1763." // "Vincenzo, son of Jacopo Viviani Franchi, Florentine nobleman, Mathematician of the Grand Dukes of Tuscany, most renowned geometrician and philosopher. | Born on April 5, 1622; dead on September 22, 1703. | To the noble youth Sinibaldo Nelli, Florentine nobleman. | From a painting of Justus Sustermans, at Giovanni Battista Nelli's, Heir of Viviani | Drawn by Giuliano Traballesi and engraved by Francesco Allegrini in 1763." In *Serie di ritratti d'uomini illustri toscani con gli elogj istorici dei medesimi*, Florence: Giuseppe Allegrini, 1766–1773, Vol. II (1768), no. XLV, f. 156r; the picture accompanies the annotated Italian translation of Bernard de Fontenelle, "Eloge de Monsieur Viviani," in Id., *Eloges des Academiciens. Avec l'histoire de l'Academie Royale des Sciences*, The Hague: Isaac van der Kloot, 1731, Vol. I, pp. 29–46 ("Elogio di Vincenzo Viviani," ff. 157r–161v). Florence: Kunsthistorisches Institut in Florenz—Max-Planck-Institut (by kind permission).

FIGURE 5. Anonymous. Portrait of Girolamo Ghilini; engraving. Inscription: "HIERONYMVS GHILINVS ABBAS. PROT[ONOTARIVS]. APOST[OLICVS]. CANON. PRÆB[ENDARIVS] DOCT[ORIS] S. AMBR[OSII]. MEDIOLANI. ÆTAT[IS] ANN[ORVM] XLVIII." // "Abbot Girolamo Ghilini, protonotary apostolic, canon, and doctoral prebendary of [the Collegiate of the Basilica of] Sant'Ambrogio, in Milan. 48 years old." In *Le glorie de gli Incogniti O vero gli Hvomini Illustri dell'Accademia de' Signori Incogniti di Venetia*, Venice: Francesco Valvasense, 1647, p. 268; the picture accompanies a short biography of Ghilini and a list of his works, both published and unpublished ("Girolamo Ghilini Alessandrino," pp. 269–271). Florence: Biblioteca Nazionale Centrale (image in the public domain).

FIGURE 6. Johann Azelt (1654–1692). Portrait of Leo Allatius; engraving. Inscription: "LEO ALLATIUS Græc[æ] Linguæ Prof[essor] Romæ" // "Leo Allatius, Professor of Greek at Rome." In Paul Freher, *Theatrvm virorum eruditione clarorum*, Nuremberg: Heirs of Andreas Knorz, at the expense of Johann Hofmann, 1688, Vol. II, Part IV, Plate 81 (detail). San Marino, CA: The Huntington Library (by kind permission).

FIGURE 7. Anonymous. Portrait of Gian Vittorio Rossi; engraving. In Gian Vittorio Rossi, *Epistolæ ad Tyrrhenvm*, Cologne: Jost Kalckhoven, 1645, f. *2v. Florence: Biblioteca Nazionale Centrale (image in the public domain).

MESSER NICCOLÒ DI FRANCESCO GHERARDINI
PATRIZIO FIORENTINO
CANONICO DELLA METROPOLITANA
E AMICO INTRINSECO DEL GRAN GALILEO

nato il dì 4. Marzo MDCVII.   morto il dì 5. Maggio MDCLXXVII.

Preso da un Quadretto a olio esistente appresso un Affine
della Famiglia Gherardini

Cos. Zocchi del.   F. Allegrini sc. 1768.

FIGURE 8. Cosimo Zocchi (1711–1767), draftsman; Francesco Allegrini (1624–1684), engraver. Potrait of Niccolò Gherardini with his family's crest; burin and etching. Inscription: "MESSER NICCOLÒ DI FRANCESCO GHERARDINI | PATRIZIO FIORENTINO | CANONICO DELLA METROPOLITANA | E AMICO INTRINSECO DEL GRAN GALILEO | nato il dì 4. Marzo MDCVII. morto il dì 5. Maggio MDCLXXVII. | Preso da un Quadretto a olio esistente appresso un Affine | della Famiglia Gherardini | Cos[imo] Zocchi del[ineavit] F[rancesco] Allegrini sc[ulpsit] 1768." // "Sir Niccolò Gherardini, son of Francesco | Florentine nobleman | canon metropolitan | a very close friend of the great Galileo | born on March 4, 1607; dead on May 5, 1677 | From a small oil painting owned by a relative | of the Gherardini family | Drawn by Cosimo Zocchi and engraved by Francesco Allegrini in 1768." In *Serie di ritratti d'uomini illustri toscani con gli elogj istorici dei medesimi*, Florence: Giuseppe Allegrini, 1766–1773, Vol. II (1768), no. XLIII, f. 149r; the picture accompanies the annotated biography of Gherardini by Salvino Salvini ("Elogio del Can. Niccolò Gherardini," ff. 150r–152v). Florence: Kunsthistorisches Institut in Florenz—Max-Planck-Institut (by kind permission).

FIGURE 9. Johann Ulrich Mayr (1629–1704), painter; Philipp Kilian (1628–1693), engraver. Potrait of Joachim von Sandrart with his family's crest; burin and etching. Inscriptions (*in the oval*): "ACADEMIA T[ED]ES[CA]" // "GERMAN ACADEMY"; (*around the oval*): "VIVRE POUR MOURIR | ET MOURIR POUR VIVRE" // "LIVE TO DIE | AND DIE TO LIVE"; (*below*): "Nobilissimo ac Præstrenuo D[omi]n[o] Ioachimo à Sandrart in Stockau, Serenissimi Principis Comitis Palatini Neoburg[i] Consiliario gravissimo, Viro undequaque Excellentissimo, Seculi nostri Apelli famigeratissimo, Antiquitatum & Elegantiarum technicarum Promo-condo consummatissimo, hanc Imaginem, æternitati sacram, omni Observantia | D[at] D[onat] D[edicat] | Philipp Kilian Chalcographus | I. Vlrich Mair pinxit." // "To the most noble and shrewd Joachim von Sandrart, in Stockau, most eminent adviser of the Most Serene Prince, the Count Palatine of Neuburg, a most excellent man in all respects, most celebrated Apelles of our time, most accomplished steward of antiquities and of the arts of elegance, | the engraver Philipp Kilian | offers, bestows, and dedicates, with the utmost reverence, this portrait, consecrated to eternity. | Painted by J. Ulrich Mayr." In Joachim von Sandrart, *L'Academia Todesca della Architectura, Scultura & Pittura: Oder Teutsche Academie der Edlen Bau- Bild- und Mahlerey-Künste*, Nurenberg: by Jacob von Sandrart, and Frankfurt: by Matthaeus Merian, printed by Johann-Philipp Miltenberger, 1675–1679, Vol. I, Tome I (1675), f. [6r]. San Marino, CA: The Huntington Library (by kind permission).

FIGURE 10. Joachim von Sandrart (1606–1688), engraver. Potrait of Paul Freher; line engraving. Inscription (*in the oval*): "Paulus Freherus Doctor, Collegii Medici Noribergensis Senior. natus a. c. mdcxi. denatus. mdclxxxii." // "Paul Freher, Senior Physician of the Nuremberg Medical School, born in 1611, deceased in 1682."; (*below*): "Dum vixit, tali spectabilis ore Freherus | Sanus erat, quamvis canus, et artis honor. | Inter eos, quorum Calamo servavit honores, | Nunc Hominum latè docta per ora volat. | Est aliquid, Vitam proferre salubribus Herbis: | Æternam Verbis reddere, majus erit. | Filialis amoris ergo f[ilius] c[arissimus] ex fratre nepos. Carolus Ioach[imus] Freherus M[edicinæ] D[octor] | Sandrart sculpsit." // "When he was alive, judging by his look, | Freher was healthy, albeit white-haired, and brought credit to the art. | Among those whose distinction he preserved in writing, | now [his name] spreads widely on the lips of learned men. | To extend someone's life with healing herbs is something, | to make it immortal with words is something greater. | With the love of a son, the dearest son, his nephew Karl Joachim Freher, physician. | Engraved by Sandrart." London: The Wellcome Library (image in the public domain).

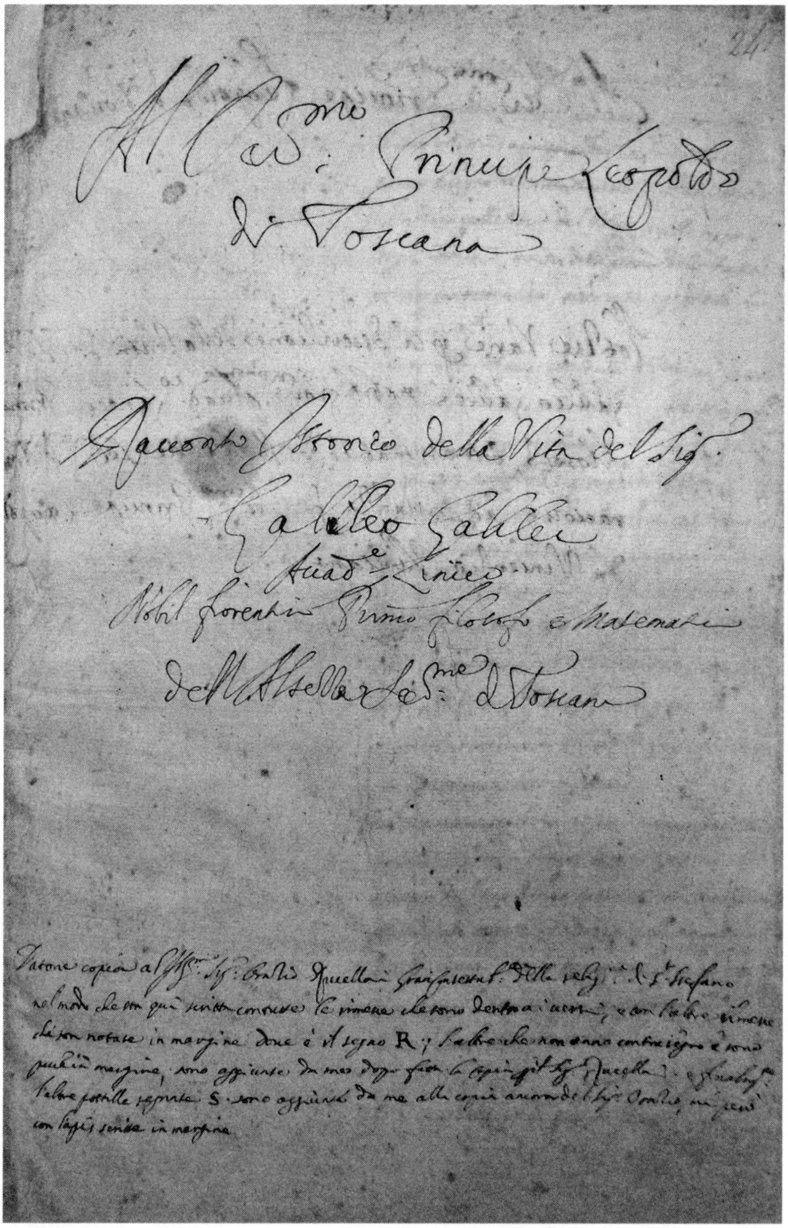

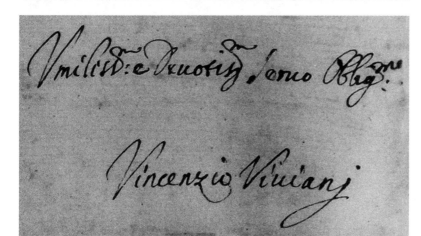

FIGURE 11A. Vincenzo Viviani (1622–1703). *Racconto istorico* (version B), title page; manuscript, post 1654. Text (*top, center*): "Al Ser.ᵐᵒ Principe Leopoldo | di Toscana | Racconto Istorico della Vita del Sig.ʳ | Galileo Galilei | Accad[emico] Linceo | Nobil fiorentino Primo Filosofo & Matematico | dell'Altezze Ser.ᵐᵉ di Toscana." // "To the Most Serene Prince Leopoldo | of Tuscany | Historical Account of the Life of | Galileo Galilei | Lyncean Academician | Florentine Nobleman, Chief Philosopher and Mathematician | of the Most Serene Highnesses of Tuscany." Florence: Biblioteca Nazionale Centrale, *Gal.* 11, f. 24*r* (by kind permission).

FIGURE 11B. Vincenzo Viviani (1622–1703). Signature at the end of the *Racconto istorico* (version B); manuscript, post 1654. Text: "Vmiliss:ᵐᵒ e Deuotiss.ᵐᵒ Seruo Oblig:ᵐᵒ | Vincenzio Vivianj" // "The most humble and devout, obliged servant | Vincenzo Viviani" Florence: Biblioteca Nazionale Centrale, *Gal.* 11, f. 68*r* (by kind permission)

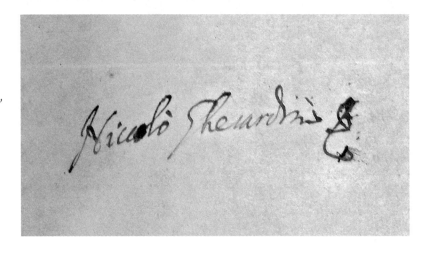

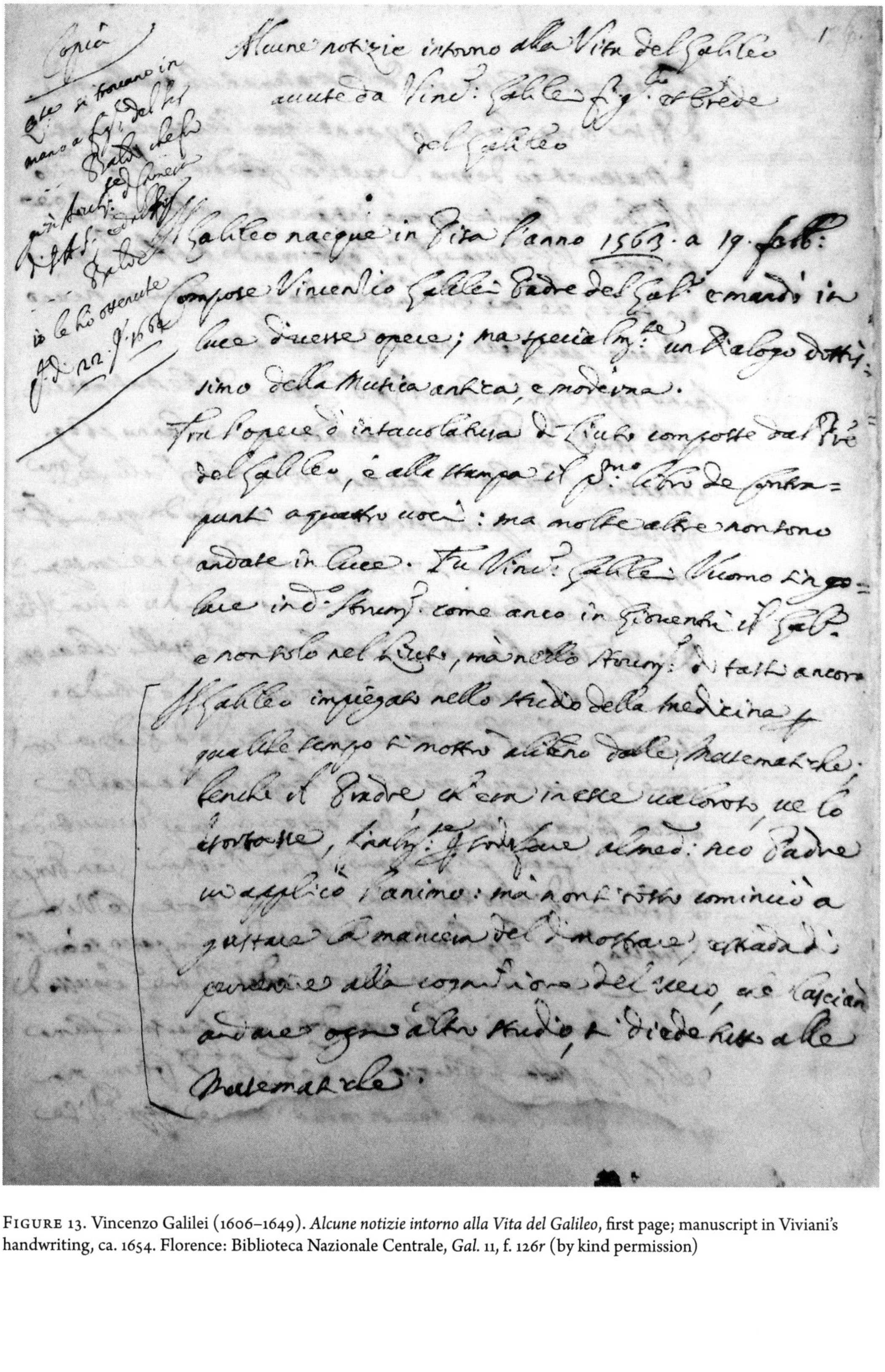

FIGURE 13. Vincenzo Galilei (1606–1649). *Alcune notizie intorno alla Vita del Galileo,* first page; manuscript in Viviani's handwriting, ca. 1654. Florence: Biblioteca Nazionale Centrale, *Gal.* 11, f. 126r (by kind permission)

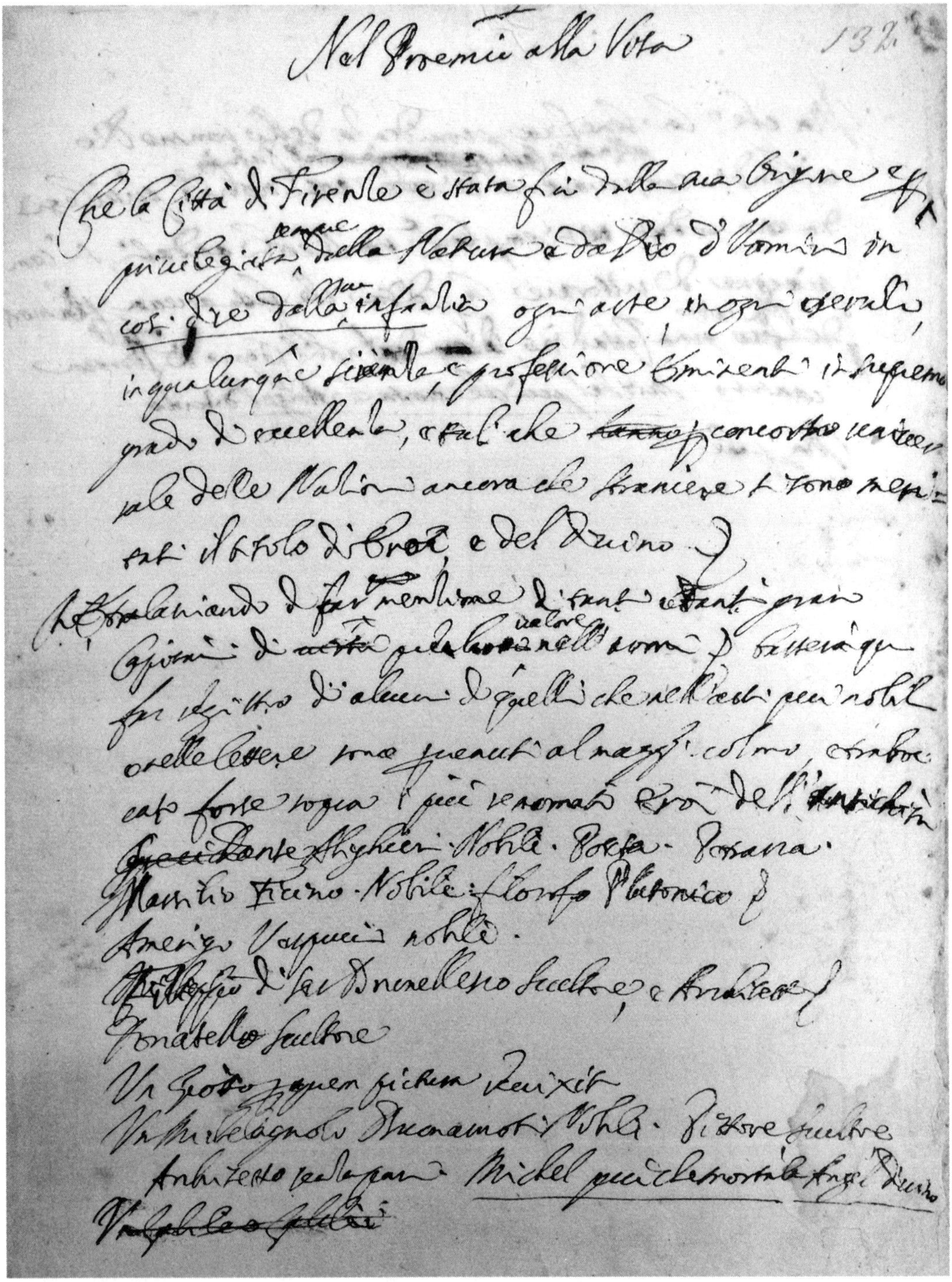

FIGURE 14. Vincenzo Viviani (1622–1703). *Proemio* to the *Racconto istorico*, first page; manuscript, post 1654. Florence: Biblioteca Nazionale Centrale, *Gal.* 11, f. 132*r* (by kind permission)

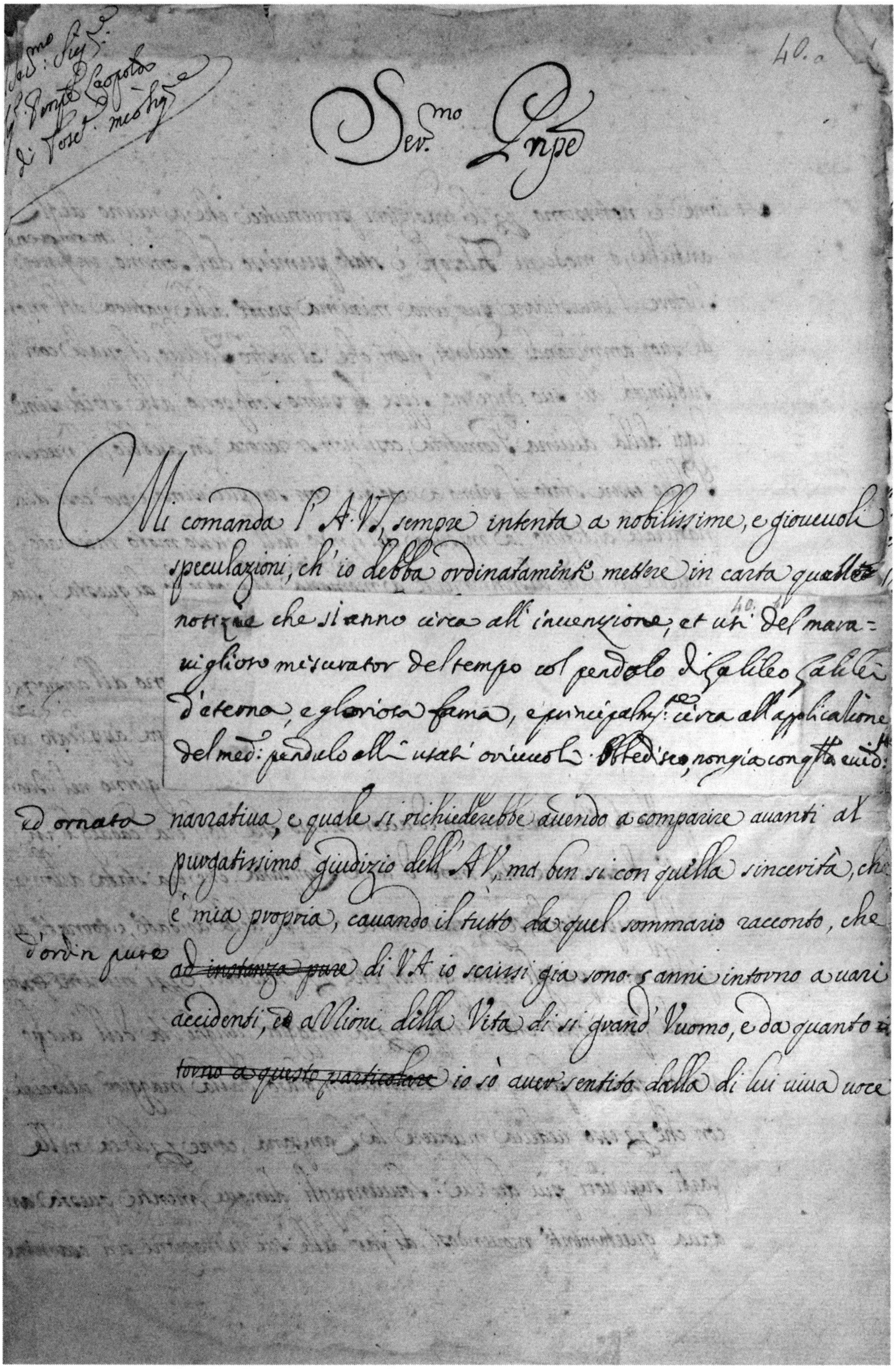

Sereniss.mo Prencipe

Mi comanda l'A.V., sempre intenta a nobilissime, e giovevoli speculazioni, ch'io debba ordinatamente mettere in carta quelle notizie che si anno circa all'invenzione, et uso del maraviglioso misurator del tempo col pendolo di Galileo Galilei d'eterna, e gloriosa fama, e principalm.te circa all'applicazione del med: pendolo alli usati oriuoli. Obbedisco, non già con gta evid. d'ornato narrativa, e quale si richiederebbe avendo a comparire avanti al purgatissimo giudizio dell'A.V., ma ben si con quella sincerità, che è mia propria, cavando il tutto da quel sommario racconto, che d'ord'n pure ad instanza pure di V.A. io scrissi già sono 5 anni intorno a varij accidenti, et attioni della vita di sì grand'uomo, e da quanto intorno a questo particolare io so aver sentito dalla di lui viva voce

FIGURE 15. Vincenzo Viviani (1622–1703). *Lettera al Principe Leopoldo de' Medici intorno all'applicazione del pendolo all'orologio*, first page; manuscript, 1659. Florence: Biblioteca Nazionale Centrale, *Gal.* 85, f. 40*r* (by kind permission)

FIGURE 16. Palazzo dei Cartelloni, façade. Florence: via S. Antonino, 11 (photograph by Stefano Gattei). See *OGA* 1, S5, pp. 420–423; see also ibid., S6, pp. 428–429, and D60, pp. 133–134.

FIGURE 16. DETAIL 1. Giovan Battista Foggini (1652–1725). Palazzo dei Cartelloni, façade (detail): bust of Galileo (bronze) and the two lateral bas-reliefs (scagliola). Florence: via S. Antonino, 11 (photograph by Stefano Gattei). See *OGA* 1, S5, pp. 420–423, and S5e, p. 428.

FIGURE 16. DETAIL 2. Palazzo dei Cartelloni, façade (detail): scagliola bas-relief on the left of the bust of Galileo, and inscription *G*. Florence, via S. Antonino, 11 (photograph by Stefano Gattei). See *OGA* 1, S5e, p. 428.

FIGURE 16. DETAIL 3. Palazzo dei Cartelloni, façade (detail): scagliola bas-relief on the right of the bust of Galileo, and inscription *H*. Florence: via S. Antonino, 11 (photograph by Stefano Gattei). See *OGA* 1, S5e, p. 428.

FIGURE 17. Giovan Battista Foggini (1625–1725). Bronze medal, ca. 1680. Obverse: right profile of Galileo; reverse (*top, left to right*): Jupiter's satellites, the Moon, with valleys and mountains, and a comet; (*center, left to right*): an object falling from a tower (possibly, a lighthouse) with accelerated motion, a floating ship, the Sun setting, with sunspots, and Venus orbiting around it, showing phases; (*bottom, left to right*): the parabolic trajectory of a projectile fired from a cannon, a broken column, a telescope, a compass, and a pendulum. Inscriptions: "GALILEVS LYNCEVS" // "Galileo, Lyncean" (obverse); "NATVRAMQVE NOVAT | MEMORIÆ OPTIMI PRÆCEPTORIS VINC[ENTIVS] VIVIANIVS" // "And he transforms nature | In memory of the best teacher, Vincenzo Viviani" (reverse). Grassina (Florence): Collection Alessandro Bruschi (with kind permission) See *OGA* 1, M4, pp. 520–521; see also ibid., D56 and D74-D74a, pp. 127–128 (II), 155 (V), and 156–157, respectively.

FIGURE 18A. (*opposite*) Temporary burial place of Galileo. Bust of Galileo; plaster (anonymous, after Foggini). Upper inscription (Vincenzo Viviani, 1674): "GALILEVS GALILÆI | GALILEO GALILÆI FLORENTINO PHILOSOPHO ET GEOMETRÆ VERE LYNCEO NATVRÆ OEDIPO | MIRABILIVM SEMP[ER] INVENTORVM MACHINATORI QVI INCONCESSA ADHVC MORTALIBVS GLORIA CŒLORVM PROVINCIAS AVXIT ET VNIVERSO DEDIT INCREMENTVM NON ENIM VITREOS SPHAERARVM ORBES FRAGILESQVE STELLAS CONFLAVIT SED ÆTERNA MVNDI CORPORA MEDICEÆ BENEFICENTIÆ DEDICAVIT CVIVS INEXTINTA GLORIÆ CVPIDITAS VT OCVLOS NATIONVM SECVLORVMQVE OMNIVM VIDERE DOCERET PROPRIOS IMPENDIT OCVLOS CVM IAM NIL AMPLIVS HABERET NATVRA QVOD IPSE VIDERET CVIVS INVENTA VIX INTRA RERVM LIMITES COMPREHENSA FIRMAMENTVM IPSVM NON SOLVM CONTINET SED ETIAM RECIPIT QVI RELICTIS TOT SCIENTIARVM MONIMENTIS PLVRA SECVM TVLIT QVAM RELIQVIT GRAVI ENIM SED NONDVM EFFECTA [*lege*: EFFŒTA] SENECTVTE NOVIS CONTEMPLATIONIBVS MAIOREM GLORIAM AFFECTANS INEXPLEBILEM SAPIENTIA ANIMAM IMMATVRO NOBIS OBITV EXHALAVIT IN ARCETRI SVBVRBANO · ANNO M · DC · XLII MENSE IANVARII DIE IX ÆTATIS SVE LXXVIII FRATER GABRIEL PIEROZZI NOVITIORVM RECTOR ET MAGISTER TANTI HEROIS ADMIRATOR VIRTVTVM POSVIT KALENDIS SEPTEMBRIS MDCLXXIIII." // "Galileo Galilei | To Galileo Galilei, Florentine Philosopher and Geometer, truly Lyncean, Nature's Oedipus | To the creator of ever new wonderful inventions, Who, blessed with a glory never before granted to any mortals, increased the heavenly provinces and added to the size of the universe. He did not assemble the crystalline orbs of the spheres, nor the frail stars, but dedicated eternal bodies of the universe to the Medicean generosity. Due to his boundless desire for glory, he sacrificed his own eyes, so that the eyes of all nations and generations might learn to see, because Nature did not have anything else he could see. His discoveries are hardly confined within the limits of the universe; the firmament itself not only contains them, but it welcomes them. And although he bequeathed us so many scientific monuments, he took with him more than he left. For, due to the burdensome yet always productive old age, wishing to achieve more glory with new researches, on January 9, 1642, aged 78 years, in the suburban estate of Arcetri, he returned his soul, always eager for knowledge, in what was for us a premature death. Father Gabriele Pierozzi, rector and teacher of the novices, admirer of the virtues of such an illustrious man, placed on September 1, 1674." Florence: Santa Croce Basilica, vestry to the Chapel of the Novitiate (photograph by Stefano Gattei). See *OGA* 1, S9, p. 439; see also ibid., S1, S21, S32, S43 and S46, pp. 413–414, 465, 479, 490–491, and 494, respectively.

FIGURE 18B. Temporary burial place of Galileo. Lower inscription (Simone di Bindo Peruzzi, 1737): "TANTI VIRI CORPVS CVIVS ANIMI PRÆCLARA MONIMENTA VBIQVE MORTALES SVSPICIVNT TOTO FERE SECVLO HIC JACERE SINE HONORE NON SINE LACRYMIS CONSPEXERVNT ERVDITI CIVES ET HOSPITES QVOTQVOT FLORENTIÆ FVERE ANNO DENIQVE CIƆDCCXXXVI IV IDVS MARTIAS VESPERE HINC TRASLATVM DECENTIORI LOCO TVMVLANDVM BONI OMNES GRATVLATI SVNT." // "The body of such a great man, the most remarkable testimonies of whose mind are everywhere for all to see, the learned citizens of Florence and all those who visited the city saw lying here, without honors but not without tears, for nearly one entire century. Finally, in the evening of March 12, 1736 [Florentine style, that is, 1737], it was moved away from here to be buried in a more proper place. All respectable people were pleased about it." Florence: Santa Croce Basilica, vestry to the Chapel of the Novitiate (photograph by Stefano Gattei); placed after the bodies of Galileo and Viviani were moved to Galileo's funerary monument, on March 12, 1737.

FIGURE 19. (*opposite*) Giovanni Battista Foggini (1625–1725), Giulio Foggini (d. 1741), Vincenzo Foggini (d. 1755), Anton Maria Fortini (1680–1738), Girolamo Ticciati (1676–1744). Funerary monument of Galileo, tomb of Galileo and Vincenzo Viviani (1737). Description (*top center*): Galilei family crest; (*center*): marble bust of Galileo, with representation of Jupiter and the Medicean Planets; (*left*): marble statue, personification of Astronomy; (*right*): marble statue, personification of Geometry. Inscription (Simone di Bindo Peruzzi, 1737): "GALILAEVS GALILEIVS PATRIC[IVS] FLOR[ENTINVS] GEOMETRIAE ASTRONOMIAE PHILOSOPHIAE MAXIMVS RESTITVTOR NVLLI AETATIS SVAE COMPARANDVS HIC BENE QVIESCAT VIX[IT] A[NNOS] LXXVIII. OBIIT A[NNO] CIƆ. IƆ. C. XXXXI. CVRANTIBVS AETERNVM PATRIAE DECVS X. VIRIS PATRICIIS SACRAE HVIVS AEDIS PRAEFECTIS MONIMENTVM A VINCENTIO VIVIANIO MAGISTRI CINERI SIBIQVE SIMVL TESTAMENTO F[IERI] I[VSSVM] HERES IO[HANNES] BAPT[ISTA] CLEMENS NELLIVS IO[HANNIS] BAPT[ISTAE] SENATORIS F[ILIVS] LVBENTI ANIMO ABSOLVIT. AN[NO] CIƆ. IƆ. CCXXXVII." // "Let Galileo Galilei, Florentine nobleman, greatest restorer of geometry, astronomy and philosophy, unrivaled by any of his contemporaries, rest here in peace. He lived 78 years and died in 1641 [Florentine style, that is, 1642]. Under the supervision of ten noblemen, appointed by this holy sanctuary for the perpetual glory of their homeland, heir Giovanni Battista Clemente Nelli, the son of Senator Giovanni Battista, willingly brings to completion the funeral monument that was to be erected, as Vincenzo Viviani stated in his will, for the remains of his master and his own. 1737." Florence: Santa Croce Basilica, left nave (photograph by Stefano Gattei). See *OGA* 1, S8, pp. 432–436, see also ibid., S8a–c, pp. 437–438.

GALILAEVS GALILEIVS PATRIC. FLOR.
GEOMETRIAE ASTRONOMIAE PHILOSOPHIAE MAXIMVS RESTITVTOR
NVLLI AETATIS SVAE COMPARANDVS
HIC BENE QVIESCAT
VIX. A. LXXVIII. OBIIT. A. CIƆ. IƆ. C. XXXXI.
CVRANTIBVS AETERNVM PATRIAE DECVS
X. VIRIS PATRICIIS SACRAE HVIVS AEDIS PRAEFECTIS
MONIMENTVM A VINCENTIO VIVIANO MAGISTRI CINERI SIBIQVE SIMVL
TESTAMENTO F. I.
HERES IO. BAPT. CLEMENS NELLIVS IO. BAPT. SENATORIS F.
LVBENTI ANIMO ABSOLVIT.
AN. CIƆ. IƆ. CCXXXVII.

FIGURE 19. DETAIL 1. Giovanni Battista Foggini (1625–1725). Funerary monument of Galileo, tomb of Galileo and Vincenzo Viviani (detail). Marble bust of Galileo (ca. 1680, after the bust cast by Giovanni Caccini, in 1613). Description (*top*): right hand holding a telescope, left hand holding a compass; left arm resting on a globe and a few books; (*bottom*): colored marble and bronze representation of Jupiter and the Medicean Planets. Florence: Santa Croce Basilica, left nave (photograph by Stefano Gattei).

FIGURE 19. DETAIL 2. Girolamo Ticciati (1676–1744). Funerary monument of Galileo, tomb of Galileo and Vincenzo Viviani. Marble statue, personification of Astronomy (detail): scroll with the Sun and sunspots. Florence: Santa Croce Basilica, left nave (photograph by Stefano Gattei).

FIGURE 19. DETAIL 3. Girolamo Ticciati (1676–1744). Funerary monument of Galileo, tomb of Galileo and Vincenzo Viviani. Marble statue, personification of Geometry (detail): scroll with diagrams of parabolic motion and accelerated motion along an inclined plane. Florence: Santa Croce Basilica, left nave (photograph by Stefano Gattei).

VRBANVS VIII BARBERINVS PONTIFEX.
Maximus Anno ætatis suæ 69.
Peter Aubrÿ Excudit.

Figure 20. Peter Aubry II (1596–1666). Portrait of Pope Urban VIII; copper engraving. Inscription: "Vrbanvs VIII Barberinvs Pontifex Maximus Anno ætatis suæ 69. Peter Aubrÿ Excudit." // "Pope Urban VIII Barberini, aged 69. Engraved by Peter Aubry." Münster: LWL-Museum für Kunst und Kultur (by kind permission).

46 CARD. BARBERINI

Premens coruſcanti micabis
Luce nouum decus inter aſtra.

ADVLATIO PERNICIOSA.

ODE.

CVM luna cælo fulget, et aëris
Pompam ſereno pandit in ambitu
Ignes coruſcantes, voluptas
Mira trahit, retinetq; viſus.
Hic emicantem ſuſpicit Heſperum,
Dirumq; Martis ſydus, et orbitam
Lactis coloratam nitore,
Ille tuam Cynoſura lucem,
Seu ſcorpij cor, ſiue canis facem.
Miratur alter, vel Iouis aſſeclas,
Patríſue Saturni, repertos
Docte tuo Galilaë vitro:
At prima Solis cum reſerat diem
Lux orta, puro Gangis ab æquore

FIGURE 21. Beginning of the *Adulatio perniciosa*, first edition. Maffeo Barberini, *Poemata*, Paris: Antoine Estienne, 1620, p. 46. Naples: Biblioteca Nazionale (image in the public domain).

FIGURE 22. Nicolas III de Larmassin (ca. 1640–1725). Portrait of Tommaso Campanella; woodcut, 1682. Inscriptions (*top right*): "PROPTER SION NON TACEBO" // "FOR ZION'S SAKE I WILL NOT KEEP SILENT [Isaiah 62:1]"; (*bottom*): "THOMAS CAMPANELLA | De Larmessin sculp[sit]" // "Tommaso Campanella | Engraved by de Larmassin" In Isaac Bullart, *Académie des Sciences et des Arts*, Paris: Jacques Ignace Bullart, 1682, Vol. II, Book II, p. 125. San Marino, CA: The Huntington Library (by kind permission).

COMMENTVM

in Oden,

cuius Titulus

ADVLATIO PERNICIOSA·

FIGURE 23. Tommaso Campanella (1568–1639). *COMMENTVM in Oden, cuius Titutuls ADVLATIO PERNICIOSA,* title page. Vatican City: Vatican Library, *Barb. lat.* 1918, f. 17r (by kind permission).

FIGURE 24. Bartholomäus II Kilian (1630–1696). Portrait of Galileo. Inscription: "GAL[LILÆI] GALILÆVS MAT[HEMATICVS]" // "Galileo Galilei, Mathematician." In Joachim von Sandrart, *Academia nobilissimæ artis pictoriæ*, Nuremberg: Christian Sigmund Froberger, 1683, Part II, Book III, Chapter XXVIII, Plate 7 (detail). Los Angeles, CA: Getty Research Institute (image in the public domain). See *OGA* 1, D34, pp. 93–94.

JOACHIM VON SANDRART

GALILÆUS GALILÆI

GALILEO GALILEI

(1683)

Galilæus Galilæus (*Jano Nicio Erithræo* teste) *Florentiæ* nobili ac vetere prosapia, clarissimum solis jubar adiit, atque umbris similes in eo maculas deprehendit; latos in Luna campos & colles & valles inspexit, novas etiam in æthereis siderum orbibus *stellas* invenit; quas *Mediceas*, ex Principum suorum cognomine, appellavit. Cum olim *Patavio*, ubi octingentorum aureorum stipendio mathematicas disciplinas juventuti tradebat, venisset *Venetias*, admonitus est à quodam patricii ordinis viro, in Germania inventum esse oculare, quod res quantumvis remotissimas ea, qua essent magnitudine, aspicienti subjiceret, ille, simul ac domum se recepit, fistulæ plumbeæ, ex organo detractæ, vitreos varii generis orbes ad certum intervallum accommodavit; unde eventum sibi ex sententia processisse cognoscens, ejus ocularis periculum in turri campanaria S. Marci fecit. Hinc notum erat omnibus, ocularis à se inventi ope, novas in cœlo stellas orbesque omnibus antè seculis, obstructas ac reconditas, esse detectas; quare & elegantissimus ille *Nuncii Siderei* titulo inscriptus ab eo liber exierat. Hac fama compulsus Magnus Etruriæ Dux accersivit eum *Pisas*; ut mathematicarum artium Doctoris nomine, centenos singulis mensibus nummos [390]argenteos magnos, quos laminas vocant, acciperet; alterum verò artis suæ vicarium sibi substitueret: Quapropter nullius rei egens, complures pertinentes ad Mathematicam libros composuit; & in his *Dialogum de Systemate mundi*, in quo præclaram suam de terræ, circa cœli orbes, nullo agitatos motu, conversione sententiam aperuit; quamvis postea Romam evocatus, à Quæsitoribus fidei palinodiam canere coactus sit. Quatuor fermè & octuaginta complevit annos, quorum postremos, luminibus orbatus, in tenebris vixit: Obiit in villa agri Florentini, cujus fines ne excederet, Quæsitorum fidei sententia cautum fuit. *Part. I. Pinacothecæ imaginum illustrium, num. CLIII.* Magni igitur Viri illius memoria me admonet, quàm familiariter ac benignè hoc ipso, Romæ, cum Inquisitionis negotio inibi vacaret, usus sim, in *Palatio Mediceo*; quod omnis antiquitatis verum armamentarium, & rarissimarum rerum singulare theatrum tunc erat: Hîc enim *Opticæ* simul ac *Geometriæ* studiis summoperè oblectatus, ab illustri doctore ac magistro ea didici, quæ universus orbis splendidè mecum ignorabat; Quid multa? per tubum, in cubiculo suo ad Lunam haud difficulter directum, montes & valles, & sylvas, & regiones, & lucem & umbram, & omnia ad oculum ostendit. Hic mihi habitus ab Eo honor, ut imaginem Ipsius debitis vicissim honoribus colerem, stimulator & concitator fuit.

Galileo Galilei (according to Gian Vittorio Rossi),[1] from a noble and ancient Florentine family, approached the brightest sunshine and snatched its spots, which resemble shadows. He examined the broad fields, the mountains, and the valleys of the Moon, and also discovered new planets in the heavenly orbs of the stars, which he named Medicean, after the family name of his princes. One day, when he traveled to Venice from Padua, where he taught mathematics to young students for a salary of 800 golden coins, a certain nobleman told him that in Germany a spyglass had been invented that offered to the beholders objects at their actual size, however very distant they actually were. As soon as he got back home, Galileo arranged a few glass disks of various kinds at given distances in a leaden pipe, which he had taken from an organ. Having realized that he had achieved the result he had hoped for, he tested his spyglass on the bell tower of San Marco. From this, everybody learned that, thanks to the spyglass he had invented, Galileo discovered new stars and orbs in the heavens, which had always remained hitherto hidden and concealed. Subsequently, he also published that fine book titled *Sidereal Messenger*. Driven by Galileo's fame, the Grand Duke of Tuscany invited him back to Pisa, so that he could receive a monthly salary of 100 large silver coins [390](which they call scudi), with the title of Professor of Mathematics, and appoint someone of his own choice as his substitute teacher.[2] Thus released from any need, he wrote several books on mathematics; among them, the *Dialogue on the System of the World*, in which he disclosed his brilliant theory on the revolution of the earth around the heavens, which are at rest. However, Galileo was later summoned to Rome by the Inquisitors, and forced to recant. He lived approximately eighty-four years,[3] the last few of these in the dark, deprived of his eyesight. He died in a villa on the outskirts of Florence, whose boundaries the sentence of the Inquisitors had prohibited him to cross (see *Pinacotheca imaginum illustrium*, Part I, no. 153).[4] Now, the biography of that great man reminds me of how I was on friendly and kind terms with him, in the Medicean Palace in Rome,[5] while he was dealing with the Inquisition. The palace was a true repository of the whole of antiquity, and an extraordinary museum displaying the rarest objects; there, I enjoyed to the highest degree the study of optics and geometry, and from that eminent scholar and teacher I learned what the world and I plainly ignored. What more [can I say]? By means of his [optical] tube, effortlessly aimed at the Moon, from his bedroom he showed to [my] eye mountains, valleys, forests, lands, lights, shadows, and all sorts of other things. This honor which I received from him led and encouraged me to pay homage to his portrait, by paying, in turn, due tribute.

1. See Rossi, pp. 114–115.

2. Actually, Galileo asked the Grand Duke to be appointed his Chief Mathematician and Philosopher: see *Racconto istorico*, pp. 24–25; and Vincenzo Galilei, footnote 3.

3. Rossi is once again the source for von Sandrart, here (see pp. 116–117): in fact, Galileo lived 78 years.

4. Von Sandrart used the second edition of Rossi's *Pinacotheca* (1645), in which all biographies are numbered with a Roman numeral: Galileo's is no. CLIII.

5. Villa Medici, an architectural complex with gardens, contiguous with the Villa Borghese gardens, on the Pincian Hill next to the church of Trinità dei Monti, in Rome. The property, whose site was part of the gardens of Lucullus in ancient times, was acquired by Cardinal Ferdinando de' Medici in 1576. The Villa Medici became the first among the Medici properties in Rome and was intended to exemplify the rise of the Medici among Italian princes and assert their presence in Rome. Upon the request of Cardinal Ferdinando, the architect Bartolomeo Ammannati (1511–1592) incorporated into the designs the Roman bas-reliefs and statues that were unearthed in the process of construction, with the result that the façades

of the villa became open-air museums. The grand gardens recalled the botanical gardens created by Grand Duke of Tuscany Cosimo I in Pisa and Florence, and the fountain in front of the villa, designed by Annibale Lippi in 1589, is formed from a red granite vase from ancient Rome. The Medici Villa assembled a striking collection of Roman sculptures, combining the Capranica and Valle collections, which had come together through marriage. To these, Cardinal Ferdinando added the *Niobe Group* and the *Wrestlers*, both discovered in 1583, and the *Arrotino*. The *Medici Lions* (a pair of marble sculptures of lions, one of which is of ancient origin) were completed in 1598, and the *Medici Vase* (a monumental marble krater sculpted in Athens in the second half of the first century AD as a garden ornament for the Roman market) was added to the collection at the villa in the 1630s, together with the *Venus de' Medici* (a Hellenistic marble sculpture). In the eighteenth century, the antiquities at the villa were moved to Florence and formed the nucleus of the collection of antiquities of the Uffizi Gallery. Until the male line of the Medici died out in 1737, the Villa Medici housed the Grand Duke of Tuscany's embassy to the Holy See. Both during and after the trial of 1633, Ambassador Francesco Niccolini and his wife, Caterina Riccardi, closely assisted Galileo, as is documented by Niccolini's letters to Galileo and his daughter, Sister Maria Celeste, as well as in the ambassador's reports to Andrea Cioli, the Grand Duke's secretary, in 1630–1634.

FIGURE 25. Johann Azelt (1654–1692). Portrait of Galileo; engraving. Inscription: "GALILÆUS GALILÆI Mathemat[icus] Florentinus." // "Galileo Galilei, Florentine Mathematician." In Paul Freher, *Theatrvm virorum eruditione clarorum*, Nuremberg: Heirs of Andreas Knorz, at the expense of Johann Hofmann, 1688, Vol. II, Part IV, Plate 81 (detail). San Marino, CA: The Huntington Library (by kind permission). See *OGA* 1, D35, pp. 95–96.

13

PAUL FREHER

GALILÆUS GALILÆI

GALILEO GALILEI

(1688)

[1536]GALILÆUS GALILÆI

Mathesin docuit Pisis & Patavii.

Florentiæ natus, Patrem habuit *Vincentium Galileum*, nobilem eruditum, & operibus Musicis editis celebrem. Mathematicis imprimis studiis delectatus, Mathesin *Pisis* per triennium, rogante *Ferdinando I.* Magno *Florentiæ* Duce, & in Academiâ *Patavinâ* 18. annis, magno studiosorum etiam nobilium virorum concursu, publicè docuit. In-

Venetiis perspicillum singulare Senatui obtulit.

tereà *Venetias* aliquando excurrens, legit ibi inter alias res novas ex *Flandriâ* scriptas, quod Opticus quidam *Mauritio* Principi *Auraico* perspicillum obtulerit, quod res longè distantes tanquàm proximè sitas oculis repræsentaret. Mox domum reversus, tale proprio *Marte* fabricavit instrumentum, quod cum inclyto Senatui *Veneto* dono obtulisset, cum admiratione receptum, eique publico decreto stipendium Lecturæ

Maculas Solis & Lunæ primus ostendit.

in Academ. *Patav.* duplicatum ad dies vitæ constitutum est. Cum hoc instrumento mirabili praeter alias stellas novas 4. planetas Joviales detexit, quibus *Mediceorum* nomen dedit. Eodem *Lunæ* & *Solis* maculas artificiosè demonstravit. Tandem *Co-*

Florentiæ magno stipendio vixit.

smus II. Magnus *Florentiæ* Dux eum, cum jam 73. ætatis annum attigisset, ad sese amplo dato stipendio accersivit, ubi vitam quietam traduxit, & vix Princeps aliquis

Moritur.

eò venit, qui celebrem hunc Mathematicum videre & alloqui non desiderâvit. Ibidem vitam beatâ morte clausit A. C. 1642.

Scripta.

*SCRIPTA:

Sidereus Nuntius. De Systemate Mundi. Difesa contra le calunnie di Baltasar Capra. L'uso del compasso geometrico, & militare, da lui ritrovato. De Luce & Lumine Disputatio. De Proportionibus. De S. Scripturæ Testimoniis. Tractatus de Proportionum instrumento à se invento. Discorso delle cose che stanno in su l'aqua, ò che in quella si muovono. Istoria & dimonstrazione intorno alle macchie Solari,[a] & loro accidenti compresi in tre lettere scritte à Marco Velsero. Il Saggiatore, nel quale con bilancia esquisita & giusta si ponderano le cose contenute nella libra Astronomica & Filosofica di Lotario Sarsi. I Dialoghi divisi in 4. giornate intorno à i due massimi Sistemi, Ptolomaico & Copernicano.

Ex Theatro Italico Viror. Eruditorum Hieron. Ghilini.

a Solari *scripsi*] Solare *F*

Born in Florence, his father was *Vincenzo Galilei*, a learned nobleman, famous for his published works on music. Attracted to mathematics since his early studies, Galileo publicly taught it for three years in *Pisa*, upon the request of *Ferdinando I*, the Grand Duke of *Florence*, and [later] at the University of Padua, for eighteen years, attracting very many scholars and noblemen. Meanwhile, on one of his quick trips to Venice, Galileo read, among other reports of novelties from *Flanders*, that one optician had presented *Maurice*, Prince of *Orange*, with a spyglass that showed distant objects as if they were close to the eyes. Galileo immediately went back home, built the instrument all by himself, and presented it as a gift to the eminent *Venetian* Senate, which received it with admiration and decided, by public decree, to grant him a lifelong lectureship at the University of *Padua*, with a doubled salary. By way of that amazing instrument, besides other stars, Galileo discovered four new planets orbiting Jupiter, which he named after the *Medici* family name; also with it, he skilfully showed the existence of spots on the *Moon* and the *Sun*. Finally, *Cosimo II*, Grand Duke of *Florence*, when Galileo was nearly 73 years old,[1] invited him to his court, with a generous salary. There he conducted a quiet life, and hardly any prince ever passed there who did not wish to see the eminent mathematician and talk to him. There he ended his life with a peaceful death, in AD 1642.

Taught mathematics in Pisa and Padua.

Offered a remarkable spyglass to the Senate in Venice.

First showed spots of the Sun and the Moon.

Lived in Florence with a generous salary.

Dies.

*Works

Works.

Sidereal Messenger. On the System of the World. A Defense Against the Libels of Baldassarre Capra. The Use of the Geometric and Military Compass He Invented. Disputation on Light and Brightness. On Proportions. On the Evidence of the Holy Scripture. Treatise on the Proportional Instrument He Invented. Discourse on Bodies That Stay atop Water, or Move in It. History and Demonstration Concerning Sunspots and Their Properties, in Three Letters to Marc Welser. The Assayer, in Which with an Exquisite and Fair Balance Are Weighted the Things Contained in Lotario Sarsi's Astronomical and Philosophical Balance. Dialogues, in Four Days, Concerning the Two Chief [World] Systems, Ptolemaic and Copernican.

*From Girolamo Ghilini's *Italian Theater of Learned Men*.

Note

1. In fact, Galileo left the University of Padua to become the Chief Mathematician and Philosopher of Grand Duke Cosimo II in 1610, when he was 46 years old. Freher's entry, explicitly based on Ghilini's, also reports his mistakes: see Ghilini, pp. 96–97.

FIGURE 26. Giovanni Antonio Lorenzini (1665–1740). Portrait of Galileo; burin and etching. Inscription: "GALILÆVS LYNCEVS ÆTATIS ANNORVM IIL. | Quem Astra, Mare, ac Terras complexum mente profunda Credibile in Solo cernere cuncta DEO | Fra[ter] Ant[onius] Lorenzini F[ecit]." // "GALILEO, LYNCEAN, AGED 48, | who embraced the stars, the sea and the Earth with his profound mind, we may believe he perceives all these in God alone. | By Father Antonio Lorenzini." In: Vincenzo Viviani, *De locis solidis secunda divinatio geometrica*, Florence: His Royal Highness' Press at Pietro Antonio Brigonci, 1701; and *Grati animi monumenta*, Florence: His Royal Highness' Press at Pietro Antonio Brigonci, [1702]. San Marino, CA: The Huntington Library (by kind permission). See *OGA* 1, D43, p. 110.

14

VINCENZO VIVIANI

GRATI ANIMI MONUMENTA

TESTIMONY OF A GRATEFUL SOUL

(1702)

[1]INSCRIPTIONES *QUÆ LEGUNTUR* IN FRONTE ÆDIUM A DEO DATARUM VINCENTII VIVIANI

Florentiæ extructarum in Via Amoris,
quæque sunt in spatiis notatis his characteribus A, B, C, D, E, F, G, H.[1]

A

ÆDES A DEO DATÆ
LUDOVICI MAGNI Inclyti Regis Christianissimi
honorificis munificentiis comparatæ, ac denuò constructæ.

B

D. O. M.
VIATOR
Qui sapientiæ amore percelleris, dum per hanc viam incedis,[2] cui fatidico quodam instinctu Amoris nomen majores fecere, siste parùm ad hoc (humile[3] quidem), sed grati, verique amoris monumentum, erga Sapientissimum PRÆCEPTOREM. Serenissimos[4] MAGNOS DUCES. Et LUDOVICUM MAGNUM Christianissimum Galliæ, & Navarræ[5] Regem; & quæ has Ædes exornant, dominique mentem demonstrant, perlege.[6]

C

GALILÆUS LYNCÆUS ætatis Annorum IIL. quem
Astra, Mare, ac Terras complexum[7] mente profunda,
Credibile[8] in solo cernere cuncta Deo.

---

1 INSCRIPTIONES—H *LM*] *deest in W*
2 incedis *LM*] incædis *W*
3 humile *V LM*] umile *W*
4 Serenissimos *LM*] Benignissimos *W*
5 & Navarræ *LM*] *om. W*
6 perlege *V LM*] per lege *W*
7 Terras complexum *LM*] Tellus complessum *W*
8 Credibile *LM*] Par est *W*

[1]INSCRIPTIONS ON THE FAÇADE OF VINCENZO VIVIANI'S HOUSE GIVEN BY GOD

In Florence, in Via dell'Amore,
and which are found in the areas designated by in the letters A, B, C, D, E, F, G, H.

A

A HOUSE GIVEN BY GOD
purchased and renovated thanks to the honorable generosity
of the Glorious and Most Christian King LOUIS THE GREAT[1]

B

To God the Almighty
Greetings, WAYFARER,
traveling this road, which the ancients prophetically named after Love![2] If you are possessed by a love of wisdom, tarry a while before this memorial—humble, indeed, but a testimony of grateful and true love, to the most wise PRECEPTOR,[3] Most Serene GRAND DUKES, and the Most Christian King of France and Navarre, LOUIS THE GREAT—and read through what adorns this house, and reveals the intention of its owner.

C

GALILEO, LYNCEAN, aged 48,
who embraced the stars, the sea and the earth with his profound mind,
we may believe he perceives all these in God alone.

AETERNÆ MEMORIÆ VIRO

GALILÆO DE[9] GALIÆIS, Patriæ, Etruriæ, Italiæ, imò Europæ totius delicio. Philosophiæ renascentis faci. Qui [2]veritatis propiùs intuendæ desiderio adeò exarsit, ut longè ultrà, tum veterum, tum recentiorum Philosophorum placita progressus, & posthabitis debilioribus humanarum mentium cogitatis, unico Geometriæ (quàm ad Cælum veritatis ducem vocabat) auxilio fretus, viam ad veritatem certiùs indagandam alios primus docuit,[10] feliciterque peregit,[11] comitante semper, per tam arduum[12] iter, pietate; itaut,[13] quæ de Maris æstu, Philolaique systemate exercendi tantùm ingenii causa,[14] (quod præsertim Epistola ad Christinam Lotharingiam[15] demonstrat) excogitaverat; religioni libens animo litaverit.

Qui[16] dum Patavii Matheseos Cathedram occuparet, vix audita Anno 1609. optici[17] tubi fama, ingenii, & dioptricæ viribus rem assecutus, instrumenti[18] structuram invenit, Senatuique Veneto dicavit, quem docti Viri meritò Galilæi[19] nomine donarunt, ut qui primus invenerit ingenio, non casu.

Novo hoc fretus auxilio, quasi Terra ejus ingenio satis non esset, Æthera[20] reclusit, novosque veluti Orbes Philosophis, & Astronomis aperuit.

In Luna montes, Valles, planities, periodicam ejus disci librationem.

In Sole nitidissimo lucis fonte, nubium, ac densarum caliginum instar nascentes, & renascentes maculas, ejus circa proprium centrum, ferè menstruam ab occasu in ortum vertiginem primus animaduertit.

Veneris sydus, ac etiam[21] Mercurii varias Lunæ facies æmulari, ac utrumque ob id proprio motu ab occasu pariter in ortum, veluti Mars, Juppiter, ac Saturnus, Solis globum circumire, tutò Astronomos docuit.

Altissimum Planetarum in variis cum Sole aspectibus tergeminâ specie, modò rotundum, modò oblongum, modò ansatum; Martemque Perigæum in quadraturis cum Sole, nonnihil[22] mutilum apparere, ante alios admonuit.

9   DE *LM*] dem *W*
10 docuit *W LM*] *deb.* docuerit
11 peregit *W LM*] *deb.* peregerit
12 arduum *V LM*] ardum *W*
13 itaut *LM*] ita ut *W*
14 causa *LM* causâ *V*] causam *W*
15 Lotharingiam *LM*] Lotaringiam *W*
16 Qui *V LM*] Quid *W*
17 optici *V LM*] opticij *W*
18 instrumenti *V LM*] instrumentis *W*
19 Galilæi *V LM*] Galilaeo *W*
20 Æthera *LM*] Aetera *W*
21 etiam *LM*] proinde *W*
22 nonnihil *LM*] non nihil *W*

TO GALILEO GALILEI,

A MAN WORTHY OF ETERNAL MEMORY, the delight of his homeland, of Tuscany, of Italy, and indeed of the whole of Europe, the light at the rebirth of philosophy; a man [2]so inflamed by the desire to contemplate the truth, that he went far beyond the accepted views of both ancient and more recent philosophers, and, disregarding weaker opinions of human minds, relying only on the help of geometry (which he called the truth's guide to heaven), he was the first to teach others a more certain way to truth, and happily pursued it, accompanied throughout his arduous way by piety; so that his discoveries about the ocean tides and Philolaus's system[4] (which he demonstrates especially in the Letter to Christina of Lorraine), and which he thought out only to engage his mind, he gladly sacrificed to religion.

A man who, in 1609, while he was holding the chair of mathematics in Padua, when news of optical tubes had not spread far, pursued the matter with the powers of his mind and of dioptrics, discovered how to construct the instrument, and dedicated it to the Venetian Senate; and so learned men deservedly named it after Galileo, since he was the first to discover it, and by reason, not by accident.

With the help of this new instrument, as if Earth were not enough to occupy his mind, he opened the ether, and disclosed new worlds, so to say, to philosophers and astronomers:[5]

The Moon, [he found] has mountains, valleys, plains, and periodic librations of its disk.

The Sun, the brightest source of light, has spots forming and re-forming like clouds and dense fog, as he was the first to see; and also revolves in nearly a month around its own center, from west to east.

The Star of Venus, and likewise that of Mercury, imitates[6] the various appearances of the Moon, as he reliably taught astronomers; consequently, both equally travel around the globe of the Sun, by their own motion, from west to east, just as Mars, Jupiter, and Saturn do.

The farthest planet, in its various aspects with respect to the Sun, has a threefold appearance, as he was the first to notice: at times it is round, at times oblong, at times it has handles; and perigean Mars, in quadrature with the Sun, appears somewhat maimed.

Inerrantes stellas, quas numero pauciores noverant prisci, ac veluti clavos unico, soli-doque Orbi fixas, quasi auxit, dum novas, & ante se nunquam[23] visas, in Orionis ense, in Plejadibus, in nebulosis, in lacteo circulo, & undique per ³Cælum, detexit, & ad Dei omnipotentiam magis, magisque declarandam, infinitas veluti lampadas, perpetuò ardentes, per immensa fluidorum Cælorum[24] spatia localiter immobi-les, sed ad instar Solis, circa propria centra revolubiles, ad primarios, & secun-darios propriorum Systematum planetas vivificandos, creatas,[25] arbitratus est.

Jovis Satellites Patavii VII. Idus Januarii Anni 1609. ante omnes primùm, & post tres tantummodò observationes[26] à se peractas, detectos, perpetuæ MEDICEORUM PROCERUM gloriæ dicavit, quorum concitatissimi motus aspectu,[27] jam-diu frustrà quæsitum problema de Locorum longitudinibus noctu captandis, proposuit; itaut[28] novis GENTIS MEDICEÆ auspiciis Geographia,[29] & Idrographia[30] corrigi, restitui, ac perfici datum sit; dum Medicearum Stellarum motus periodicos, & ab Jove distantias, improbo trienni labore assecutus, ad earum citissimè abeuntes aspectus prænunciandos,[31] Canones, & Tabulas con-fecit; spretisque amplissimis præmiis[32] iis, qui tantùm problema[33] enodarent, promissis; proprias etiam Theoricas, Tabulas, & Ephemeridas;[34] proprios opticos tubos; propriumque[35] Horologium[36] Oscillatorium à se jam a pluribus Annis Pisis excogitatum, ac insuper Viros horum instrumentorum usum probè callentes Anno 1615. Catholico[37] primùm Regi PHILIPPO TERTIO; postmodum Anno 1635. confœderatis Hollandiæ[38] Provinciis, hæroicâ[39] sanè magnanimitate obtu-lit; sed Dei Omnipotentis decreto tam generosa oblatio, ac nobile tentamentum utrinque evanuit, ut maximum opus Nauticæ, & Geographiæ bono, LUDOVICI MAGNI Terra, Marique potentissimi munificentia; & summi Astronomi Cassini labore, per ipsa Medicea Sydera inciperet, & perficeretur.

Cometarum denique Generationem, Incrementa, Motus, Interitum, explicavit.

---

23 nunquam *LM*] numquam *W*
24 Cælorum *LM*] Coelorum *W*
25 creatas *LM*] creatos *W*
26 observationes *V LM*] osseruationes *W*
27 motus aspectu *V LM*] motus, aspectus *W*
28 itaut *LM*] ita ut *W*
29 Geographia *W*] Geograhia *LM*
30 Idrographia *W LM*] *deb.* Hydrographia
31 prænunciandos *LM*] prenunciandos *W*
32 præmiis *V LM*] praemis *W*
33 problema *W*] probeblema *(sic) LM*
34 Ephemeridas *V LM*] effemeridas *W*
35 Propriumque *V LM*] propiumque *W*
36 Horologium *LM*] Orologium *W*
37 Catholico *LM*] Catolico *W*
38 Hollandiæ *V LM*] Hollandiæ, et *W*
39 hæroicâ *LM*] Hæroicâ *W deb.* heroicâ

The fixed stars, which the ancients thought to be few in number, and driven like nails into a single solid orb, he increased in number, disclosing new ones, never seen before, in the sword of Orion, in the Pleiades, in nebulas, in the Milky Way, everywhere throughout [3]the heavens; and he believed that they declare God's omnipotence more and more,[7] an infinite number of lights, glowing forever throughout the immense spaces of the fluid heavens; at rest locally, but, just like the Sun, revolving around their own centers, in order to give life to the primary and secondary planets of their own systems.[8]

The satellites of Jupiter, which he discovered first and before all others in Padua, on 7 January 1609,[9] after completing only three observations, he dedicated to the everlasting glory of the MEDICEAN PRINCES. And, due to the observation of their very quick motion, he set forth the question, vainly investigated for a long time, of how to obtain the longitude of places at night, so that under the new auspices of the Medicean House, geography and hydrography could be corrected, renewed, and refined. And while he worked out over three years of persistent work the periods of the motion of the Medicean Stars and their distances from Jupiter, he prepared canons and tables to predict their very rapidly changing aspects. Disdaining the most generous rewards promised to those who solved such difficult problem, with truly heroic magnanimity he offered his set of theories [of Jupiter's satellites], tables, and ephemerides,[10] his optical tubes, and his pendulum clock, which he had devised several years earlier in Pisa, and also men skilled in the use of these instruments: first to PHILIP III, the Catholic King,[11] in 1615, and then to the confederate Provinces of Holland, in 1635. However, by the decree of Omnipotent God the generous offer and noble attempt came to nothing in both cases, so that this essential work for the good of navigation and geography was begun by means of the Medicean Stars and carried through owing to the liberality of LOUIS THE GREAT, the most powerful on land and sea, and to the labors of that most excellent astronomer [Giovanni Domenico] Cassini.[12]

And lastly, comets generate, grow, move, and disappear, as he explained.

Qui verò cœlestia, & longinqua Dei opera aperuit, idem, ut summum Opificem, in minimis etiam operibus laudandum proponeret, humanæ Philosophiæ secretioria[40] penetralia reseravit; dum Microscopii ope[41] ex unicâ, & ex duplici[42] lente à se primùm excogitati, & confecti, ac jam Anno 1612. [4]instanti CASIMIRO POLONORUM REGI, dono missi, humano obtutui minima subjecit, & naturæ ipsius quamdam veluti anatomen instituit.

Et sicut Geometriam Philosophiæ nutricem vocabat, ita exemplo, & inventis demonstravit; siquidem novâ Methodo scientiam Centrobarycam quorumdam solidorum, vix etiam initiatis in Geometria aperuit.

Archimedis doctrinam de iis, quæ innatant fluidis, & eorum libramenta, ob vim alternarum pressionum primus indigitavit, innumeraque scriptis suis sparsit semina, è quibus plurimorum tractatuum seges præsenti ætate accrevit, & in dies Posteris accrescet.

Ante alios, vim percussionis infinitam suapte naturâ animadvertit.

Novas scientias omnibus usque ad ejus ætatem seculis intactas animadvertit; de solidorum resistentia; de motibus gravium, tum æquabiliter incedentium, tum naturaliter descendentium, tum projectorum (è quibus præcipuè bellicorum missilium artem elicuit) primus Philosophiæ Sacrario, intulit, promovit, ac Geometricè demonstravit.

Tantis rerum humanarum bono inventis, fama celeberrimi Viri in Æternitatem permansura, oblivionis,[43] temporumque victrix triumphabit.[44]

Hoc monumento hujus Ædis Dominus gratum animum erga eximiam virtutem, ob auctas, illustratas, perfectas naturales scientias, tantum testatum[45] in futuras ætates voluit.

E

GALILÆO INQUAM DE GALILÆIS

Patritio Florentino, Serenissimorum Etruriæ Magnorum Ducum, FERDIN. I.,[46] COSMI II.,[47] ac FERDINANDI II.[48] primario Philosopho, ac Mathematico. Academico, verè Lynceo, Geographiæ, Hydrographiæ, Cosmographiæ, Mechanices, Phisices, Astrorum scientiæ, opitulante Geometria, felicissimo instauratori. Inanis Artis Genethliacæ[49] perpetuo insectatori.

40 secretiora *LM*] secretioria *W*
41 ope *LM*] *om. W*
42 duplici *V LM*] duplice *W*
43 *ante* oblivionis *scr.* se ipsa *W* CAMBIARE TRADUZIONE
44 triumphabit *V LM*] triumphauit *W*
45 testatum *LM*] textatum *W*
46 FERDIN. I. *LM*] *Cosmi* I. *V Cosmi* primi *W*
47 II. *V LM*] secundi *W*
48 II. *V LM*] secundi *W*
49 Genethliacæ *V LM*] genetliacae *W*

In fact, the man who disclosed God's distant heavenly works was the same who unlocked to human philosophy the most secret recesses, showing that the supreme artificer deserves praise even in his most minute works; when, as early as 1612, aided by the microscope, which he first devised and produced, with a single lens or with two lenses, and sent it as a gift ⁴to the POLISH KING CASIMIR,¹³ who had asked for it, he brought the tiniest things to human contemplation, and established something like an anatomy of nature itself.

He justified his calling geometry the wet nurse of philosophy, through examples and discoveries, for he was able by his new method to introduce even those barely initiated into geometry to the centrobaric science of certain solid bodies.

He was the first to proclaim Archimedes's doctrine concerning bodies floating in fluids and their buoyancies, arising from the force of reciprocal pressures, and scattered innumerable seeds with his writings, from which a harvest of many treatises has grown in the present time, and will grow day by day in the future.

Before others, he called attention to the force of percussion, which is infinite by its very nature.

He called attention to new sciences, which had remained untouched until his own age: he was the first to introduce into the sanctuary of philosophy, and develop and prove geometrically, the resistance of solids, the motion of heavy bodies, whether proceeding uniformly or falling naturally, or else of projected bodies (from which he drew mainly the technique of military artillery).

Because of such great discoveries for the good of humanity, the fame of this most distinguished man will remain forever, and will triumph, victorious, over oblivion and time.

With this monument, the owner of this house wished to give to future ages a proper testimony of his grateful soul for [Galileo's] exceptional ability in extending, casting light upon, and perfecting natural sciences.

E¹⁴

TO GALILEO GALILEI, I SAY,¹⁵

Florentine Patrician, Chief Philosopher and Mathematician of the Most Serene Grand Dukes of Tuscany FERDINANDO I, COSIMO II, and FERDINANDO II;¹⁶ truly Lyncean Academician, most productive renovator, with the help of geometry, of geography, hydrography, cosmography, mechanics, physics, and the science of stars; tireless critic of the vain art of horoscopes,

[5]NOVISSIMUS TANTI VIRI DISCIPULUS.

Quòd ob aurea, Civilis, Moralis, & Christianæ sapientiæ monita; ob exemplum vitæ[50] viam veritatis eligere curaverit, ac pro virili, prosecutus fuerit, judicia Dei non sit oblitus. Nonnulla[51] ex infinitis abditis[52] vera, ex immensis Geometriæ Thesauris deprompserit,[53] & per ea homines ad ipsum Deum propiùs accedere senserit.

Quòd hinc veritatem, & justitiam esse fortiter propugnandas. Mendacium, assentationem, & hypocrisin, veluti pestes defugiendas. A segni otio potissimùm abhorrendum. Beneficia in ære, Maleficia in aere incidenda. Benemeritis quantum fieri potest, aut grato saltem animo satisfaciendum. Unicuique promissa religiosè exolvenda,[54] datamque fidem integrè servandam. Honestè acquisita pro se, suisque honestè impendenda. Avaritiæ sordes, & turpia lucra reiicienda.[55] Nihil in perniciem ingrati animi vitio laborantium cumulandum. Reliqua, omni prius ære alieno dissoluto, ingenuis potius, & bene merentibus[56] læto animo dandum perceperit.

Quòd præceptis hujusmodi juvenili, tum primum suo in animo à natura, à Genitoribus, à studiis, & à Præceptoris doctrina impressis, suavissimis propriorum Principum imperiis, nutibusque, se planè devoverit; atque hinc ab ingenita Serenissimi FERD. II. benignitate plura sibi ultrò, graviaque munera, maximis cum honoribus, ac stipendiis fuerint collata, certatimque à Serenissimo COSMO III.[57] incomparabili[58] clementia, denuo impartita, in quibus is[59] deditissimus cliens per quinquaginta ferè Annos, semper totus fuerit, iisque (veritate, & justitia ducibus) eximia sedulitate, & constanti fide ad extremum usque responderit.

Quòd denique, ob hæc omnia, LUDOVICI MAGNI, Galliarum, & Navarræ[60] invictissimi REGIS CHRISTIANISSIMI, tanquam[61] Numinis sui, judicium, ac voluntatem promeritus, amplissima ejus augustæ[62] liberalitatis dona diutissimè sit consecutus.

---

50 vitæ *V LM*] uitæ sanctitatem redolentis *W*
51 Nonnulla *LM*] Nonnullas *W*
52 abditis, *V LM*] abditis, ac Deo coaeuis *W*
53 deprompserit *V LM*] depromerit *W*
54 exolvenda *LM* exsoluenda *V*] essoluenda *W*
55 reiicienda *LM*] reicienda *W*
56 merentibus *LM*] smærentibus *W*
57 III. *V LM*] tertio *W*
58 incomparabili *LM*] incoparabili *W*
59 is *V LM*] his *W*
60 Navarræ *V LM*] Nauarra *W*
61 tanquam *LM*] tamquam *W*
62 augustæ *V LM*] auguste *W*

For he cared to choose the way of truth and pursued it to the utmost of his ability, following the golden commandments of civil, moral, and Christian wisdom, and the example of [Galileo's] life. He was never oblivious to God's will. He brought some truths to light out of the infinite secrets, from the immense treasuries of geometry, and discerned that by means of them men draw closer to God Himself.

For, in virtue of this, he himself perceived that truth and justice are to be fought for vigorously, and that falsehood, flattery, and hypocrisy must be avoided like the plague; that the laziness of an idle life is abhorrent; that good deeds are to be carved in bronze, evil ones in air;[18] that the deserving are to be given satisfaction as much as possible, or at least regarded with a grateful mind; that promises must be conscientiously fulfilled, and plighted faith irreproachably kept; that what a man has honestly acquired may be spent on himself and his kin; that the dirt of greed and filthy gain are repellent; that nothing is to be gained at the expense of those who are tainted with ingratitude; and that what remains, when all previous debts have been settled, should be given away joyfully, preferably to the honest and deserving.[19]

For, thanks to such principles instilled in his young mind first by nature, then by his parents, by studies, and by the teachings of his preceptor, he submitted himself wholly to the most agreeable commands and pleasure of his Princes; the innate benevolence of the Most Serene FERDINANDO II then entrusted several important duties to him, rewarded by the greatest honors and stipends; and these were renewed, with incomparable benignity, by the Most Serene COSIMO III;[20] and in carrying out these duties, as a most devoted subject for nearly fifty years, he was always faithful, and to which he responded (guided by truth and justice) with extraordinary earnestness and constant loyalty, until his very last moments.

Finally, for he deserved, for all these things, the trust and benevolence of LOUIS THE GREAT, most invincible and MOST CHRISTIAN KING of France and Navarre, whom he regarded almost as a divinity, and from whom he obtained the most generous gifts of august liberality for a very long time;

SIMULACRUM HOC ÆNEUM

Præceptoris sui perpetua[63] veneratione dignissimi ex Proto⁶plasmate[64] à celebri Sculptore Joanne Caccinio coram Serenissimo COSMO II.[65] Anno 1610. ad vivum efformato, exiguum uti Minerval, & grati animi pignus, ingenuique Amoris monumentum, tot, tantorumque beneficiorum Autoris æternùm memor, Serenissimorum eorumdem MM. DD.

Primarius Mathematicus.

Ætatis annorum LXXII.

Anno à Salut.[66] MDCXCIII.

A Galilæi ortu CXXX.

Ab interitu LII.

Primus publicè posuit.

F

FLORENTIA præ aliis Urbibus DEO nimis cara.

Exurge grata, & gratulabunda.

Ut enim non interruptam illustrium, divinorumque Virorum seriem videres; eodem Anno, mense, ac die, quo Mundi Conditor substulit nobiliorum artium penè deperditarum Picturæ, Sculpturæ, atque Architecturæ ad summum usque reparatorem, perfectoremque Patritium[67] tuum MICHAELEM ANGELUM, eodem ipso Anno, Mense, Die, ac propemodum Horâ hanc dolendam decoris tui jacturam ipsemet Deus refecit, & ut tu adhuc per nova lustra possis Civium tuorum virtuti,[68] Orbi universo prodesse, fastos tuos, Patritii tui GALILÆI ortu auxit, Philosophiæ, Geometriæ, atque Astronomiæ felicissimi Instauratoris, Patris, Principis, Ducis.

---

63 perpetua *LM*] prepetua *W*
64 Protoplasmate *LM*] Prothoplasmate *W*
65 II. *V LM*] secundo *W*
66 *post* Salut. *scr.* Inc. *W*
67 Patritium *LM*] Patricium *W*
68 virtuti *W LM*] *deb.* virtute

THIS BRONZE BUST

of his Preceptor, most worthy of everlasting veneration, from the original prototype
[6]molded from life by the famous sculptor Giovanni Caccini, in 1610,[21] in the presence of the Most Serene COSIMO II, as inadequate Minerval,[22] and token of a grateful soul, and a monument of sincere love, eternally preserving the memory of the author of so many and such important services, the Chief Mathematician

of the Most Serene Grand Dukes,

at the age of 72 years,

in the Year of Salvation 1693,

130 years since the birth of Galileo

52 years since his death,

first publicly set up.

F

FLORENCE, immeasurably dear to GOD, more than any other cities,

rise, grateful and rejoicing.

In order that you should not see the series of your illustrious and divine men interrupted, in the same year, month, and day in which the Creator of the world took away your Patrician MICHELANGELO,[23] supreme restorer and perfecter of the almost lost noble arts of Painting, Sculpture, and Architecture—in that very year, month, and day, and almost hour, God Himself remedied this painful loss to your fame, and, in order that you may promote, thanks to the excellence of your citizens, the whole world for many new lustra, enriched your splendour with the birth of your Patrician GALILEO, most fortunate renovator, father, prince, and guide of Philosophy, Geometry, and Astronomy.

Hic enim, cœlestis planè ingenii Vir (longè secus, ac Encomiastes quidam, invido-
rum Antagonistarum[69] fidei malè nixus, falsò conscripserat) Imperante inclyto
COSMO I. Pisis legittimè nascitur ex patre Vincentio Michaelis Angeli Joannis
de Galilæis, Patritio[70] Florentino, (qui de vetere, ac recentiore Theorica Musices
pereruditos Dialogos conscripsit) & ex honestissima ejusdem Vincentii uxore
egregia Julia Cosmi Venturæ è vetustissima, ac eminentissima Pistoriensi familia
de Am⁷mannatis, tunc Pisis cum eodem Vincentio commorante, Anno à Christi
Incarnatione 1563.[71] stylo Florentino, mense Februarii,[72] die decima octava,[73] &
hora ab occasu vigesima prima, & s. qui quidem Annus, Mensis, Dies, Hora tamen
23 &[74] s. itidem ab occasu, Pisis Galilæo nostro natalis; eidem Michaeli Angelo
Bonarrotio Romæ lethalis[75] fuit, ut ipsi legimus in domesticis Commentariis
Leonardi Bonarrotæ Michaelis Angeli fratris filii propria manu conscripti;[76] non
verò die 17.[77] ut à Vasario in ejus vita enarratur.

Exurge ergo, grata, & gratulabunda Florentia, & Summo Conditori illustres toto
Orbe Cives donanti demississima gratiarum actione, obsequia repende. Non
defuturos enim semper tibi nobilissimos, insignesque filios, illustria duorum
Virorum æternùm mansura, & semper futura fœcunda exempla promittunt.

Sed sicuti in Galilæi ortu, ejusque præclarè[78] gestâ vitâ meritò lætaris,[79] ita in ipso
ejusdem religiosissimo obitu, Pietatis Christianæ exemplum Civibus monstratura
pone luctum, imò exulta.

---

69 Antagonistarum *LM*] Antogonistarum *W*
70 Patritio *LM*] Patricio *W*
71 1563 *LM*] 1564 *W*
72 Februarii *LM*] Februari *W*
73 octava *LM*] ottaua *W*
74 23 & *LM*] XXIII. *W*
75 lethalis *LM*] læthalis *W deb.* letalis
76 conscripti *LM*] conscriptis *recte W*
77 17. *LM*] XVII. *W*
78 præclarè *LM*] preclare *V* precabo (?) *W*
79 lætaris *LM*] lætharis *W*

For this man, of quite celestial genius (far from what some biographer wrote, dis-honestly relying on the testimony of envious adversaries),[24] was born legiti-mately in Pisa, under the rule of the glorious COSIMO I, of Vincenzo, [son] of Michelangelo, [son] of Giovanni Galilei, Florentine Patrician (who composed the most learned Dialogues on the Ancient and Modern Theory of Music),[25] and of the most honorable wife of Vincenzo, the excellent Giulia [daughter] of Cosimo [son] of Ventura, belonging to the very ancient and eminent Ammannati family, [7]from Pistoia,[26] who at the time was living in Pisa together with Vincenzo. The year 1564 of Christ's Incarnation, Florentine style, in the month of February, on the eighteenth day, and at the twenty-first hour and a half, from sunset, that gave birth to Galileo in Pisa—that very same year, month, and day, though at the twenty-third hour and a half from sunset, Michelangelo Buonarroti died in Rome, as we ourselves read in the household notebooks by Lionardo Buonarroti, son of Michelangelo's brother,[27] written in his own hand—in fact, not on the seventeenth day, as Vasari recounts in his life.[28]

Rise, therefore, Florence, grateful and rejoicing, and reward, with an act of most humble thanksgiving, the Supreme Creator, who presented you with citizens il-lustrious throughout the world. For the bright examples of these two men, which will remain forever, and will always be fruitful, ensure that you will never be short of the noblest and distinguished sons.

But just as you rightly rejoice in Galileo's birth and gloriously conducted life, so set aside the sorrow for his passing, in a most religious manner, so as to show an example of Christian piety to your citizens—rather, rejoice!

Postquam enim[80] de rerum abditis[81] nihil pro mentis humanæ captu non conspexisset, ut meliùs in Creatorem animum intenderet, Deo permittente, oculis orbatus, per postremum vitæ quinquennium Divinæ Voluntati pius obsecundavit, quod fortiori animo præstitisse agnoscitur, quo amantissimo[82] eo sensu in nova semper detegenda fuerat usus. Lenta tandem correptus febre (quum[83] bonorum Virorum instituto vixisset, æs proprium,[84] non alienum, in pauperes occultè,[85] effusèque erogando, & multa singularis pietatis exempla edidisset) sensim deficiens, petitis sæpiùs salutaribus Ecclesiæ præsidiis,[86] ac piè susceptis, Pontificia Urbani VIII. benedictione munitus, optimus Philosophus, invocato sæpius Jesu, immortalem spiritum Creatori suo reddidit pacatissimè, Anno à Christi Incar. MDCXLI.[87] die Mercurii VIII. Januarii,[88] hora quarta n. s. Annos agens LXXVII. Menses X. Dies X. in Suburbano Martellinorum Arcetri Rure, ubi plusquam triginta annos scientiis vacauerat.

Tanti Viri postremæ invaletudini adstarunt[89] assiduè, & postremas voces accepere, Doctor Vincentius filius, Nurus, [8]Proximiores, Sacerdos Parœciæ;[90] duoque alii singulari doctrina, & pietate præstantes ad expiandam animam à Galilæo jam pridem delecti; duoque hospites jam & socii Mensæ,[91] alter Evangelista Torricellius acutissimus Geometra per postremum trimenstre. Alter per ultimum triennium novissimus Discipulus ter felix, GALILÆO à Sereniss. FERDIN.[92] II. sollicitè commendatus, qui memoranda hæc[93] posuit, ut à se in Præceptore conspecta,[94] vel à cognatis, amicis, famulis, sedulò, & tutè audita,[95] nepotibus, & posteris ad Christianos Philosophos edocendos, fideliter aperiret; assentiente, & jubente præsertim Serenissimo FERDINANDO Principe Etruriæ Primogenito, Artium, & Scientiarum Cultore,[96] ac Mœcenate munificentissimo.

---

80 enim W^{sl} LM] uero W
81 abditis V LM] additis W
82 amantissimo W LM] deb. amantissime
83 quum LM] cum W
84 æs proprium V LM] æx propium W
85 occultè LM] obcultè W
86 præsidiis V LM] Præsidibus W
87 Christi Incar. MDCXLI. LM] Christo Nato MDCXLII. W
88 Januarii LM] Ianuari W
89 adstarunt LM] astarunt W
90 Parœeciæ LM] Paroecie W
91 Mensæ V LM] Mense W
92 FERDIN. LM] Ferd: W
93 memoranda hæc LM] memoriam hanc W
94 conspecta LM] conspectam W
95 audita LM] auditam W
96 Cultore V LM] cultorem W

For, no natural secrets compatible with human capacity to understand remaining for
him to disclose, and God having allowed the deprivation of his eyesight during the
last five years of his life so that he could better reach out his mind to the Creator,
he devoutly yielded to the Divine Will; for which renunciation he showed a stronger spirit the more passionately he appealed to that sense to discover ever new
things. In the end, consumed by a persistent fever[29]—having lived according
to the custom of good men, secretly and generously contributing his own (not
others') money to the poor, and having offered several examples of remarkable
piety—gradually passing away, having repeatedly required the salutary aids of
the Church, which he devoutly received, strengthened by the Pontifical bene-
diction of Urban VIII,[30] the greatest of philosophers, having frequently invoked
Jesus, most peacefully returned his immortal soul to his Creator, in the year 1642
of Christ's birth, on Wednesday, 8 January, at the fourth hour, in the new style,
aged 77 years, 10 months and 10 days, in the suburban estate of the Martellini in
Arcetri,[31] where for over thirty years[32] he occupied himself with the sciences.

During the last illness, those who constantly assisted the great man and received
his last words were his son Doctor Vincenzo,[33] his daughter-in-law, [8]closest relatives, the parish priest, and two others of extraordinary doctrine and piety, long
before selected by Galileo so that they could purify his soul; and two guests who
were then his table companions: one, the sharpest of mathematicians, Evangelista
Torricelli,[34] for the last three months; the other, for the last three years, the last
and thrice fortunate disciple, who was warmly recommended to GALILEO by
the Most Serene FERDINANDO II,[35] and has placed these inscriptions in
order to convey faithfully to his descendants and to posterity the memory of
what he had seen in his preceptor, and diligently and safely heard about him
from his relatives, friends, and attendants, for the education of Christian philosophers. With the assent and the command particularly of the Most Serene
FERDINANDO, Prince of Tuscany by primogeniture, lover of the arts and the
sciences, and most munificent patron.

In Diaglyptico Phrenoschemate   G

> Este Duces, ô si qua via est.
> VIRGIL. ÆNEID. LIB. VI.

In Diaglyptico Phrenoschemate   H

> In Sole, quis credat? retectas
> Arte tua, Galilæe, labes.
> URB. VIII. P.M.[97]

APPENDIX

[120]MONITUM LECTORI.

EN Tibi, Amice Lector, hoc Anno 1702. cum suis Epigrammatis ære incisam Orthographiam Ædium A DEO DATARUM, unde tandem in lucem prodit Secunda Geometrica hæc Divinatio, quæ post sex & quinquaginta Annos ab Auctore conscripta fuit; & cujusmodi novem ac viginti ab hinc Annis fuerat typis impreßa. Habes hìc ejusdem Auctoris grati animi monumenta: tum erga potentissimum Galliarum Regem LUDOVICUM MAGNUM, cujus amplissimis Honorariis Ædes ipsæ comparatæ sunt, & instauratæ: tum erga Celsitudines Regias Mediceæ Gentis, Patronos clementissimos, quorum profusam liberalitatem ab Anno ætatis suæ XVI. est expertus: tum erga Præceptorem amantissimum GALILÆUM, cui, quantulumcumque id est, quod in Geometria progreßus est Auctor, totum se debere profitetur. Tantas ergo beneficentias, quum apud Posteros testatas ipse relinquere cuperet, & ingravescente ætate, afflictaque valetudine, ac ingruente mortis periculo, omnes alias vias præclusas eße animadverteret; Anno Sal. CIƆ DC LXXXXIII. Elogia hæc in fronte earumdem Ædium, quàm citissimè fieri potuit, inscribi jussit. Nunc ut ad exteros etiam, qui non peregrinantur, sempiternò propagetur hæc sua grati animi significatio, typis ea, ut vides, mandari curavit; ut (si fortè in posterum hæc ipsa, aut temporis edacis culpa, aut succeßorum in Ædibus voluntate, ad alia substituenda, fuerint abrasa) in indelebili Eruditorum memoria perpetuò maneant.

---

97 *In Diaglyptico Phrenoschemate* G—URB. VIII. P.M. *LM*] *deest in W*

In the carved emblem[36]   G

> Be my guides, if any way there be.
> VIRGIL, *AENEIS*, VI[37]

In the carved emblem   H

> in the Sun (who would believe it?)
> laid bare by your art, Galileo.
> POPE URBAN VIII[38]

APPENDIX

[120]ADVICE TO THE READER

Behold, dear Reader, in this year 1702, the engraving of the façade and the inscriptions of the House GIVEN BY GOD, whence is brought to light this Second Geometric Divination, which the author wrote fifty-six years ago, and now appears as it was printed twenty-nine years ago. Here you can find the memorials of a grateful soul: to the most powerful King of France LOUIS THE GREAT, thanks to whose most generous gifts this house was built and restored; to the Royal Highnesses of the Medici Family, most gentle patrons, whose lavish generosity the author has experienced ever since he was sixteen years old; and to GALILEO, most beloved teacher, to whom the author declares he owes the results he achieved in geometry, however little they are. Hence, wishing to leave a testimony of such great benevolence to posterity, and realizing that, due to advancing age, debilitating health conditions and the approaching threat of death, all other ways are hindered, in AD 1693, he [the author] ordered that these elogia be inscribed, as quickly as possible, on the façade of this very house. Now, in order that this expression of his grateful soul may forever also reach foreigners, who do not travel, he published it, as you can see—so that (in case these elogia might be scratched off, either by time all-consuming, or by decision of those who inherit this house, in order to replace them with others) they might remain indelible in the memory of educated people.

1. Louis XIV (1638–1715), also known as Louis the Great (*Louis le Grand*) or the Sun King (*le Roi Soleil*), King of France and Navarre from 1643. He was named Louis Dieudonné (Louis the God-given), hence Viviani's pun. Viviani appeals to the same pun twice in the first few pages of *De locis solidis* (1701), ff. ✠4r and 5r); and again on the title page of the *Grati animi monumenta* (1702).

2. Viviani's palace is in via Sant'Antonino 11, formerly via dell'Amore ("Street of Love") 13, in Florence.

3. Rheticus refers to Copernicus by calling him "D. Praeceptor meus," too: see *Narratio prima* (1540), f. Aijr, and *passim*.

4. That is, the Copernican system. Philolaus of Croton (ca. 470–ca. 385 BC) was the first to suggest displacing Earth from the center of the cosmos and making it a planet, setting it in motion around a central fire. The disguised reference to Copernicus (whose *De revolutionibus* was suspended *donec corrigatur* in 1616), and to the ebb and flow of the sea, clearly points to Galileo's *Dialogue*, banned by the Inquisition in 1633 and listed in the *Index* of forbidden books in 1634. This is the only implicit reference to Galileo's *Dialogue* and his defense of the Copernican hypothesis in the *Grati animi monumenta*.

5. This passage is reminiscent of a famous section from the first Dialogue of Giordano Bruno's *Ash Supper Wednesday*: "Now here is he who has pierced the air, penetrated the sky, wandered among the stars, traversed the boundaries of the world, dissolved the fictitious walls of the first, eighth, ninth, tenth spheres, and the many more you could add to these according to the theories of vain mathematicians and the blind vision of popular philosophers. Thus, aided by all senses and reason, with the key provided by his most diligent inquiry, he laid open those enclosures of truth that it is possible for men to open, and laid bare shrouded and veiled nature. He gave eyes to moles and light to the blind, who could not fix their gaze and behold their own image reflected in so many mirrors which surround them on every side. He untied the tongues of the mute, who could not and dared not express their entangled thoughts. He restored strength the lame, who were unable to make that progress in spirit which the ignoble and dissolvable compound [i.e., the body] cannot make. He provided them with no less a presence [vantage point] than if they were the actual inhabitants of the Sun, of the Moon, and of other known stars [i.e., wandering stars, that is, planets]" ("Or ecco quello ch'ha varcato l'aria, penetrato il cielo il cielo, discorse le stelle, trapassati gli margini del mondo, fatte svanir le fantastiche muraglia de le prime, ottave, none, decime, et altre che vi s'avesser potute aggiongere sfere per relazione de vani matematici e cieco veder di filosofi volgari. Cossì al cospetto d'ogni senso e raggione, co la chiave di solertissima inquisizione aperti que' chiostri de la verità che da noi aprir si posseano, nudata la ricoperta e velata natura: ha donati gli occhi a le talpe, illuminati i ciechi che non possean fissar gli ochi e mirar l'imagin sua in tanti specchi che da ogni lato gli s'opponeno. Sciolta la lingua a muti, che non sapeano e non ardivano esplicar gl'intricati sentimenti; risaldati i zoppi che non valean far quel progresso col spirto, che non può far l'ignobile e dissolubile composto: le rende non men presenti, che si fussero proprii abitatori del sole, de la luna, et altri nomati astri": Bruno, *La cena de le Ceneri* (1584), pp. 47, 49; English translation, p. 61).

6. By the term *aemulari* ("imitate") Viviani implicitly refers to the anagram with which Galileo announced the discovery of the phases of Venus to the Grand Duke's ambassador to Rudolph II, in Prague: see Galileo to Giuliano de' Medici, 11 December 1610, in *OG* 10, no. 435, p. 483. Once decoded, three weeks later (Galileo to Giuliano de' Medici, 1 January 1611, in *OG* 11, no. 451, p. 12), it read: *Cynthiae figuras aemulatur mater amorum*, that is, "the mother of love (i.e., Venus) imitates the faces of Cynthia (i.e., the Moon)."

7. See *Psalms* 18:22 (= KJV 19:1): "caeli enarrant gloriam Dei et opera manuum eius adnuntiat firmamentum" ("The heavens declare the glory of God, and the firmament shows His handiwork").

8. In the *Dialogue*, Salviati (the spokesperson for Galileo) says fixed stars are suns; see p. 327: "fixed stars, which are many suns, agree with our Sun in enjoying perpetual rest" ("le stelle fisse, che sono tanti Soli, conforme al nostro Sole godono una perpetua quiete": *OG* 7, p. 354). Surprisingly, however, Viviani ascribes to Galileo the idea (daringly close to Giordano Bruno's, although no reference is made to possible inhabitants) of fixed stars with their own planetary systems, of which there is no trace in Galileo's works. Perhaps, the two of them talked about this in private conversations.

9. Without further specification, the date is computed according to the Florentine style (*ab Incarnatione*). According to our calendar, the discovery took place on January 9, 1610.

10. Theories, tables, and ephemerides were the three standard genres of astronomical works in the Renaissance and early modern times; they belonged to distinct genres, although they often happened to be different parts of the same work. Theoretical works included Ptolemy's *Almagest*, Copernicus's *De revolutionibus*, Tycho's *De Mundi Aetherei Recentioribus Phaenomenis*, and Kepler's *Astronomia Nova*. They dealt with cosmological theories. Astronomical tables were designed to facilitate the computation of the positions of the Sun, the Moon, and the planets relative to the fixed stars, lunar phases, eclipses, and calendrical information. Ptolemy's *Almagest* and Copernicus's *De revolutionibus*, for instance, also included astronomical tables; when independently published, they often included explanations of astronomical instruments. Most important were the *Toledan Tables*, completed around the year 1080 by a group of Arabic astronomers in Toledo; the

Alfonsine Tables (first published in 1483), named after Alfonso X of Castile, who supported their creation; the *Prutenic Tables*, compiled by Erasmus Reinhold on the basis of Books 2–6 of Copernicus's *De revolutionibus*, and published in 1551; and Kepler's *Rudolphine Tables*, published in 1627, using the observational data gathered by Tycho over four decades. Finally, ephemerides (literally, "calendars") tabulated the positions of naturally occurring astronomical objects in the sky at regular intervals of date or time.

11. Philip III of Spain (1578–1621), who reigned from September 13, 1598.

12. Giovanni Domenico Cassini (1625–1712), Italian (naturalized French) mathematician, astronomer, and engineer.

13. As Nelli correctly remarked, here Viviani is actually referring to King Sigismund III Vasa (1566–1632), who was crowned on December 27, 1587: see Nelli, *Vita e commercio letterario di Galileo Galilei* (1793), vol. 2, p. 861. Possibly, Viviani confused King Sigismund with his son John II Casimir (1609–1702), King of Poland and Grand Duke of Lithuania, who reigned from 1649 to 1669.

14. The rather convoluted structure of E is as follows: "The very last disciple of so great a man" (i.e., Viviani) "first publicly set up" "this bronze bust" "To Galileo Galilei" "For . . . For . . . For . . . Finally, for." Viviani is talking about himself, listing the reasons why he is dedicating the bust to his master, Galileo.

15. The word *inquam* ("I say") resumes the text of D, which starts with a dedication to Galileo, too.

16. Ferdinando I de' Medici (1549–1609), third Grand Duke of Tuscany (1587–1609); Cosimo II de' Medici (1590–1621), fourth Grand Duke of Tuscany (1609–1621); and Ferdinando II de' Medici (1610–1670), fifth Grand Duke of Tuscany (1621–1670).

17. That is, Vincenzo Viviani himself. We find this epithet also in the title pages of Viviani's *Qvinto libro degli Elementi d'Evclide* (1674), *De locis solidis* (1701), and *Formazione, e misvra di tvtti i cieli* (1692), as well as in his will, *Testamento* (1735), p. 3. On April 4, 1692, he published a mathematical challenge, "Ænigma Geometriccm de miro opificio Testudinis Quadrabilis Hemisphæricæ," under the name "D. Pio Lisci Posillo Geometra," that is, "Postremo Galilaei Discipulo" ("Galileo's last pupil"); see also *Gal.* 70, 11ᵃr and *Gal.* 179, 1r–2v (where he engages in several anagrams); *Gal.* 183, 17r, where he offers different anagrams of "Ultimo scolare di Galileo," the title he chose for himself; and *Gal.* 243, 199r, where he comes up with "L'umido e gelato Carlo Lilles" ("The wet and cold Carlo Lilles") as a transposition of "L'ultimo scolare del Galileo." When addressing the "Noble beginner geometers" ("Nobili geometri principianti"), in the *Qvinto libro degli Elementi d'Evclide*, f. *5r, Viviani wrote: "There might be someone who will blame my self-description as the last pupil of Galileo, on the title page of this work, on excessive yearning; but many more will be those who will crave for it. As a matter of fact, I was very fortunate to be his very last pupil, as he was my teacher continuously, throughout the last three years of his life; and of those who were present when he breathed his last (besides two priests, Torricelli, Doctor Vincenzo Galilei his son, and other relatives of his), I alone—although the least to take advantage from this—have outlived them all and am nearly the last of those who assiduously frequented him" ("FORSE alcuno vi sarà che m'attribuirà a soverchia ambizione il palesarmi in fronte di quest'Opera per ultimo Discepolo del Galileo; ma però molti più saranno quei, che me n'invidieranno. Il fatto si è che, per mia grande ventura, io son l'ultimo suo Discepolo, perché egli mi fu continuo Maestro per gli ultimi tre anni di sua Vita, e di quanti ci trovammo presenti all'ultimo suo respiro, (che oltre a due Sacerdoti, v'interuennero il Torricelli, il Dottor Vincenzio Galilei suo figliuolo, e gli altri di sua Casa) io solo, (benché l'ultimo, nell'essermene approfittato) sono a tutti sopravvissuto, e quasi anche rimasto l'ultimo di quanti più intimamente lo praticarono"). Indeed, all other pupils of Galileo's died shortly after him, or at any rate before 1674, when Viviani first used the epithet: Benedetto Castelli in 1643; Michelangelo Buonarroti the Younger and Mario Guiducci in 1646; Bonaventura Cavalieri, Vincenzo Renieri, and Evangelista Torricelli in 1647; Clemente Settimi probably in the late 1640s or early 1650s; Braccio Manetti in 1652; Giovanni Battista Rinuccini in 1653; Pier Francesco Rinuccini in 1657; Antonio Santini in 1662; Famiano Michelini in 1666; Andrea Arrighetti in 1672. Galileo's son, Vincenzo, died in 1649, and his daughter Livia in 1659; Galileo's other daughter, Virginia, had already died in 1634. Galileo outlived some of his pupils, too: Niccolò Aggiunti died in 1635, Niccolò Arrighetti in 1639, Dino Peri in 1640, and Iacopo Soldani in 1641. No doubt, of all of them Viviani was the most devoted disciple, and spent his entire life struggling to have Galileo's works and correspondence published, as well as his memory celebrated.

18. Inevitably, the pun—*aes* ("bronze") and *aer* ("air")—is lost in translation.

19. Viviani is appealing to the literary *topos* of the wise person's good use of wealth, a recurrent topic in ancient philosophy and literature: see, for example, Seneca, *De vita beata*, 21–26 (esp. 23, 1–3).

20. Cosimo III de' Medici (1642–1723), sixth and second to last Grand Duke of Tuscany (1670–1723).

21. The bronze bust of Galileo was cast by Giovanni Battista Foggini (1652–1725) from a terracotta model, made by Giovanni Battista Caccini (1556–1613) on behalf of Grand Duke Cosimo II: see Viviani to Foggini, 7 December 1691, in *Gal.* 159, f. 256r, as well as Nelli, *Vita e commercio letterario di Galileo Galilei* (1793), vol. 2, pp. 855 and 871. Foggini is also the author of the bas-reliefs in the scagliola cartouches on the two sides of Galileo's bust, above the entrance of the palace: see Lorenzo Bellini to Viviani, 8 February 1693, in *Gal.* 257, f. 120r.

22. *Minerval* is a satirical word referring to the fee paid by pupils for their instruction: see Varro, *Rerum rusticarum libri*, III 2.18, and Juvenal, *Saturae*, X 116. Here, it is a learned reference to the gift, or homage, paid by Viviani to his preceptor, Galileo.

23. Michelangelo Buonarroti (1475–1564). In fact, Michelangelo died on February 18, 1564, in Rome, and was buried in the church of Santi Apostoli: see *Eseqvie del Divino Michelagnolo Buonarroti* (1564), f. A2v. His body was later moved to Florence by his nephew, Lionardo Buonarroti, on March 11; he was buried in the Santa Croce Basilica the next day: see Girolamo Ticciati, "Supplemento alla Vita di Michelagnolo Buonarroti" (1746), p. 62. Galileo's remains were eventually moved to their present location, opposite the tomb of Michelangelo, in 1737, on the same day (12 March) on which Michelangelo was buried: see Giovanni Camillo Piombanti, *Instrumento Notarile*, in the public records of the city of Florence (Archivio di Stato di Firenze, Notarile Moderno, Prot. 25439, 12 March 1737). Whereas in the version of Viviani's *Racconto istorico* reproduced in this volume the date of Galileo's death is stated as February 19, 1564, in the first printed edition (1717), p. 397, the date given is February 15, 1564.

24. See Rossi, pp. 114–115.

25. Vincenzo Galilei, *Dialogo della musica antica, et della moderna* (1581).

26. Giulia Venturi degli Ammannati (1538–1620).

27. Lionardo Buonarroti's (Michelangelo's nephew) household notebooks are available in Florence, Casa Buonarroti, Archivio Buonarroti, vol. 38: *Debitori, Creditori e Ricordi. Cod. segnato B,* c. CXIIII: "1563 [i.e., 1564]. I record that today, 18 February, Friday, at 23:30, Michelangelo, son of Ludovico, son of Lionardo Buonarroti Simoni, passed away. He was 88 years, 11 months, and 14 days old. His body was deposited at Santi Apostoli, in Rome, on Saturday, at 7 pm; and it remained in Rome until the following 2 March, after which it was moved to Florence by Simone, son of Bernardo, a carrier. It was put in San Pietro Maggiore, where it remained for two days; on 22 February, the painters and sculptors of the Florentine academies carried the body to Santa Croce, where it was put in a walled depository so as to preserve it while a sepulcher was prepared. I, Lionardo Buonarroti, went to Rome on 19 February 1564, and got there on 24 February. I found the above-mentioned Michelangelo dead, and sent his body to Florence, as said above: I remained in Rome until 6 May 1564, when I left for Florence; I got there on 12 February" ("An. 1563. Ricordo come oggi questo di xviij di febraio in uenerdi a ore 23² passo di questa presente uita Michelangelo di Ludouico di Lionardo Buonarroti Simoni quale mori in roma auente anni .88. messi 11. dj 14. fu meso in deposito in santi aposto<li> il sabato alli 19 detto<.> in roma stetteui infino alli .2. di marzo prossimo dipoi si fece portare affire<n>ze per le mani di simone di bernardo uetturale. ariuo in fire<n>ze alli 20 di marzo detto e deposito in santo piero magiore doue stette giorni due dipoi alli 22. fu portato in santa croce dalli aca[me]demici di pittori e scultori fiorentini doue si fece uno deposito murato per saluarl[l]o per farli uno sepulcro. Io Lionardo Buonaroti andai a roma alli 19 di febraio 1564 et ariuai alli 24 detto<.> trouai il sudetto michelangelo morto e lo mandai affire<n>ze come di sopra si dice e vi stei fino alli 6 di magio 1564 et mi parti detto di<.> ariuai affire<n>ze alli 12 detto"). See also ibid., c. XXXXVI, as well as *Gal.* 11, ff. 169r–171r.

28. See Giorgio Vasari, "Vita di Michelagnolo Buonarruoti Fiorentino" (1568), p. 774: "And so on February 17, 1563, at 11 pm, Florentine style (1564, in the Roman style) [Michelangelo] breathed his last, and passed away" ("[. . .] & cosi a di 17. di Febraio l'anno 1563. a hore 23. a uso Fiorentino, che al Romano sarebbe 1564. spiro per irsene a miglior uita"). In the *Racconto istorico,* which was written earlier than the text for the inscriptions on the façade of his palace, Viviani chose a different "parallel life," placing Galileo side by side with another eminent Florentine, Amerigo Vespucci: see *Racconto istorico,* pp. 46–47, as well as Viviani, *Racconto istorico* (1717), p. 393. In the preface, Viviani added a parallel with Christopher Columbus: see *Racconto istorico,* pp. 60–61.

29. To further emphasize the parallel between Michelangelo and Galileo, Viviani uses almost identical expressions: Viviani's "Lenta tandem correptus febre" is a loan from Vasari's "in fine della uita sua, [. . .] quando amalatosi Michelagnolo di una lente febbre" (Vasari, "Vita di Michelagnolo Buonarruoti Fiorentino" (1568), p. 744. See also Viviani, *Historical Account,* p. 000, and *Racconto istorico* (1717), p. 423.

30. Maffeo Barberini (1568–1644), elected Pope Urban VIII in 1623.

31. Esaù Martellini (1580–1650), the owner of the Villa Il Gioiello, in Pian de' Giullari, Arcetri, where Galileo spent his last years. Galileo rented it on September 22, 1631 (see *OG* 20, *Supplementi,* no. XLbis (b), pp. 587–588), so that he could live closer to his daughter Virginia (Sister Maria Celeste, 1600–1634), who was in the nearby Convent of San Matteo; see Virginia to Galileo, 12 August 1631, in *OG* 14, no. 2198, p. 288.

32. Giovanni Battista Clemente Nelli corrected Viviani twice: "In fact, eight" ("Imo octo"), he noted in his *Vita e commercio letterario di Galileo Galilei* (1793), vol. 2, p. 866; and "In fact, only nine" ("immo novem tantum"), in Viviani, *Grati animi monumenta,* ed. Nelli (1791), p. 15. Francesco Albèri, in turn, corrected Nelli: "In fact, ten" ("Imo decem"), in *Opere 1842–1856,* vol. 15, p. 380. In fact, whereas Galileo did live in Arcetri from September 1631 to January 1642 (that is, for a little over ten years), here Viviani is not referring to the time Galileo actually lived on the estate of Esaù Martellini but to the time he spent "occupying himself with the sciences." Moreover, Viviani is not likely referring specifically to Galileo's last residence (Villa Il Gioiello) in Arcetri but used this to refer, more generally, to the time Galileo spent in Florence and its surroundings, practicing science and writing about it. Indeed, during the years he spent in Padua (December 1592–September 1611) Galileo often returned to Florence for the summer, living either in one of the summer residences of the Grand Duke (such as the Medici villa at Pratolino, or La Ferdinanda, at Artimino) or with his sister Virginia (in the neighborhood of the Church of the Carmine). And after he took up his position as the Chief

Mathematician and Philosopher of the Grand Duke (save for his journeys, of course, especially those to Rome), he spent "over thirty years" in or near Florence, working on a number of scientific issues—first, in a house he rented on the south shore of the river Arno; then (from August 15, 1617) at the Villa L'Ombrellino, at Bellosguardo; and finally (from September 9, 1631) at the Villa Il Gioiello in Arcetri (in 1638, he was granted permission to move temporarily to his house in Costa San Giorgio, in Florence, next to his son Vincenzo's). Galileo also spent time at the Medici villas La Petraia and in Marignolle, on the outskirts of Florence; at the so-called Villa Michelangelo, in Settignano, hosted by Michelangelo Buonarroti the Younger; and he spent long and fruitful periods at Filippo Salviati's Villa Le Selve, in Lastra a Signa. In Florence, Galileo published *On Floating Bodies* (1612) after extensive discussions documented in several letters; see also the 1625 correspondence on hydraulics (*OG* 13, pp. 291–294). In Florence, he observed the Moon, Jupiter's satellites, lunar eclipses, "tricorporeal" Saturn, the phases of Venus, and sunspots; he also engaged in long negotiations with representatives of the Spanish Court about using the period of Jupiter's satellites to calculate longitudes. In Bellosguardo, Galileo worked on the microscope, on his reply to Francesco Ingoli, and on his theory of tides and other parts of the *Dialogue*, and made several telescopic observations; there he worked on fluid mechanics, too, and offered suggestions for the design of the façade of the cathedral of Florence. During his frequent stays at Salviati's Villa Le Selve, Galileo conceived and wrote the first and third letter on sunspots, and made many telescopic observations of sunspots and Jupiter's satellites, as well as performed several other scientific activities (such as the study of the center of gravity of solids). Finally, in Arcetri, Galileo worked for several years on the *Two New Sciences*, performed many astronomical observations (he discovered the librations of the Moon, for example), responded to objections, and discussed at length with Benedetto Castelli about geometric issues. He also spent considerable time on negotiations about his solution to the problem of longitude with representatives of the United Provinces of Holland, and engaged in exchanges with Fortunio Liceti.

33. Vincenzo Galilei (1606–1649), third born of Galileo and Marina Gamba (1570–1612), after Virginia and Livia, and the only one he recognized; he graduated in law at the University of Pisa, on June 5, 1628 (*OG* 19, no. 27c, pp. 427–430), and married Sestilia Bocchineri (d. 1669), on January 28, 1629.

34. Evangelista Torricelli (1608–1647), of Faenza, one of Galileo's best pupils. He is best known for the invention of the barometer, and his advances in optics.

35. See Viviani to Abbott Salviati, 5 April 1697, in Angelo Fabroni (ed.), *Lettere inedite di uomini illustri*, (1773–1775), no. 2, pp. 6–7: "Ever since I was 17 years old, on his own initiative, the Most Serene Grand Duke Ferdinando began to support me, with a salary out of his own pocket, so that I could provide myself with books on pure mathematics. From then on, he destined me as his own mathematician. And again, on the Most Serene Grand Duke's own initiative, and by his own words, upon the occasion of paying a visit to our great Galileo in Arcetri (as I used to do), the Grand Duke suggested that he welcome me as his pupil. Thanks to this most generous initiative, Galileo immediately and most lovingly accommodated me, as he would do with his own son. There I lived until he lived, following his instructions, for about three years, the last three months of which in the company of Torricelli. With Torricelli, three priests, Galileo's own son, and all his relatives, I attended to the blessed passing of Galileo's great soul to his Creator" ("Cominciai di 17. anni ad esser di proprio moto assistito dal Serenissimo Gran Duca Ferdinando con provvisione dal suo stipo, perché io mi provvedessi de' libri di Matematica speculativa, e fin d'allora mi destinò per suo Matematico. Dal medesimo Serenissimo fui di proprio moto e dalla sua propria bocca raccomandato al nostro gran Galileo in occasione d'esser a visitarlo in Arcetri, come spesso così onorar lo soleva, a ricevermi per suo scolare. Mediante questo benignissimo ufizio, subito il Galileo amorosissimamente mi ci accolse, trattandomi come figliuolo: quivi dimorai finché ei visse sotto la di lui disciplina per lo spazio di tre anni in circa, e negli ultimi tre mesi col Torricelli, con cui presenti tre Sacerdoti, il proprio figliuolo, e tutti di sua famiglia, intervenni al felice passaggio di quella grand'anima al suo Creatore").

36. The adjective *diaglypticus* is a loan from the Greek διάγλυπτος, "carved." For the translation of *phrenoschema*, see Alessandro Donati, *Ars Poetica* (1631), pp. 378–388 (Book 3, Chapter XXXII: "Figurata Epigrammata, vulgò Impresiæ"); and Athanasius Kircher, *Œdipvs Ægyptiacvs* (1652–1654), vol. 2, pp. 7–8 (Chapter II: "De Emblemate, & Impresia, siue Phrenoschemate").

37. Inscriptions G and H appear in the two scagliola cartouches on either side of Galileo's bust. The cartouche on the left presents a bas-relief of a man looking at Jupiter's satellites with the telescope, on the stern of a ship (a reference to Galileo's telescopic discoveries, published in the *Sidereal Messenger*, and to their possible applications in the determination of longitude). The source is Vergil, *Aeneis*, VI 194. See also next footnote.

38. The second cartouche, on the right of the bust above the entrance, represents a man observing sunspots with a telescope, a man watching the (parabolic) course of a cannonball, and a beam breaking under its own weight. The source is Pope Urban VIII's *Adulatio perniciosa*, lines 21–22 (see the appendix). The two bas-reliefs, designed by Foggini, are also cast in a bronze medal, ca. 1680: on one side, Galileo's profile; on the other, besides the symbolic images in the second cartouche, a representation of a pendulum and of free fall, as well as of Jupiter's satellites, the phases of the Moon and Venus, and a comet. On the side with Galileo's profile, we read the inscription "GALILEVS LYNCEVS" ("Galileo, Lyncean"); on the opposite side, the inscriptions "NATVRAMQVE NOVAT" ("Renovator of Nature") and "MEMORIÆ OPTIMI PRÆCEPTORIS VINC. VIVIANIVS" ("To the memory of the Greatest of Preceptors, from Vincenzo Viviani").

FIGURE 27A. Giovanni Antonio Lorenzini (1665-1740). Façade of Palazzo dei Cartelloni; burin and etching. Inscription: "ÆDES A DEO DATÆ LVDOVICI MAGNI HONORIFICIS MVNIFICENTIIS COMPARATÆ AC DENVO CONSTRVCTÆ FLORENTIÆ A NOVISSIMO M. GALILÆI DISCIPVLO ANNO SAL[VTIS] MDCXCIII. | HARVM ÆDIVM PROSPECTVS EST PEDVM REGIORVM PARISIENSIVM IN LONG[ITVDINE] LXXXV. IN ALT[ITVDINE] LI. | F[rater] A[ntonius] Lor[enzini] F[ecit]." // "A house given by God, purchased and renovated

in Florence, thanks to the honorable generosity of Louis the Great, by the very last disciple of Galileo's, in 1693 A.D. | The prospect of this house is 85 Paris Royal Feet [= 27.6 meters] in length, and 51 [= 16.56 meters] in height. | By Father Antonio Lorenzini." In: Vincenzo Viviani, *De locis solidis secunda divinatio geometrica*, Florence: His Royal Highness' Press at Pietro Antonio Brigonci, 1701; and *Grati animi monumenta*, Florence: His Royal Highness' Press at Pietro Antonio Brigonci, [1702]. San Marino, CA: The Huntington Library (by kind permission). See *OGA* 1, S5a, p. 424; see also ibid., S5c, p. 426.

FIGURE 27B. Giovanni Antonio Lorenzini (1665-1740). Diagram of the inscriptions on the façade of Palazzo dei Cartelloni; burin and etching. Inscription: "F[rater] A[ntonius] Lor[enzini] F[ecit]." // "By Father Antonio Lorenzini." In: Vincenzo Viviani, *De locis solidis secunda divinatio geometrica*, Florence: His Royal Highness' Press at Pietro Antonio Brigonci, 1701; and *Grati animi monumenta*, Florence: His Royal Highness' Press at Pietro Antonio Brigonci, [1702]. San Marino, CA: The Huntington Library (by kind permission). See *OGA* 1, S5b, p. 425; see also ibid., S5d, p. 427.

APPENDIX

MAFFEO BARBERINI'S
ADULATIO PERNICIOSA

In late September or early October 1611 Cardinal Maffeo Barberini and Galileo (who had conversed in Rome earlier that year) were invited to attend a banquet hosted by Grand Duke Cosimo II de' Medici. Galileo was requested to present his views on hydrostatics against an opponent (possibly, Flaminio Papazzoni, who had been appointed professor of philosophy at Pisa), who upheld the established Aristotelian view. As the dinner progressed, Barberini sided with Galileo, while another eminent cardinal, Ferdinando Gonzaga, supported Galileo's opponent. The following year, Galileo sent Barberini a copy of his book *On Floating Bodies*, and the two exchanged a few letters on Galileo's observations of sunspots, and on the *querelle* that ensued between him and Christoph Scheiner-Apelles.[1] Later on, in 1613, they had a brief exchange about the Medicean Planets, too.[2]

AN UNEXPECTED POETICAL GIFT

The two must have kept on very good terms, as some seven years later Barberini sent this letter to Galileo:

> Most Honorable Sir,
> [T]he esteem I have always had of Your Lordship and your talents provided me with the subject matter of the poem I am enclosing. In case you find it wanting of any parts that might suit it, please consider only my affection, so long as I intend to dignify it [the poem] with your name. Whence, without anymore lingering in further apologies (I pass on them, given the familiarity I have with Your Lordship), I ask you accept the little demonstration of goodwill I show to you. And by sending my heartfelt regards, I beg God might grant whatever wish you may have.
>
> Rome, 28 August 1620
> As the brother of Your Lordship, Galileo Galilei,
> Cardinal Maffeo Barberini[3]

---

1 See Galileo to Barberini, 2 June 1612, in *OG* 11, no., 684*, pp. 305–311. See also Barberini to Galileo, 5 June 1613, ibid., no. 690, p. 318; Galileo to Barberini, 9 June 1612, ibid., no. 694*, pp. 322–323; Barberini to Galileo, 13 June 1612, ibid., no. 697, p. 325.

2 See Galileo to Barberini, 14 April 1613, ibid., no. 861*, pp. 494–495; and Barberini to Galileo, 20 April 1613, ibid., no. 862, pp. 495–496.

3 *OG* 13, no. 1479, pp. 48–49: "Molto Ill. S.^{re}, / La stima che ho fatta sempre della persona di V. S. et delle virtù che concorrono in lei, ha dato materia al componimento che viene incluso; il quale se mancherà

The enclosed poem was the *Adulatio perniciosa* ("Perilous Flattery"), in which the name of Galileo is explicitly mentioned twice, with references to his telescopic discoveries of Jupiter's satellites, "three-bodied" Saturn, and sunspots. Galileo is the only period figure to be named in the poem (the others being historical or mythical ones), and the only person the author addresses directly, in the second person, thereby indicating that the poem is dedicated to him.[4]

Galileo very much appreciated the kind letter, signed with unusual warmth, and especially the enclosed poem. On September 7, he penned his reply. After a

---

di quelle parti che se gli convengono, havrà ella da notarvi solamente il mio affetto, mentre io intendo d'illustrarlo col puro suo nome. Onde, senza prolongarmi più in altre scuse, che rimetto alla confidentia che io ho in V. S., la prego che gradisca la picciola dimostratione della volontà grande che le porto; et con salutarla di tutto cuore, le desidero dal Signor Iddio qualunque contento. / Di Roma, li 28 di Agosto 1620. / Di V. S. S.ᵉ Galileo Galilei / Come fratello / M. Card.[1] Barberino".

4 The *Adulatio perniciosa* first appeared in print in 1620, in the first edition of Maffeo Barberini's *Poemata*, published in Paris and including 31 poems. Most likely, it was written shortly before Barberini sent it to Galileo. Although Galileo's discoveries date to 1610–1611, and they are mentioned in the previous exchanges between him and Galileo, there are no references to the poem prior to the letter of August 28, 1620. A selection of nine of Barberini's Latin poems was first published in the 1606 posthumous reprint of Aurelio Orsi's poems (originally published in 1589), edited by Giuliano Castagnacci, which also included poems by Melchiorre Crescenzi, Claudio Contulio, Giovan Battista Lauri, Vincenzo Palettari, and Marco Antonio Bonciari: see Orsi et al., *Academicorum Insensatorum Carmina* (1606), pp. 191–200. As soon as Barberini became pontiff, there began a sort of competition to republish his poems. The 1620 edition was reprinted in 1623, in revised form (one of the poems was replaced with a new one, for a total of 31) and supplemented with two letters between Lauri and the Scottish poet John Barclay. This 1623 edition was the reference edition for all subsequent ones, until 1628 (at least six of them appeared): Palermo (1624); Cologne (1626); Vienna (1627); Venice, Bologna, and Codogno (1628). Behind them were the great European powers: the king of France, the archbishop and prince-elector of Cologne, and the Viennese court; or various Italian cities and their local authorities: the Republic of Venice, the cardinal of Palermo, Bologna's Accademia della Notte (Cardinal Barberini governed the legation of Bologna from 1611 to 1614). A new, entirely revised edition—almost certainly edited (or supervised) by the author himself, and prompted by the expiration of the privilege for the Paris 1620 edition, which had been granted for ten years: see Barberini, *Poemata* (1620), p. 103—appeared in Rome, in 1631 (83 poems). It was edited by the Jesuits of the Roman College, and included 35 previously unpublished poems, as well as a description of their various meters. It was published in two versions: a luxury *in quarto* edition, printed in red and black throughout, with woodcut papal arms on the *verso* of the title page, and an engraved title page and portrait of the author by Claude Mellan, after Gian Lorenzo Bernini; and *in octavo*, printed in black and white. The 1631 *editio maior* was the model for the Antwerp 1634 edition (94 poems), elegantly printed by Balthasar Moretus, heir of Christophe Plantin, and enriched with an engraved title page and portrait of Urban VIII by Cornelius Galle the Younger, after a drawing of Peter Paul Rubens. The 1631 edition was also the base for further editions, all published in Rome by The Printing Office of the Reverend Apostolic Chamber: 1635 (107 poems; for the first time, 73 poems in the vernacular, *Poesie toscane*, are appended, in the second half of the volume), 1637 (121 poems), and 1638 (129 poems). The fullest edition, including 144 Latin poems and 81 vernacular ones, was published in Rome, in 1640. There followed an edition published in Dillingen in 1641, with imprint 1640 (144 poems), to which an appendix (with six *Analecta*) was added in 1643. In 1642, King Louis XIII of France honored Pope Urban VIII with an elegant edition of both the *Poemata* (107 poems, based on the Antwerp 1634 edition) and the *Poesie toscane* (73 poems, based on the Rome 1635 edition). In 1726, Joseph Brown published the only edition of the *Poemata* to appear after the pope's death (Oxford: Clarendon Press). Based on the Antwerp 1634 edition, it includes 107 poems. Little can be said about the date of the various poems. According to some indications, the *Poemata* (at least those included up to the Rome 1631 edition) were written *dum in minoribus*, that is, before Barberini received the major orders (1592–1601): see Marco Aurelio Maraldi, Secretary of Secret Briefs, "Vrbanvs Papa VIII. Ad fvtvram rei memoriam," in Barberini, *Poemata* (1631) (page not numbered). In fact, however, the Latin poems were written at different times, throughout Barberini's entire life, and a more precise dating is possible only for some poems, whenever references to specific events or persons are offered. Nor does the order of the poems in the various editions follow a chronological order; rather, the poems are arranged according to their subject. On the publication history of the *Poemata*, see Castagnetti, "I *Poemata* e le *Poesie toscane* di Maffeo Barberini" (1982).

beautifully crafted paragraph in which Galileo expresses his joy and gratitude for the most unexpected gift with which the cardinal had honored him and makes himself available if there is any way to reciprocate Barberini's consideration and esteem, Galileo adds:

> Your Most Eminent and Reverend Lordship's Ode seemed admirable to all those who are knowledgeable about poetry. I do not add my opinion to their own, as it is by its very nature imperfect, and still too much enthralled by the huge favor you made to me by mentioning my name twice in your most learned poem. I shall not say that in order to show the brightness of your mind you decided to cast light on darkness, but I will say that, by an excess of kindness, you decided to disclose to the world your affection to me. This I deem the highest honor that could have possibly been bestowed on me—and for this honor, as I am unable to do anything to reciprocate it, I thank you profusely, and by humbly bowing in front of you, I kiss your hem, and pray God that the greatest felicity shall be yours.[5]

Galileo knew very well how to appreciate the singular and generous gift from an eminent Prince of the Church, and perfectly understood the proper deference that was expected of him. But amidst the carefully worded expressions of gratitude and devotion, he called attention to the cardinal's renewed support for his work, against all the resistance he was facing. The explicit mention of Galileo's name and reference to his discoveries, Galileo says, "casts light on darkness," the "dark side" being all those—either fellow scientists, or representatives of the Church hierarchy (or both, as in the case of Jesuit scholar Christoph Scheiner)—who strenuously opposed his criticism of the established view.

The *Adulatio perniciosa*

Like most of Barberini's poems, the *Adulatio perniciosa* is a mere display of learning and exercise of technical skills. It may have some kind of formal elegance, but it is utterly devoid of poetic inspiration. In fact, the author manages to insert a couple of explicit references to Galileo's telescopic discoveries, amidst a long series of rather dull commonplaces. Still, the double tribute Barberini paid to Galileo and his 1610 controversial telescopic discoveries was significant, and did not go unnoticed.

The first reference occurs at lines 10–12 (11–12 in the translation), where Barberini praises Galileo for his discovery of Jupiter's satellites (and three-bodied Saturn). Here Barberini takes a strong stance in support of the announcements made in the *Sidereal Messenger* and, by implication, against all those who attacked (at times

---

5 Galileo to Maffeo Barberini, 7 September 1620, in *OG* 13, no. 1481*, p. 50: "La Ode di V. S. Ill.ma e Rev.ma è parsa ammirabile a tutti gl'intendenti, con i giudizii de i quali non porto in schiera il mio, come per sé stesso imperfetto et hora troppo affascinato dalla grandezza del favore usatomi da lei nel nominarmi ben due volte nella sua dottissima composizione. Io non dirò che per mostrar l'eminenza del suo ingegno ella habbia voluto illustrar le tenebre, ma dirò bene che un trabocco di gentilezza habbia voluto scoprire al mondo l'affezione che ella mi porta; e questo reputo io per il maggior honore che già mai avvenir mi potesse: del quale, non potendo altro, le rendo grazie infinite, e con humiltà inchinandomegli le bacio la veste, e dal S. Dio gli prego il colmo delle felicità."

viciously) Galileo in 1610, particularly the Bohemian mathematician Martin Horký, who dismissed Galileo's telescopic discoveries as either flaws in the instrument or cynical attempts on Galileo's part to make easy money; and Lodovico delle Colombe, who clashed swords with Galileo on the 1604 nova and later opposed him in Florence, for his support of Copernicanism.

Even more important, however, is the ninth stanza (lines 34–36), in which Barberini refers to Galileo's observations of sunspots. Galileo announced the discovery in 1610, but German mathematician and astronomer Christoph Scheiner claimed priority over it; there ensued a bitter quarrel, culminating, in 1612, in two epistolary treatises by Scheiner (under the pseudonym Apelles) and in Galileo's *On Sunspots*. The quarrel dragged on for several years and involved other Jesuits at the Roman College, of which Scheiner was a prominent member. By explicitly mentioning Galileo's name and discovery not only did Barberini support the latter's claim of priority but clearly took sides with Galileo about his criticism of Scheiner's interpretation of the phenomenon. By calling attention to Galileo's "art" (*retectas arte tua Galilæe labes*), Barberini took a clear stance for Galileo against Scheiner and the Jesuits. This was of great significance, as in *On Sunspots* Galileo openly advocated heliocentrism, and this caused an uproar of Aristotelian philosophers, especially at the Jesuit Roman College, which eventually led to Cardinal Bellarmino's decree (March 5, 1616) and the inclusion—"until corrected" (*donec corrigatur*)—of Copernicus's *De revolutionibus* in the *Index* of prohibited books (the "salutary edict," as Galileo sarcastically referred to it in the *Dialogue*),[6] not to mention the injunction issued against Galileo himself.

It is easy to see what Galileo read in Barberini's lines: an encouragement to pursue his researches further, to go on making the best use of the "art" that had allowed him to achieve such amazing results—and he could count on the backing of an eminent Prince of the Church. Galileo's feelings and confidence could not but grow stronger three years later, when the author of the *Adulatio perniciosa* rose to the Throne of Saint Peter, and the poem was reprinted in the second edition of the *Poemata* (1623).

"Amazing circumstance"

Cardinal Barberini was very fond of his nephew Francesco, the son of his brother Carlo. After Galileo helped Francesco obtain a doctorate *in utroque iure* at the University of Pisa, in June 1623, Maffeo Barberini wrote to Galileo from Rome, expressing his appreciation and inquiring about Galileo's health conditions. At the bottom of the letter, he hurriedly added a postscript:

> I am very much in Your Lordship's debt for your continuing goodwill toward myself and the members of my family, and I look forward to the opportunity of

---

6 See *OG* 19, pp. 322–323 (Bellarmino's decree followed the report of eleven consultants appointed by the Holy Office, dated 24 February 1616: see ibid., pp. 321–322); and the *Dialogue*, p. 5: "salutary edict" (*OG* 7, p. 29: "salutifero editto").

reciprocating. I assure that I will be more than willing to be of service, in consideration of your great merit, and the gratitude I owe you.[7]

Less than two months later, on August 6, 1623, Maffeo Barberini became Pope Urban VIII; owing to illness, he was crowned on September 29.[8] Shortly thereafter, Galileo wrote to Prince Federico Cesi, the head of the Academy of the Lynxes:

I keep going over and over things of momentous relevance for the Republic of Letters: if we do not make them happen at the time of this amazing circumstance, we cannot hope to see another one like it ever again, at least as far as I am concerned.[9]

Galileo immediately seized the opportunity and dedicated to the pontiff his new book, *The Assayer* (published in October 1623), which had been lingering at the publisher's for some time. The richly engraved title page, redesigned at the very last minute, shows, on the plinth at bottom, the Lyncean flower, a lynx, and two telescopes, calling attention to the role of the Academy of the Lynxes in the printing of the book, as well as indicating the observational basis of Galileo's philosophy, and the source of his reputation. But it also prominently displays, at the top center, a tiara (symbol of the papacy) and the coat of arms of the Barberini family: in a single image, the new pope and the new science were united.

Formally a public letter to Virginio Cesarini—a close friend, fellow Lyncean academician, and newly appointed lord chamberlain to the pontiff—the work was dedicated to Urban VIII. A learned humanist of many intellectual interests, and a

7 "Io resto tenuto molto a V. S. della sua continuata affetione verso di me et li miei, et desidero occasione di corrisponderle, assicurandola che troverrà in me prontissima dispositione d'animo in servirla, rispetto al suo molto merito et alla gratitudine che le devo": Barberini to Galileo, 24 June 1623, in *OG* 13, no. 1561, p. 119.

8 The next day, Francesco Barberini joined the Academy of the Lynxes and was given copies of *On Floating Bodies* and *On Sunspots*. On reporting the news to Galileo, Francesco Stelluti and Federico Cesi also announced that Francesco Barberini will be created a cardinal in the consistory of October 2: "whence we will have in him an eminent cardinal and patron, who, we believe, should be also our benefactor" ("onde haveremo un protettore porporato e principale, che possiamo credere debbia anco essere nostro benefattore": 30 September 1623, in *OG* 13, no. 1579, p. 133). Galileo shared Stelluti's and Cesi's satisfaction and hopes (Galileo to Francesco Barberini, 9 October 1623, ibid., no. 1580*, pp. 133–134), and his feelings were reciprocated by the newly appointed cardinal (Francesco Barberini to Galileo, 18 October 1623, ibid., no. 1584, p. 137). Francesco, who had just turned 27, became his uncle's right hand and was to remain supportive of Galileo, even at the time of the trial, and in the years that followed. In 1633, as the newly appointed Grand Inquisitor, he summoned Galileo to Rome and was part of the Inquisition tribunal investigating him (Francesco did not share Galileo's Copernican views: see Benedetto Castelli to Galileo, 9 February 1630, in *OG* 14, no. 1984, p. 78) but was one of the three members of the tribunal (the other two being Cardinal Gaspar de Borja and Cardinal Laudivio Zacchia) who did not sign the sentence, on 22 June 1633: see *OG* 19, pp. 403 and 406. Soon after Urban VIII's death, Francesco (together with the Barberini family) fell into disgrace with the new Pope, Innocent X (Giovanni Battista Pamphilj), and in 1646 fled to Paris, where he remained under the protection of Cardinal Mazarin. Two years later, he was pardoned, and his properties were restored to him. On his return to Rome, Francesco resumed his role as a patron of the arts, although on a reduced scale. He died in 1679.

9 "Io raggiro nella mente cose di qualche momento per la republica litteraria, le quali se non si effettuano in questa mirabil congiuntura, non occorre, almeno per quello che si aspetta per la parte mia, sperar d'incontrarne mai più una simile": Galileo to Federico Cesi, 9 October 1623, in *OG* 13, no. 1581, p. 135.

patron of the arts and sciences, the pope had befriended many Lyncean Academicians in Rome through the years. Indeed, all the Lynxes signed the dedication collectively:

> In this universal jubilee of humanities, and of virtue itself, when the whole City and especially the Holy See is more resplendent than ever from the placing there of Your Holiness by celestial and divine disposition, and every heart is inflamed to praiseworthy studies and worthy actions, in order to venerate, by imitating it, so eminent an example, we come to appear before you, mindful of infinite obligation for the benefits continually received from your benign hand, and full of joy and contentment to see exalted, on so sublime a throne, such a great Patron. As evidence of our devotion, and as a tribute from our true fealty, we bring you *The Assayer* of our Galileo, the Florentine discoverer not of new lands but of hitherto unseen portions of the heavens, containing investigations of those celestial splendors that usually attend great wonders. This we dedicate and present to Your Holiness as to one who has filled his soul with true ornaments and splendors and has turned his heroic mind to the highest undertakings, and we hope that this discourse upon unusual torches in the sky will be to you a sign of our love and ardent passion to serve Your Holiness and to merit your grace. Meanwhile, humbly inclining ourselves at your feet, we supplicate you to continue favoring our studies with the gracious rays and vigorous warmth of your most benign protection.[10]

The purpose of *The Assayer* was to respond to the *Libra astronomica ac philosophica* (1619) by Orazio Grassi (a prominent Jesuit mathematician at the Roman College, who used the pseudonym Lotario Sarsi), and offer a much better "weighing" of the different arguments about the nature of comets. The *saggiatore* ("assayer") alluded to the fine balance used by goldsmiths for weighing gold and precious metals, in contrast with the ordinary, much less sensitive steelyard balance (*libra*) implied by Grassi's title.[11] But *The Assayer* went well beyond the astronomical issues that fueled

---

10 *The Assayer*, p. 153; *OG* 6, p. 201: "In questo universal giubilo delle buone lettere, anzi dell'istessa virtù, mentre la Città tutta, e spezialmente la Santa Sede, più che mai risplende per esservi la Santità Vostra da celeste e divina disposizione collocata, e non vi è mente alcuna che non s'accenda a lodevoli studi ed a degne operazioni per venerare, imitando, essempio sì eminente, vegniamo noi a comparirle davanti, carichi d'infiniti oblighi per li benefizii sempre dalla sua benigna mano ricevuti, e pieni di contento e d'allegrezza per vedere in così sublime seggio un tanto Padrone essaltato. Portiamo, per saggio della nostra devozione e per tributo della nostra vera servitù, il Saggiatore del nostro Galilei, del Fiorentino scopritore non di nuove terre, ma di non più vedute parti del cielo. Questo contiene investigazioni di quegli splendori celesti, che maggior maraviglia sogliono apportare. Lo dedichiamo e doniamo alla Santità Vostra, come a quella c'ha l'anima di veri ornamenti e splendori ripiena, e c'ha ad altissime imprese l'eroica mente rivolta; desiderando che questo ragionamento d'inusitate faci del cielo sia a lei segno di quel più vivo ed ardente affetto che è in noi, di servire e di meritare la grazia di Vostra Santità. Ai cui piedi intanto umilmente inchinandoci, la supplichiamo a mantener favoriti i nostri studi co' cortesi raggi e vigoroso calore della sua benignissima protezzione."

11 The imprimatur was granted under Pope Gregory XV, Urban VIII's predecessor, who died while the printing of *The Assayer* was slowly progressing. In it, the Dominican Niccolò Riccardi, consultor for the Congregation of the Holy Office (who would later, in his capacity of Master of the Sacred Palace, grant the imprimatur for Galileo's *Dialogue*), wrote: "I have read, by order of the Most Reverend Father, Master of the Sacred Palace, this work *The Assayer*; and besides having found here nothing offensive to morality, nor anything which departs from the supernatural truth of our faith, I have remarked in it so many fine considerations pertaining to natural philosophy that I believe our age is to be glorified by future ages not

the controversy with Grassi. In order to counter his opponent's arguments about rotating solids and fluids (which Grassi offered in connection with the structure of the heavens), Galileo advanced remarks about ships in water, river currents, stirred fluids, adhesion and cohesion, and commented on the Aristotelian conception of motion as the cause of heat, contemplating instances in which heating results from friction. In fact, *The Assayer* was a new attack on Aristotelian science and its stubborn upholders, as well as a methodological manifesto of Galileo's natural philosophy.[12]

In the Vatican gardens

At the beginning of April 1624, Galileo left from Florence and headed toward Rome. After spending a couple of weeks in Acquasparta, hosted by Cesi, he arrived in Rome in the late evening of April 23. The next day, he was granted an audience with the pope, with whom he conversed for one hour. On April 25, he had an equally enjoyable meeting with Cardinal Francesco Barberini.[13] In the next few days, he met with the pope five more times, enjoying extensive exchanges on different subject

---

only as the heir of works of past philosophers but as the discoverer of many secrets of nature which they were unable to reveal, thanks to the deep and sound reflections of this author in whose time I count myself fortunate to be born, when the gold of truth is no longer weighed in bulk and with the steelyard, but is assayed with so delicate a balance" (*The Assayer*, p. 152; *OG* 6, p. 200: "Ho letto per ordine del Reverendissimo P. Maestro del Sacro Palazzo quest'opera del Saggiatore; ed oltre ch'io non ci trovo cosa veruna disdicevole a' buoni costumi, né che si dilunghi dalla verità sopranaturale di nostra Fede, ci ho avvertite tante belle considerazioni appartenenti alla filosofia naturale, ch'io non credo che 'l nostro secolo sia per gloriarsi ne' futuri di erede solamente delle fatiche de' passati filosofi, ma d'inventore di molti segreti della natura ch'eglino non poterono scoprire, mercè della sottile e soda speculazione dell'autore, nel cui tempo mi reputo felice d'esser nato, quando non più con la stadera ed alla grossa, ma con saggi sì delicati, si bilancia l'oro della verità").

12 Besides the famous passage in which Galileo claims that the book of nature is written in the language of mathematics, and those who wish to read it must know its language (and not simply act as philosophers *in libris*, relying on Aristotle's authority: see *Racconto istorico*, footnotes 178 and 92), much of Galileo's epistemological manifesto is offered in the engraved title page. The two central standing figures—Natural Philosophy (left), holding the philosopher's sphere, the symbol of the real heavens; and Mathematics (right), holding the mathematical astronomer's armillary sphere—are both represented on the title page, and are both given the same relevance. For the Jesuits, natural philosophy and mathematics were distinct disciplines: philosophy was the business of philosophers, and not of mathematicians (at the time, astronomy was part of mathematics; it was supposed to provide models that could account for the observations of the planets and allow for the computation of their future positions, whereas it was the task of natural philosophy to provide a true description of the actual structure of the universe). In Galileo's eyes, by contrast, cosmology and mathematics are simply two faces of the same coin, and have a lot to say to each other. To borrow Kepler's words from the back of the title page of the *Astronomia nova* (in which he first revealed that it was the Lutheran theologian Andreas Osiander, and not Copernicus, who was the author of the unsigned preface to the *De revolutionibus*, arguing that mathematics, and not physics, should be the basis for properly understanding the new theory: the task of the astronomer is to devise hypotheses, and these need not be true or probable but simply have to provide computations consistent with observations): "Copernicus does not mythologize [i.e., he does not offer fictitious models] but seriously provides paradoxes [i.e., seemingly absurd theories, apparently contradicting common sense and observations], that is, he philosophizes [i.e., he performs the task of the natural philosopher]—which is what you would expect of an astronomer" ("Non igitur μυθολογεῖ Copernicus, sed serio παραδοξολογεῖ, hoc est, φιλοσοφεῖ: quod tu [i.e., Petrus Ramus] in Astronomo desiderabas": Kepler, *Astronomia nova* (1609), p. 4). This was one of Galileo's most deep-seated beliefs. When he was appointed at the Medici court, in 1610, he insisted on the dual title of the Grand Duke's Chief Mathematician *and* Philosopher. See Galileo to Belisario Vinta, 7 May 1610, in *OG* 10, no. 307, p. 353.

13 See Galileo's report to Curzio Picchena, 27 April 1624, in *OG* 13, no. 1628*, p. 175.

matters—except for the Copernican hypothesis, which was never mentioned. As Galileo later reported to Cesi:

Most importantly, I received great honors and favors from Our Lordship [Urban VIII]: I was granted six audiences, during which we conversed at length. And yesterday, when I took leave from him, he steadfastly promised a pension for my son: Monsignor Ciampoli will pursue this matter on my behalf, as ordered by His Holiness.[14] And three days earlier he presented me with a beautiful painting and two medals, one of gold, the other of silver, and several *Agnus Dei*.[15] I found the usual benevolence in Cardinal [Francesco] Barberini, just as I did in his father [Carlo] and his brothers [Antonio and Taddeo].

Among the other cardinals, I greatly enjoyed various meetings with Santa Susanna [Scipione Cobelluzzi], [Francesco] Boncompagni, and [Friedrich Eutel von] Zollern, who left yesterday for Germany. He told me he spoke with Our Lordship about Copernicus, and told him that all heretics [i.e., the Protestants] accept the Copernican opinion and hold it as most certain; consequently, one has to be very cautious in arriving at any resolution about it. To this, His Holiness replied that the Holy Church had not condemned it, nor was about to condemn it, as heretical but only as temerarious. Nor, he added, should one fear that it could ever be proved to be necessarily true.

However far from being able to delve into the details of astronomical speculations as one might wish, Father Monster [Niccolò Riccardi][16] and [Caspar] Schoppe are nonetheless firmly of the opinion that this is not a matter of faith, nor that Scriptures should be brought in. And as far as truth and falsehood are concerned, Father Monster does not uphold Ptolemy or Copernicus but rests content with his own simple way of having angels move the heavenly bodies as these go, without the slightest difficulty or puzzle. And that's got to be enough for us.[17]

14 The pension (60 scudi) was eventually granted to Galileo's son, Vincenzo Galilei, on March 20, 1627: see *OG* 19, pp. 460–462, and Francesco Barberini to Galileo, 12 May 1627, in *OG* 13, no. 1816, p. 356. Upon Galileo's request, the pension was later transferred to Galileo's nephew (Vincenzo Galilei, the son of Michelangelo, Galileo's younger brother) and subsequently revoked upon Galileo's request, and passed on to himself on September 9, 1628: see *OG* 19, pp. 462–464. On February 12, 1630, Urban VIII granted a pension (40 scudi) to Galileo himself: see ibid., pp. 465–468; this was later increased to 100 scudi: see Giovanni Ciampoli to Galileo, 10 August 1630, in *OG* 14, no. 2046, p. 133. A second pension (60 scudi) was granted to him by the pope on November 4, 1632: see *OG* 19, pp. 471–472.

15 Cakes of wax made from Easter candles and stamped with an impression of a lamb bearing a cross, or a flag, emblematic of Christ ("the Lamb of God").

16 Niccolò Riccardi, who succeeded Niccolò Ridolfi as Master of the Sacred Palace in 1629, was nicknamed "Father Monster," either for his extraordinary memory and eloquence, or his unusual obesity.

17 Galileo to Federico Cesi, 8 June 1624, in *OG* 13, no. 1637, pp. 182–183: "[. . .] ho principalmente ricevuti grandissimi honori e favori da N. S., essendo stato fin a 6 volte da S. Santità in lunghi ragionamenti; et hieri, che fui a licentiarmi, hebbi ferma promessa di una pensione per mio figliuolo, per la quale resta mio sollecitatore, di ordine di Sua Santità, Mons. Ciampoli; e 3 giorni avanti fui regalato di un bel quadro e 2 medaglie, una d'oro e l'altra di argento, e buona quantità d'*Agnus Dei*. Nel Sig. Cardinal Barberino ho trovato sempre la sua solita benignità, come anco nell'Eccellentiss. Sig. suo padre e fratelli. Tra gli altri Signori Cardinali, sono stato più volte con molto gusto in particolare con Santa Susanna, Buoncompagno e Zoller, il quale partì hieri per Alemagna, e mi disse haver parlato con N. S. in materia del Copernico, e come gli heretici sono tutti della sua opinione e l'hanno per certissima, e che però è da andar molto circospetto nel venire a determinatione alcuna: al che fu da S. Santità risposto, come Santa Chiesa non l'havea dannata né era per dannarla per heretica, ma solo per temeraria, ma che non era da temere che alcuno fosse mai per dimostrarla necessariamente vera. Il P. Mostro e 'l Sig. Scioppio, benché sieno assai lontani

The pope's words were likely to be much less than what Galileo had hoped for. He was unable even to broach the matter in his own conversations with the pope and had to rest content with a secondhand report, however trustworthy. In Urban VIII's eyes, the Sun-centered universe remained an unproven hypothesis and one without any prospect of proof in the future: astronomers cannot in any way attain to true causes, and their models are, by their very nature, mere conjectures. The pope's position was the same Andreas Osiander had stated in the preface to Copernicus's *De revolutionibus*, in 1543, and all Galileo knew for sure was that in February 1616 Cardinal Bellarmino had ordered Copernicus's work to be listed in the *Index* of prohibited books.[18]

Bellarmino had further admonished Galileo that the Copernican opinion was erroneous and had enjoined him to abandon it completely, and henceforth not to hold, teach, or defend it in any way whatsoever, either orally or in writing; otherwise, the Inquisition would proceed against him. Galileo acquiesced to Bellarmino's injunction and promised to obey.[19] Eight years later, in 1624, in what was doubtless a very successful journey to Rome, during which Galileo was granted six extraordinary audiences with the newly elected pope, it was Urban VIII himself who said, in not unclear terms, that although the Church had not condemned the proposition that the earth moves and the Sun stands still at the center of the world as formally heretical, such a proposition was "temerarious." It asserts, that is, a thesis that contrasts

---

dal potersi internar quanto bisognerebbe in tali astronomiche speculazioni, tuttavia tengono ben ferma opinione che questa non sia materia di fede, né che convenga in modo alcuno impegnarci le Scritture. E quanto al vero o non vero, il Padre Mostro non aderisce né a Tolomeo né al Copernico, ma si quieta in un suo modo assai spedito, di mettere angeli che, senza difficoltà o intrico veruno, muovano i corpi celesti così come vanno, e tanto ci deve bastare".

18 On February 24, 1616, eleven consultants to the Holy Office declared that the proposition that the Sun is motionless at the center of the world is "foolish and absurd in philosophy, and formally heretical, as it explicitly contradicts in many places the sense of Holy Scripture, according to the literal meaning of the words and according to the established interpretation and understanding of the Holy Fathers and the doctors of theology" ("Omnes dixerunt, dictam propositionem esse stultam et absurdam in pbilosophia, et formaliter haereticam, quatenus contradicit expresse sententiis Sacrae Scripturae in multis locis secundum proprietatem verborum et secundum communem expositionem et sensum Sanctorum Patrum et theologorum doctorum": *OG* 19, p. 321); and that the proposition that the earth is not at the center of the world, and it moves both around the Sun and around its own axis, is equally foolish and absurd in philosophy, and "in regard to theological truth, it is at least erroneous in faith" ("Omnes dixerunt, hanc propositionem recipere eandem censurarm in philosopbia; et spectando veritatem tbeologicam, ad minus esse in Fide erroneam": ibid.). On March 5, Copernicus's *De revolutionibus* (1543), together with Diego de Zúñiga's *In Iob Commentaria* (1591) and Paolo Antonio Foscarini's *Lettera... Sopra l'Opinione de' Pittagorici, e del Copernico* (1615), were ordered to be listed in the *Index*: see ibid., pp. 322–323.

19 On February 25, 1616, Cardinal Giovanni Garzia Millini ordered Bellarmino to summon Galileo and warn him: see *OG* 19, p. 321. The next day, Galileo met with Bellarmino and instructed him to abandon the Copernican opinion. Galileo promised to obey: "[...] the aforementioned Father Commissary [i.e., Cardinal Bellarmino], in the name of His Holiness the pope and the whole Congregation of the Holy Office, ordered and enjoined the said Galileo, who was himself still present and in attendance, to abandon completely the abovementioned opinion that the Sun is at the center of the world and the Earth moves, and henceforth not to hold, teach, or defend it in any way whatever, either orally or in writing; otherwise the Holy Office would proceed against him. The same Galileo acquiesced in this injunction and promised to obey" ("[...] supradictus P. Commissarius praedicto Galileo adhuc ibidem praesenti et constituto praecepit et ordinavit [proprio nomine] S.^mi D. N. Papae et totius Congregationis Officii, ut supradictam opinionem, quod sol sit centrum mundi et immobilis et terra moveatur, omnino relinquat, nec eam de caetero, quovis modo, teneat, doceat aut defendat, verbo aut scriptis; alias, contra ipsum procedetur in S.^to Officio. Cui praecepto idem Galileus aquievit [*sic*] et parere promisit" (ibid., p. 322).

with the current view of the Church, as passed down from the Fathers, even though it does not touch upon issues that are articles of faith.

Before departing from Rome, Galileo paid a last visit to Cardinal Francesco Barberini, who gave him letters for the Grand Duke Ferdinando II and the Archduchess Maria Maddalena.[20] Urban VIII gave him an ornate letter for the Grand Duke, too, written in Latin by his secret chamberlain, Giovanni Ciampoli. Once again, the pope expressed his support and benevolence toward Galileo:

> Recently [. . .], our beloved son Galileo entered the ethereal spaces, cast light on unknown stars, and plunged into the inner recesses of the planets. Wherefore, while the favorable star of Jupiter will twinkle in the sky accompanied by its four new attendants, it will bring along, as company for life, the praise of Galileo. So great a man, whose fame shines in the heavens and spreads through Earth, we have embraced with paternal affection for a long time. We know that in him [lie] not only the glory of humanities but also a profound sense of devotion; he is very good at the arts the pope greatly enjoys. Now, however, after he came to Rome to congratulate on our papacy, we embraced him most affectionately, and were pleased to hear that he keeps contributing to the ornament of the Florentine eloquence in learned disputations. Accordingly, we could not let him return to his homeland, whither he is drawn by the benefices granted by your nobility, without bidding farewell, commensurately with the pope's affection. We have considered the benefices the Grand Dukes rewarded him with, for his wonderful discoveries, he who placed among the stars the glory of the Medici family. Actually, many people say they are not at all amazed that virtues abound so much in this city [i.e., Florence], in which the magnanimity of its rulers fosters them with such extraordinary benefices. However, in order that you know how dear he is to the Pontiff, we wanted to give him this honourable testimony of virtue and devotion. Accordingly, we wish to openly state that all the benefices your nobility will grant him—by not merely imitating the generosity of a father but also increasing on it—will give us comfort.[21]

A few days later, Galileo was back in Florence. Cardinal Bellarmino had died in 1621, and Urban VIII had just renewed his goodwill and support, both of him and the

20 See Francesco Barberini to Maria Maddalena d'Austria and Ferdinando II de' Medici, 8 June 1624, in *OG* 13, nos. 1639 and 1640, pp. 184–185.

21 Urban VIII to Ferdinando II, 8 June 1624, in *OG* 13, no. 1638, p. 184: "Nuper autem dilectus filius Galilaeus, aethereas plagas ingressus, ignota sydera illuminavit, et planetarum penetralia reclusit. Quare, dum beneficum Iovis astrum micabit in coelo quatuor novis asseclis comitatum, comitem aevi sui laudem Galilaei trahet. Nos tantum virum, cuius fama in coelo lucet et terras peragrat, iamdiu paterna charitate complectimur. Novimus enim in eo non modo literarum gloriam, sed etiam pietatis studium; iisque artibus pollet, quibus Pontificia voluntas facile demeretur. Nunc autem, cum illum in Urbem Pontificatus nostri gratulatio reduxerit, peramanter ipsum complexi sumus, atque iucundi idemtidem audivimus Florentinae eloquentiae decora doctis disputationibus augentem. Nunc autem non patimur eum sine amplo Pontificiae charitatis commeatu in patriam redire, quo illum Nobilitatis tuae beneficentia revocat. Exploratum est, quibus praemiis Magni Duces remunerentur admiranda eius ingenii reperta, qui Medicei nominis gloriam inter sydera collocavit. Quin immo non pauci ob id dictitant, se minime mirari tam uberem in ista civitate virtutum esse proventum, ubi eas dominantium magnanimitas tam eximiis beneficiis alit. Tamen ut scias quam charus Pontificiae menti ille sit, honorificum hoc ei dare voluimus virtutis et pietatis testimonium. Porro autem significamus, solatia nostra fore omnia beneficia, quibus eum ornans Nobilitas tua paternam munificentiam non modo imitabitur, sed etiam augebit."

Academy of the Lynxes. No doubt, Galileo had good reason to believe the moment had come to go back to Copernicanism. True, he could not expect the Church to endorse it, but he could at least be sure of its impartiality. The fruitful journey and extensive exchanges with Urban VIII gave him new confidence; he felt he could finally resume work on the project he had been toying with since the publication of the *Sidereal Messenger*, in 1610, and write what would eventually become the *Dialogue Concerning the Two Chief World Systems*.

Meanwhile, Galileo's friends and supporters were growing in strength and influence. In Rome, the group included the Grand Duke's uncle and official representative in the Curia, Cardinal Carlo de' Medici; many of the names in the guest lists of Caterina Riccardi, the wife of the Florentine ambassador, Francesco Niccolini; Caterina's distant cousin, Niccolò Riccardi (Father Monster), who was in charge of licensing books for publication; Giovanni Ciampoli, the pope's secret chamberlain and secretary for briefs; the whole Magalotti family, of whom Lorenzo had been vice-legate of Cardinal Maffeo Barberini, in Bologna, and had just been elevated to the cardinalate by Urban VIII; and Lorenzo's sister, Costanza, was married to the pope's brother, Carlo Barberini. The "Galileans" also included the Rinuccini family, particularly Giovanni Battista, Archbishop of Fermo, and his brother Tommaso, who had worked for six years for the pope's nephew, Francesco Barberini, before becoming consul of the Accademia Fiorentina in 1631; and, to some extent, also the pope's own family, the Barberini, who was of Florentine descent. Galileo had the admiration of Cardinal Pietro Sforza Pallavicino, the tacit approval of the Dominican Niccolò Ridolfi, and the unequivocal support of the Scolopians, the Fathers of the "Pious Schools."

No doubt, Galileo's and his supporters' strong confidence was based on an understanding of what was at stake that was very different from the pope's own. Whereas for Urban VIII the Copernican opinion was not only an unproven hypothesis but one that had no real prospect of proof in the future, for Galileo it was far from temerarious, as he felt sure he could provide solid and compelling confirmations for it. Chief among them were the tides, which Galileo saw as the most solid argument in favor of the motion of the earth—so much that the *Dialogue* was supposed to be titled *Dialogue on Ebb and Flow of the Sea* (*Dialogo sul flusso e reflusso del mare*).[22] As Galileo would later have Salviati, his spokesperson, say toward the end of the *Dialogue*:

> [. . .] the ordering of the world bodies and the integral structure of that part of the universe recognized by us was in doubt up to the time of Copernicus, who finally supplied the true arrangement and the true system according to which

22 On January 8, 1616, Galileo sent to Cardinal Alessandro Orsini the *Discorso del flusso e reflusso del mare* (in *OG* 5, pp. 377–395), which remained unpublished but was the basis for the Fourth Day of the *Dialogue*, some passages of which are very similar to, if not verbatim transcriptions of, Galileo's 1616 *Discorso*. Galileo makes the issue very clear in a letter to Cesare Marsili, on 7 December 1612: "[. . .] I am working on my *Dialogue on the Ebb and Flow*, of which the Copernican system is a consequence" ("[. . .] vo tirando avanti il mio Dialogo del flusso e reflusso, che si tira in consequenza il sistema copernicano": *OG* 13, no. 1688*, p. 236; see also Galileo to Élie Diodati, 20 October 1625, in *OG* 13, no. 1733*, p. 282). Indeed, the Inquisitors required that Galileo remove all mentions of tides from the title and to change the preface, because granting approval to such a title would look like approval of his theory of the tides using the motion of the earth as proof: see Niccolò Riccardi to Clemente Egidi, 24 May 1632, in *OG* 19, p. 327.

these parts are ordered, so that we are certain that Mercury, Venus, and the other planets revolve about the Sun and that the Moon revolves around the earth.[23]

BACK TO ROME, SEEKING THE IMPRIMATUR

Galileo completed the *Dialogue* in December 1629,[24] and in the spring of 1630 set out for Rome to push it through the censorship. He arrived on May 3, 1630,[25] and there he received encouraging news from Castelli, sent a few weeks earlier:

A few days ago, Father Campanella spoke with Our Lordship[26] and told him he had met with some German gentlemen, in order to convert them to the Catholic faith. They were willing to do so, but when they knew about the prohibition of Copernicus etc., they became distraught, and he could do nothing more. Our Lordship told him exactly the following words: *We never meant [to suspend Copernicus]; and had we had our say, we would have never issued that decree.* All this I was told by Prince Cesi [...]. Moreover, as I wrote in another letter of mine,[27] Father Master Monster is very well-disposed toward you, and Monsignor Ciampoli is certain that, upon your arrival in Rome, all problems will be solved. Cheer up, then, and come with a light heart; you will end up most comforted![28]

Things, however, had changed considerably since 1624, when Urban VIII had granted Galileo six audiences in seven weeks. The Thirty Years' War, which had begun in 1618 as a clash between German Catholic and Protestant princes, had spiraled out of control, involving many other countries, including Italy, France, Spain, and Portugal, and by 1630 only a few of the various causes that fueled the conflict still pertained to genuinely religious issues. As the leader of Christendom the pope was expected to try to reconcile the French Bourbons and the Spanish Habsburgs, but his overt sympathy for King Louis XIII irked the Spanish cardinals,

23 *Dialogue*, p. 455 (*OG* 7, p. 480: "qual sia l'ordine solamente de i corpi mondani e la integrale struttura delle parti dell'universo da noi conosciute, è stata dubbia sino al tempo del Copernico, il quale ci ha finalmente additata la vera costituzione ed il vero sistema secondo il quale esse parti sono ordinate; sì che noi siamo certi che Mercurio, Venere e gli altri pianeti si volgono intorno al Sole, e che la Luna si volge intorno alla Terra").

24 See, for example, Galileo to Federio Cesi, 24 December 1629, in *OG* 14, no. 1971, p. 60, and 13 January 1630, ibid., no. 1978, p. 67; and Galileo to Cesare Marsili, 12 January 1630, ibid., no. 1977, p. 66. See also Giovanni Ciampoli to Galileo, 5 January 1630, ibid., no. 1975, p. 64.

25 See Francesco Niccolini to Andrea Cioli, 4 May 1630, in *OG* 14, no. 2004*, p. 97.

26 Pope Urban VIII.

27 See Castelli to Galileo, 9 February 1630, in *OG* 14, no. 1984, pp. 77–78.

28 "Il Padre Campanella, parlando a' giorni passati con Nostro Signore, li hebbe a dire che haveva hauti certi gentiluomini Tedeschi alle mani per convertirli alla fede Catolica, e che erano assai ben disposti; ma che havendo intesa la prohibizione del Copernico etc., che erano restati in modo scandalizati, che non haveva potuto far altro: e Nostro Signore li rispose le parole precise seguenti: *Non fu mai nostra intenzione; e se fosse toccato a noi, non si sarebbe fatto quel decreto.* Tutto questo ho inteso dal Sig.ʳ Principe Cesi [...]. Di più, come ho scritto in un'altra mia, il P. Maestro Mostro è benissimo disposto a servirla, e Mons.ʳ Ciampoli tiene per fermo, che venendo V. S. a Roma, supererà qual si voglia difficoltà: però si faccia buon animo e venga allegramente, chè restarà consolatissima": Benedetto Castelli to Galileo, 16 March 1630, in *OG* 14, no. 1993, pp. 87–88.

who began to denounce his policy. Nepotism was a way of ensuring that higher officials remained loyal, and discontents with the pope's external policy were fueled by resentment against the promotion and pensions he showered on members of his family. Eventually, the growing dissatisfaction with the pope and the Barberini family found expression in astrological forecasts prophesying the early demise of the pontiff.

During the eight weeks he spent in Rome, Galileo was granted a single meeting, which probably took place on May 18. The meeting was good, as Galileo himself reported to the Grand Duke's private secretary, Geri Bocchineri.[29] On that very day, in the *Avvisi di Roma* (printed news-sheets), Antonio Badelli spread the news that Galileo was in Rome trying to publish a book in which he attacked the Jesuits, and that he prophesied, among other things, the death of the pope.[30] The rumor was pure libel, but soon after he dined with his friend Orazio Morandi (who presided over a center for astrological computations in the Vallombrosan convent of Santa Prassede, in Rome), the pope ordered Morandi's arrest for presuming to calculate the chances of the cardinals likely to succeed him. Galileo had no part in Morandi's astrological research, but he took care, with Michelangelo Buonarroti's help, to clear himself of every shred of suspicion.[31]

Galileo left Rome for Florence on June 26. He was very happy about his fifth journey to The Eternal City, though. As the Grand Duke's ambassador reported: "The pope was pleased to meet him, and treated him most kindly, just as Cardinal [Francesco] Barberini did, who also invited Galileo for supper. The whole Court held Galileo in high esteem, and honored him, as due."[32] Still, the pope's position

---

29 See Geri Bocchineri to Galileo, 21 May 1630, in *OG* 14, no. 2014, p. 105: "I reported to Balì Cioli [Andrea Cioli, the secretary of state] what Your Lordship wrote to me in his letter of [May] 18. He was very pleased to know about the kindness His Holiness showed to you in his first and long audience" ("Ho fatto sentire al S.ʳ Balì Cioli quanto V. S. mi ha scritto con la sua de' 18; et egli ha havuto molto gusto di intendere la benignità che le ha dimostrata S. B.ⁿᵉ nella sua prima et lunga audienza").

30 See *OG* 14, no. 2009**, p. 103.

31 See Michelangelo Buonarroti the Younger to Galileo, 3 June 1630, in *OG* 14, no. 2022, p. 111, in which he reports a meeting with Cardinal Francesco Barberini: "When I met face to face with the cardinal in his room, [. . .] I had the opportunity to talk about the libel invented against Your Lordship. He cut me off, and spoke before I could say anything. He told me that someone—you see, Your Lordship, what evil people dare to do—had come to see him and talked about Your Lordship, just like Your Lordship had known from another source. The Cardinal cut him off, too, and told him that Galileo had no greater friend than the pope himself: the pope knew who Galileo was, and that he was far from thinking of these things. The cardinal was furious with that man, who was baffled by this. And while I was highlighting the recklessness of those heinous people who do these things, he told me he perfectly knew that such deeds were not in the least meant to damage Your Lordship, but the pope himself. He who slandered [Galileo], he continued, must have thought that since a great mathematician had come to Rome, he must have been a great astrologer—and on this he grounded the whole of his fiction" ("Trovandomi poi testa testa col Sig.ʳ Cardinale in camera, [. . .] ebbi campo lì di trattar della calunnia inventata contro a V. S. Mi tagl[iò] la parola e s'espresse prima di me, e dissemi essere stato un tale (guardi V. S. se gli sciagurati s'avventano) che gli era entrato a parlar di V. S. nella istessa maniera che V. S. per altra via ha saputo; a cui tagliando pur il parlare, disse il S. Ca[rdinale] che il S.ʳ Galileo non aveva il maggior amico che sè e che 'l Papa stesso, e che sapeva chi egli era, e che sapeva che egli non haveva queste cose in testa; e se li mostrò controverso del tutto, e colui rimase brutto. E mentre che io ostentavo la ribalderia di persone sì sciagurate e che fan tali uffizi, mi si dichiarò penetrare che e' non eran fatti per offender di punta V. S., ma lui stesso, e che chi malignò dovette far conto, che essendo venuto a Roma un gran matematico, argomentasse: Adunque un grande astrologo; e sopra di lui fondasse la macchina della sua favola").

32 "Il Papa l'ha visto volontieri, gli ha fatto moltissime carezze, come il S.ʳ Card.ᵉ Barberino, che l'ha anco tenuto seco a desinare; e da tutta la Corte è stato stimato et honorato come l'era dovuto": Francesco

had not changed—nor, unfortunately, had Galileo's confidence. Most important, he failed to recognize the profound change the Church had undergone in structure and administration around 1600. Galileo and his friends still looked for backing from the bishops and cardinals, overlooking the fact that Clement VIII had imposed a new *modus governandi* by gradually depriving the consistory of its authority, and by appointing as cardinals two of his nephews. The offices previously assumed by the consistory were divided into different congregations, and all decisions of the congregations required pontifical approval. The pope's *plenitudo postestatis* was further increased by Urban VIII, who forbade individual members of the congregations to speak unless specifically invited to do so.[33] All initiative was concentrated solely in the hands of the pope and the bureaucrats to whom he delegated whatever he could not attend to in person. Worse still, not only did Galileo and his supporters misjudge the structure and function of the Church, but they misjudged the personality of its head, Pope Urban VIII, in whose eyes the arts and sciences served the ultimate end of exalting the papacy—and the papacy, in turn, served the end of exalting the Barberini family and the image it had of itself.[34]

As to the Copernican hypothesis, Urban VIII never changed his mind: he rested convinced that Galileo had to treat it as a mere mathematical hypothesis. Had Galileo done that, he would have had all the intellectual freedom and support he could possibly hope for. Galileo, by contrast, rejected the Aristotelian distinction between philosophy (i.e., philosophy of nature, that is, physics) and mathematics, and its implied subordination of the former to the latter. Moreover, for Galileo, physics could not be confined to mere philosophical speculation, but needed to include sensory experience. And it is the continuous interplay between sensory experience and necessary demonstration—a key tenet of Galileo's methodology—that leads to the truth.[35]

The trial, the sentence, Galileo's abjuration, and house arrest for life would soon follow.

THE SCIENTIST, THE FRIAR, AND THE POPE

Their contrasting views notwithstanding, Galileo was determined to take his chance. He trusted he could do that—just like the Dominican friar Tommaso Campanella (1568–1639). Indeed, from this respect, Galileo's and Campanella's personal trajectories share interesting aspects—not in the least because Campanella wrote an

---

Niccolini to Andrea Cioli, 29 June 1630, in *OG* 14, no. 2034*, p. 121. See also Giovanni Ciampoli to Galileo, 13 July 1630, ibid., no. 2037, p. 123: "Our Lordship [the pope] talks about you with words of great esteem and fondness" ("N. S.^{re} parla di lei con parole di grande stima et affetto").

33 Indeed, when the crucial moment came, only three out of the several member cardinals of the Congregation of the Holy Office sympathetic to Galileo dared to silently express their dissent by avoiding to affix their signatures to the condemnation, on 22 June 1633: see *OG* 19, pp. 402–403 and 406.

34 See Cochrane, *Florence in the Forgotten Centuries 1527–1800* (1973), pp. 189–194. Furthermore, while the pope was unwilling to give Galileo and his supporters the backing they needed, Grand Duke Ferdinando II—who represented the second great power they were counting on—was unable to do so, faltering where his predecessors Ferdinando I and Cosimo II had succeeded.

35 See, for example, *Letter to Castelli*, pp. 50–51 (*OG* 5, p. 283). See also Benedetto Castelli's remarks in *OG* 4, p. 653.

extensive grammatical and philosophical commentary on the pope's *Poemata*, giving the *Adulatio perniciosa* pride of place, at the beginning of his *Commentaria*.[36] The story is worth telling.

As soon as he got to know about Bellarmino's 1616 decree, Campanella wrote a passionate defense of Galileo (*Apologia pro Galilaeo*), which was eventually edited by Lutheran Tobias Adami and published in Germany, in 1622. This was a brave act, as in 1616 Campanella was in prison, in Naples, where he had been sentenced (for life) since 1599. The *Apologia* was dedicated to Cardinal Bonifacio Caetani (1567–1617), by whose order, he says, Campanella investigated whether the proposition stating the motion of the earth should be deemed heretical.[37] By the time the *Apologia* appeared in print, however, Caetani had died and could not either confirm or deny Campanella's claim. Possibly, Campanella dedicated the *Apologia* to him in order to gain official credit for his work. At any rate, Campanella's passionate defense of Copernicus and Galileo in the *Apologia* fell on deaf ears, and not even Galileo paid much attention to it, possibly since its author was in prison by order of the Holy Office.[38]

Ten years after Bellarmino's injunction and Campanella's *Apologia*, the latter was brought from Naples to Rome. In order to win the goodwill of the pope, and possibly obtain his release, Campanella seized the opportunity to appeal to Urban VIII's narcissism, especially as a poet. In 1627 he began to write a voluminous commentary on the pope's Latin poems, a first part of which he completed by the first half of 1629; beginning with 1631, he worked on a second part, of which only a few sections survive.[39] After some three decades in prison, Campanella managed to get the pope's attention, and was eventually released.

Amidst all sorts of ornate, learned, and lavish praises of Urban VIII's poetic talent—Campanella refers to him as "divine poet" (*diuinus Poeta*), and exalts him over Moses, King David, Orpheus, Pindar, Plato, Catullus and Horace, Dante Alighieri and Ludovico Ariosto, not to mention St. Ambrose—the Dominican friar struggles to get the new Pope on his own (and Galileo's) side and at the same time set him in contrast with Aristotle, at least as far as poetry is understood. Whereas Aristotle, a blind leading the blind (*coecus coecorum Dux*), stated that the end and essence of poetry is fiction,[40] the subject matter of poetry, which is supposed to

36 On the *Commentaria*, see Bolzoni, "I *Commentaria* di Campanella ai *Poëmata* di Urbano VIII" (1988). See also Amabile, *Fra Tommaso Campanella ne' castelli di Napoli in Roma ed in Parigi* (1887), vol. 1, pp. 412–420; and Bolzoni, "La restaurazione della poesia nella Prefazione dei *Commentaria* campanelliani" (1971).

37 Campanella, *Apologia pro Galileo* (1622), p. 39. At the time of Pope Paul V, Cardinals Caetani and Barberini (later to become Pope Urban VIII) stood in opposition to the proposal that the Copernican opinion be declared contrary to the faith: see Giovanfrancesco Buonamici's letter of 2 May 1633 (in *OG* 15, no. 2492*, p. 111) and July 1633 report (in *OG* 19, pp. 408–409). Caetani was charged with correcting Copernicus's *De revolutionibus*: see Galileo to Curzio Picchena, 6 March 1616, in *OG* 12, no. 1187, p. 244. On Campanella's *Apologia*, see Firpo "Campanella e Galileo" (1968) and Ernst, *Tommaso Campanella* (2002), pp. 159–166.

38 See Pietro Bartolini to Galileo, 6 September 1616, in *OG* 12, no. 1223, p. 277.

39 Whatever is extant of Campanella's *Commentaria* is preserved at the Vatican Library, in Rome: *Barb. lat.* 1918 (over 1000 pages) contains the full first part of the *Commentaria*; *Barb. lat.* 2037 and 2048 contain what is left of the second part.

40 See Aristotle, *Poetics*, 1, 1447 a 8–10: "We are to discuss both poetry in general and the capacity of each of its genres: the canons of plot construction needed for poetic excellence" ("Περὶ ποιητικῆς αὐτῆς τε καὶ τῶν εἰδῶν αὐτῆς, ἥν τινα δύναμιν ἕκαστον ἔχει, καὶ πῶς δεῖ συνίστασθαι τοὺς μύθους εἰ μέλλει καλῶς ἕξειν ἡ ποίησις"); see also 9, 1451 b 27–28: "It is clear from these points, then, that the poet should be more

have an educational aim, is, in fact, philosophy, harmony, and beauty—as the pope's *Poemata* most clearly and cleverly testify.[41] Therefore, all arts should be led back to their origins, under the great and wise new pontiff, so that they may finally abandon the Iron Age, with its discords and divisions, and enter a new Golden Age.[42] As we read in the preface, §7:

> We should not be afraid of the term "restorer," by which the Fathers, in holy councils, wisely designate heretics, and which contemporary wiseasses, rolling over in mud and in the scums, foolishly refer to the renovators of the sciences; nor should we be afraid of ending up in the lake that burns with sulfur, together with the sexually immoral, the detestable, and the faithless, as the holy book of *Revelation* threatens.[43] Indeed, by "restorer" we mean he who brings forth new ideas in order to demolish old and well-defined dogmas, as St. Stephen, Pope and martyr, warns in writing to St. Cyprian; [. . .] whence St. Jerome, in awe of being labeled "restorer," in the prologue to the [translation of] the Holy Scriptures,[44] often arguing against those who contrasted him with the Holy Ghost (who had spoken through the 72 interpreters who, isolated in 72 different cells, offered identical translations of the Bible from Hebrew to Greek), says: "We make new things not in order to destroy old ones but in order to fix them". [. . .] For, just like a new translation of the Holy Scriptures was required, given the light shed by the Gospel, as Jerome says, so a new interpretation of nature is required, given Columbus's discovery of a new world, and Galileo's discovery of new stars.[45]

"Let zeal without science stop," continues Campanella, "and let science blush, if not accompanied by zeal."[46] The author presents his commentary as a defense of the pope-poet against traditional Aristotelians.

---

a maker of plots than of verses" ("δῆλον οὖν ἐκ τούτων, ὅτι τὸν ποιητὴν μᾶλλον τῶν μύθων εἶναι δεῖ ποιητὴν ἢ τῶν μέτρων"). In Aristotle's text, however, μῦθος (*fabula*) does not mean "fiction," as Campanella—or possibly, his source—understood it, but "plot." According to Aristotle, the essential feature of all forms of poetry is that they are all modes of imitation, or mimesis. Imitation is not mere mimicry of nature, though, but the thing that is reproduced in us and for us by the poet's art.

41 See *Barb. lat.* 1918, ff. 2r-5r; Campanella paraphrases *Matthew* 15, 14: "[. . .] they are blind guides. And if the blind leads the blind, both will fall into a pit." The *Commentaria* were meant to be a companion to the *Poemata*, for students at the Pious Schools (held by the Scolopi Fathers, or Piarists).

42 See *Barb. lat.* 1918, ff. 8v–9r.

43 See *Revelation* 19, 20.

44 See Jerome, *Praefatio in Pentateuchum*, in *Patrologia Latina*, vol. 28, col. 177 A–184 A.

45 "Nec nomen Nouatoris, quod sapienter hæreticis Patres in Concilijs sacris, et insipienter scientiarum Instauratoribus Recentiores illi scioli qui, in coeno, et fæcibus [scientiarum] (scientiarum *delevi*) uolutati, adscribunt, formidandum est, ne cum fornicatoribus, et execratis, et formidolosis pars nostra in stagno sulfuris, veluti sancta Apocalipsis comminatur, siet. Siquidem Nouator est, qui noua promit ad destructionem veterum. benè præiactorum dogmatum, vti S. Stephanus Papa et Martyr ad S. Cyprianum scribendo admonet: [. . .], vnde et S. Hieronimus Nouatoris metuens nomen in prologis Hagiographorum sepè clamitans contrà eos qui Hieronimum spiritui sancto opponebant per 72. Interpretum in Cellis 72. Separatorum, eisdem uerbis sacra Biblia Hebraica grecè uertentium, ora locuto, ipse nouam exorsus versionem, ait. Non ita cudimus noua, vt destruamus vetera, sed vt statuamus. [. . .] sicuti enim corruscante Euangelio vt ait Hieronimus, noua interprætatio: Scripturis Sacris debebatur, Ità et rerum naturae tum ex [Euangelio tum ex *delevi*] detecto per Columbum orbe nouo, [et] (et *delevi*) <tum ex> (tum ex *addidi*) nouis per Galileum stellis, noua debetur Interprætatio": *Barb. lat.* 1918, ff. 9r–10v.

46 *Barb. lat.* 1918, f. 14v: "Cesset Zelus sine scientia, erubescat scientia sine zelo"; see also Campanella *Apologia pro Galileo* (1622), p. 51.

After the preface, the first poem Campanella annotates is the *Adulatio perniciosa*.[47] In the various editions of the *Poemata*, the poem never appears in a prominent position; usually, it is grouped with other poems with a moral content. Campanella's decision to have it at the very opening of his work is very interesting, as he obviously wished to call particular attention to it. The poem that celebrates Galileo's discoveries is given pride of place, making it appear as if it were the most important one in Urban VIII's very eyes.

Section III of Campanella's commentary annotates the second stanza (lines 9–12), which includes references to Galileo's discovery of Jupiter's satellites and three-bodied Saturn:

In AD 1611, the Florentine G. Galilei, the glory of our age, and friend of mine since 1592 (when he moved to Padua, where I was staying, to teach mathematics at the university, due to his talent), whom I got to know on the occasion of a letter of Ferdinand, the Grand Duke of Tuscany, which he [Galileo] gave to me,[48] shrewdly detected four little stars orbiting Jupiter, and two orbiting Saturn, as is well known from the *Sidereal Messenger* and his three letters on sunspots. I sent him a letter at once, congratulating [him] on what philosophy was to be gained from that discovery, and what was left to be done to complete the observations. And when, in Rome, Galileo was sentenced because he inclined toward the Copernican opinion about the motion of the earth and the stars around the Sun (an opinion he seemed to advocate, on the basis of his discovery of the stars orbiting Jupiter and Saturn, I wrote the *Apologia*, showing that, perhaps, this opinion—which I disproved as false, as far as physics is concerned, does not contradict all the Doctors of the Church, as may be shown appealing to Chrysostom, Justin, Methodius, Theodore of Tharsus, Augustine, and Procopius of Gaza (after the Church's decree, however, I rejoiced, as I was right to write against Copernicus). The book [*Apologia pro Galileo*], published through the good offices of the most learned Cardinal Bonifacio Caetani, and by whose order was written, was printed in Germany. Our divine poet, when he was elected Supreme Pontiff—just like, when he was a cardinal, being himself a very learned man, he used to favor learned men with great talent—saw that in the new *Index*, the Copernican opinion, purged from error, was upheld understood as a hypothesis, and [in so doing], with commendable providence, he took care of the interest of the philosophers and of the safety of the State, and did not allow for the Muses to leave Italy.[49]

47 Of the many commentaries on selections from Urban VIII's *Poemata*—either the few that were published during Urban VIII's pontificate or the many that were still unpublished at the time of his death (and remain so up to this day) and are preserved in Rome, at the Vatican Library—only two include the *Adulatio perniciosa* (the other being Giulio Cesare Capaccio, *In Odas ... Cardinalis olim Barberini nvnc ... Vrbani VIII.* (1633), no. 10, pp. 157–169), and Campanella's is the only one that opens with his annotations on the *Adulatio perniciosa*: see Barb. lat. 1918, ff. 17r–57v. Although extensive parts of Campanella's *Commentaria* have been variously published in the past few decades, the commentary on the *Adulatio perniciosa* has not; it is now forthcoming, edited, and translated by the author of the present work.

48 Campanella had met Galileo in Padua, in 1592. See Campanella to Galileo, 13 January 1611, in Campanella, *Lettere* (2010), no. 37, p. 193; see also Campanella to Ferdinando II de' Medici, 6 July 1638, ibid., no. 165, p. 510.

49 "Anno Christi Dei 1611. G. Galilæus Florentinus, nostri æui decus, amicus noster ab anno 1592. ex quo Patauium conductus, ubi ego morabar, ad Mathematicarum lecturam, eius virtutis gratia, et literarum Magni Ducis Hetruriæ Ferdinandi ab ipso mihi redditarum occasione notus, exiguæ quantitatis circa

Now, the whole manuscript of the *Commentaria* was ready to be sent to the printer and published, and was in the handwriting of a professional copyist. By contrast, the words I underlined in this excerpt were added in between lines by a different hand; and, by the same hand, the word *sustinendam* ("upheld") was crossed out and replaced with *legendam* ("understood"). These were Campanella's own corrections, penned at the very last minute in the manuscript of the *Commentaria*. We shall see a possible explanation for this in a minute.

The end of Section III is worth reading in full, too:

[...] the pronoun "your"[50] is an elegant addition, by right of discovery, and it is by this very right that he named the stars "Medicean," after the name of the Medici, Princes of Florence, patrons of talents and the sciences, who broke the yoke of the sophists into pieces. However, if my own opinion might be of any value for a friend, I would name the stars that circle Saturn after the Barberini family, because this family adds luster to the tutelage, promotion, and reform of the divine Muses. And in Plato's *Statesman*, Saturn, father of the gods, indicates the highest king, who is, at the same time, the highest priest, under whose rule the circular paths of celestial bodies move not from the region we now call east, but from the west, and things, tells Plato, revert from old age to youth.[51] Let the reader understand what I mean.[52]

Section IX of Campanella's commentary annotates lines 33–36, and includes mention of Galileo's discovery of sunspots. Most strikingly, in the philosophical part of the commentary (which, as in every section, follows the grammatical one) Campanella openly refers to authors whose work had been notoriously banned by the Church: Bernardino Telesio (1509–1588), whose views are mentioned twice, and *quidam Nolanus*, namely, Giordano Bruno (1548–1600), who was burned at the stake

---

Iouem stellas quatuor sagaciter deprehendit; duasque circa Saturnum, vt ex Nuncio Sidereo, et epistolis tribus de maculis solaribus, nemo ignorat. Ad quem statim ego epistolam misi, quid ex hac inuentione philosophia lucratura sit, gratulatus, et quid ab ipso desideretur ad obseruationum complementum. Cumque Romæ taxaretur, quòd ad Copernici inclinaret sententiam de Telluris, et planetarum circa Solem rotatione, quam ex Stellarum repertarum circa Iouem, ac Saturnum, gyro vagantium declaratione approbare visus est. Ego Apologeticum scripsi, ostendens, quod ista forsan opinio <u>à me in physicis tanquam falsa reprobata</u> cum sanctis <u>omnibus</u> non pugnat Doctoribus, vt ex Chrysostomo, Iustino, Methodio, Theodoro Tarsensi, Augustino et Procopio Gazæo probari posset (<u>sed post decretum ecclesiæ gavisus sum, quod ego contra Copernicum rectè scripsissem</u>). Qui libellus cura Bonifacii Caetani Cardinalis doctissimi, cui obediens illum composueram, in Germania impressus est. Noster autem diuinus Poeta summum Pontificatum adeptus, quemadmodum in Cardinalatu Sapientes non vulgares ipse Sapientissimus fouere consueuerat, in nouo indice opinionem Copernici <u>errore purgatam</u> hypotheticè ~~sustinendam~~ legendam cum philosophorum commodo, ac Reipublicæ incolumitate, mira prouidentia curauit Musasque ab Italia exulare non patitur": *Barb. lat.* 1918, ff. 29r–29v.

50 *Adulatio perniciosa*, line 12: "through your telescope, learned Galileo" (*Docte tuo Galilæe vitro*).

51 Plato, *Statesman*, 270 B 3–271 A 1.

52 "Eleganter ergo additur pronomen 'Tuo,, iure inuentionis, quo etiam nomen stellis fecit Mediceis à Principibus Florentiæ Mediceis virtutum, ac scientiarum patronis, sophistici iugi fractoribus. Siquid autem mea apud amicum ualet autoritas, eas, quæ Saturnum circumeunt, Barberinas vocarem: quoniam tutela, promotio, et reformatio diuinarum Musarum ab hac domo nobis illucescit. Ac Saturnus pater Deorum summum Regem Antistitem summum in libro Platonis de Regno indicat, sub cuius regimine cœlestium circumuolutio non hac à plaga, quam nunc Orientalem vocamus, sed ab Occidentali reuertit, et res à senio ad pueritiam remeare a Platone narratur; qui legit, intelligat rectè": *Barb. lat.* 1918, ff. 29v–30r.

after being declared an impenitent and pertinacious heretic by the Roman Inquisitors (among whom was also Cardinal Bellarmino).[53] Even more strikingly, shortly thereafter Campanella declares: "suffice it [. . .] to praise the genius of the divine poet, who philosophizes at the highest level, because he reproaches incompetent Peripatetics by saying, 'Who would believe it?'"[54] One might think that, amidst the renewed celebration of the pope's genius and Galileo's famous discoveries, Campanella wished to "smuggle" the names of two well-known heretics such as Telesio and Bruno.

As can be seen from the two sections of the commentary dealing with Barberini's praise of Galileo, Campanella seems to be using Galileo's name and achievements to play his own game, winning the pope over to his own side and for his own ends—and, perhaps, turning him into an advocate of the new science against the dogmatic Aristotelians from within the Church. However, in a way similar to Galileo's, he crossed a line he was not supposed to cross: on June 10, 1628, Campanella was informed that the pope had read his commentary and got annoyed at his references to the Copernican theory.[55] That is why, as was pointed out in the passage from Section III quoted above, Campanella rushed to revise the text, inserting a few sentences or replacing words here and there, in an effort to tone down his remarks.

The last-minute attempt to reconcile with the pope and take advantage of his vanity succeeded only for a short period. As Campanella was at work on the *Commentaria*, rumors spread of the pope's imminent death, in view of unfavorable celestial signs.[56] Urban VIII had a profound interest in astrology and rose increasingly alarmed by the rumors, as he feared there might be something true in the insistent predictions. He turned therefore to Campanella, whose astrological skills were well known, and the friar reassured him, on the basis of an attentive analysis of the pope' birth chart. From the beginning of the summer 1628, Urban VIII repeatedly and secretly called Campanella to the Papal Palace, asking him to put into practice the theories of natural magic he described in the treatise *De siderali fato vitando* ("On avoiding sidereal destiny"). The pope followed Campanella's advice, and consideration for the Dominican friar grew considerably in the Roman circles. At the beginning of 1629, he was definitively cleared of the charges that had led to his long incarceration, and on June 2 the General Chapter of the Dominican Order bestowed on him the title of

---

53 See *Barb. lat.* 1918, f. 39v.

54 "hic satis est [. . .] admirari diuini Poetae ingenium non vulgariter philosophantis, proptereà exclamat ad inertes Peripateticos, Quis credat?": ibid.

55 See Campanella to Urban VIII, 10 June 1628, in Campanella, *Lettere* (2010), no. 74, p. 304.

56 Gian Vittorio Rossi hinted at this famous incident in his *Evdemia*: some crewmen on board a vessel poked fun at some foreign princes (namely, some French Cardinals, who had hurriedly travelled to Rome in view of a new conclave) who were traveling back home after they had been "persuaded by mathematicians [= astrologers], who had predicted, on the basis of the positions of the stars, that the reigning king [= the pope] was going to die. However, as they saw that one month had already passed since the prediction, and there was no hope that Minos would draw from the urn the sentence for the king, gloomy, and not without [the crewmen's] laugh, they returned back home" ("nam, cum respexissent, triremes quatuor viderant, quæ dynastas aliquot in regnum Geryonis reportabant, unde ad Regis sacrorum comitia velis remisque contenderant, persuasi à mathematicis qui ex positione astrorum prædixerant, intra paucos dies Regem, qui tunc rerum potiebatur, orco destinari. sed, cum vidissent, mensem ab ea prædictione jam elapsum esse, neque spem aliquam ostendi, ut apud inferos Regis dicam ex urna Minos educeret, tristes, nec sine risu, eodem unde discesserant, revertebantur": Rossi, *Evdemiæ libri decem* (1645), Book VIII, p. 157).

magister theologiae; there were rumors about his possible appointment as consultor to the Holy Office, and perhaps even higher honors. Then, in the autumn of 1629, a scandal erupted: the *De siderali fato vitando* was printed in Lyon as the seventh and last book of the *Astrologicorum Libri VI*. [57] The circumstances surrounding the publishing are not entirely clear, although it seems that the booklet was added without Campanella's knowing. Possibly, Campanella's sudden rise bothered fellow members of the Curia or the Dominican Order (also owing to his previous charges of heresy), who wished to rob him of the pope's benevolence. [58]

The publication unleashed a violent anger in the pope, who was worried at the risk of being publicly compromised and suspected of magical and superstitious practices (astrology was officially banned by the Church, as a consequence of Pope Sixtus V's Bull *Coeli et terrae*, promulgated in January 1586). Campanella's appointment as consultor to the Holy Office was suspended, and he hurried to write the *Apologeticus ad libellum De siderali fato vitando*, in which he argued that the practices suggested in the booklet, far from being superstitious rituals, were in fact completely licit natural remedies. [59] But he could see the gathering of the storm over astrology that was to come, culminating in the trial of astrologers in the summer of 1630, and in the promulgation of the Papal Bull *Inscrutabilis Iudiciorum Dei* on April 1, 1631, banning all sorts of astrological and divination practices, as well as threatening harsh punishments for authors of any prognostications pertaining to the life of the pope and his relatives. Campanella wrote a *Disputatio contra murmurantes citra et ultra montes in Bullas SS. Pontificum Sixti V. et Urbani VIII. adversus judiciarios editas*, in which, under the pretext of replying to the criticisms of Italian and foreign adversaries, he purported to offer a mitigating interpretation of Urban VIII's and Sixtus V's Bulls. His efforts notwithstanding, the short period of papal favor enjoyed by Campanella came to an end: his works, especially the *De sensu rerum et magia* (1620) and *Atheismus triumphatus* (1631), came under new attacks. The *Commentaria*—which had already passed a preliminary revision by the Dominican theologian Vincenzo Candido, on 10 July 1629, and later a second revision by the Somascan Father Francesco Tontoli—were at a very advanced stage around 1632, but were never published. [60]

---

57 *De siderali fato vitando*, printed (with distinct page numbers) in Campanella, *Astrologicorum Libri VI.* (1629), after f. Gg4v. On the whole incident, see Amabile, *Fra Tommaso Campanella ne' castelli di Napoli in Roma ed in Parigi* (1887), vol. 1, pp. 347–363; Firpo, "Il Campanella astrologo e i suoi persecutori romani" (1939); and Ernst, *Tommaso Campanella* (2002), pp. 220–226.

58 See Campanella to Urban VIII, 14 February 1630, in Campanella, *Lettere* (2010), no. 79, pp. 318–319; 2 November 1634 (from Aix-en-Provence), ibid., no. 103, pp. 353–356; 25 February 1635, ibid., no. 109, pp. 376–377; and 9 April 1635, ibid., no. 114, pp. 387–396. See also Campanella's long letters to Francesco Barberini, 24 March 1630, 4 December 1634, and 3 July 1635, ibid., nos. 80, 105 and 123, pp. 320–321, 360–364, and 414–415; and to Antonio Barberini, 1 February 1635, ibid., no. 107, pp. 369–373.

59 The text was published in Amabile, *Fra Tommaso Campanella ne' castelli di Napoli in Roma ed in Parigi* (1887), vol. 2, Documenti, pp. 172–179.

60 In August 1633, Tommaso Pignatelli, a former disciple of Campanella's, was jailed in Naples, on the charge of having plotted a conspiracy against the Spanish monarchy. Renewed suspicions began to circulate about Campanella, too. In the early 1630s he struck a friendship with French scholars Gabriel Naudé and Pierre Gassendi, and when the situation hopelessly deteriorated, in October 1634, Campanella fled to France, aided by the French ambassador François de Noailles. He was welcomed at the court of Louis XIII, who granted him a pension, and spent the rest of his days in Paris, at the reformed Dominican convent in Rue St.-Honoré, under the protection of Cardinal de Richelieu.

Campanella blamed his downfall on members of the Curia and fellow Dominican friars. For sure, many were jealous of the growing benevolence he enjoyed from the pope.[61] And no doubt it was someone in the papal court, envious of the rapidity of Campanella's rising to power, who called Urban VIII's attention to the friar's philosophical inclinations.[62] But by fulsomely adulating the pope, Campanella clearly tried to present him as an opponent of Aristotelianism, a supporter of Galileo's new science, and an advocate of Campanella's own philosophical views (which he shared with such heretics as Bernardino Telesio and *quidam Nolanus*).[63]

The fall of Campanella and the (non-)publication history of the *Commentaria* go hand in hand with the success enjoyed by Urban VIII's *Poemata*, whose fate— beginning with its 1631, considerably reorganized *maior* and *minor* editions, published by the Roman College—was taken care of by the members of the Society of Jesus. The same years proved crucial for Galileo, too, who published the *Dialogue* in 1632, was found vehemently suspect of heresy by the Tribunal of the Inquisition in 1633, forced to abjuration, and sentenced to house arrest for life; once again, the Jesuits played a major role in the events, also blaming on the Dominicans the *imprimatur* granted to the *Dialogue*. And Urban VIII's rage at his argument's being adopted by Simplicio, at the end of Galileo's book,[64] was not so distant, in time, from the

---

61 See Campanella, *Syntagma* (1642), pp. 29–30: "And the book *De siderali fato vitando*, which was borrowed by a slimy friar, who gave it to a printer in Lyon and published it together with [my] six books on astrology, in order to provoke against me the hate of the most learned Pope Urban VIII, who, because of his very learnedness, despises astrology. Hence, he [i.e., the slimy father] at once accused me of disobedience about that publication, and of superstition for its content; on which account, since I had not published it, I refused to accept it as mine and wrote an apology that was approved by two selected censors" ("Item librum de siderali fato vitando, quem mutuatum insidiosus frater, vna cum sex libris Astrologicis, vt odium Vrbani VIII. Sapientissimi Pontificis eoque nomine Astrologiam detestantis, mihi conciliaret, Typographo dedit Lugdunensi, à quo publicatus est; ac simul me accusauit inobedientiæ circa impressionem, & superstitionis circa impressas res; quare hunc, vt à me minimè impressum, pro meo arripere nolui, & apologiam feci quam delecti duo Censores approbauere". The "slimy friar" was likely the Dominican Niccolò Riccardi, who instigated the plot along with the General of the Dominican Order, Niccolò Ridolfi. See also Campanella to Urban VIII, 2 November 1634: "The Most Eminent Cardinal [Francesco] Barberini always cautiously tried his best to prevent me from visiting the [Apostolic] Palace too much, dedicating books to Your Beatitude, and publishing my *Commentaria*" ("Il Card. Em.^mo Barberino prudentemente ha cercato sempre che io praticassi poco il Palazzo e che non dedicassi libri a V. B. e che non si stampassero i miei Commenti").

62 See Campanella to Urban VIII, 29 September 1631, in Campanella, *Lettere* (2010), no. 85, pp. 326–328.

63 For this reading, see Spini, "Galileo uomo del Seicento" (1995), and *Galileo, Campanella e il «Divinus Poeta»* (1996), ch. 3.

64 See *Dialogue*, p. 464 (*OG* 7, pp. 488–489). The argument was explicitly mentioned in the proceedings of the trial, as one of the eight textual points offensive to the Church: see *OG* 19, p. 326: "[. . .] that he put the 'medicine of the end' in the mouth of a simpleton, and in a place where it can only be found with difficulty; and then he had it approved coldly by the other speaker, by merely mentioning but not elaborating the positive things, which he seems to utter against his will" ("[. . .] et aver posto la medicina del fine in bocca di un sciocco, et in parte che né anche si trova se non con difficoltà, approvata poi dall'altro interlocutore freddamente, e con accennar solamente e non distinguer il bene, che mostra dire di mala voglia"). In fact, in order to obtain the *imprimatur*, Galileo was explicitly required to include Urban VIII's argument on God's omnipotence at the end of the work: see *OG* 19, p. 330: "At the end, [. . .] Galilei should add the argument by God's infinite omnipotence Our Lordship [i.e., Urban VIII] offered to him, which must put our mind at rest, however compelling the Pythagorean arguments might result" ("Nel fine si dovrà fare la perorazione delle opere (*sic*) in conseguenza di questa prefazione, aggiongendo il Sr Galilei le raggioni della divina onnipotenza dettegli da Nostro Signore, le quali devono quietar l'intelletto, ancorché da gl'argomenti Pittagorici non se ne potesse uscire"). On the whole issue, see Bianchi, "'Mirabile e veramente angelica dottrina': Galileo e l'argomento di Urbano VIII" (2011).

---

annoyance he must have felt when he realized (or someone had him realize) that in Campanella's *Commentaria* his vanity had been leveraged to ascribe to him positions, and convey philosophical ideas, he was far from advocating.

Truth, we read in the *Adulatio perniciosa*, escapes powerful people, blinded as they are by their self-confidence and boundless approval from their courtiers; that is how flattery perniciously injects its deadly venom.[65] Argus Panoptes, at the end of the poem, provides "a terrible example of failure": intent on listening to Mercury's song, he closed all his hundred eyes, and was slain with a stone by the messenger of the Olympian gods. For sure, the *Adulatio perniciosa* twice approvingly mentions Galileo's name and discoveries, and it was written and sent to him to show appreciation and support—but it is not, as it is all too often presented, a poem in praise of Galileo (besides, wouldn't it be a very strange title for a laudatory poem?). It is a poem with a moral content, and as such it is grouped with other poems of the same kind in the various editions and reprints of Urban VIII's *Poemata*. Had it been a praise of Galileo, it would have been excluded from the editions published after the 1633 trial, whereas, by contrast, it is always included in all editions since its first appearance in 1620 (the change of structure, order, and selection of poems throughout almost a quarter of a century notwithstanding).

In 1620, ten years after the publication of the *Sidereal Messenger*, Galileo was enjoying the success and fame his works had earned him; his very well paid position as Chief Mathematician and Philosopher at the Grand Duke's court was envied by all. As years passed by, new discoveries complemented previous ones, and he was looked up to as the new (celestial) Amerigo or Columbus: his glory seemed never-ending. Himself a powerful and influential Prince of the Church, Maffeo Barberini presented Galileo with a friendly warning, inviting him to keep his feet firmly on the ground, not to be seduced by too much success and pride, and remain a humble devotee of the Church.[66] And he did so by showing examples of the risks into which he would run should he be blinded by the dangerous flattery of his supporters. *Mutatis mutandis*, at the 1633 trial the Inquisition did something similar: the very idea that he might undergo torture[67] was enough to make him recant his Copernican sins.

Sadly, Cardinal Barberini was right: overconfident in his ability to change a century-old state of affairs, trusting that the support and goodwill the pope had

65 See *Adulatio perniciosa*, line 53: "Truth avoids the dwellings of the powerful" ("Fugit potentum limina veritas"). Interestingly, the line was quoted in Campanella to Urban VIII, 29 September 1631, in Campanella, *Lettere* (2010), no. 85, p. 327.

66 This was Christoph Scheiner's own advice in the engraved frontispiece of his optical treatise, *Ocvlvs hoc est: fvndamentvm opticvm* (1619), in which a blind man groping in the dark is accompanied by the statement "Without the eye, the hand does not see anything" ("Manus Nil videt absque oculo"); opposite, an armless man vainly gazes at various tools: "Without the hand, the eye cannot accomplish anything" ("Oculus Nil valet absque manu"). The meaning is clear: observation is not sufficient to establish truth and must be complemented by (divinely inspired) reason. Similarly, the telescope allows us to overcome long distances, but its glasses may alter what we see ("Non integer intrat", that is, "[Light] does not enter intact"): we cannot trust it as a completely reliable source of knowledge. A big peacock (the bird whose tail bears a thousand eyes) at the center of the engraving symbolizes optics but also pride, one of the seven cardinal sins; above it we read "I am raised on the ground" ("Attollor in imo"). Hence, Scheiner has the perceptive reader conclude, aided by two slogans in capital letters: we can safely trust what comes from God, and need to remain gloriously humble ("SUBLIMITAS SECURA—HUMILITAS GLORIOSA").

67 See *OG* 19, p. 360; see also pp. 283, 360, 399, and 419.

showed him on many occasions would shield him against any criticisms, in the *Dialogue* Galileo crossed a line he was not supposed to cross. And so did Campanella, presenting his patron as a supporter of Campanella's own views and covering it all up in lavish encomiastic words. The pope, moreover, was himself blinded by too much adulation and was prevented from seeing what was happening in front of his eyes. He fell victim to the dangers Cardinal Barberini had poetically described years earlier. Unfortunately for both Galileo and Campanella, Pope Urban VIII reconsidered.

Cvm luna cœlo[1] fulget, & auream[2]
    Pompam sereno pandit in ambitu
        Ignes coruscantes, voluptas
        Mira trahit, retinetque visus.
5   Hic emicantem suspicit Hesperum,
    Dirumque Martis sydus, et orbitam
        Lactis coloratam nitore,
        Ille tuam Cynosura lucem,
    Seu Scorpij cor, siue Canis facem
10   Miratur alter, vel Iouis asseclas,
        Patrisue Saturni, repertos
        Docte tuo Galilæe vitro.
    At prima Solis cum reserat diem
    Lux orta, puro Gangis ab æquore
15   [47]Se sola diffundit, micánsque
        Intuitus radijs moratur.
    Non vna vitæ sic ratio genus
    Mortale ducens pellicit; horrida
        Hic bella per flammas, et enses
20   Lætus init meditans triumphos.
    Est pacis ambit qui bonus artibus
    Ad clara rerum munia prouehi.
        Illum Peruanas ad oras
        Egit amor malesuadus auri.
25   Hunc, sumptuosus dum Siculæ iuuat
    Mensæ paratus spes alit aleæ
        Mendacis, ac fundis auitis
        Exuit, et laribus paternis.
    Nil esse regum sorte beatius,
30   Mens, et cor æquè concipit omnium,
        Quos larua rerum, quos inani
        Blanda capit specie cupido.
    Non semper extra quod radiat iubar
    Splendescit intra: respicimus nigras
35   In sole (quis credat?) retectas
        Arte tua, Galilæe, labes
    [48]Sceptri coruscat gloria Regij
    Ornata gemmis; turba satellitum
        Hinc inde præcedit, colentes

---

1 cœlo *ed. 1620*] Cælo *ed. 1631*
2 auream *ed. 1631*] aëris *ed. 1620*

When the Moon shines in the sky, and lays open
    the glittering stars, as in a golden parade,
    along its serene course, wondrous delight
    draws and restrains the sight.
5   Some people look up at the rising evening star,
    And the dreadful star of Mars, and the track
    colored with the brightness of milk.
    Others look up at your light, Cynosura.[1]
Still others marvel at the heart of the Scorpion,
10   or the brightest star of the Dog,[2]
    or Jupiter's attendants, or his father Saturn's, which you discovered
    through your telescope, learned Galileo.
But when the first light of the Sun
    discloses the day, it spreads itself, alone,
15   [47]from the pure water of the Ganges,
    And, shining, captures every onlooker with its rays.
By contrast, no single idea of life
    entices mortals: one sort goes in for savage wars,
    and through flames and swords
20   joyfully contemplates triumphs.
Then there is the good man who, by the arts of peace,
    seeks votes for high public office.
    There is he whom
    the seductive love of gold draws to Peru.
25   Yet another, while enjoying the sumptuous
    Sicilian table,[3] nourishes the hope
    of gambling, and loses his ancestral lands
    and his father's house.
The minds and the hearts of all men equally believe
30   that nothing is happier than the destiny of kings;
    they are prisoners of the deceitful semblance of things
    and of passion for worthless ornament.
That which outwardly radiates brightness
    does not always shine within: we gaze at black spots
35   in the Sun (who would believe it?)
    laid bare by your art, Galileo.
[48]The glory of the royal scepter, adorned with gems,
    glitters: a crowd of attendants from everywhere
    leads the way, and courtiers

40   Officijs comites sequuntur;
Luxu renidet splendida, personat
     Cantu, superbit delicijs domus:
     Sunt arma, sunt arces, et aurum:
     Iussa libens populus capeßit.[3]

45   At si recludas intima, videris
     Vt sæpe curis gaudia suspicax
     Mens icta perturbet. Promethei
     Haud aliter laniat cor ales.
Cui sensa mentis prouidus abdita

50   Rex credat? aut quos cauerit? omnium
     Sincera, seu fallax, eodem
     Obsequio tegitur voluntas,[4]
Fugit potentum limina veritas:[5]
     Quamquam salutis Nuncia. Nauseam

55   Invisa proritat, vel iram.
     Sæpe magis iuuat hostis hostem:
Ictus sagittâ Rex Macedo videt
     Non esse prolem se Iouis. Irrita
     [49]Xerxem[6] tumentem spe, trecentis

60   Thermopylæ[7] cohibent sarißis,
Docéntque fractum clade quid aulici
     Sint verba plausus. Vt nocet vt placet
     Stillans adulatrix latenti
     Lingua fauos madidos veneno!

65   Hæc in theatri puluere barbarum
     Infecit atro sanguine Commodum,
     Probrísque fœdauit Neronem, ac
     Perdidit illecebris vtrumque;
Artes nocendi mille tegit dolis

70   Imbuta. quis tam lynceus aspicit
     Quod vitet? Intentus canentis
     Mercurij numeris, sopore
Centena claudens lumina, sensibus
     Abreptus, aures dum vacuas melos

75   Demulcet, exemplum peremptus
     Exitij, graue præbet Argus.

---

3 capeßit *ed. 1620*] capescit *ed. 1642*
4 voluntas *ed. 1620*] voluptas *ed. 1726*
5 See Virgil, *Aeneis*, XII 519-520; and Horace, *Epodon Liber*, II 7-8
6 Xerxem *ed. 1623*] Xerxen *ed. 1620*
7 Thermopylæ *ed. 1623*] Thermopilæ *ed. 1620*

40   seeking offices follow.

The house shines with splendor, resounds
    with music, and prides itself in pleasures.
    There are weapons, strongholds, and gold:
    the people willingly obey orders.

45 But if you disclose the innermost secrets,
    you will see how suspicious minds often
    spoil joys with cares. In the same way,
    the eagle tears Prometheus's heart into pieces.[4]

To whom could the cautious king reveal
50   his innermost feelings? Whom should he fear?
    Everybody's intentions, sincere or misleading,
    are concealed by the same deference.

Although the herald of prosperity,
    truth avoids the dwellings of the powerful,
55   detested, it brings out revulsion, or even rage.
    Enemies often help each other:

the Macedonian King, struck by an arrow,
    realizes that he is not the progeny of Jupiter.[5]
    [49]With three hundred soldiers, the Thermopylae
60   stop Xerxes, pompous with vain hope,[6]

and teach him, broken by defeat,
    what the courtiers' words of approval are worth.
    How harmful and pleasant is the flattering tongue,
    exuding honey filled with hidden venom!

65 This smeared brutal Commodus with black blood,
    in the dust of the circus;[7]
    this defiled Nero with abominations,[8]
    and plunged them through enticements into ruin.

Imbued with a thousand artifices,
70   [the lying tongue] hides the arts of harming.
    Who has eyes sharp enough[9] to see what is to be avoided?
    Attentive to the harmonies of Mercury's song,

closing his hundred eyes with sleepiness
    while the music caresses his empty ears,
75   Argus in death offers
a terrible example of ruin.[10]

Notes

1. That is, the constellation Ursa Minor (the "Little Bear"), in the northern sky, whose ancient name was Cynosura (Κυνόσουρα, "dog's tail").

2. Sirius, the brightest star of Canis Major, a constellation of the Southern Hemisphere.

3. See Athenaeus, *Deipnosophistae*, XII 518c; see also 527c–e and 536b–c.

4. Prometheus was the son of the Titan Iapetus: the god of forethought and crafty counsel. He was given the task of molding mankind out of clay. His attempts to better the lives of men brought him into conflict with Zeus. He tricked the gods out of the best portion of the sacrificial feast, acquiring the meat for men; then, when Zeus withheld fire, he stole it from heaven and delivered it to mortals. As punishment for these rebellious acts, Zeus ordered the creation of Pandora (the first woman) as a means to deliver misfortune into the house of men. Meanwhile, Prometheus was bound to a stake on Mount Caucasus, where an eagle was set to feed upon his ever-regenerating liver (or, some say, heart). See Hesiod, *Opera et Dies*, 42–105, and *Theogonia*, 520–525; see also Hyginus, *Astronomica*, II, 15, 1–3.

5. Philip II (382–336 BC), the king of Macedon from 359 to his assassination; he was the father of Alexander the Great. In 355–354, during the siege of Methone (the last city on the Thermaic Gulf controlled by Athens), he was struck by an arrow and lost his right eye. See Demosthenes, *De corona*, 67; Strabo, *Geographia*, VIIa 1, 22, and VII 6, 15; and Lucian, *Quomodo historia conscribenda sit*, 38. See also Diodorus Siculus, *Bibliotheca Historica*, XVI 34, 5; Pliny, *Naturalis Historia*, VII 37, 124; and Justin, *Epitome*, VIII 6, 14.

6. The Battle of Thermopylae was fought between an alliance of Greek city-states, led by King Leonidas of Sparta, and the Persian Empire of Xerxes I, during the second Persian invasion of Greece. The battle took place at the narrow coastal pass of Thermopylae (the "Hot Gates"), over the course of three days, in August or September 480 BC. Xerxes had amassed a huge army and navy, and set out to conquer all of Greece; the vastly outnumbered Greeks held off the Persians for seven days, including three of battle, before the rear guard was annihilated. After the second day, Leonidas, aware that his force was being outflanked, dismissed the bulk of the Greek army and remained to guard the retreat with 300 Spartans, fighting to the death. The Persians overran Boeotia and captured Athens, but the Greek fleet defeated the invaders at the Battle of Salamina (late 480 BC), and Xerxes withdrew with much of his army to Asia. The following year, what remained of the Persian army was decisively defeated at the Battle of Plataea, thereby ending the Persian invasion. See Herodotus, *Historiae*, VII 201–233, especially 202 and 224–228.

7. See *Historia Augusta*, Aelius Lampridius (*Vita Commodi Antonini*), XVII 1–2.

8. See Tacitus, *Annales*, XV 33–47.

9. Literally, "as sharp as a lynx" (*tam lynceus*): lynxes were believed to be the animals with the sharpest eyesight. Federico Cesi chose the lynx as the symbol of the Lyncean Academy (i.e., the Academy of the Lynxes), which he founded in Rome, in 1603, and whose aim was the scrupulous and constant observation of nature. It was the first scientific society; Galileo was inducted on April 25, 1611.

10. Argus Panoptes (the "all-seeing") was a many-eyed giant, endowed with prodigious strength. Io was a young Argive girl of whom Hera was jealous, as she suspected Zeus was in love with her. In order to save Io from his wife's jealousy, Zeus transformed her into a white heifer. Zeus swore to Hera that he never loved that animal, but Hera demanded custody of it, and enjoined Argus to watch it. Argus tethered Io to an olive tree in the sacred wood at Mycenae, and thanks to his many eyes, of which only half were ever shut at one time, he could fulfill his task. Zeus had Hermes (Mercury) disguise himself as a shepherd, put all of Argus's eyes asleep by playing to him on the pipes of Pan, and slay him with a sword. In order to give immortality to her faithful servant, Hera moved his eyes to the tail of the bird that was sacred to her, the peacock: see Apollodorus, *Bibliotheca*, II 1 2–3; Ovid, *Metamorphoses*, I 588–721; and Valerius Flaccus, *Argonautica*, IV 366–422. In Barberini's poem, Argus represents the powerful and prominent individual ruined by flattery, which—just like Mercury's song—numbs his senses, thereby making him an easy target for enemies.

BIBLIOGRAPHY

With the exception only of Viviani, because more than one work of his is included here, references to the works in the present collection are simply made by author and page. In the case of Viviani, the date—1659, 1674, or 1701—precedes the page reference; references to the *Racconto istorico* are made just by its title and the corresponding page.

Abbreviations

Gal. Collection of Galilaean manuscripts at the Biblioteca Nazionale Centrale, in Florence; followed by volume number and leaf numbers.

KGW *Johannes Kepler gesammelte Werke*, edited by Max Caspar et al., Munich: C. H. Beck, 1937–2009, 21 vols.

ODG *Le Opere dei Discepoli di Galileo Galilei: Carteggio*, edited by Paolo Galluzzi and Maurizio Torrini, Florence: Giunti-Barbèra, 1975 and 1984, 2 vols.

OG *Le Opere di Galileo Galilei: Edizione Nazionale*, Antonio Favaro editor-in-chief, Florence: Giunti-Barbèra, 1890–1909, 20 vols. (new enlarged edition, 1929–1939; reprint, 1968).

OGA *Le Opere di Galileo Galilei: Appendice*, vol. 1: *Iconografia galileiana*, edited by Federico Tognoni, Florence: Giunti, 2013; vol. 2: *Carteggio*, edited by Michele Camerota and Patrizia Ruffo, with the assistance of Massimo Bucciantini, Florence: Giunti, 2015; vol. 3: *Testi*, edited by Andrea Battistini, Michele Camerota, Germana Ernst, Romano Gatto, Mario Otto Helbing and Patrizia Ruffo, Florence: Giunti, 2017 (one more volume is forthcoming).

Opere *Opere di Galileo Galilei Linceo nobile fiorentino Già Lettore delle Matem-*
1655–1656 *atiche nelle Vniversità di Pisa, e di Padoua, di poi Sopraordinario nello Studio di Pisa. Primario Filosofo, e Matematico del Serenissimo Gran Duca di Toscana*, Bologna: Heirs of Evangelista Dozza, 1655–1656, 2 vols.

Opere 1718 *Opere di Galileo Galilei Nobile Fiorentino Primario Filosofo, e Mattematico del Serenissimo Gran Duca di Toscana*, [edited by Tommaso Buonaventuri, in collaboration with Luigi Guido Grandi], Florence: Giovanni Gaetano Tartini & Santi Franchi, 1718, 3 vols.

Opere 1744 *Opere di Galileo Galilei divise in quattro tomi, In questa nuova Edizione accresciute di molte cose inedite*, [edited by Giuseppe Toaldo], Padua: Stamperia del Seminario & Giovanni Manfrè, 1744, 4 vols.

Opere
1842–1856
Le Opere di Galileo Galilei: Prima edizione completa condotta sugli autentici manoscritti palatini, Eugenio Albèri editor-in-chief, Florence: Società Editrice Fiorentina, 1842–1856, 15 vols. + Supplement.

PRIMARY SOURCES

Galileo's Works

Balance
"Discorso . . . intorno all'arteficio che vsò Archimede nel scoprir il furto dell'Oro nella Corona di Hierone con la fabrica d'vn nuouo Strumento, detto dall'Autore, Bilancetta", in Giovanni Battiasta Odierna, Archimede redivivo con la stadera del momento, Palermo: Decio Cirillo, 1644, pp. 1–8; La Bilancetta, in OG 1, pp. 215–220.

English translation, The Ballance; In which, in imitation of Archimedes in the Problem of the Crown, he sheweth how to find the proportion of the Alloy of Mixt-Metals; and how to make the said Instrument, in Thomas Salusbury, Mathematical Collections and Translations, vol. II, Part 1, London: William Leybourn, 1665, pp. 303–310.

Compass
Le operazioni del compasso geometrico et militare, Padua: Pietro Marinelli, 1606; reprinted in OG 2, pp. 365–424.

English translation in Galileo Galilei, Operations of the Geometric and Military Compass, translated, with an introduction by Stillman Drake, Washington: Smithsonian Institution Press, 1978, pp. 37–92.

Dialogue
Dialogo doue ne i congressi di quattro giornate si discorre sopra i due massimi del mondo, tolemaico e copernicano, Florence: Giovanni Battista Landini, 1632; reprinted in OG 7, pp. 21–520.

English translation by Stillman Drake, Dialogue Concerning the Two Chief World Systems—Ptolemaic & Copernican, Foreword by Albert Einstein, Berkeley-Los Angeles: University of California Press, 1962; second edition, 1967.

See also Dialogo sopra i due massimi sistemi del mondo tolemaico e copernicano, critical edition and commentary by Ottavio Besomi and Mario Helbing, Padua: Antenore, 1998, 2 vols.

Difesa
Difesa Contro alle Calunnie & imposture di Baldessar Capra Milanese, Vsategli sì nella Considerazione Astronomica sopra la nuoua Stella del MDCIIII. come (& assai più) nel publicare nuouamente come sua inuenzione la fabrica, & gli vsi del Compasso Geometrico, & Militare, sotto il titolo di Vsus & fabrica Circini cuiusdam proportionis, &c, Venice: Tommaso Baglioni, 1607; reprinted in OG 2, pp. 513–601.

Letter to
Castelli
in Pierre Gassendi, Apologia in Io. Bap. Morini librum, Cui titulus, Alæ Tellvris fractæ: Epistola IV. de Motu impresso à Motore translato, Lyon: Guillaume Barbier, 1649, pp. 65–78; reprinted in OG 5, pp. 281–288.

English translation in Maurice Finocchiaro, The Galileo Affair: A Documentary History, Berkeley: University of California Press, 1989, pp. 49–54.

Letter to Christina	"Ad Serenissimam Dominam, Magnam-Dvcem Hetruriæ, Magni-Ducis Matrem / Alla Serenissima Madama La Gran Duchessa Madre", in *Nov-Antiqua Sanctißimorum Patrum, & Probatorum Theologorum Doctrina, De Sacræ Scripturæ Testimoniis, in Conclusionibus mere Naturalibus, quæ sensatâ experientiâ, & necessariis demonstrationibus evinci possunt, temere non usurpandis: In gratiam Serenißimæ Christinæ Lotharingæ, Magnæ Ducis Hetruriæ, privatim ante complures annos, Italico idiomate conscripta*, Strasbourg: Elzevier (publisher) and David Hautt (printer), 1636, pp. 1–60 (Latin translation by Robertus Robertinus Borussus, i.e., Élie Diodati); reprinted in *OG* 5, pp. 309–348.

English translation in Maurice Finocchiaro, *The Galileo Affair: A Documentary History*, Berkeley: University of California Press, 1989, pp. 87–118.

See also *Lettera a Cristina di Lorena*, critical edition by Ottavio Besomi, with the collaboration of Daniele Besomi, edited by Giancarlo Reggi, Rome-Padua: Antenore, 2012.

On Floating Bodies	*Discorso Intorno alle cose, che Stanno in su l'acqua, ò che in quella si muouono*, Florence: Cosimo Giunta, 1612; reprinted in *OG* 4, pp. 58–141.

English translation in Stillman Drake, *Cause, Experiment and Science: A Galilean Dialogue Incorporating a New English Translation of Galileo's "Bodies That Stay atop Water, or Move in It,"* Chicago-London: The University of Chicago Press, 1981.

On Mechanics	*Le mecaniche* (ca. 1600), in *OG* 2, pp. 155–190.

English translation by Stillman Drake, *On Mechanics*, in Galileo Galilei, *On Motion and On Mechanics*, Madison: The University of Wisconsin Press, 1960, pp. 147–186.

See also *Le Mechaniche*, edited by Romano Gatto, Florence: Leo S. Olschki, 2002.

On Motion	*De motu* [*antiquiora*] (ca. 1590), in *OG* 1, pp. 251–340 (1), 341–366 (2), 367–408 (3), 409–417 (4), and 418–419 (5).

English translation by Israel E. Drabkin, in Galileo Galilei, *On Motion and On Mechanics*, translated and annotated by Israel E. Drabkin and Stillman Drake, Madison: The University of Wisconsin Press, 1960, pp. 13–114 (1, "On Motion"), 118–123 (2, "The Reworking of Parts of the *De Motu*"), and 130–131 (5, "The Memoranda on Motion"); and by Israel E. Drabkin, in *Mechanics in Sixteenth-Century Italy*, translated and annotated by Stillman Drake and Israel E. Drabkin, Madison: The University of Wisconsin Press, 1969, pp. 331–377 (3, "Dialogue on Motion") and 378–387 (4, "Memoranda on Motion").

On Sunspots	*Istoria e dimostrazioni matematiche intorno alle macchie solari e loro accidenti*, Rome: Giacomo Mascardi, 1613; reprinted in *OG* 5, pp. 72–149.

English translation in *On Sunspots*, translated and introduced by Eileen Reeves and Albert Van Helden, Chicago-London: The University of Chicago Press, 2010, pp. 87–105, 107–168, and 251–304.

Sidereal Messenger	*Siderevs nvncivs Magna, longeqve admirabilia Spectacula pandens, suspiciendaque proponens vnicuique, præsertim verò philosophis, atque astronomis*, Venice: Tommaso Baglioni, 1610; reprinted in *OG* 3.1, pp. 52–96.

English translation, *Sidereal Messenger*, in Galileo Galilei, *Sidereus Nuncius or The Sidereal Messenger*, translated, with introduction, conclusion, and notes by Albert Van Helden, Chicago-London: The University of Chicago Press, 1989, 2016[2], pp. 27–88.

Critical edition and annotated French translation, *Sidereus nuncius. Le messager céleste*, edited by Isabelle Pantin, Paris: les Belles Lettres, 1992.

The Assayer	*Il saggiatore. Nel quale Con bilancia esquisita e giusta si ponderano le cose contenute nella Libra astronomica e filosofica di Lotario Sarsi Sigensano*, Rome: Giacomo Mascardi, 1623; reprinted in *OG* 6, pp. 199–372.

English translation by Stillman Drake, *The Assayer: In which with a delicate and precise scale will be weighed the things contained in The Astronomical and Philosophical Balance of Lotario Sarsi of Siguenza*, in *The Controversy on the Comets of 1618: Galileo Galilei, Horatio Grassi, Mario Guiducci, Johann Kepler*, edited by Stillman Drake and Charles D. O'Malley, Philadelphia: University of Pennsylvania Press, 1960, pp. 151–336.

See also *Il saggiatore*, critical edition and commentary by Ottavio Besomi and Mario Helbing, Rome-Padua: Antenore, 2005, pp. 85–320.

Two New Sciences	*Discorsi e dimostrazioni matematiche, intorno à due nuoue scienze Attenenti alla Mecanica & i Movimenti Locali, Con vna Appendice del centro di grauità d'alcuni Solidi*, Leiden: Elsevier, 1638; reprinted in *OG* 8, pp. 41–318 and *OG* 1, pp. 187–208.

English translation: *Two New Sciences: Including Centers of Gravity and Force of Percussion*, translated, with introduction and notes by Stillman Drake, Madison: The University of Wisconsin Press, 1974.

See also *Discorsi e dimostrazioni matematiche intorno a due nuove scienze*, edited by Adriano Carugo and Ludovico Geymonat, Turin: Paolo Boringhieri, 1958.

Other Primary Sources

ACCADEMICO INCOGNITO
Considerazioni sopra il Discorso del Sig. Galileo Galilei Intorno alle cose, che stanno in su l'Acqua, o che in quella si muouono. Fatte a difesa, e dichiarazione dell'opinione d'Aristotile da Accademico Incognito, Pisa: Giovanni Battista Boschetti & Giovanni Fontani, 1612; reprinted (with Galileo's remarks and rejoinder) in *OG* 4, pp. 145–196.

ALLATIUS, Leo
Apes Vrbanae, siue De viris illvstribvs, Qui ab Anno MDCXXX. per totum MDCXXXII. Romæ adfuerunt, ac Typis aliquid euulgarunt, Rome: Lodovico Grignani, 1633.

ARCHIMEDES

Arenarius, in *Archimedis Opera Omnia cum Commentariis Eutocii*, edited by Johann L. Heiberg, Leipzig: B. G. Teubner, 1881, pp. 242–291.

De iis quae vehuntur in aqua libri duo. A Federico Commandino Vrbinate in pristinum nitorem restituti, et commentariis illustrati, Bologna: Alessandro Benacci, 1565.

De insidentibus aquae, Venice: Curzio Troiano Navò, 1565.

Opera Archimedis Syracusani philosophi et mathematici ingeniosissimi, edited by Niccolò Tartaglia, Venice: Venturino Ruffinelli, 1543.

ARGELATI, Filippo

Bibliotheca Scriptorum Mediolanensium, seu acta, et elogia virorum omnigena eruditione illustrium, qui in Metropoli Insubriæ, oppidisque circumjacentibus orti sunt; additis literariis monumentis post eorundem obitum relictis, aut ab aliis memoriæ traditis, Milan: Tipografia Palatina, 1745, 2 vols. in 4 parts.

ARIOSTO, Ludovico

Orlando fvrioso, Ferrara: Giovanni Mazzocchi, 1516 (in 40 cantos); revised edition, Ferrara: Giovanni Battista Della Pigna, 1521; definitive edition, Ferrara: Francesco Rosso, 1536.

BARBERINI, Maffeo (Pope URBAN VIII)

Poemata, Paris: Antoine Estienne, 1620; second edition, 1623.

Poemata, Rome: Vatican Printing Office, 1631 (*editio maior*); and The Printing Office of the Reverend Apostolic Chamber, 1631 (*editio minor*).

Poemata, Rome: The Printing Office of the Reverend Apostolic Chamber, 1640.

Poemata, Paris: The Royal Press, 1642.

Poemata, edited by Joseph Brown, Oxford: Clarendon Press, 1726.

BELLINI, Lorenzo

Discorsi di anatomia, Florence: Francesco Moücke, 1741.

Gratiarvm actio ad Serenissimos Hetrvriæ Principes. Quædam Anatomica in Epistola ad Serenissimvm Ferdinandvm II. Magn. Hetr. Dvc. et propositio mechanica, Pisa: Giovanni Ferretti, 1670.

"Lettera di Lorenzo Bellini al Sig. Antonio Vallisneri, intorno all'ingresso dell'aria dentro il nostro sangue," *Giornale de' letterati d'Italia*, IV, 1710, pp. 152–164.

BENEDETTI, Giovanni Battista

Demonstratio proportionvm motvvm localivm contra Aristotelem et omnes Philosophos, Venice: [Bartolomeo Cesano], 1554; second edition, with identical imprint.

Diversarvm specvlationvm Mathematicarum, & Physicarum Liber, Turin: Heir of Nicolò Bevilacqua, 1585.

Resolvtio omnivm Evclidis problematvm aliotumque ad hoc necessario inuentorum vna tantummodo circini data apertura, Venice: [Bartolomeo Cesano], 1553.

BOREL, Pierre

De vero telescopii inventore, Cum brevi omnium conspiciliorum historia, The Hague: Adriaen Vlacq, 1655.

Borelli, Giovanni Alfonso
De motv animalivm, Rome: Angeli Bernabò, 1680–1681, 2 vols.
De vi percvssionis liber, Bologna: Giacomo Monti, 1667.

Bruno, Girordano
La cena de le Ceneri, [s.l.]: [s.e.], 1584; in Giordano Bruno, *Œuvres complètes*, vol. II:
    Le souper des cendres, edited by Giovanni Aquilecchia, Preface of Adi Ophir,
    translated by Yves Hersant, Paris: Les Belles Lettres, 1994; English translation by
    Stanley L. Jaki, *The Ash Supper Wednesday: La cena de le Ceneri*, translated, with
    and Introduction and Notes by Stanley L. Jaki, The Hague-Paris: Mouton, 1975.

[Brusoni, Girolamo]
*Le glorie de gli Incogniti O vero gli hvomini illvstri dell'Accademia de' Signori Incogniti
    di Venetia*, Venice: Francesco Valvasense, 1647.

Bullart, Isaac
*Académie des Sciences et des Arts, Contenant les Vies, & les Eloges Historiques des Hom-
    mes Illustres, Qui ont excellé en ces Professions depuis environ quatre Siécles parmy
    diverses Nations de l'Europe: Avec leurs Pourtraits tirez sur des Originaux au Naturel,
    & plusieurs Inscriptions funebres, exactement recueïllies de leurs Tombeaux*, Paris:
    Jacques Ignace Bullart, 1682, 2 vols.

Buonamici, Francesco
*De motv libri X. Qvibvs generalia natvralis philosophiae principia summo studio collecta
    continentur*, Florence: Bartolomeo Sermartelli, 1591.

Calcagnini, Celio
Opera aliqvot, Basel: Hieronymus Frobenius & Nikolaus Episcopus, 1544.

Campanella, Tommaso
*Apologia pro Galileo, mathematico florentino. Vbi disqviritvr, vtrvm ratio philosophandi,
    qvam Galilevs celebrat, faueat sacris scripturis, an aduersetur*, Frankfurt: Erasmus
    Kempfer, at the expense of Gottfried Tampach, 1622; English translation by Rich-
    ard J. Blackwell, *A Defense of Galileo the Mathematician from Florence: Which Is an
    Inquiry as to Whether the Philosophical View Advocated by Galileo Is in Agreement
    with, or Is Opposed to, the Sacred Scriptures*, Notre Dame-London: University of
    Notre Dame Press, 1994.
*Astrologicorum Libri VI. In qvibvs Astrologia, omni superstitione Arabum, & Iudæorum
    eliminata, physiologicè tractatur*, Lyon: Jacques, André & Mathieu Prost, 1629.
De Libris propriis & recta ratione studendi. Syntagma, Paris: Guillaume veuve Pelé, 1642.
Lettere, edited by Germana Ernst on the basis of preparatory work by Luigi Firpo,
    with the collaboration of Laura Salvetti Firpo and Matteo Salvetti, Florence:
    Leo S. Olschki, 2010.

Capaccio, Giulio Cesare
*In Odas eminentissimi Cardinalis olim Barberini nvnc Sanctissimi Summi Pontificis
    Vrbani VIII*, Naples: Lazzaro Scoriggio, 1633.

CAPRA, Baldassarre

*Consideratione astronomica Circa la noua, & portentosa Stella che nell'anno 1604. adi
10. Ottobre apparse*, Padua: Lorenzo Pasquati, 1605; reprinted (with Galileo's annotations) in *OG* 2, pp. 287–305.

*Vsvs et fabrica circini cvivsdam proportionis, Per quem omnia ferè tum Euclidis, tum
Mathematicorum omnium problemata facili negotio resoluuntur*, Padua: Pietro Paolo
Tozzi, 1607; reprinted in *OG* 2, pp. 427–510.

CASTELLI, Benedetto

Della misvra dell'acqve correnti, Rome: Stamperia Camerale, 1628; English translation by Thomas Salusbury, *Of the Mensuration of Running Waters*, in *Mathematical
Collections and Translations*, London: William Leybourne, 1661, vol. I, Part 2,
pp. 1–33.

*Risposta alle opposizioni del S. Lodovico delle Colombe, e del S. Vincenzio di Grazia,
contro al Trattato del Sig. Galileo Galilei, delle cose che stanno su l'Acqua, ò che in
quella si muouono*, Florence: Cosimo Giunta, 1615; reprinted in *OG* 4, pp. 451–789.

CECCO DI RONCHITTI [Girolamo Spinelli]

Dialogo in perpvosito De La Stella Nvova, Padua: Pietro Paolo Tozzi, 1605; reprinted
in *OG* 2, pp. 309–334; English translation, *Dialogue Concerning the New Star*, in
Galileo Against the Philosophers, translated, with introduction and notes by Stillman Drake, Los Angeles: Zeitlin & Ver Brugge, 1976, pp. 33–53.

CHIARAMONTI, Scipione

*Antitycho in qvo contra Tychonem Brahe, & nonnullos alios rationibus eorum ex opticis,
& geometricis principijs solutis Demonstratur cometas esse svblvnares non coelestes*,
Venice: Evangelista Deuchino, 1621.

*Apologia … pro antitychone svo adversvs Hyperaspistem Ioannis Kepleri. Confirmatur in
hoc Opere, Rationibus ex Parallaxi Præsertim ductis, contrarijsquè omnibus reiectis,
cometas svblvnares esse Non Cælestes*, Venice: Evangelista Deuchino, 1626.

*De tribvs novis stellis qvæ Annis 1572. 1600. 1604. Comparuere libri tres In quibus demon-
stratur rationibus, ex Parallaxi præsertim ductis Stellas eas fuiße Sublunares, & non
Cælestes adversvs Tychonem, Gemmam, Mestlinum, Digesseum, Hagecium, Santu-
cium, Keplerum, aliosque plures qvorvm Rationes in Contrarium adductæ soluuntur*,
Cesena: Giuseppe Neri, 1628.

*Difesa … al svo Antiticone, e Libro delle tre nuoue Stelle dall'oppositioni dell'avtore de'
Due massimi Sistemi Tolemaico, e Copernicano. Nella qvale Si sostiene, che la nuoua
Stella del 72. Non fù celeste: Si difende Arist. ne' suoi principali Dogmi del Cielo: si
rifiutano i principij della nuoua Filosofia, e l'addotto in distesa, e proua del Sistema
Copernicano*, Florence: Giovanni Battista Landini, 1633.

Discorso della cometa pogonare dell'anno MDCXVIII, Venice: Pietro Ferri, 1618.

COLLETET, Guillaume

Epigrammes, avec vn discovrs de l'epigramme, Paris: Jean Baptiste Loyson, 1653.

COMMANDINO, Federico

Liber de centro gravitatis solidorum, Bologna: Alessandro Benacci, 1565.

COPERNICUS, Nicolaus
De revolvtionibvs orbium cælestium, Libri VI, Nuremberg: Johann Petreius, 1543; second edition, Basel: Sebastian Henric Petri, 1566; critical edition and annotated French translation, *De revolvtionibvs orbivm cælestivm. Des révolutions des orbes célestes*, edited by Michel-Pierre Lerner, Alain-Philippe Segonds, and Jean-Pierre Verdet, Paris: Les Belles Lettres, 2015, 3 vols.

CORESIO, Giorgio
Operetta intorno al galleggiare de corpi solidi, Florence: Bartolomeo Sermartelli, 1612; reprinted in *OG* 4, pp. 199–244.

CRASSO, Lorenzo
Elogii d'hvomini letterati, Venice: Sebastiano Combi & Giovanni La Noù, 1666, 2 vols.

DANESI, Luca
Opere, Ferrara: Giulio Bolzoni Giglio, 1670.

DELLA PORTA, Giovanni Battista
De refractione optices parte: Libri Nouem, Naples: Orazio Salviani, at Giovanni Giacomo Carlino & Antonio Pace, 1593.
Magiae natvralis, siue De miraculis rerum naturalium libri IIII, Naples: Mattia Cancer, 1558; revised edition, *Magiae natvralis libri XX*, Naples: Orazio Salviani, 1589.

DELLE COLOMBE, Lodovico
Discorso apologetico d'intorno al Discorso di Galileo Galilei, Circa le cose, che stanno sù l'Acqua, ò che in quella si muouono, Florence: Zanobi Pignoni, 1612; reprinted in *OG* 4, pp. 313–369.
Discorso Nel quale si dimostra, che la nuoua Stella apparita l'Ottobre passato 1604. nel Sagittario non è cometa, ne Stella generata, ò creata di nuouo, ne apparente: ma vna di qvelle che fvrono da principio nel Cielo; e ciò eßer conforme alla vera Filosofia, Teologia, e Astronomiche dimostrazioni, Florence: Giunta, 1606
Risposte piacevoli, e cvriose alle considerazioni di certa Maschera saccente nominata Alimberto Mauri, fatte sopra alcuni luoghi del discorso del medesimo Lodouico dintorno alla stella apparita l'anno 1604, Florence: Giovanni Antonio Caneo & Raffaello Grossi, 1608.

DEL MONTE, Guidobaldo
Mechanicorvm liber, Pesaro: Girolamo Concordia, 1577.
Perspectivae libri sex, Pesaro: Girolamo Concordia, 1600.
Planisphaeriorvm vniversalivm theorica, Pesaro: Girolamo Concordia, 1579.

DE PINEDA, Juan
In Ecclesiasten Commentariorum Liber unus, Seville: Gabriel Ramos Bejarano, 1619.

DE ZÚÑIGA, Diego
In Iob Commentaria. Qvibvs triplex eivs editio vulgata Latina, Hebræa, & Græca Septuaginta interpretum, necnon & Chaldæa explicantur, Rome: Francesco Zanetti, 1591.

DI GRAZIA, Vincenzo

Considerazioni sopra 'l Discorso di Galileo Galilei Intorno alle cose che stanno su l'acqua, e che in quella si muouono, Florence: Zanobi Pignoni, 1613; reprinted in *OG* 4, pp. 373–440.

DONATI, Alessandro

Ars Poetica, Rome: Guglielmo Facciotti, 1631

Esequie del Divino Michelangelo Bvonarroti Celebrate in Firenze dall'Accademia de Pittori, Scultori, & Architettori, Florence: Giunti, 1564.

EUCLID

Elementa, in *Euclidis Opera Omnia*, edited by Johan L. Heiberg and Heinrich Menge, vols. 1–5, Leipzig: B. G. Teubner, 1883–1888.

The Thirteen Books of Euclid's Elements, translated, with introduction and commentary by Thomas L. Heath, Cambridge: Cambridge University Press, 1908, 3 vols.

FONTANA, Francesco

Novae cœlestivm terrestrivmqve rervm observationes. Et fortasse hactenus non uulgatæ . . . specillis à se inventis, et ad summam perfectionem perductis editæ, Naples: Giacomo Gaffaro, 1646.

FABRONI, Angelo (ed.)

Lettere inedite di uomini illustri, Florence: Francesco Moücke, 1773–1775, 2 vols.

FOSCARINI, Paolo Antonio

Lettera . . . Sopra l'Opinione de' Pittagorici, e del Copernico. Della mobilità della Terra e stabilità del Sole, E del nuouo Pittagorico Sistema del Mondo, Naples: Lazzaro Scoriggio, 1615.

FREHER, Paul

Theatrvm virorum eruditione clarorum, Nuremberg: Heirs of Andreas Knorz, at the expense of Johann Hofmann, 1668–1688, 2 vols.

GALILEI, Vincenzo

Canto de contrapvnti a dve voci, Florence: Giorgio Marescotti, 1584.

Dialogo della musica antica, et della moderna, Florence: Giorgio Marescotti, 1581; English translation by Claude V. Palisca, *Dialogue on Ancient and Modern Music*, New Haven, CT: Yale University Press, 2003.

Discorso intorno all'opere di messer Gioseffo Zarlino da Chioggia, et altri importanti particolari attenenti alla musica, Florence: Giorgio Marescotti, 1589.

Fronimo, Dialogo nel quale si contengono le vere et necessarie regole del intavolare la musica nel liuto, Venice: Girolamo Scoto, 1568.

Tenore de contrapvnti a dve voci, Florence: Giorgio Marescotti, 1584.

GASSENDI, Pierre

Animadversiones in decimvm librvm Diogenis Laertii, qvi est de Vita, Moribus, Placitisque Epicvri, Lyon: Guillaume Barbier, 1649.

De vita et moribvs Epicvri libri octo, Lyon: Guillaume Barbier, 1647.

Epistolica exercitatio, In qua Principia Philosophiæ Roberti Flvddi medici reteguntur; Et ad recentes illius Libros, aduersus R. P. F. Marinvm Mersennvm Ordinis Minimorum Sancti Francisci de Paula scriptos, Respondetur. Cum Appendice aliquot Obseruationum Cœlestium, Paris: Sébastien Cramoisy, 1630.

Exercitationes paradoxicæ adversvs Aristoteleos. In quibus totius Peripateticæ doctrinæ fundamenta excutiuntur, Grenoble: Pierre Verdier, 1624.

Institvtio astronomica, Iuxta Hypotheseis tam vetervm, qvam Copernici et Tychonis, Paris: Louis de Heuqueville, 1647; second edition, London: Jacob Flesher, 1653.

Mercvrivs in Sole visvs, et Venvs invisa Parisiis, Anno 1631. Pro voto, & Admonitione Keppleri, Paris: Sébastien Cramoisy, 1632.

Parhelia, sive Soles qvatvor, Qvi circa Vervm apparuerunt Romæ, Die xx Mensis Martij, Anno 1629. et . . . ad Henricvm Renervm epistola, Paris: Pierre Vitré, 1630.

GILBERT, William

De magnete, magneticisqve corporibvs, et de magno magnete tellure; Physiologia noua, plurimis & argumentis, & experimentis demonstrata, London: Peter Short, 1600.

GRASSI, Orazio [Lotario Sarsi]

De tribvs cometis anni M. DC. XVIII. Dispvtatio astronomica, Rome: Giacomo Mascardi, 1619; reprinted in *OG* 6, pp. 23–35; English translation by Charles D. O'Malley, *On the Three Comets of the Year MDCXVIII: An Astronomical Disputation*, in *The Controversy on the Comets of 1618: Galileo Galilei, Horatio Grassi, Mario Guiducci, Johann Kepler*, edited by Stillman Drake and Charles D. O'Malley, Philadelphia: University of Pennsylvania Press, 1960, pp. 3–19.

Libra astronomica ac philosophica qva Galilaei Galilaei Opiniones de Cometis a Mario Gvidvcio In Florentiina Academia expositæ, atque in lucem nuper editæ, examinantur, Perugia: Marco Naccarini, 1619; reprinted in *OG* 6, pp. 111–180 (with Galileo's annotations); in Galileo Galilei, *Il saggiatore*, critical edition and commentary by Ottavio Besomi and Mario Helbing, Rome-Padua: Antenore, 2005, pp. 365–433 and 321–364; English translation by Charles D. O'Malley, *The Astronomical and Philosophical Balance: on which the Opinions of Galileo Galilei regarding Comets are weighed, as well as those presented in the Florentine Academy by Mario Guiducio and recently published*, in *The Controversy on the Comets of 1618: Galileo Galilei, Horatio Grassi, Mario Guiducci, Johann Kepler*, edited by Stillman Drake and Charles D. O'Malley, Philadelphia: University of Pennsylvania Press, 1960, pp. 67–132.

Ratio pondervm libræ et simbellæ: in qva qvid è Lotharii Sarsii Libra astronomica, qvidqve è Galilei Galilei simbellatore, De Cometis statuendum sit, collatis vtriusque rationum momentis, Philosophorum arbitrio proponitur, Paris: Sébastien Cramoisy, 1626.

GUALDO, Paolo

Vita Ioannis Vincentii Pinelli, Patricii Genvensis, Augsburg: Christoph Mang, 1607.

GUIDUCCI, Mario

Discorso delle comete, Florence: Pietro Cecconelli, 1619; reprinted in *OG* 6, pp. 39–108; English translation by Stillman Drake, *Discourse on the Comets*, in *The Controversy* on the Comets of 1618: Galileo Galilei, Horatio Grassi, Mario Guiducci, Johann Kepler, edited by Stillman Drake and Charles D. O'Malley, Philadelphia: University of Pennsylvania Press, 1960, pp. 21–65.

Lettera al M. R. P. Tarqvinio Gallvzzi. Della Compagnia di Giesv, Florence: Zanobi Pignoni, 1620; reprinted in *OG* 6, pp. 183–196; English translation by Stillman Drake, *Letter To the Very Reverend Father Tarquinio Galluzzi of the Society of Jesus*, in *The Controversy on the Comets of 1618: Galileo Galilei, Horatio Grassi, Mario Guiducci, Johann Kepler*, edited by Stillman Drake and Charles D. O'Malley, Philadelphia: University of Pennsylvania Press, 1960, pp. 133–150.

HORKÝ, Martin

Brevissima peregrinatio contra Nvncivm siderevm nvper ad omnes philosophos et mathematicos emissvm, Modena: Giuliano Cassini, 1610; reprinted in *OG* 3.1, pp. 129–145.

HUYGENS, Christiaan

Horologivm, The Hague: Adriaan Vlacq, 1658.

Horologivm oscillatorivm. Sive de motv pendvlorvm ad horologia aptato demonstrationes geometricæ, Paris: François Muguet, 1673; English translation by Richard J. Blackwell, *The Pendulum Clock or Geometrical Demonstrations Concerning the Motion of Pendula as Applied to Clocks*, introduction by Henk J. M. Bos, Ames, IA: The Iowa State University Press, 1986.

Œuvres complètes, The Hague: Martinus Nijhoff, 1888–1950, 22 vols.

Systema Satvrnivm, Sive De causis mirandorum Satvrni Phænomenôn, Et Comite ejus Planeta Novo, The Hague: Adriaan Vlacq, 1659.

INCHOFER, Melchior

Tractatvs syllepticvs, In quo qvid de Terrae, Solisqve motv, vel statione, secundùm S. Scripturam, & Sanctos Patres sentiendum, quauè certitudine alterutra sententia tenenda sit, breuiter ostenditur, Rome: Lodovico Grignani, 1633; English translation by Richard J. Blackwell, in Id., *Behind the Scenes at Galileo's Trial: Including the First English Translation of Melchior Inchofer's* Tractatus syllepticus, Notre Dame, IN: University of Notre Dame Press, 2006, pp. 105–206.

KEPLER, Johannes

Ad Vitellionem Paralipomena, Quibus Astronomiæ pars optica traditvr; Potißimùm de artificiosa observatione et æstimatione diametrorvm deliquiorumque Solis et Lunæ. Cvm exemplis insignivm eclipsivm, Frankfurt: Claude de Marne and Heirs of Johann Aubry, 1604; reprinted in *KGW* 2: *Astronomiae pars optica*, edited by Franz Hammer, Munich: C. H. Beck, 1939, pp. 5–391; English translation by William H. Donahue, *Optics: Paralipomena to Witelo & Optical Part of Astronomy*, Santa Fe, NM: Green Lion Press, 2000.

Astronomia nova ΑΙΤΙΟΛΟΓΗΤΟΣ, sev physica coelestis, tradita commentariis de motibvs stellæ Martis, Ex observationibus G. V. Tychonis Brahe, [Heidelberg: Gotthard Vögelin], 1609; reprinted in *KGW* 3: *Astronomia nova*, edited by Max Caspar, Munich: C. H. Beck, 1937, pp. 5–424; English translation by William H. Donahue, *Astronomia Nova*, new revised edition, Santa Fe, NM: Green Lion Press, 2015.

De cometis libelli tres, Vol. I: *Astronomicus*, Augsburg: Andreas Aperger & Sebastian Müller, 1619; vol. II: *Cometarvm physiologia nova et παραδόξος*, Augsburg: Andreas Aperger, 1619; vol. III: *De significationibvs cometæ qvi anno MDCVII. Conspectvs est*, Augsburg: Andreas Aperger, 1620; reprinted in *KGW* 8: *Mysterium cosmographicum, editio altera cum notis. De cometis. Hyperaspistes*, edited by Franz Hammer, 1963, pp. 131–262.

De stella nova in pede Serpentarii, et qui sub ejus exortum de novo iniit, Trigono Igneo, Prague: Pavel Sessius, 1606.

Dioptrice sev Demonstratio eorum quæ visui & visibilibus propter Conspicilla non ita pridem inventa accidunt, Augsburg: David Franck, 1611; reprinted in *KGW* 4: *Kleinere Schriften 1602/1611. Dioptrice*, edited by Max Caspar and Franz Hammer, Munich: C. H. Beck, 1941, pp. 329–414.

Dissertatio Cum Nvncio Sidereo nuper ad mortales misso à Galilæo Galilæo, Prague: Daniel Sedlčanský, 1610; reprinted in *KGW* 4: *Kleinere Schriften 1602/1611. Dioptrice*, edited by Max Caspar and Franz Hammer, Munich: C. H. Beck, 1941, pp. 283–311; English translation by Edward Rosen, *Kepler's Conversation with Galileo's Sidereal Messenger*, New York-London: Johnson Reprint Corporation, 1965.

Gründtlicher Bericht Von einem ungewohnlichen Newen Stern, wellicher im October ditz 1604. Jahrs erstmahlen erschienen, Prague: Schuman, 1604.

Narratio de observatis a se quatuor Iouis satellitibus erronibus, qvos Galilævs Galilævs mathematicus Florentinus iure inuentionis Medicæa sidera nuncupauit, Frankfurt: Zacharias Palthen, 1611; reprinted in *KGW* 4: *Kleinere Schriften 1602/1611. Dioptrice*, edited by Max Caspar and Franz Hammer, Munich: C. H. Beck, 1941, pp. 315–325.

Tychonis Brahei Dani Hyperaspistes, adversvs Scipionis Claramontii Cæsennatis Itali, Doctoris & Equitis Anti-Tychonem, Frankfurt am Main: Gottfried Tampach, 1625; reprinted in *KGW* 8: *Mysterium cosmographicum, editio altera cum notis. De cometis. Hyperaspistes*, edited by Franz Hammer, 1963, pp. 265–437.

KɪʀCHER, Athanasius
Œdipvs Ægyptiacvs. Hoc est Vniuersalis Hieroglyphicæ Veterum Doctrinæ temporum iniuria abolitæ instavratio. Opus ex omni Orientalium doctrina & sapientia conditum, nec non viginti diuersam linguarum authoritate stabilitum, Rome: Vitale Mascardi, 1652–1654, 3 vols.

LaGALLA, Giulio Cesare
De phænomenis in orbe Lvnæ novi telescopii vsv nvnc itervm svscitatis Physica disputatio ... necnon De lvce, et lvmine Altera disputatio, Venice: Tommaso Baglioni, 1612; reprinted in *OG* 3.1, pp. 311–393, with Galileo's annotations (pp. 393–399).

LaɪʀD, Walter R.
The Unfinished Mechanics of Giuseppe Moletti: An Edition and English Translation of his Dialogue on Mechanics, 1576, Toronto: Toronto University Press, 2000.

Lauʀɪ, Giovanni Battista
Theatri Romani Orchestra. Dialogvs De Viris sui Aeui doctrina illustribus, edited by Josse Rycke, Rome: Andrea Fei, 1625.

Leibniz, Gottfried W.
Sämtliche Schriften und Briefe, Series III: *Mathematischer naturwissenschaftlicher und technischer Briefwechsel*, under the supervision of the Akademie der Wissenschaften in Gottingen, edited by the Leibniz-Archiv der Niedersachsischen Landesbibliothek Hannover, Berlin: Akademie, 1988–2015, 8 vols.

Liceti, Fortunio
De lvnae svbobscvra lvce Prope Coniunctiones, & in Eclipsibus obseruata libros tres: . . . Fortvnivs Licetvs . . . Serenissimo Principi Leopoldo Tvsciæ dedicat, Udine: Nicola Schiratti, 1642.
Litheosphorvs, sive De lapide Bononiensi Lucem in se conceptam ab ambiente claro mox in tenebris conseruante liber, Udine: Nicola Schiratti, 1640.

Locher, Johann Georg
Disqvisitiones mathematicæ, de controversiis et novitatibvs astronomicis. Quas svb præsidio Christophori Scheiner . . . pvblice dispvtandas posvit, propvgnavit, Ingolstadt: Wolfgang Eder & Elisabeth Angermaier, 1614.

Magalotti, Lorenzo
Delle lettere familiari del Conte Lorenzo Magalotti e di altri insigni uomini a lui scritte, Florence: Gaetano Cambiagi, 1769, 2 vols.
Saggi di natvrali esperienze fatte nell'Accademia del Cimento sotto la protezione del Serenissimo Principe Leopoldo di Toscana e descritte dal segretario di essa Accademia, Florence: Giuseppe Cocchini, 1666 (in fact, 1667); reprinted in *Le Opere dei Discepoli di Galileo Galilei*, vol. I: *L'Accademia del Cimento*, Part I, Florence: S. A. G. Barbèra, 1942, pp. 77–270; English translation by Richard Waller, *Essayes of Natural Experiments Made in the Academie del Cimento Under the Protection of the Most Serene Prince Leopold of Tuscany*, London: Benjamin Alsop, 1684.

Marchetti, Alessandro
Fvndamenta vniversæ scientiæ De motu vniformiter accelerato à Galileo Galilei primum iacta, ab Euangelista Torricellio, alijsque celeberrimis Mathematicis confirmata; nunc vero demum euidentibus demonstrationibus stabilita, Pisa: Giovanni Ferretti, 1674.

Marino, Giovan Battista
L'Adone, Paris: Olivier de Varennes, 1623.
La Galeria, Distinta in Pittvre, & Sculture, Venice: Giovanni Battista Ciotti, 1620.

Mästlin, Michael
Disputatio de eclipsibvs solis et lvnæ, Tübingen: Georg Gruppenbach, 1596.

Mauri, Aliberto
Considerazioni sopra alcvni lvoghi del Discorso di Ludouico delle Colombe intorno alla stella apparita 1604, Florence: Giovanni Antonio Caneo, 1606; English translation, *Considerations On Some Places in the Discourse of Lodovico Delle Colombe about the*

Star which appeared in 1604, in *Galileo Against the Philosophers*, translated, with Introduction and Notes, by Stillman Drake, Los Angeles: Zeitlin & Ver Brugge, 1976, pp. 73–130.

MAYR, Simon [Simon Marius]
Mundus Iovialis Anno M. DC. IX. Detectus ope Perspicilli Belgici. Hoc est, Quatuor Iovialium Planetarum, cum theoria, tum tabulæ, propriis observationibus maxime fundatæ, ex quibus situs illorum ad Iovem, ad quodvis tempus datum promptissimè & facilimè supputari potest, Nuremberg: Johann Lauer, 1614.

MAZZONI, Jacopo
Della difesa della Comedia di Dante. Distinta in sette libri. Nella quale si risponde alle oppositioni fatte al Discorso di M. Iacopo Mazzoni, e si tratta pienamente dell'arte Poetica, e di molt'altre cose pertenenti alla Philosophia, & alle belle lettere. Parte prima, che contiene li primi tre libri, Cesena: Bartolome Raverio, 1587.
Della difesa della Comedia di Dante. Distinta in sette libri, Nella quale si risponde alle oppositioni fatte al discorso di M. Iacopo Mazzoni, e si tratta pienamente dell'Arte Poetica, e di molte altre cose pertenenti alla Filosofia, & altre belle lettere. Parte Seconda Posthuma, che contiene gli vltimi qvattro libri, non piv' stampati, Cesena: Severo Verdoni, 1688.
Discorso in difesa della Comedia Del Diuino Poêta Dante, Cesena: Bartolomeo Raverio, 1573.
In vniversam Platonis, et Aristotelis Philosophiam Præludia, siue de Comparatione Platonis, & Aristotelis. Liber primvs, Venice: Giovanni Guerigli, 1597.
Mechanics in Sixteenth-Century Italy: Selections from Tartaglia, Benedetti, Guido Ubaldo, & Galileo, translated and annotated by Stillman Drake and Israel E. Drabkin, Madison: The University of Wisconsin Press, 1969.

MILTON, John
Areopagitica; A Speech For the Liberty of Vnlicenc'd Printing, To the Parlament of England, London: [s.e.], 1644.

MOLETI, Giuseppe
Tabvlae Gregorianae Motuum Octauæ Sphæræ ac Luminarium ad vsum Calendarij Ecclesiastici, & ad Vrbis Romæ Meridianum supputatæ, Venice: Pietro Deuchino, 1580.

NELLI, Giovanni Battista Clemente
Vita e commercio letterario di Galileo Galilei, Losanna: s.n., 1793 [yet Florence: Francesco Moücke, 1791], 2 vols.

ODIERNA, Giovanni Battista
Archimede redivivo con la stadera del momento, Palermo: Decio Cirillo, 1644.

ORSI, Aurelio et al.
Academicorum Insensator. Carmina, Perugia: Accademici Augusti, 1606.

PATRICIUS, Franciscus
Nova de vniversis philosophia. In qva aristotelica methodo non per motum, sed per lucem, & lumina, ad primam causam ascenditur. Deinde propria Patricii methodo; Tota in contemplationem venit Diuinitas: Postremo methodo Platonica, rerum vniversitas, à conditore Deo deducitur, Ferrara: Benedetto Mammarello, 1591.

RHETICUS, Georg Joachim
De Libris Revolutionum . . . Nicolai Copernici . . . Narratio Prima [Gdańsk: s.e., 1540].

RICCIOLI, Giovanni Battista
Almagestvm Novvm, astronomiam veterem novamqve complectens observationibvs aliorvm, et propriis Nouisque Theorematibus, Problematibus, ac Tabulis promotam, in tres tomos distribvtam, Bologna: Heirs of Vittorio Benacci, 1651.

ROCCO, Antonio
Esercitationi filosofiche . . . Le qvuali versano in considerare le Positioni, & Obiettioni, che si contengono nel Dialogo del Signor Galileo Galilei Linceo contro la Dottrina d'Aristotile, Venice: Francesco Baba, 1633; reprinted in *OG* 7, pp. 571–712, with Galileo's annotations (pp. 712–750).

ROFFENI, Giovanni Antonio
Epistola apologetica contra cæcam peregrinationem Cuiusdam furiosi Martini, cognomine Horkij editam aduersus nuntium sidereum, Bologna: Heirs of Giovanni Rossi, 1611; reprinted in *OG* 3.1, pp. 193–200.

ROSSI, Gian Vittorio
Epistolæ ad Tyrrhenvm, Amsterdam: Joan Blaeu, 1645.
Epistolae ad Tyrrhenvm et ad diuersos, edited by Johann C. Fischer, Amsterdam: Joan Blaeu, 1649.
Evdemiæ libri octo, Leiden: Bonaventura & Abraham Elzevier, 1637; second edition, *Evdemiæ libri decem*, Cologne: Cornelius von Egmondt, 1645.
Pinacotheca imaginvm illvstrivm, doctrinæ vel ingenii laude, virorvm, qui, auctore superstite, diem suum obierunt, Cologne: Cornelius von Egmondt, 1643.

SALUSBURY, Thomas
Mathematical Collections and Translations, London: William Leybourn, 1661–1665, 2 vols.

SALVINI, Salvino
"Elogio del Can. Niccolò Gherardini," in *Serie di ritratti d'uomini illustri toscani con gli elogj istorici dei medesimi*, Florence: Giuseppe Allegrini, 1766–1773, vol. II (1768), no. XLIII, ff. 149r–152v.

SALVINI, Salvino (ed.)
Fasti consolari dell'Accademia Fiorentina, Florence: His Royal Highness's Press at Gaetano Tartini & Santi Franchi, 1717.

Sandrart, Joachim von

Academia nobilissimæ artis pictoriæ. Sive De veris & genuinis hujusdem proprietatibus, theorematibus, secretis atque requisitis aliis; nimirum de Inventione, Delineatione, Evrythmia & Proportione corporum: de Picturis in albario recente, sive fresco, in tabulis item, atque linteis; de pingendis historiis, imaginibus humanis, iconibusque viventium; de subdialibus & nocturnis; de subactu colorum oleario & aquario, de affectibus & perturbationibus animi exprimendis; de lumine & umbra; de vestibus, deque colorum proprietate, efficacia, usu, origine, natura atque significatione Instructio Fundamentalis, Multarum industria lucubrationum, & plurimorum annorum experientia conquisita, Nuremberg: Christian Sigmund Froberger, and Frankfurt: Heirs of Michael & Johann Friedrich Endter, and Johan Jacob von Sandrart, 1683.

Santorio, Santorio

Methodi vitandorvm errorum omnium qui in arte Medica contingunt Libri Qvindecim, Quorum principia sunt ab auctoritate Medicorum & Philosophorum principum desumpta, eaque omnia experimentis, & rationibus analyticis comprobata, Venice: Francesco Bariletti, 1603.

Sarpi, Paolo

Historia del Concilio tridentino, London: John Bill, 1619 (under the pseudonym Paolo Soave); second edition, Geneva: [s.e.], 1629.

Scheiner, Christoph

De macvlis solaribvs et stellis circa Iouem errantibus, accvratior disqvisitio, Augsburg: Ad insigne pinus, 1612; reprinted in *OG* 5, pp. 37–70; English translation in *On Sunspots*, translated and introduced by Eileen Reeves and Albert Van Helden, Chicago-London: The University of Chicago Press, 2010, pp. 183–230.

Ocvlvs hoc est: fvndamentvm opticvm, Innsbruck: Daniel Paur, 1619.

Pantographice, sev ars delineandi res qvaslibet per parallelogrammvm lineare sev cavvm, mechanicvm, mobile; Libellis suobus explicata, & Demonstrationibus Geometricis illustrata: quorum prior Epipedographicen, siue Planorum, posterior Stereographicen, seu Solidorum aspectabilium viuam imitationem atque proiectionem edocet, Rome: Lodovico Grignani, 1631.

Prodromvs pro Sole mobili et terra stabili, contra Academicvm Florentinvm Galilævm a Galilæis, s.l.: s.n., 1651.

Refractiones Coelestes, sive Solis Elliptici Phænomenon illvstratvm; in qvo variæ atqve antiqvæ astronomorvm circa hanc materiam difficultates enodantur, dubia multiplicia dissoluuntur, via ad multa recondita eruenda sternitur, Ingolstadt: Eder, for Elisabeth Angermaier, 1617.

Rosa Vrsina sive Sol ex admirando Facvlarvm & Macularum suarum Phænomenis varivs, necnon Circa centrum suum & axem fixum ab occasu in ortum annua, circaq. alium axem mobilem ab ortu in occasum conuersione quasi menstrua, super polos proprios, Libris quatuor mobilis ostensus, Bracciano: Andrea Feo, 1630.

Sol Ellipticus hoc est Nouum & perpetuum Solis contra hi soliti Phænomenin, quodnouiter inuentum, Augsburg: Christoph Mang, 1615.

Tres epistolae de macvlis solaribvs. Scriptæ ad Marcvm Velservm, Augsburg: Ad insigne pinus, 1612; reprinted in *OG* 5, pp. 23–33; English translation in *On Sunspots*, translated and introduced by Eileen Reeves and Albert Van Helden, Chicago-London: The University of Chicago Press, 2010, pp. 59–73.

SCHÖNBERGER, Johann Georg
Exegeses Fvndamentorvm Gnomonicorvm Quas In Alma Ingolstadiensi Academia, Præside Christophoro Scheinero . . . Publicæ Disputationi exponebat . . . mense Septembri, die 26, Ingolstadt: Eder, for Elisabeth Angermaier, 1615.

SIRI, Vittorio
Del Mercvrio Ouero Historia De' correnti tempi, Casale: Cristoforo della Casa, 1644–1682, 15 vols.

SIZZI, Francesco
ΔΙΑΝΟΙΑ astronomica, optica, physica, Qua Syderei Nuncij rumor de Quatuor Planetis à Galilæo Galilæo Mathematico Celeberrimo recens perspicilli cuiusdam ope conspectis, vanus redditur, Venice: Pietro Maria Bertano, 1611; reprinted (with Galileo's annotations) in *OG* 3.1, pp. 203–250.

STELLUTI, Francesco
Persio tradotto in verso sciolto e dichiarato, Rome: Giacomo Mascardi, 1630.

STENO, Nicolas
De musculis & glandulis observationum specimen Cum Epistolis duabus Anatomicis, Copenhagen: Matthias Jorgenson Godiche, 1664.
Elementorvm myologiæ specimen, sev Musculi descriptio Geometrica, Florence: Insegna della Stella, 1667.

TARDE, Jean
Borbonia Sidera, id est planetæ qvi solis limina circvmvolitant motv proprio ac regulari, falsò hactenus ab helioscopis maculæ Solis nuncupati, Paris: Jean Gesselin, 1620.

TARGIONI TOZZETTI, Giovanni
Notizie degli aggrandimenti delle scienze fisiche accaduti in Toscana nel corso degli anni LX. del secolo XVII, Florence: Giuseppe Bouchard, 1780, 3 vols.

TASSO, Torquato
Gervsalemme liberata, edited by Angelo Ingegneri, Parma: Erasmo Viotti, 1581.

TICCIATI, Girolamo
"Supplemento alla Vita di Michelagnolo Buonarroti," in Ascanio Condivi, *Vita di Michelagnolo Buonarroti pittore scultore architetto e gentiluomo fiorentino*, Florence: Gaetano Albizzini, 1746, pp. 59–63.

Tondini, Giambattiatsa (ed.)
Delle lettere di uomini illustri pubblicate ora per la prima volta, Macerata: Bartolomeo Capitani, 1782, 2 vols.

Torricelli, Evangelista
Lezioni accademiche, Florence: Jacopo Guiducci & Santi Franchi, 1715.
Opere di Evangelista Torricelli, edited by Gino Loria and Giuseppe Vassura, vols. 1–3, Faenza: G. Montanari, 1919; vol. 4, Faenza: Lega, 1944.
Opera Geometrica, Florence: Amadore Massi & Lorenzo Landi, 1644.

Vasari, Giorgio
"Giotto Pittor Fiorentino," in Id., *Le vite de piv eccellenti Architetti, Pittori, et Scvltori italiani, da Cimabve insino a' tempi nostri*, Florence: [Lorenzo Torrentino], 1550, Part I, pp. 138–149; revised edition, "Vita di Giotto Pittore, Scvltore, et Architetto Fiorentino," in Id., *Le vite de' piv eccellenti Pittori, Scvltori, e Architettori*, Florence: Giunti, 1568, Part I, pp. 119–133.
"Michelangelo Bonarroti Fiorentino. Pittore Scvltore et Architetto," in Id., *Le vite de piv eccellenti Architetti, Pittori, et Scvltori italiani, da Cimabve insino a' tempi nostri*, Florence: [Lorenzo Torrentino], 1550, Part III, pp. 947–991; revised edition, "Vita di Michelagnolo Buonarruoti Fiorentino Pittore, Scultore, & Architetto," in Id., *Le vite de' piv eccellenti Pittori, Scvltori, e Architettori*, Florence: Giunti, 1568, Part III, vol,. 2, pp. 715–796.

Viviani, Vincenzo
"Ænigma Geometriccm de miro opificio Testudinis Quadrabilis Hemisphæricæ propositum die 4 April. A. 1692," *Acta Eruditorum*, June 1692, pp. 274–275.
De locis solidis secunda divinatio geometrica in quinque libros iniuria temporum amissos Aristæi senioris geometræ, Florence: His Royal Highness's Press at Pietro Antonio Brigonci, [1701].
Formazione, e misvra di tvtti i cieli Con la struttura, e quadratura esatta dell'intero, e delle parti di un nuovo Cielo ammirabile, e di uno degli antichi delle Volte regolari degli Architetti. Curiosa esercitazione matematica, Florence: Piero Matini, 1692.
Grati animi monumenta . . . Uti fuerunt conscripta Florentiæ in Fronte Ædium A DEO DATARUM Anno Salutis 1693, in Viviani, *De locis solidis* [1701], pp. 121–128 (of the second half of the book); Florence: His Royal Highness's Press at Pietro Antonio Brigonci, [1702]; edited by Giovanni Battista Clemente Nelli, Florence: Francesco Moücke, 1791.
Qvinto libro degli Elementi d'Evclide, ovvero scienza vniversale delle proporzioni spiegata colla dottrina del Galileo, Florence: alla Condotta, 1674.
Racconto Istorico della vita del Sig. Galileo Galilei Fiorentino Accademico Linceo Primo Filosofo, e Matematico Sopraordinario del Serenissimo Granduca di Toscana, in *Fasti consolari dell'Accademia Fiorentina*, edited by Salvini (1717), pp. 397–431.
Testamento dell'Illustrissimo Signor Vincenzo Viviani. Rogato da Ser Simone Mugnai, Florence: Pietro Gaetano Viviani, 1735.

WALLACE, William A.
Galileo's Early Notebooks: The Physical Questions, Notre Dame-London: University of Notre Dame Press, 1977.

WEDDERBURN, John
Qvatvor problemavm qvæ Martinvs Horky Contra Nuntium Sidereum de qvatvor planetis novis Disputanda proposuit. Confvtatio, Padua: Pietro Marinelli, 1610; reprinted in *OG* 3.1, pp. 149–178.

SECONDARY SOURCES

AFFÒ, Ireneo.
Memorie degli scrittori e letterati parmigiani, Parma: Stamperia Reale, 1789–1797, 5 vols.

AMABILE, Luigi
Fra Tommaso Campanella ne' castelli di Napoli in Roma ed in Parigi. Narrazione con molti documenti e 10 opuscoli del Campanella inediti, Naples: Antonio Morano, 1887, 2 vols.

BIANCHI, Luca
"'Mirabile e veramente angelica dottrina': Galileo e l'argomento di Urbano VIII," in *Il caso Galileo: Una rilettura storica, filosofica, telogica*, edited by Massimo Bucciantini, Michele Camerota, Franco Giudice, Florence: Leo S. Olschki, 2011, pp. 213–233.

BOLZONI, Lina
"La restaurazione della poesia nella Prefazione dei *Commentaria* campanelliani," *Annali della Scuola Normale Superiore di Pisa. Classe di Lettere e Filosofia*, Series III, 1, 2, 1971, pp. 307–344.
"I *Commentaria* di Campanella ai *Poëmata* di Urbano VIII: Un uso infedele del commento umanistico," *Rinascimento*, 28, 1988, pp. 113–132.

BONELLI, Maria Luisa
"L'ultimo discepolo: Vincenzo Viviani," in *Saggi su Galileo Galilei*, edited by Carlo Maccagni, vol. 2 (the only one published), Florence: G. Barbèra, 1972, pp. 656–688.

BOSCHIERO, Luciano
"Post-Galilean Thought and Experiment in Seventeenth-Century Italy: The Life and Work of Vincenzio Viviani," *History of Science*, 43, 2005, pp. 77–100.
"Robert Southwell and Vincenzio Viviani: Their Friendship and an Attempt at Italian-English Scientific Collaboration," *Parergon*, 26, 2, 2009, pp. 87–108.

BUCCIANTINI, Massimo
Contro Galileo. Alle origini dell'affaire, Florence; Leo S. Olschki, 1995.

CASTAGNETTI, Marina
"I Poemata e le Poesie toscane di Maffeo Barberini. I—Stampe e problemi di cro-
nologia," *Atti della Accademia di Scienze Lettere e Arti di Palermo*, Series IV, 39, 2,
1982, pp. 283–388.

CERBU, Thomas, and LERNER, Michel-Pierre
"La disgrâce de Galilée dans les *Apes Urbanae*: sur la fabrique du texte de Leone
Allacci," *Nuncius*, 15, 2, 2000, pp. 589–610.

COCHRANE, Eric
*Florence in the Forgotten Centuries 1527–1800: A History of Florence and the Florentines
in the Age of the Grand Dukes*, Chicago-London: The University of Chicago Press,
1973.

COOPER, Lane
Aristotle, Galileo, and the Tower of Pisa, Ithaca, NY: Cornell University Press, 1935.

CROCE, Benedetto
Nuovi saggi sulla letteratura italiana del Seicento, Bari: Laterza, 1931.

DE SEPIBO, Giorgio
Romani Collegii Societatus [*sic*] *Jesu Musæum celeberrimum*, Amsterdam: Johannes
van Waesbergen, 1678.

DRABKIN, Israel E.
"Two Versions of G. B. Benedetti's *Demonstratio Proportionum Motuum Localium*,"
Isis, 54, 2, 1963, pp. 259–262.

DRAKE, Stillman
"Galileo Gleanings II. A Kind Word for Salusbury," *Isis*, 49, 1, 1958, pp. 26–33.
"Introduction," in Thomas Salusbury, *Mathematical Collections and Translations, in
Two Tomes, London 1661 and 1665*, facsimile reprint with an analytical and biobib-
liographical introduction by Stillman Drake, London: Dawsons of Pall Mall and
Los Angeles: Zeitlin & Ver Brugge, 1967, vol. 1, pp. 1–27.

ERNST, Germana
Tommaso Campanella: Il libro e il corpo della natura, Rome-Bari: Laterza, 2002; En-
glish translation by David L. Marshall, *Tommaso Campanella: The Book and the
Body of Nature*, Dordrecht-Heidelberg-London-New York: Springer, 2010.

FAVARO, Antonio
Amici e corrispondenti di Galileo [1894–1919], edited by Paolo Galluzzi, Florence:
Salimbeni, 1983, 3 vols.
"Amici e corrispondenti di Galileo Galilei: XXIX. Vincenzio Viviani," *Atti del
Reale Istituto Veneto di scienze, lettere ed arti*, 72, 2, 1912–1913, pp. 1–155; reprinted

in Antonio Favaro, *Amici e corrispondenti di Galileo Galilei* (1983), vol. 2, pp. 1009–1163.

"Di alcune inesattezze nel *Racconto Istorico della Vita di Galileo* dettato da Vincenzio Viviani," *Archivio storico italiano*, 74, 2, 1917, pp. 127–150.

"Miscellanea galileiana inedita. Studi e ricerche," *Memorie del Reale Istituto Veneto di scienze, lettere ed arti*, 22, 1882, pp. 701–1034; reprinted as Antonio Favaro, *Miscellanea galileiana inedita. Studi e ricerche*, Venice: Tipografia di Giuseppe Antonelli, 1887.

"Rarità bibliografiche Galileiane: III. Sopra una traduzione inglese di alcune opere di Galileo," *Rivista delle biblioteche*, 2, 18–19, 1889, pp. 86–91.

Scampoli galileiani [1886–1915], edited by Lucia Rossetti e Maria Laura Soppelsa, Trieste: LINT, 1992, 2 vols.

"Serie decimottava di scampoli galileiani: CXVIII. In qual giorno del febbraio '64 dovrebbe celebrarsi la ricorrenza del natalizio di Galileo?" *Atti e memorie della R. Accademia di scienze, lettere ed arti in Padova*, 24, 1908, pp. 6–8; reprinted in Favaro, *Scampoli galileiani* (1992), vol. 2, pp. 564–566.

"Serie sesta di scampoli galileiani: XLII. Ulteriori notizie intorno alla traduzione inglese di alcune Opere di Galileo," *Atti e memorie della R. Accademia di scienze, lettere ed arti in Padova*, 292, 1891, pp. 46–48; reprinted in Favaro, *Scampoli galileiani* (1992), vol. 1, pp. 164–166.

"Serie ventesimaterza di scampoli galileiani: CXLIX. Un mancato biografo di Galileo," *Atti e memorie della R. Accademia di scienze, lettere ed arti in Padova*, 304, 1914, pp. 70–77; reprinted in Favaro, *Scampoli galileiani* (1992), vol. 2, pp. 732–739.

"Sulla veridicità del *Racconto istorico della Vita di Galileo* dettato da Vincenzo Viviani," *Archivio storico italiano*, 73, 1, 1915, pp. 323–380.

"Vincenzo Viviani e la sua *Vita di Galileo*," *Atti del Reale Istituto Veneto di scienze, lettere ed arti*, 72, 2, 1902–1903, pp. 683–703.

FINDLEN, Paula
"Living in the Shadow of Galileo: Antonio Baldigiani (1647–1711), a Jesuit Scientist in Late Seventeenth-Century Rome," in *Conflicting Duties: Science, Medicine and Religion in Rome, 1550–1750*, edited by Maria Pia Donato and Jill Kraye, London: Warburg Institute, and Turin: Nino Aragno Editore, 2009, pp. 211–254.

"Rethinking 1633: Writing the Life of Galileo after the Trial," in *Nature Engaged: Science in Practice from the Renaissance to the Present*, edited by Mario Biagioli, Jessica Riskin, London: Palgrave Macmillan, 2012, pp. 205–226.

FINOCCHIARO, Maurice
The Galileo Affair: A Documentary History, Berkeley: University of California Press, 1989.

Retrying Galileo, 1633–1992, Berkeley: University of California Press, 2005.

FIRPO, Luigi
"Campanella e Galileo," in Tommaso Campanella, *Apologia di Galileo*, edited by Luigi Firpo: Turin: Utet, 1968, pp. 7–26.

"Il Campanella astrologo e i suoi persecutori romani," *Rivista di filosofia*, 30, 1939, pp. 200–215.

Galluzzi, Paolo
"I Sepolcri di Galileo: Le Spoglie 'Vive' di un Eroe della Scienza," in *Il Pantheon di Santa Croce a Firenze*, edited by Luciano Berti, Florence: Cassa di Risparmio di Firenze, 1993, pp. 145–182; partial English translation by Michael J. Gorman, "The Sepulchers of Galileo: The 'Living' Remains of a Hero of Science,, in *The Cambridge Companion to Galileo*, edited by Peter Machamer, Cambridge: Cambridge University Press, 1998, pp. 417–447.

Gattei, Stefano
"From Banned Mortal Remains to the Worshipped Relics of a Martyr of Science: Vincenzo Viviani and the Birth of Galileo's Mythography," in *Savant Relics: Brains and Remains of Scientists*, edited by Marco Beretta, Maria Conforti, Paolo Mazzarello, Sagamore Beach, MA: Science History Publications, 2016, pp. 67–92.
"Galileo's Legacy: A Critical Edition and Translation of the Manuscript of Vincenzo Viviani's *Grati Animi Monumenta*," *The British Journal for the History of Science*, 50, 2, 2017, pp. 181–228.

Gerboni, Luigi
Un umanista nel Seicento. Giano Nicio Eritreo: Studio biografico critico, Città di Castello: S. Lapi, 1899.

Giachino, L.
"*Cicero libertinus*. La satira della Roma barberinia na nell'*Eudemia* dell'Eritreo," *Studi secenteschi*, 43, 2002, pp. 185–215.

Gingerich, Owen
"The Censorship of Copernicus's *De Revolutionibus*," *Annali dell'Istituto e Museo di Storia della Scienza*, 6, 2, 1981, pp. 45–61.

Giusti, Enrico
Euclides reformatus: La teoria delle proporzioni nella scuola galileiana, Turin: Bollati Boringhieri, 1993.

Herklotz, Ingo
Apes Urbanae: Eruditi, mecenati e artisti nella Roma del Seicento, Città di Castello: LuoghInteriori, 2017.

Kris, Ernst, and Kurz, Otto
Die Legende vom Künstler: Ein geschichtlicher Versuch, Wien: Kristall, 1934; English translation by Alastair Laing, revised by Lottie M. Newman, *Legend, Myth, and Magic in the Image of the Artist: A Historical Experiment*, Preface by Ernst H. Gombrich, New Haven-London: Yale University Press, 1979.

LERNER, Michel-Pierre
"Le panégyrique différé ou les aléas de la notice 'Thomas Campanella' des *Apes Urbanae*," *Bruniana & Campanelliana*, 7, 2, 2001, pp. 413–451.
"Premessa," in Leone Allacci, *Apes Urbanae*, facsimile reprint of the 1633 edition, Lecce: Conte, 1998, pp. VII-XXX.

LUNARDI, Roberto, and SABBATINI, Oretta (eds.)
Il rimembrar delle passate cose. Una casa per Memoria: Galileo e Vincenzo Viviani, Florence: Polistampa, 2009.

PANOFSKY, Erwin
Galileo as a Critic of the Arts, The Hague: Martinus Nijhoff, 1954; abridged version, "Galileo as a Critic of the Arts: Aesthetic Attitude and Scientific Thought," *Isis*, 47, 1, 1956, pp. 3–15.

ROWLAND, Ingrid
"The Lost *Iter Hetruscum* of Athanasius Kircher (1665–78)," in *New Perspectives on Etruria and Early Rome: In Honor of Richal Daniel De Palma*, edited by Sinclair Bell and Helen Nagy,Madison, WI: The University of Wisconsin Press, 2009, pp. 274–289.

RUFFO, Patrizia
"Da Firenze a Lima: origine e fortuna della Vita di Galileo di Vittorio Siri," *Galilæana*, 2, 2005, pp. 181–192.

SEGRE, Michael
"Galileo, Viviani, and the Tower of Pisa," *Studies in History and Philosophy of Science*, 20, 4, 1989, pp. 435–451.
In the Wake of Galileo, New Brunswick, NJ: Rutgers University Press, 1991.
"Viviani's Life of Galileo," *Isis*, 80, 2, 1989, pp. 206–231.

SIEBERT, Harald
"Kircher and His Critics: Censorial Practice and Pragmatic Disregard in the Society of Jesus," in *Athanasius Kircher: The Last Man Who Knew Everything*, edited by Paula Findlen, New York-London: Routledge, 2004, pp. 79–104.

SPINI, Giorgio
Galileo, Campanella e il «Divinus Poeta», Bologna: Il Mulino, 1996.
"Galileo uomo del Seicento," in *Galileo a Padova 1592–1610*, vol. 4: *Tribute to Galileo in Padua*, Trieste: LINT, 1995, pp. 69–88.

TOGNONI, Federico
Galileo: il mito tra Otto e Novecento, Pisa: Pacini, 2014.
Il carteggio Cigoli-Galileo, 1609–1613, Preface by Lucia Tomasi Tongiorgi, Pisa: ETS, 2009.

Tognoni, Federico (ed.)
Il carteggio Cigoli-Galileo, 1609–1613, Introduction by Lucia Tomasi Tongiorgi, Pisa: Edizioni ETS, 2009.

Vaccalluzzo, Nunzio
Galileo Galilei nella poesia del suo secolo. Raccolta di poesie edite e inedite scritte da' contemporanei in lode di Galileo, Milan-Palermo-Naples: Sandron, [1910].

Wilding, Nick
"The Return of Thomas Salusbury's *Life of Galileo* (1664)," *The British Journal for the History of Science*, 41, 2, 2008, pp. 241–265.

Wohlwill, Emil
"Galilei-Studien", *Mitteilungen zur Geschichte der Medizin und der Naturwissenschaften*, 4, 2, 1905, pp. 229–248 ("I. Die Pisaner Fallversuche"); 5, 3, 1906, pp. 439–464, and 6, 3, 1907, pp. 231–242 ("II. Der Abschied von Pisa").
Galilei und sein Kampf für die copernicanische Lehre, Hamburg-Leipzig: Leopold Voss, 1909 and 1926, 2 vols.
"Über einen Grundfehler aller neueren Galilei-Biographien," *Münchener medizinische Wochenschrift*, 50, 1903, pp. 1850–1851; expanded reprint in *Verhandlungen der Gesellschaft deutscher Naturforscher und Ärzte*, edited by Albert Wangerin, Leipzig: Leopold Voss, 1904, vol. 2, Part 2, pp. 100–101.

Zeitlin, Jacob
"Salusbury Discovered," *Isis*, 50, 4, 1959, pp. 455–458.

Further Readings

The secondary literature on the various aspects of Galileo's life and works, and especially his trial, is enormous. The *Galilean Bibliography*, constantly updated by the Museo Galileo, in Florence—listing books, articles, book chapters, and critical reviews—includes a total of approximately 23,000 entries. It can be accessed online at https://galileoteca.museogalileo.it/AmicusSearch/AmicusSearch?type=bibliografia&lang=en.

In what follows, I offer a selection of works (not already referred to) that may be useful to the reader seeking additional information or context about the topics dealt with in this volume.

Biographies

Camerota, Michele
Galileo Galilei e la cultura scientifica nell'età della Controriforma, Rome: Salerno Editrice, 2004.

DRAKE, Stillman
Galileo at Work: His Scientific Biography, Chicago-London: The University of Chicago Press, 1978.

FANTOLI, Annibale
Galileo per il copernicanesimo e per la Chiesa, Città del Vaticano: Libreria Editrice Vaticana, 1993, 1997^2, 2010^3; English translation by George V. Coyne, *Galileo for Copernicanism and for the Church*, Città del Vaticano: Libreria Editrice Vaticana, 1994, 1996^2, 2003^3.

HEILBRON, John L.
Galileo, Oxford-New York: Oxford University Press, 2010; 2012^2.

OLSCHKI, Leonardo
Galilei und seine Zeit, Halle: Max Niemeyer, 1927.

SHEA, William R., and ARTIGAS, Mariano
Galileo in Rome: The Rise and Fall of a Troublesome Genius, Oxford: Oxford University Press, 2003.

WOOTTON, David
Galileo: Watcher of the Skies, New Haven-London: Yale University Press, 2010.

Other Works

ARDISSINO, Erminia
Galileo: la scrittura dell'esperienza. Studi sulle lettere, Pisa: Edizioni ETS, 2010.

BATTISTINI, Andrea
Galileo e i Gesuiti. Miti letterari e retorica della scienza, Milan: Vita e Pensiero, 2000.

BELLINI, Eraldo
Stili di pensiero nel Seicento italiano. Galileo, i Lincei, i Barberini, Pisa: Edizioni ETS, 2009.
Umanisti e Lincei. Letteratura e scienza a Roma nell'età di Galileo, Padua: Editrice Antenore 1997.

BERETTA, Francesco
Galilée devant le Tribunal de l'Inquisition. Une relecture des sources, Fribourg: [s.e.], 1998.

BERETTA, Francesco (ed.)
*Galilée en procès, Galilée réhabilité ?*Saint-Maurice: Éditions Saint-Augustin, 2005.

BIAGIOLI, Mario
Galileo, Courtier: The Practice of Science in the Culture of Absolutism, Chicago-London: The University of Chicago Press, 1993.
Galileo's Instruments of Credit: Telescopes, Images, Secrecy, Chicago-London: The University of Chicago Press, 2006.

BLACKWELL, Richard J.
Behind the Scenes at Galileo's Trial: Including the First English Translation of Melchior Inchofer's Tractatus Syllepticus, Notre Dame, IN: University of Notre Dame Press, 2006.
Galileo, Bellarmine, and the Bible: Including a Translation of Foscarini's Letter on the Motion of the Earth, Notre Dame, IN: University of Notre Dame Press, 1991.

BOSCHIERO, Luciano
Experiment and Natural Philosophy in Seventeenth-Century Tuscany: The History of the Accademia del Cimento, Dordrecht: Springer, 2007.

BUCCIANTINI, Massimo, CAMEROTA, Michele, and GIUDICE, Franco
Il telescopio di Galileo. Una storia europea, Turin: Einaudi, 2012; English translation by Catherine Bolton, *Galileo's Telescope: A European Story*, Cambridge, MA-London: Harvard University Press, 2015.

BUCCIANTINI, Massimo, CAMEROTA, Michele, and GIUDICE, Franco (eds.)
Il caso Galileo. Una rilettura storica, filosofica, teologica, Florence: Leo S. Olschki, 2011.

CLAVELIN, Maurice
Galilée copernicien. Le premier combat, 1610–1616, Paris: Albin Michel, 2004.
La philosophie naturelle de Galilée. Essai sur les origines et la formation de la mécanique classique, Paris: Colin, 1968; English translation by A. J. Pomerans, *The Natural Philosophy of Galileo: Essay on the Origins and Formation of Classical Mechanics*, Cambridge, MA-London: The MIT Press, 1974.

DAMANTI, Alfredo
Libertas philosophandi. Teologia e filosofia nella Lettera alla Granduchessa Cristina di Lorena *di Galileo Galilei*, Rome: Edizioni di Storia e Letteratura, 2010.

DRAKE, Stillman
Essays on Galileo and the History and Philosophy of Science, edited by Noel M. Swerdlow and Trevor H. Levere, Toronto: University of Toronto Press, 1999, 3 vols.
Galileo: Pioneer Scientist, Toronto-Buffalo-London: University of Toronto Press, 1990.
Galileo Studies: Personality, Tradition, and Revolution, Ann Arbor: The University of Michigan Press, 1970.

DREYER, John L. E.

History of the Planetary Systems from Thales to Kepler, Cambridge: Cambridge University Press, 1906; revised edition, *A History of Astronomy from Thales to Kepler*, Foreword by William H. Stahl, New York: Dover Publications, 1953.

FANTOLI, Annibale

Galileo e la Chiesa. Una controversia ancora aperta, Rome: Carocci, 2010; English translation by George V. Coyne, *The Case of Galileo: A Closed Question?* Notre Dame, IN: University of Notre Dame Press, 2012.

FAVARO, Antonio

Adversaria galilaeiana [1916–1923], edited by Lucia Rossetti e Maria Laura Soppelsa, Trieste: LINT, 1992.

Galileo Galilei a Padova. Ricerche e scoperte, insegnamento, scolari, Padua: Editrice Antenore, 1968.

Galileo Galilei e lo Studio di Padova [1883], Padua: Editrice Antenore, 1966.

"Nuovi studi Galileiani", *Memorie del Reale Istituto Veneto di scienze, lettere ed arti*, 24, 1891, pp. 7–430; reprinted as Antonio Favaro, *Nuovi studi Galileiani*, Venezia: Tipografia Antonelli, 1891.

Oppositori di Galileo [1892–1921], edited by Stefano Gattei, Lanciano: Rocco Carabba, 2018.

FEINGOLD, Mordechai (ed.)

The New Science and Jesuit Science: Seventeenth Century Perspectives, Dordrecht: Springer, 2003.

Jesuit Science and the Republic of Letters, Cambridge, MA: The MIT Press, 2003.

FERRONE, Vincenzo, and BRAVO, Gian Mario (eds.)

Il processo a Galileo Galilei e la questione galileiana, Rome: Edizioni di Storia e Letteratura, 2010.

FREEDBERG, David

The Eye of the Lynx: Galileo, His Friends, and the Beginnings of Modern Natural History, Chicago: The University of Chicago Press, 2002.

GALLUZZI, Paolo

Libertà di filosofare in naturalibus. I mondi paralleli di Cesi e Galileo, Rome: Scienze e Lettere, 2014; English translation by Peter Mason, *The Lynx and the Telescope: The Parallel Worlds of Cesi and Galileo*, Leiden: Brill, 2017.

GALLUZZI, Paolo (ed.)

Novità celesti e crisi del sapere. Atti del Convegno internazionale di studi galileiani, Florence: Giunti-Barbera, 1984.

GARCIA, Stéphane

Élie Diodati et Galilée. Naissance d'un réseau scientifique dans l'Europe du XVIIe siècle, Florence: Leo S. Olschi, 2004.

GEBLER, Karl von
Galilei und die römische Kurie, Stuttgart: J. G. Cotta, 1876–1877, 2 vols.; English translation by George Sturge, *Galileo Galilei and the Roman Curia*, London: C. K. Paul & Co., 1879; reprint, Merrick, NY: Richwood, 1977.

GUERRINI, Luigi
Cosmologie in lotta. Le origini del processo di Galileo, Florence: Edizioni Polistampa, 2010.
Galileo e gli aristotelici. Storia di una disputa, Rome: Carocci, 2010.
Galileo e la polemica anticopernicana a Firenze, Florence: Edizioni Polistampa, 2009.

HALL, Crystal
Galileo's Reading, Cambridge-New York: Cambridge University Press, 2013.

KOYRÉ, Alexandre
Études galiléennes [1939], Paris: Hermann, 1966; English translation by Joseph Mepham, *Galileo Studies*, Hassocks: The Harvester Press, 1978.

MACHAMER, Peter (ed.)
The Cambridge Companion to Galileo, Cambridge: Cambridge University Press, 1998.

MACCAGNI, Carlo (ed.)
Saggi su Galileo Galilei, Florence: G. Barbera, 1972.

McMULLIN, Ernan (ed.)
Galileo: Man of Science, New York: Basic Books, 1967.
The Church and Galileo, Notre Dame, IN: University of Notre Dame Press, 2005.

MONTESINOS, José, and SOLÍS, Carlos (eds.)
Largo campo di filosofare. Eurosymposium Galileo 2001, La Orotava: Fundación Canaria Orotava de Historia de la Ciencia, 2001.

PAGANO, Sergio (ed.)
I documenti vaticani del processo di Galileo Galilei (1611–1741), new, revised and annotated edition, Città del Vaticano: Archivio Segreto Vaticano, 2009.

PALMIERI, Paolo
Reenacting Galileo's Experiments: Rediscovering the Techniques of Seventeenth-Century Science, Lewiston-Queenston-Lampeter: The Edwin Mellen Press, 2008.

PETERSON, Mark A.
Galileo's Muse: Renaissance Mathematics and the Arts, Cambridge, MA-London: Harvard University Press, 2011.

REDONDI, Pietro
Galileo eretico, Turin: G. Einaudi, 1983, 1988[2], 2004[3]; new edition, Rome-Bari: Laterza, 2009; English translation by Raymond Rosenthal, *Galileo Heretic*, Princeton, NJ: Princeton University Press, 1987.

REEVES, Eileen
Galileo's Glassworks: The Telescope and the Mirror, Cambridge, MA-London: Cambridge University Press, 2008.
Painting the Heavens: Art and Science in the Age of Galileo, Princeton, NJ: Princeton University Press, 1997.

RENN, Jürgen (ed.)
Galileo in Context, Cambridge: Cambridge University Press, 2001.

RICCI, Saverio
Campanella, Rome: Salerno Editrice, 2018.

RIGHINI-BONELLI, Maria Luisa, and SHEA, William R.
Galileo's Florentine Residences, Florence: Istituto e Museo di Storia della Scienza, 1979.

ROSEN, Edward
The Naming of the Telescope, Foreword by Harlow Shapley, New York: H. Schuman, 1947.

SHEA, William R.
Galileo's Intellectual Revolution, London: Macmillan, 1972.

STRANO, Giorgio (ed.)
Galileo's Telescope: The Instrument that Changed the World, Florence: Giunti, 2008.

TOGNONI, Federico
I volti di Galileo. Fortuna e trasformazione dell'immagine galileiana tra XVII e XIX secolo, Lugano: Agorà&Co., 2018.

VALLERIANI, Matteo
Galileo Engineer, Dordrecht-Heidelberg-London-New York: Springer, 2010.

VAN HELDEN, Albert
Measuring the Universe: Cosmic Dimensions from Aristarchus to Halley, Chicago: The University of Chicago Press, 1985.
The Invention of the Telescope, Philadelphia: American Philosophical Society, 1977, 2008[2].

Van Helden, Albert, Dupré, Sven, van Gent, Rob, and Zuidervaart, Huib (eds.)
The Origins of the Telescope, Amsterdam: KNAW Press, 2010.

Wallace, William A.
Galileo and His Sources: The Heritage of the Collegio Romano in Galileo's Science, Princeton, NJ: Princeton University Press, 1984.
Prelude to Galileo: Essays on Medieval and Sixteenth-Century Sources of Galileo's Thought, Dordrecht-Boston-London: D. Reidel, 1981.

Wilding, Nick
Galileo's Idol: Gianfrancesco Sagredo and the Politics of Knowledge, Chicago-London: The University of Chicago Press, 2014.

Willach, Rolf
The Long Route to the Invention of the Telescope, Philadelphia: American Philosophical Society, 2008.

Wohlwill, Emil
Galilei und sein Kampf für die copernicanische Lehre, Hamburg-Leipzig: L. Voss, 1909–1926, 2 vols.

INDEX

Captions and bibliographical references have not been indexed.